The Photostability of Drugs and Drug Formulations

The Photostability of Drugs and Drug Formulations

EDITED BY H. HJORTH TØNNESEN

Taylor & Francis
Publishers since 1798

UK Taylor & Francis Ltd, 1 Gunpowder Square, London EC4A 3DE
USA Taylor & Francis Inc., 1900 Frost Road, Suite 101, Bristol, PA 19007

British Library Cataloguing in Publication Data
A catalogue record for this book is available from the British Library
ISBN 0-7484-0449-X (cased) (formerly 013 127564 X)

Library of Congress Cataloging Publication Data are available

Cover design by Jim Wilkie

Typeset in Times 10/12pt by Keyset Composition, Colchester, Essex

Printed in Great Britain by T. J. Press (Padstow) Ltd

Contents

List of Contributors

N. H. ANDERSON
Sanofi Research Division, Alnwick, Northumberland, UK

J. BOXHAMMER
Heraeus Xenotest GmbH, Hanau, Germany

P. L. CARTER
Smithkline Beecham Pharmaceuticals, Worthing, Sussex, UK

D. CLAPHAM
Smithkline Beecham Pharmaceuticals, Worthing, Sussex, UK

R. S. DAVIDSON
The Chemical Laboratory, University of Kent, Canterbury, Canterbury, Kent, UK

J. V. GREENHILL
Department of Chemistry, University of Florida, USA

J. C. HUNG
Mayo Clinic, Rochester, Minnesota, USA

J. KARLSEN
Department of Pharmaceutics, University of Oslo, Oslo, Norway

S. KRISTENSEN
Department of Pharmaceutics, University of Oslo, Oslo, Norway

D. R. MERRIFIELD
Smithkline Beecham Pharmaceuticals, Worthing, Sussex, UK

J. MOAN
Institute for Cancer Research, Oslo, Norway

D. E. MOORE
Department of Pharmacy, University of Sydney, Sydney, Australia

K. NORD
Department of Pharmaceutics, University of Oslo, Oslo, Norway

T. OPPENLÄNDER
Fachhochschule Furtwangen (Schwarzwald) Schwenningen, Germany

J. E. ROBERTS
Division of Science and Mathematics, Fordham University, New York City, New York, USA

S. A. SANDE
Department of Pharmaceutics, University of Oslo, Oslo, Norway

F. D. SANDERSON
Smithkline Beecham Pharmaceuticals, Worthing, Sussex, UK

K. THOMA
Ludwig-Maximilians-Universität, Munchen, Munich, Germany

H. HJORTH TØNNESEN
Department of Pharmaceutics, University of Oslo, Oslo, Norway

Preface

Photochemical degradation of drugs and drug formulations is an area of interest that recently has developed into an important field of research. This can partly be ascribed to the demand for harmonized guidelines for photochemical stability studies of drugs and drug products from regulatory bodies dealing with drug registration worldwide. This was also the background for the 1st International Meeting on Photostability of Drugs held in Oslo, Norway, June 1995. As a result of our efforts, this book contains the full text of the plenary lectures held at the symposium in addition to a part discussing a test case and a review of literature on photostability of actual drug compounds. In my opinion the chapters in the book cover the most important aspects of the meeting and would be a valuable tool for anybody interested in this exciting topic.

The organizing committee realized that studies on photochemical degradation implies collaboration between groups of scientists with different background and that the plenary lectures should reflect this multidisciplinary approach. I am therefore happy to be able to present a book to the audience that covers many different approaches to the subject of photostability.

Information about the stability of compounds and drug products is needed to predict the shelf-life. Particularly in the field of photodecomposition there is limited help to be obtained from the existing literature. Usually this is caused by the use of non-standardized experimental conditions since there are no set guidelines for light stability testing of pharmaceuticals. The testing procedures therefore vary widely among pharmaceutical laboratories. Important research groups in Europe, Japan and the US are now discussing the problems of standardization of experimental conditions for photochemical stability testing.

This book is intended to highlight the different aspects of the combination of drugs and light, i.e. with respect to stability and quality control, toxicity and as a tool in drug targeting. It is my hope that this book will be of help to anyone interested in this field and that it also will initiate projects using light impact to change molecules for practical applications.

Preface

The intention of the chapters published in this book is to bridge a gap between different areas of research. The views and data expressed in the chapters are those of the authors and the topics have been covered on the initiative of the editor.

Hanne Hjorth Tønnesen
Oslo

1

Introduction: Photostability Testing of Drugs and Drug Formulations — Why and How?

H. HJORTH TØNNESEN

It is well known that light can change the properties of different materials and products. This is often observed as bleaching of coloured compounds like paint and textiles or as a discolouration of colourless products. Photostability has for many years been a main concern within several fields of industry, e.g. the textile, paint, food, cosmetic and agricultural industries. In the field of pharmacy, photostability has played a less important role. Meanwhile, the number of drugs found to be photochemically unstable is steadily increasing. The European pharmacopoeia prescribes light protection for about 250 medical drugs and a number of adjuvants. New compounds are frequently added to this list, although the justification of light protection requirements for certain compounds has been questioned (Reisch and Zappel, 1993).

Several points need to be clarified before developing and adopting a protocol for photostability testing of drugs.

- What is the rationale for evaluating drug photostability?
- What can be achieved by evaluating drug photostability?
- How can adequate information about drug photostability be obtained?

In this context the term 'photostability' is used to describe how a compound responds to light exposure and includes not only degradation reactions but other processes such as formation of radicals, energy transfer and luminescence.

1.1 The rationale for evaluation of drug photostability

The most obvious result of drug photodecomposition is a loss of potency of the product. In the final consequence this can result in a drug product which is therapeutically inactive. Although this is not often the case, even less severe degradation can lead to problems. Adverse effects due to the formation of minor degradation products during storage and administration have been reported (de Vries *et al.*, 1984). The drug substance can also cause light-induced side-effects

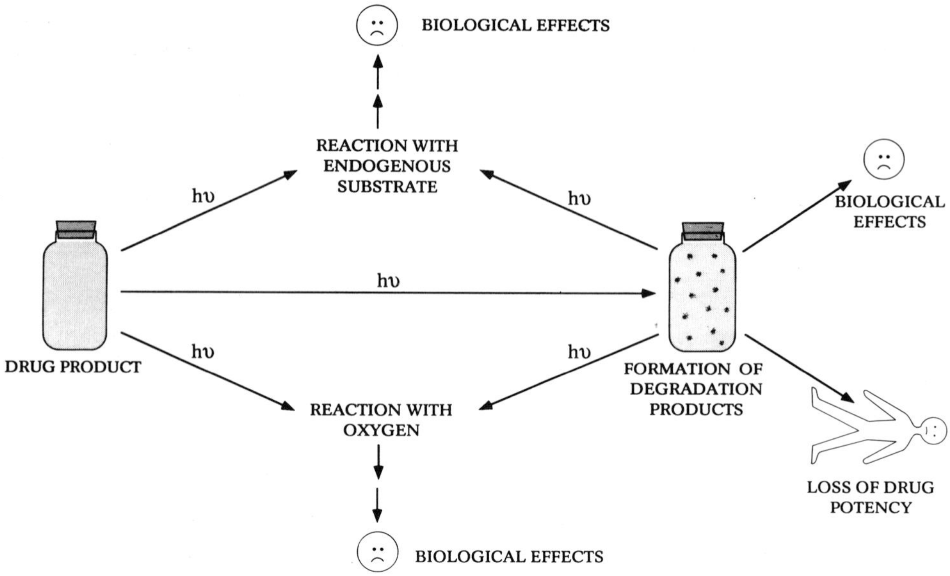

Figure 1.1 Possible consequences of drug photoinstability

after administration to the patient by interaction with endogenous substances. Therefore, two aspects of drug photostability have to be considered: *in vitro* stability and *in vivo* stability. The possible consequences of drug photoinstability are illustrated in Fig. 1.1. Independent of what we are concerned about: *in vitro* stability or *in vivo* effects, characterization of the photochemical properties of drug substances and drug formulations is a part of the formulation work and cannot be ignored.

Many drug substances and drug products are found to decompose *in vitro* under exposure to light, but the practical consequences will not necessarily be the same in all cases. Derivatives of the drug nifedipine have a photochemical half-life of only a few minutes while other drugs may decompose only a few per cent after several weeks' exposure (Squella *et al.*, 1990). They are all 'sensitive to light' but the same precautions are not required in handling of these compounds. Knowledge about the photostability of drug substances and drug products is important in order to evaluate:

- Handling, packaging and labelling;
- adverse effects and
- therapeutic aspects and new drug delivery systems

1.1.1 Handling, packaging and labelling

The ability of a drug substance to degrade or undergo a gradual change in colour upon light exposure is not an uncommon property. Polymorphs of drug substances can even exhibit different sensitivity to light (Nyqvist and Wadsten, 1986). In practice, the drug substance would mainly experience exposure to visible light (i.e. cool white fluorescent tubes) during storage and production. Many drug substances are white and essentially no visible light will be absorbed by these compounds. It is, however, important to know that all lamps, even incandescent ones, emit some radiation in the UV region of the spectrum. Light protection of the drug substance during storage and production must, therefore, be recommended in many cases, Solid-state photostability of drugs is not fully understood and it remains unclear as to whether a change in colour upon exposure necessarily is correlated with the extent of chemical stability of the material (Matsuda and Tatsumi, 1990).

A change in the selection of packing materials combined with a change in storage conditions or conditions during administration of the drug products seems to generate new stability problems *in vitro*. Most people are familiar with the traditional brown medicinal flask or the white pill-box. These containers offer adequate protection of most drug products during storage and distribution. In modern hospital pharmacies, drugs are often stored in unit-dose containers in an open shelf. The protective market pack is in many cases removed. The inner container can be made of transparent plastic materials which offer little if any protection towards UV and visible radiation (Tønnesen, 1989; Tønnesen and Karlsen, 1987). The unprotected drug product can then be exposed to fluorescent tubes and/or filtered daylight for several days or weeks (Tønnesen and Karlsen, 1995).

Infusion solutions should be stored in transparent infusion bottles or infusion bags. Long-term infusions can lead to the exposure of the drug to filtered daylight for hours. During intravenous medication of premature babies which are under treatment for hyperbilirubinemia the drug can experience radiation of high intensity. Portable drug delivery devices are often used to treat patients with severe pain. Various types of plastic materials are used in the drug reservoirs for these pumps.

The precautions taken in handling these drugs, including adequate labelling and selection of packaging, will in each case depend on the photochemical half-life of the drug substance in the actual formulation. Basic information about the photostability of the compounds is needed and evaluation of *in vitro* stability is therefore essential to ensure good quality over the entire lifespan of the drug.

1.1.2 Adverse effects

Although a drug product is shown to be photochemically inert in the sense that it does not decompose during exposure to light, it can still act as a source of free radicals or form phototoxic metabolites *in vivo* (Beijersbergen van Henegouwen, 1981). The drug will then be photoreactive after administration if the *patient* is exposed to light, causing light-induced adverse effects (Epstein and Wintroub, 1985). This emphasizes the importance of including studies of reaction mechanisms and sensitizing properties of the parent compound, its degradation products and *in vivo* metabolites in the evaluation of photostability of drugs.

The increase in number of reported adverse effects that can be ascribed to the combination of drugs and light is due to an increase in exposure to artificial light sources such as daylight lamps and solaria; a change in human leisure habits (we spend more time outdoors) and a widespread use of drugs.

Several requirements are to be met if a drug is to cause phototoxic reactions.

First, the drug or metabolites of the compound must be distributed to tissues near the body surface, e.g. skin, eye and hair, that are exposed to light. Then the absorption spectrum of the drug must overlap with the transmission spectrum of light through the actual tissue. The Federal Drugs Administration (FDA) has recently requested photocarcinogenicity testing on specific classes of drugs (Loveday and Bergman, 1994).

1.1.3 Therapeutic aspects and new drug delivery systems

In vivo photodecomposition and radical formation should not always be avoided as these properties can be advantageous from a therapeutic viewpoint. More than 3000 years ago, the Egyptians, Chinese and Indians were using photosensitization in attempts to cure such disorders as vitiligo, rickets, psoriasis, skin cancer and psychosis (Harber *et al.*, 1982; Spikes, 1985). Treatment of psoriasis by combination of psoralens and UV-A light (PUVA therapy) is now well established. Alternative photosensitizers are certainly in demand. The potential for new drug delivery systems such as light-activated liposomes or prodrugs should not be ignored. The use of fiber optics for activation of therapeutic compounds in drug targeting seems to be promising (Bayley *et al.*, 1987). New developments in the group of topical preparations would certainly have advantages, especially for localized tumours near the skin surface.

1.2 What can be achieved by evaluation of drug photoreactivity?

Great effort is taken to stabilize a formulation in such a way that the shelf-life becomes independent on the storage conditions. Photostability of drugs and excipients should be evaluated at the formulation development stage in order to assess the effects of formulation and packaging on the stability of the final product. The information obtained should also result in label storage recommendations. The purpose of label storage recommendations is to guarantee the maintenance of the quality of the product in relation to its *safety*, *efficacy* and *acceptability* throughout the proposed shelf-life, i.e during *storage*, *distribution* and *use* (including reconstitution or dilution as recommended in the labelling). Details on photostability will also be helpful for advising the patient to avoid direct sun, wear sunglasses and use sun protective creams in order to minimize side-effects.

1.3 How to obtain information on drug photoreactivity?

In contrast to the testing of thermal stability there are no established guidelines for the photostability testing of drugs. Hence, testing procedures differ widely among pharmaceutical laboratories (Anderson *et al.*, 1991). This is reflected in the large discrepancy in stability data reported and may in some cases also result in misleading information. In many cases it will be expected by regulatory reviewers that adequate studies have been conducted. The lack of any specific guidelines concerning photostability testing of pharmaceuticals has been the focus of considerable effort by members of an expert working group of the International Conference of Harmonization (ICH) and by members of an expert working group of the European Pharmacopoeia (Ph. Eur.). Their ultimate goal is to establish a standard protocol for photostability testing. A consensus on ICH guidelines was reached in Yokohama, Japan, in November 1995 (see Chapter 14) and a draft for a monograph in the European Pharmacopoeia is currently under discussion.

Photostability testing of drugs may be considered as consisting of two parts. Stress testing is undertaken to evaluate the overall photosensitivity of the *drug substance*. Such evaluation is not mandatory but should be established as a part of the preformulation work. The design of the photoassay is left to the applicant's discretion and may include a variety of exposure conditions. The photoassay should lead to the determination of degradation pathways, identification of degradation products and evaluation of sensitizing properties of the parent compound, its degradation products, impurities or *in vivo* metabolites. Accelerated testing includes a standardized photostability test for *drug substances* and *drug products* in order to determine the need for a label warning according to regulatory requirements. This test can be designed as a simple pass/fail test or it can be designed to allow a classification of photostability.

Knowledge about the photochemical and photophysical properties of the compound is essential for handling, packaging and labelling of the drug substance and drug product but is also needed in order to predict drug phototoxicity. Several *in vitro* methods for phototoxicity studies are previously described (Valenzeno *et al.*, 1991), but in many cases *in vivo* test methods will also be required (Oppenländer, 1988). A complete assay for photostability/phototoxicity is, however, time and money consuming and requires a broad spectrum of techniques. A selection of the drug compounds to undergo a full screening can be made on certain criteria:

- The drug or metabolites of the drug accumulate in tissues that are frequently exposed to light (skin, eye, hair).
- The drug is administered at a high accumulative dosage.
- The drug is photolabile *in vitro*.
- The drug forms photolabile degradation products or *in vivo* metabolites.
- The drug is administered topically.
- The drug molecule contains essential functionalities known to induce phototoxicity reactions.

Large structural variations are found among molecules that can act as photosensitizers in biological systems, and photostability is difficult to predict (Greenhill and McLelland, 1990). It is also important to be aware that the photostability of a pure compound can change when the sample is introduced into a biological system. Interactions between the drug substance and excipients in the actual formulation can further influence the photostability. Tests on the final product should therefore be included in the total evaluation of photostability.

1.4 Conclusion

Photostability testing of the drug substance is undertaken to evaluate the overall photosensitivity of the material for development and validation purposes and to provide information necessary for handling, packaging and labelling.

A photostability assay for pharmaceutical products should provide information related to the practical use of the product, i.e. the light-exposure conditions the product will experience under its normal applications. Well-designed photostability studies ensure the quality of the product throughout the shelf-life and guarantee its safety, efficacy and acceptability to the patient. Although photostability testing at present is not required by the regulating authorities this is about to change. Standardized experimental conditions must then be applied in stability testing. There is also an increasing demand for photoreactivity data in order to minimize side-effects of frequently used drugs. The evaluation of interactions between drugs and light should be a natural part of the research and development of new drug substances and drug products in the future.

References

ANDERSON, N. H., JOHNSTON, D., MCLELLAND, M. A. and MUNDEN, P., 1991, Photostability testing of drug substances and drug products in UK pharmaceutical laboratories, *J. Pharm. Biomed. Anal.*, **9**, 443–9.

BAYLAY, H., GASPARRO, F. and EDELSON, R., 1987, Photoactivatable drugs, *TIPS*, **8**, 138–43.

BEIJERSBERGEN VAN HENEGOUWEN, G. M. J., 1981, Photochemistry of drugs *in vitro* and *in vivo*, in Breimer, D. D. and Speiser, D. (eds), *Topics in Pharmaceutical Sciences*, pp. 233–56. Holland: Elsevier/North-Holland Biomedical Press.

DE VRIES, H., BEIJERSBERGEN VAN HENEGOUWEN, G. M. J. and HUF, F. A., 1984, Photochemical decomposition of chloramphenicol in a 0.25% eyedrop and in a therapeutic intraocular concentration, *Int. J. Pharm.*, **20**, 265–71.

EPSTEIN, J. H. and WINTROUB, B. U., 1985, Photosensitivity due to drugs, *Drugs*, **30**, 42–57.

GREENHILL, J. V. and MCLELLAND, M. A., 1990, Photodecomposition of drugs, in Ellis, G. P. and West, G. B. (eds), *Progress in Medicinal Chemistry*, pp. 51–121. Holland: Elsevier Science Publishers, B.V.

HARBER, L. C., KOCHEVAR, I. E. and SHALITA, A. R., 1982, Mechanisms of photosensitization to drugs in humans, in Regan, J. D. and Parrish, J. A. (eds), *The Science of Photomedicine*, pp. 323–47. New York: Plenum Press.

LOVEDAY, K. S. and BERGMAN, C. L., 1994, Trends in phototoxicity testing, *Drug Cosmetic Ind.*, **155**, 30–7.

MATSUDA, Y. and TATSUMI, E., 1990, Physicochemical characterization of frusemide modifications, *Int. J. Pharm.*, **60**, 11–26.

NYQVIST, H. and WADSTEN, T., 1986, Preformulation of solid dosage forms: Light stability testing of polymorphs as a part of a preformulation program, *Acta Pharm. Technol.*, **32**, 130–2.

OPPENLÄNDER, T., 1988, A comprehensive photochemical and photophysical assay exploring the photoreactivity of drugs. *CHIMIA*, **42**, 331–42.

REISCH, J. and ZAPPEL, J., 1993, Photostabilitätsuntersuchungen an Natrium-Warfarin in kristallinem Zustand, *Sci. Pharm.*, **61**, 283–6.

SPIKES, J. D., 1985, The historical development of ideas on applications of photosensitized reactions in the health sciences, in Bensasson, R. V., Jori, G., Land, E. J. and Truscott, T. G. (eds), *Primary Photo-Processes in Biology and Medicine*, pp. 209–77. New York: Plenum Press.

SQUELLA, J. A., ZANOCCO, A., PERNA, S. and NUÑEZ-VERGARA, L. J., 1990, A polarographic study of the photodegradation of nitrendipine, *J. Pharm. Biomed. Anal.*, **8**, 43–7.

TØNNESEN, H. H., 1989, Emballasjens betydning ved formulering av fotokjemisk ustabile legemidler, *Norg. Apot. Tidsskr.*, **97**, 79–85.

TØNNESEN, H. H. and KARLSEN, J., 1987, Studies on curcumin and curcuminoids. X. The use of curcumin as a formulation aid to protect light-sensitive drugs in soft gelatin capsules, *Int. J. Pharm.*, **38**, 247–9.

1995, Photochemical degradation of components in drug formulations. III. A discussion of experimental conditions, *PharmEuropa*, **7**, 137–41.

VALENZENO, D. P., POTTIER, R. H., MATHIS, P. and DOUGLAS, R. H., 1991, *Photobiological Techniques*, pp. 85–120, 165–78, 347–9. London: Plenum Press.

2

Photophysical and Photochemical Aspects of Drug Stability

D. E. MOORE

2.1 Absorption spectra of drugs

A photon corresponding to the ultraviolet wavelength 300 nm has an energy of 400 kJ mol^{-1}, which is of comparable magnitude to the bonding energy of organic compounds. The fact that a drug absorbs radiation in the ultraviolet or visible region of the electromagnetic spectrum means that it is absorbing energy that is sufficient to break a bond in the molecule. Thus the property of absorption is a first indication that the drug *may* be capable of participating in a photochemical process leading to its own decomposition or that of other components of the formulation. The statement is a qualified one because there are a number of processes that may occur following absorption of UV or visible light, some of which lead to no net change to the absorbing molecule or the system. The photochemical reaction must follow the basic law of photochemical absorption, established by Grotthus and Draper in 1818, that no photochemical (or subsequent photobiological) reaction can occur unless electromagnetic radiation is absorbed. The absorption spectrum of a compound is therefore an immediate way of determining the wavelength range to which the drug may be sensitive.

Some drug substances and formulation excipients are coloured, meaning that they absorb light in the visible region. The colour they display is complementary to the light they absorb, e.g., a red powder is absorbing blue light. The great majority of therapeutic substances are white in appearance, meaning that they do not absorb light in the visible region, but they may absorb in the UV region as a consequence of their chemical structure. The presence of aromatic residues and conjugated double bonds containing N, S or O is usually associated with the absorption. Two contrasting examples, ibuprofen and sulindac, chosen from the wide range of anti-inflammatory drugs, are given in Fig. 2.1. Ibuprofen is a white powder with a weak absorption centred on 265 nm, due to the aromatic chromophore, unaffected by substituents. On the other hand, sulindac is yellow in colour, and absorbs strongly across both the UV and visible regions with maxima at 280 and 327 nm due to the extended chromophore.

When each of these compounds is irradiated with wavelengths corresponding

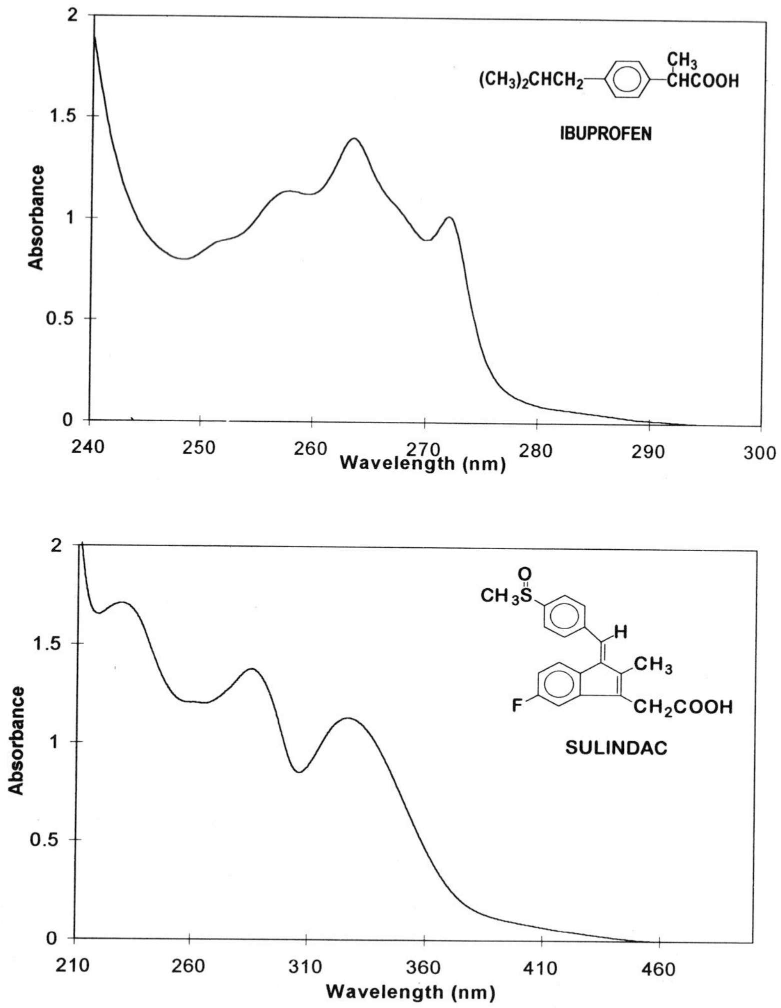

Figure 2.1 Structure and absorption spectrum of (a) ibuprofen and (b) sulindac

to their maximum absorption, photodegradation occurs, but when the extent of decomposition is equated with the amount of radiation absorbed, it transpires that ibuprofen is significantly more photoreactive than sulindac (unpublished results). The difference in the way we have to store these two substances stems, of course, from the fact that ibuprofen would experience exposure to UV radiation of around 265 nm only under the most unusual storage conditions involving germicidal lamps which emit at 254 nm, while sulindac can absorb the output from regular room lighting. Under normal storage ibuprofen does not require amber glass whereas sulindac does.

There are two important factors to ponder in relation to the potential of a drug to be degraded following absorption of electromagnetic radiation. First, the absorption spectrum is normally described by the maximum absorption wavelength and the molar absorptivity at that wavelength, but the spectrum can be broad and

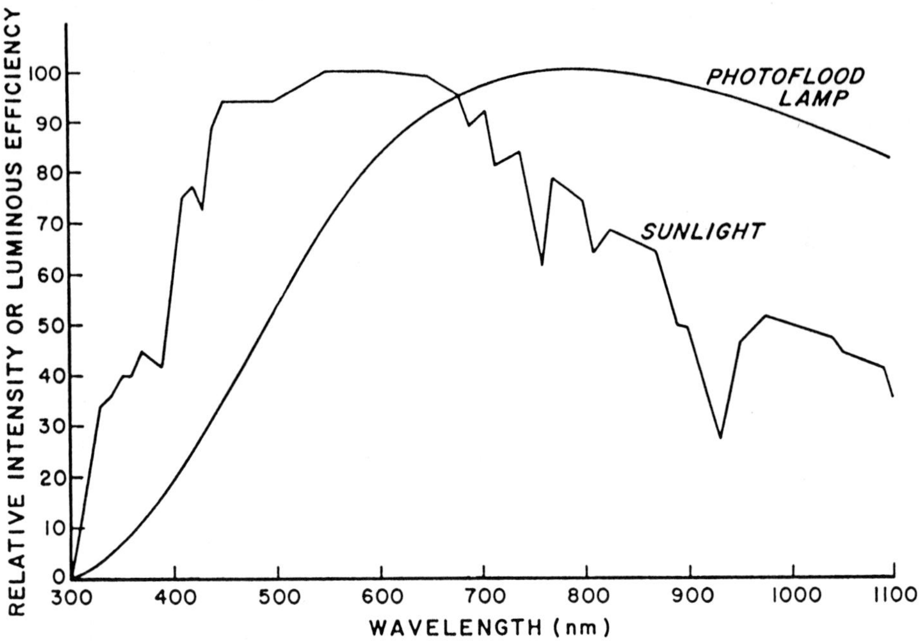

Figure 2.2 Spectral power distribution of sunlight compared with an incandescent lamp (the curves are normalized)

any overlap of the absorption spectrum with the output of the photon source impinging upon it has the potential to lead to photochemical change. Second, the decomposition may be initiated by another component of the formulation which has the absorption characteristics that overlap with the incident radiation while the therapeutic component does not. In this case, the process is called photosensitization and the absorbing component, or photosensitizer, may transfer the absorbed energy completely and not be altered itself in the process, although it is more likely that it will undergo some degradation.

2.2 Spectral characteristics of sunlight and artificial light sources

Pharmaceuticals may be exposed to electromagnetic radiation from a number of sources, ranging from direct sunlight, through filtered sunlight to a variety of artificial light conditions. In terms of the possibility of photochemical reaction, the UV component of sunlight is the most potentially damaging, but there may be long exposures to fluorescent and incandescent lighting during the various stages of manufacture, storage and use, so it is important to consider their spectral distribution as well. Relative spectral intensity curves for sunlight and an incandescent (filament) lamp are shown in Fig. 2.2. Each of these extends from near 300 nm in the ultraviolet region to beyond 3000 nm in the infrared, but with differing distribution.

Ultraviolet radiation (UV-R) has been divided into three subbands: UV-C, UV-B and UV-A (Grossweiner, 1989). The UV-C band ranges from 200 nm to 290 nm and is often called shortwave or far-UV because the wavelengths in this region are the shortest UV-R transmitted through air. Although most drugs and all cellular constituents absorb UV-C, sunlight at the earth's surface contains no UV-C because of efficient absorption by molecular oxygen and ozone in the upper atmosphere (Frederick *et al.*, 1989). Despite its absence from natural sunlight at the earth's surface, UV-C is present in artificial radiation sources such as discharge and germicidal lamps and welding arcs and can cause rapid photochemical degradation, as well as serious damage to the skin and cornea following exposure. The determination of chemical and biological effects of UV-C is still receiving much attention today partially because of the increasing knowledge of far-UV photochemistry and the specificity of the damage generated (Cadet *et al.*, 1992).

The UV-B spectral region is often defined as encompassing wavelengths from 280 to 320 nm (Grossweiner, 1989). However, no solar radiation penetrates to the ground at wavelengths between 280 and 290 nm, and this remains true even in the case of a large reduction in atmospheric ozone such as occurs over Antarctica in springtime. Therefore it has been suggested that the interval from 290 to 320 nm be adopted as a practical definition of the UV-B (Frederick, 1993). The purine and pyrimidine bases of DNA and the aromatic amino acids are the major cellular absorbers of UV-B. Although the intensity of UV-B in the solar UV-R reaching the earth's surface is relatively small (Thorington, 1985), it is abundantly clear that UV-B is the most important band since it causes sunburn, skin cancer and other biological effects, and is responsible for the direct photoreaction of many chemicals in natural sunlight (Epstein, 1989). The UV-B intensity at a particular latitude varies greatly with time of day and the season of the year, as the variation of the solar azimuthal angle varies the pathlength of the sun's rays through the stratospheric ozone layer.

UV-A is the long wavelength UV region from 320–400 nm. It is also called near-UV because of its proximity to the visible spectrum. In total energy the amount of solar UV-A reaching the earth's surface is enormously greater than that of UV-B (Gates, 1966). Chemical and biological effects induced by UV-A may involve either direct energy absorption, e.g., in the long wavelength absorption tail of proteins and DNA, or photosensitization by endogenous or exogenous substances.

Sunlight has a very high output in the visible (400–800 nm) and infrared (800–3200 nm) regions, while the incandescent lamp typifies black body radiation with a higher relative infrared output. The only importance that infrared radiation can accrue in the context of photodegradation is that the sample can be heated, thereby activating thermal decomposition. The visible region is relevant when a coloured substance is present in the formulation.

Artificial light sources can have varying spectral characteristics depending on the particular construction. The key components of a fluorescent light are the low pressure mercury discharge at 254 nm within a glass tube coated internally with a phosphor having specific emission characteristics. The spectral power distribution of several commercial fluorescent artificial light sources is shown in Fig. 2.3. While the principal output is in the visible region, there is a significant UV component. Note also the discontinuous line spectrum superimposed upon the background of continuous radiation. It has been estimated that at least 90 per cent of all lighting in the business and manufacturing sectors in the USA is achieved by 'cool-white' fluorescent tubes, while the domestic sector uses incandescent lighting for 80 per

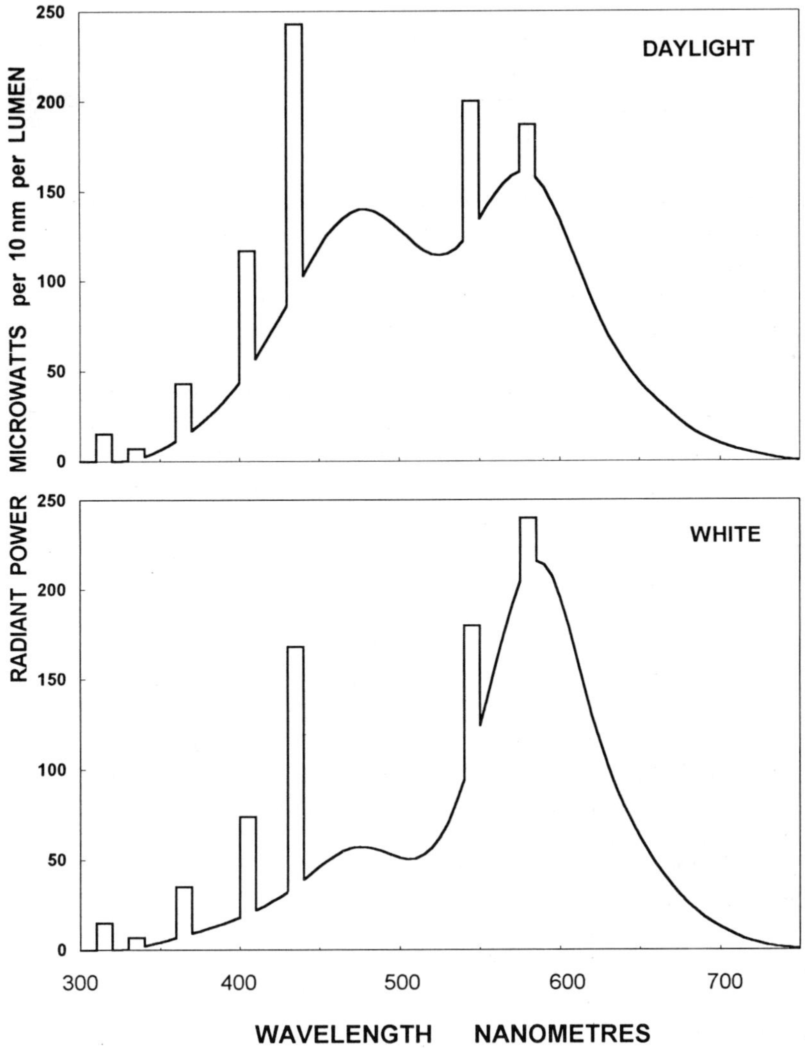

Figure 2.3 Spectral power distribution of 'Daylight' and 'White' fluorescent lamps (redrawn from Thorington, 1985)

cent of its artificial light needs (Thorington, 1985). The glass bulb or tube in an artificial light source can be said to act as the ozone layer does with respect to natural sunlight, limiting the UV-R component to about 300 nm, depending on the glass used. According to Thorington (1985), there are no criteria for the UV-R component in most commercial artificial light sources because the sole function is to provide light in the narrow definition of illuminating engineering (i.e., visible light).

2.3 Action spectrum and overlap integral

There are two rather different ways in which the term *action spectrum* has been used. The first usage is strictly incorrect in that it relates to the overlap integral

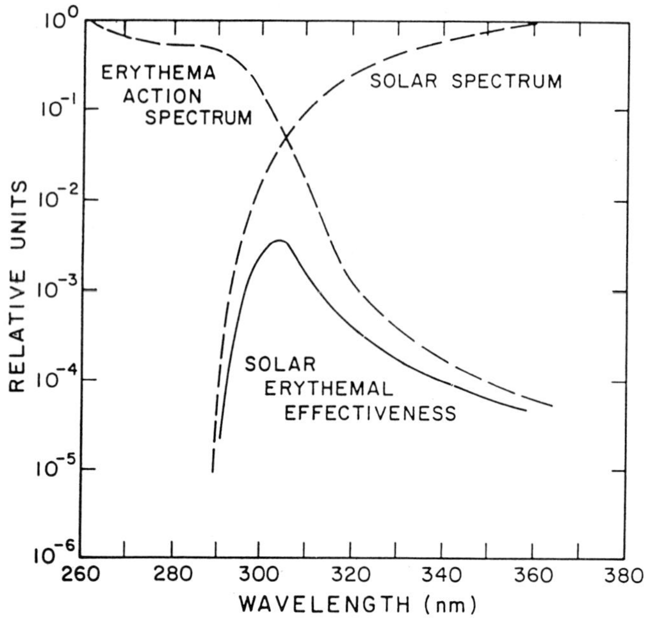

Figure 2.4 Erythemal response (sunburn) action spectrum, the midday solar spectrum and the resultant erythemal effectiveness spectrum (Parrish *et al.*, 1978)

of the spectra for the particular combination of photon source and absorbing substance. A familiar example of this definition of an action spectrum is the sunburn or *erythemal effectiveness spectrum* which is the overlap of the sunlight UV spectrum and the absorption spectra of proteins and nucleic acids as shown in Fig. 2.4 (Jagger, 1985). The sunburn response (erythema) can be elicited in human skin by an artificial light source emitting any of the wavelengths corresponding to the absorption spectra of protein and nucleic acid. Sunburn (as caused by the sun) occurs only for the narrow range of wavelength for which the overlap with the solar emission is finite. This type of overlap integral would be found for quite a large number of drugs whose absorption maxima occur at around 270 to 280 nm with a broad tail extending into the UV-B region. For examples one need look no further than the sulfonamide group of antibacterials. Sulfamethoxazole has its absorption maximum at 268 nm but is decomposed on exposure to sunlight. Indeed, such is its change in decomposition rate with time of day and season of the year that its use has been suggested for an absolute chemical reaction standard for measuring the seasonal variation in UV-B intensity (Moore and Zhou, 1994).

The second usage of the term *action spectrum* is the more correct, according to photochemists. The action spectrum is obtained by measuring the radiation dose required to evoke the same chemical or biological response at different wavelengths. It will usually coincide with the absorption spectrum of the compound when the irradiation source variation with wavelength is corrected. To parallel the absorption spectrum, an action spectrum should meet the following conditions: (1) the action mechanism is the same at all wavelengths; (2) the quantum yield is the same at all wavelengths; (3) absorption of radiation by inactive chromophores, and radiation scattering is negligible; (4) only a small fraction of the incident radiation is absorbed by the sample in the wavelength range of interest; (5) the exposure time is inversely proportional to the fluence rate for the same effect. The action spectrum for any specific photosensitized reaction would normally overlay the absorption spectrum of the sensitizer (Grossweiner, 1989). The erythemal response (sunburn) spectrum in Fig. 2.4 is an example.

In the context of drug photostability, the *action spectrum* is less important, in that the formulation developer is concerned with the overlap of the drug's spectrum with the spectral output of the incident radiation. In order to avoid confusion, the term *overlap integral* is recommended for this situation.

2.4 Penetration of UV

The extent to which UV radiation (UV-R) is able to provoke reactions is obviously dependent on its penetration of the system. For pharmaceutical formulations, this will depend on the degree of transparency of the packaging material. The most frequently used materials for which this is an issue are glass and plastic, but there is a variation in light transmission characteristics caused by different compositions. The transmission cut-off can only be clearly delineated in terms of a filter of defined composition and thickness. Thus the Corning glass filter O-53 in Fig. 2.5 corresponds to standard Pyrex glass and can be characterized as giving 50 per cent transmission at 310 nm for a sheet of 2 mm thickness. Note, however, that it still transmits 1 per cent at 280 nm. What this means is that glass will cut down the incident radiation in the UV-B region by a significant proportion but not completely. Thus, for a substance which absorbs in the UV-B region, but whose absorption spectrum does not extend above 310 nm, storage in glass containers is not sufficient to protect it. If the substance is exposed for long enough, there remains the possibility of photochemical reaction. Also shown in Fig. 2.5 is the transmission spectrum of a plastic film as used for overhead transparencies. For experimental purposes, this film provides a much sharper cut-off than does glass,

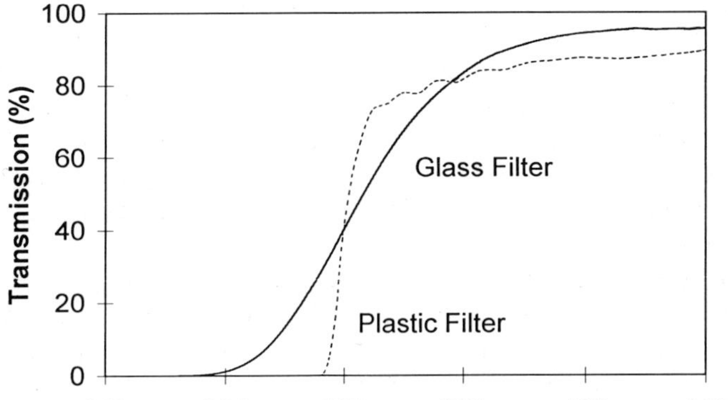

Figure 2.5 UV transmission curves of Corning O-53 glass filter and a plastic filter made from overhead transparency film

although it does not completely exclude the UV-B region. The transmission characteristics of plastics vary according to their composition.

With respect to human response to UV-R, the transmission of Caucasian skin is such that most of the UV-R shorter than 320 nm is absorbed in the stratum corneum (Epstein, 1989). To evoke a photochemical reaction in the skin, UV-R must penetrate to the site of the absorbing molecule in the peripheral blood capillaries. The penetration is governed by the optical properties of the skin and is modified by absorption by melanin and scattering processes which vary dramatically with wavelength (Diffey, 1983). The transmission of radiation through the human stratum corneum, for example, was estimated to vary from 15 per cent at 297 nm, through 33 per cent at 313 nm and 50 per cent at 365 nm, to 72 per cent at 546 nm (Bruls *et al.*, 1984). It is widely accepted that UV-A can penetrate into non-melanized skin and has the potential to cause photoreactions in the skin at a greater depth than UV-B which can only reach the viable layers of the epidermis (Lovell, 1993).

2.5 Excited states, radiative and non-radiative processes

Photochemical damage to a substance is initiated by the absorption of energy by the compound itself or by a photosensitizer. Many photochemical reactions are complex, and may involve a series of competing reaction pathways in which oxygen may play a significant role. In fact, the great majority of photoreactions in biological systems involve the consumption of molecular oxygen and are photosensitized oxidation processes (Spikes, 1989).

Consider first the photophysical processes, which can be best described by an energy level diagram (Fig. 2.6) and equations (2.1) to (2.7).

Absorption $$D_0 + h\nu \rightarrow {}^1D \quad (2.1)$$

Internal conversion $${}^1D \rightarrow D_0 \quad (2.2)$$

Fluorescence $${}^1D \rightarrow D_0 + h\nu' \quad (2.3)$$

Photoionization $${}^1D \rightarrow D^{\cdot +} + e^- \quad (2.4)$$

Intersystem crossing $${}^1D \rightarrow {}^3D \quad (2.5)$$

Intersystem crossing $${}^3D \rightarrow D_0 \quad (2.6)$$

Phosphorescence $${}^3D \rightarrow D_0 + h\nu'' \quad (2.7)$$

Any UV-R or visible light-induced process begins with the excitation of drug molecules or sensitizers from their ground state (D_0) to reactive excited states, by absorption of photons of certain wavelengths. As shown in equation (2.1), upon absorption of radiation, the drug molecule, D_0, in the ground state in which the valence electrons are paired or antiparallel (a spin singlet state) is raised to a higher energy level, as a valence electron moves to the first available outer shell corresponding to the first excited singlet state ^1D (the electron spins remain antiparallel). When the absorption spectrum shows more than one absorption band, it indicates a corresponding number of excited states which can be reached by irradiation with the appropriate excitation wavelength. For example, sulindac can be raised to the second excited state when irradiated with UV-R in the

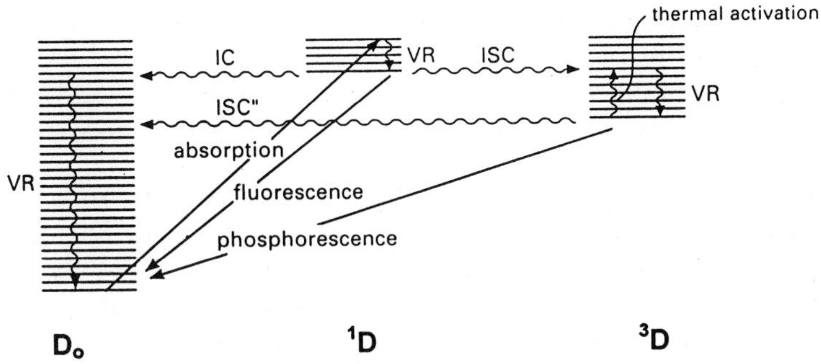

Figure 2.6 Energy levels of molecules, showing transitions involving fluorescence, phosphorescence, internal conversion and intersystem crossing

wavelength range around 280 nm, while longer wavelengths around 327 nm yield the first excited state only (Fig. 2.1).

The molecule cannot persist in an excited state indefinitely since it represents a situation which is less stable with respect to the ground state. There are a variety of competing physical processes involving energy dissipation and resulting in deactivation of the excited states. The energy dissipation may be via either internal conversion (IC) (equation (2.2)), which is a non-radiative transition between states of like multiplicity, or via photon emission (fluorescence) resulting in return to D_0 (equation (2.3)). Even if excitation occurs to an excited state higher than the first, IC will always bring the molecule to the ^1D level (within a picosecond) before fluorescence occurs. Thus the fluorescence emission wavelength is the same, irrespective of the irradiating wavelength. Any excess energy within a particular electronic state is dissipated as heat by collision with neighbouring molecules. This is referred to as vibrational relaxation (VR). As the lifetime of the excited singlet state of a molecule is generally of the order of nanoseconds (but up to microseconds for rigid molecular structures), the possibility of interaction with neighbouring molecules leading to chemical change is limited at this stage. However, in the excited singlet state, the ionization potential of the molecule is reduced, and the excited electron is more easily removed than it is from the ground state molecule, but requires that an appropriate acceptor be present. This process of photoionization (equation (2.4)) is also more likely to occur if higher energy UV-R is used (i.e., wavelengths less than 300 nm) and if the molecule is in the anionic state. Alternatively, intersystem crossing (ISC) may occur from the excited singlet state to a metastable excited triplet state ^3D (electron spins parallel) (equation (2.5)). Despite the low probability in general for transfer between states of differing multiplicity, ISC occurs with relatively high efficiency for most photochemically active molecules. The excited triplet state, because of its longer lifetime (microseconds to seconds, or even longer), may diffuse a significant distance in fluid media and therefore has a much higher probability of interaction with other molecules. If no interaction occurs, it decays back to the ground state by a further ISC event (equation (2.6)), or by phosphorescence emission (equation (2.7)).

The nature of the excited state decay processes is studied by the technique of laser flash photolysis, a description of which has been given by Bensasson, Land

and Truscott (1983). Briefly, flash photolysis involves irradiating a sample with a short (nanosecond) intense pulse from a laser, then observing by rapid response spectrophotometry the spectral changes that occur on the time scale nanoseconds to milliseconds. Several standard tests have been established to aid in the identification of the transient species. Solvated electrons generated by photoionization in a nitrogen-gassed solution, have a characteristic broad structureless absorption peak at about 700 nm depending on the solvent (720 nm in aqueous solution). Oxygen quenches this absorption and also quenches the triplet state, while nitrous oxide gassing can be used to quench the solvated electron only, thereby gaining an indication of any transient absorption which arises from the triplet state. One difficulty with flash photolysis experiments at present lies with the laser exciting source. To achieve the required pulse intensity, the source usually employed is a Nd-YAG laser emitting at 1064 nm, with frequency doubling to produce 532 nm, tripling to 355 nm, and quadrupling to 266 nm. In the majority of drugs, this provides excitation at very specific wavelengths 266 or 355 nm, leaving an unfortunate gap in the 280 to 340 nm region. Thus for many drug molecules whose absorption does not extend to 355 nm, one is forced to use the high energy 266 nm excitation, which may produce upper excited states and lead to events such as photoionization. In the context of photodegradation initiated by UV-R greater than 300 nm, some of these events may not be relevant. The efficiency, or quantum yield, of each of the processes described by equations (2.2) to (2.7), is defined as the fraction of the molecules excited by absorption (equation (2.1)) which then undergo that particular mechanism of deactivation. While the quantum yield of fluorescence is readily determined by reference to quinine fluorescence as described by Calvert and Pitts (1966), those of the other processes can only be obtained by difference. Phosphorescence is usually too weak to be observed in solution at room temperature, but can be measured if the drug is held in a glassy matrix at low temperature. The usual procedure is to dissolve the drug in ethanol and immerse in liquid nitrogen. The phosphorescence accessory of the fluorimeter incorporates a mechanical chopper enabling the phosphorescence to be observed free of interference from any fluorescence. Because of the difference in temperature and matrix, it is not possible to compare the phosphorescence yield with that of fluorescence. Nevertheless, phosphorescence is worth measuring because it is an important indicator of the capacity of a molecule to populate its triplet state.

2.6 Direct reactions from the excited states of the drug

The excited molecule has a different electronic character compared to the ground state, and is often able to form a complex (called an exciplex) with another species which will be designated as Q, i.e., the complex is $D^{\star}Q$. The symbol Q is used because, in effect, the interacting molecule is a quencher of the native fluorescence of D. Sometimes at high concentration of the absorbing molecule, this occurs with the ground state itself (in which case the $D^{\star}D$ species formed is called an excimer). The formation of the exciplex or excimer is observed as a shift in the fluorescence emission to longer wavelength, the difference in energy between exciplex and normal fluorescence reflecting the stability of the exciplex. More details of this type of interaction can be found in Gilbert and Baggott (1991).

The substances for which this phenomenon has been observed are invariably polycyclic aromatic hydrocarbon structures. No exciplex formation has been reported in the literature to involve drug molecules, but this remains a possibility in concentrated solution or perhaps in solid state mixtures. The consequences of exciplex formation are radiative or non-radiative return to the ground state without chemical change, or electron transfer leading to chemical reaction of the drug, the quencher or both. Many photoaddition processes are postulated to proceed via exciplex formation with the quencher molecule becoming chemically bound.

The electronically excited state of a molecule will act as a more powerful electron donor or acceptor than the ground state. The reactions that can occur are, respectively, oxidative or reductive quenching:

$$\text{Oxidative quenching } D^{\star} + Q \rightarrow D^{\star}Q \rightarrow D^{+\cdot} + Q^{-\cdot} \tag{2.8}$$

$$\text{Reductive quenching } \rightarrow D^{-\cdot} + Q^{+\cdot} \tag{2.9}$$

The exact nature of the reaction (oxidative vs. reductive) will depend on the redox properties of $D^{\star}$ and Q. The electron transfer process is a special case of exciplex formation which is favoured in the strongly polar solvents, such as water. The involvement of an exciplex in a photochemical reaction is generally established by studying the effects of known exciplex quenchers such as amines on both the exciplex fluorescence and the product formation. The heavy atom effect, due to the presence of substituents such as bromine or iodine either intra- or intermolecular, causes an exciplex to move to the triplet state preferentially, with a quenching of fluorescence.

2.6.1 *Photodehalogenation reactions*

In regard to drug photodegradation reactions that appear to involve exciplex formation, the most frequently observed are those in which an aromatic chlorine substituent is lost in the photoreaction. Examples of drugs which lose their chlorine substituent are chlorpromazine (Davies *et al.*, 1976), hydrochlorthiazide (Tamat and Moore, 1983), chloroquine (Moore and Hemmens, 1982), frusemide (Moore and Sithipitaks, 1983) and diclofenac (Moore *et al.*, 1990). In each case, when the drug (Aryl-Cl) is photolysed in aqueous or alcoholic (ROH) solution, HCl is liberated and a mixture of reduction (Aryl-H) and substitution (Aryl-OR) products is obtained. This is exemplified by the photodegradation of diclofenac shown in Fig. 2.7. The photodechlorination occurs for these compounds more strongly in deoxygenated solution. When oxygen is present, it promotes ISC to the triplet state and the production of singlet oxygen (see below).

The mechanism is by no means completely clear, but the photodehalogenation reaction is postulated to occur through the formation of a pair of radical ions from an exciplex resulting in the excited state (Grimshaw and de Silva, 1981). The precursor of the reduction product (Aryl-H) is suggested to be a radical anion ($\text{Aryl-Cl}^{-\cdot}$) while a radical cation ($\text{Aryl-Cl}^{+\cdot}$) is postulated as the precursor of the substitution product (Aryl-OR). In a less polar solvent, e.g., iso-propanol, direct homolysis of the C–Cl bond occurring from the triplet state has been suggested based on flash photolysis experiments with chlorpromazine (Davies *et al.*, 1976). 3,3′,4′,5-Tetrachlorosalicylanilide represents a class of antibacterial agents

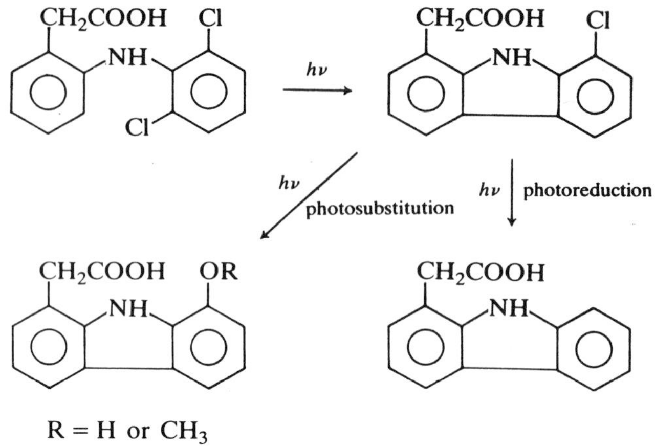

R = H or CH_3

Figure 2.7 Photodegradation of diclofenac in aqueous solution at pH 7 (from Moore *et al.*, 1990)

formerly used in cosmetics and soaps. These compounds were found to undergo sequential photodehalogenation which was presumed to be related to their capacity to induce skin rashes upon sunlight exposure (Davies *et al.*, 1975).

Not all chloroaromatic drugs appear to follow this type of reaction. For example, free chloride ion is not formed on irradiation of chlordiazepoxide for which an oxaziridine is the major photoproduct (Cornelissen *et al.*, 1979). There is a variability among reports on other drugs which contain chlorine substituents. This can arise due to differences in the irradiation conditions. If an unfiltered mercury arc source is used, the sample will receive 254 nm irradiation and the C–Cl bond will certainly break, while under longer wavelength irradiation (> 300 nm) the bond may be stable.

2.7 Photosensitized reactions

Any photochemical process in which there is a transfer of reactivity to a species other than that absorbing the radiation initially, is called a photosensitization reaction. As a result of the long lifetime and the bi-radical nature with unpaired electron spins, the excited triplet states can mediate photosensitized reactions, the most common of which are photosensitized oxidations. Due to the triplet spin nature of its ground state, oxygen is spin matched with the drug triplet state, and also is a very good scavenger of free radicals. These characteristics lead to two distinct mechanisms of photooxidation, as shown in Scheme 2.1 using AH to refer to an oxidizable substrate.

The excited triplet sensitizer can undergo its primary reaction with molecules in its vicinity by (1) electron transfer including simultaneous transfer of a proton corresponding to the transfer of a hydrogen atom resulting in free radical reactions (equations (2.10) to (2.12)), termed Type I or free radical reaction or (2) energy transfer, with spin conservation, to ground-state molecular oxygen ($^3O_2$) to form singlet oxygen (equation (2.14)), termed Type II reaction (Spikes, 1989).

Both Type I and II processes can take place simultaneously in a competitive

Free radical formation	$^3D + AH$	$\rightarrow D_0 + A^{\cdot} + H^+$	(2.10)
		or $DH^{\cdot} + A^{\cdot}$	(2.11)
		or $D_0 + A^{\cdot -} + H^+$	(2.12)
(Type I photooxidation)	$A^{\cdot}$ (or $A^{\cdot -}$) $+ O_2$	$\rightarrow AO_2^{\cdot}$	(2.13)
Energy transfer to oxygen	$^3D + {}^3O_2$	$\rightarrow D_0 + {}^1O_2$	(2.14)
Oxidation by singlet oxygen	$^1O_2 + AH$	$\rightarrow AOOH$	(2.15)
(Type II photooxidation)			

Scheme 2.1: Photosensitized Oxidation Reactions – Types I and II

fashion, as in the cases of thionine (Kramer and Maute, 1973) and chlorpromazine (Moore and Burt, 1981). The distribution between the two processes depends on the sensitizer, the substrate, the solvent, and the oxygen concentration, as well as the affinity of sensitizer and substrate (Henderson and Dougherty, 1992). One of the processes may be dominant in a specific system. For example, in an air-saturated aqueous solution at neutral pH the excited triplet of the dye Rose Bengal reacts overwhelmingly with oxygen rather than directly with DNA (Lee and Rodgers, 1987). For 2-methyl-1,4-naphtho-quinone, however, a similar study revealed that the one-electron transfer to thymine can effectively compete with singlet oxygen formation (Fisher and Land, 1983).

2.7.1 Type I photosensitization of chain reactions

The Type I mechanism of photosensitization commonly proceeds through the transfer of electrons or protons, depending on the polarity of the medium (Foote, 1968). The formed cation or neutral radical is expected to undergo further reactions which, in the absence of oxygen, means recombination, dimerization or disproportionation. When oxygen is available in sufficient concentration, there is a rapid addition of molecular oxygen to the radical. The peroxy radical which is formed is also reactive and will seek to stabilize itself by proton abstraction from neighbouring molecules. If the sample consists of a high concentration of the drug, the extent to which the reaction continues will depend on the reactivity of the drug. This sequence may be thought of as a chain reaction because the radical activity is continually transferred and kept ‘alive’. Except in very unusual structures, free radicals are considered as high reactivity species, but there is a need for a suitable donor or acceptor in the near vicinity. Secondary alcohols are examples of molecules with readily abstractable hydrogens. Thus *iso*-propanol, mannitol and ascorbic acid are very good scavengers of free radicals, and can be used to protect the therapeutic substance while themselves undergoing oxidation.

The chain reaction mechanism is frequently referred to as autocatalytic, because it starts slowly but the rate becomes faster as the reaction proceeds. There are not many known examples of drug substances that decompose by a free radical chain mechanism since the process requires the participation of a very reactive (i.e., unstable) compound. This usually means a compound susceptible to oxidation and is illustrated by the photooxidation of benzaldehyde, as shown in Scheme 2.2 (Moore, 1976):

Initiation $\text{Ph-CHO} + h\nu \rightarrow \text{Ph-CO}^{\cdot} + \text{H}^{\cdot}$

Propagation $\text{Ph-CO}^{\cdot} + \text{O}_2 \rightarrow \text{Ph-CO-OO}^{\cdot}$
peroxy radical

$\text{Ph-CO-OO}^{\cdot} + \text{Ph-CHO} \rightarrow \text{Ph-CO-OOH} + \text{Ph-CO}^{\cdot}$

peroxybenzoic acid

Termination $\text{Ph-CO-OO}^{\cdot} + \text{Ph-CO-OO}^{\cdot} \rightarrow$ Dimeric peroxy compounds

Scheme 2.2: Chain reaction mechanism for the degradation of benzaldehyde

While the peroxy products are themselves unstable and will break down potentially generating new free radical species, the faster processes are those given as the propagation steps in Scheme 2.2. Although benzaldehyde has only a weak $n \rightarrow \pi^*$ absorption at 320 nm, it is only necessary to generate one radical by dissociation of an excited state molecule. This is quite sufficient to set a chain reaction into progress resulting in the oxidation of thousands of benzaldehyde molecules (depending on temperature). The free radical chain reaction is categorized in terms of the *chain length* which means the number of propagation steps occurring for every initiation event. In this case the chain length, and also the quantum yield for the overall photochemical process, will be in the thousands.

The limit to the chain reaction is determined by the relative values of the rate constants for the propagation step and the branching or transfer reactions involving solvent or inhibitor molecules. As the concentration of the oxidizable molecule falls in the solution, the reaction rate also falls. The reaction is characterized by a 'steady state' or maximum rate, represented by the linear portion of the sigmoidal reaction progress curve. This is achieved when the rate of generation of new initiating radicals is equal to their termination rate. Here the kinetics are simplified by the 'steady state approximation' and the maximum rate is first order with respect to the benzaldehyde concentration.

Inhibition of chain processes is achieved by the addition of free radical scavengers which react by chain transfer more rapidly than the propagation step. The product of chain transfer is also a free radical, but the key to the transfer agent being a good inhibitor is that it must be a very unreactive radical, such as the sterically hindered radicals formed from the widely used antioxidants BHT and BHA.

Chain reactions are the major pathway by which hydrocarbon polymers as used in packaging are broken down, with the radicals for initiation arising from photo-induced decomposition of trace amounts of peroxide or hydroperoxide impurities. Indeed, the development of 'biodegradable packaging' is an application of this principle. Figure 2.8 shows an example of the chain reaction process leading to the breakdown of a hydrocarbon polymer backbone.

In biological systems, free radicals can react with cellular macromolecules in a variety of ways, the most important of which is hydrogen abstraction from DNA leading to chain scission or crosslinking. In proteins, tryptophan is the amino acid residue most susceptible to free radical attack. Lipid peroxidation by free

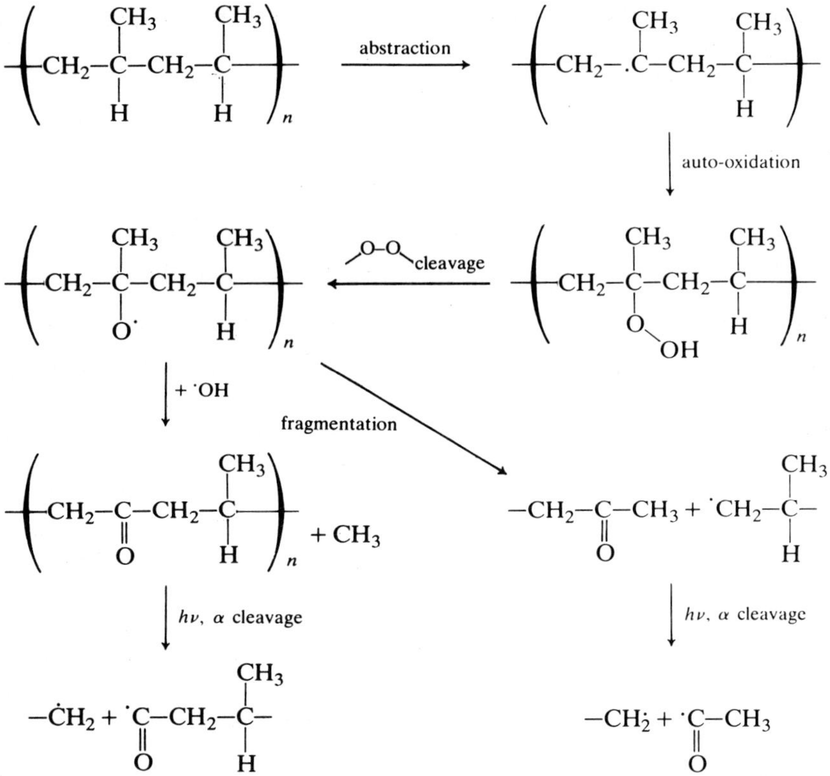

Figure 2.8 Photodegradation of a hydrocarbon polymer (from Gilbert and Baggott, 1991)

radicals in turn is liable to cause alteration in cell membranes (Grossweiner and Smith, 1989).

2.7.2 ***Electron-transfer sensitized photooxidation***

As mentioned above in the discussion of exciplex formation, electron transfer between an excited state species and a ground state molecule (equations (2.8) and (2.9)) is frequently observed in the photochemistry of systems containing an electron donor and acceptor combination. As a result, a pair of radical ions are formed, both of which react with oxygen but with different rates. The reaction of ground state oxygen with radical anions occurs rapidly and yields superoxide anion (equation (2.16)). The superoxide then adds to the radical cation forming DO_2 (equation (2.17)). When D is an olefin DO_2 is a dioxetan which is liable to cleave to yield ketones as products.

$$Q^{-\cdot} + O_2 \rightarrow Q + O_2^{-\cdot} \qquad (2.16)$$

$$D^{+\cdot} + O_2^{-\cdot} \rightarrow DO_2 \qquad (2.17)$$

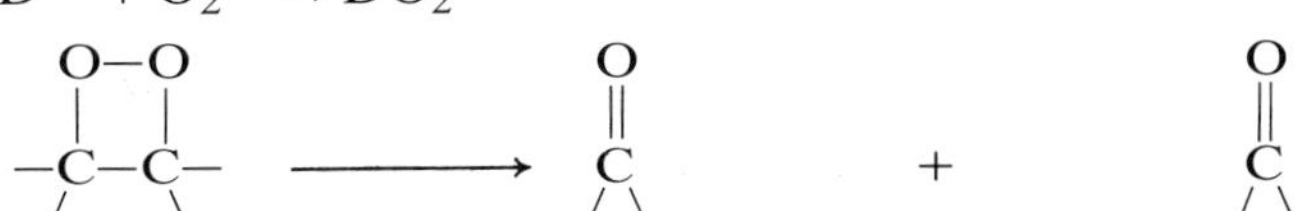

2.7.3 Detection of free radicals

The above is a simplified view of some of the processes which may occur involving free radicals generated from the excited state. Determination of the detailed reaction mechanism is a difficult task and requires knowledge of the quenching efficiency of the sensitizer excited state by the substrate, the ability of the radical anion to transfer an electron to oxygen, and the rate of reaction of the substrate radical cation with ground state oxygen. A number of techniques have been developed to enable the detection of free radical intermediates in photochemical reactions, including electron paramagnetic resonance spectroscopy (EPR). EPR is useful for radicals which are formed in relatively high concentration and persist for relatively long times. Unfortunately that is not true for the great majority of photochemical reactions, and special procedures are necessary such as rigid solution matrix isolation. Addition of free radical trapping compounds to the system (spin traps) is an alternative (Mason and Chignell, 1982; Chignell *et al.*, 1985). The superoxide anion is also readily trapped and identified by this technique.

An extremely sensitive technique able to detect the nature of radical pairs in a photochemical reaction, called chemically induced dynamic nuclear polarization (CIDNP), depends on the observation of an enhanced absorption in a nuclear magnetic resonance (NMR) spectrum of the sample irradiated *in situ* in the cavity of the NMR spectrometer. The background to and interpretation of CIDNP are discussed by Gilbert and Baggott (1991).

Probably the main technique that has been used to detect free radical intermediates in photochemical reactions is the competitive reaction rate study in which various free radical scavengers are added to the sample during irradiation, and the rate of disappearance of drug and appearance of particular products is compared with that occurring without the scavenger. Typical scavengers include ascorbic acid and glutathione for aqueous systems, and 2,6-di-t-butyl-hydroxytoluene (BHT) and α-tocopherol for lipophilic systems. However, there is some difficulty in interpreting the results of such a study, since the relative reactivity of both radicals and scavengers determine the outcome and the product profile will invariably change. If the radical intermediates are extremely reactive, they may react with the solvent before they encounter a scavenger molecule, and no change will be observed.

2.7.4 Polymerization as a detector of free radicals

The chain reaction process can be used as a diagnostic aid to determine whether free radicals are generated from a drug when irradiated. Acrylamide is an acrylic monomer which is widely used in gel electrophoresis as a polymer formed *in situ* by peroxide or UV-initiated polymerization. This monomer is a water-soluble solid, more easily handled than most other vinyl monomers, and the progress of its polymerization is readily followed by measuring the contraction in volume by dilatometry, or the increase in viscosity in a viscometer. Details of this experimental technique can be found in Moore and Burt (1981).

While this technique does not give any information as to the identity of the free radical generated by irradiation of the drug, it is a chemical amplification process in which very small concentrations of free radicals can be detected. The rate of polymerization caused by free radicals generated by the UV irradiation of a drug solution containing acrylamide is a reflection not only of the rate of radical generation, but also of their lifetime. Note that oxygen must be excluded from the system so that the polymer radicals are not scavenged and the reaction inhibited.

2.8 Singlet oxygen and its reactivity

The Type II reaction involves electronic energy transfer from the triplet excited photosensitizer to ground state molecular oxygen which is spin-matched, thereby forming excited singlet molecular oxygen while the photosensitizer is regenerated (equation (2.14)). There are two types of singlet oxygen with different spectroscopic symmetry notations, i.e., $^1\Delta_g$ and $^1\Sigma_g^+$. Their energies are, respectively, 92 kJ/mol and 155 kJ/mol higher than that of ground state oxygen $^3\Sigma_g^-$. The $^1\Delta_g$ state possesses a much longer lifetime and normally has a higher yield in biological system than does $^1\Sigma_g^+$. Consequently, the $^1\Delta_g$ state is the main consideration here. It is because of the relatively small energy difference from the ground state that many compounds are capable of acting as sensitizers for singlet oxygen formation. For example, the dyes methylene blue and Rose Bengal have a triplet state energy of about 140 and 170 kJ/mol, respectively.

The production of $^1O_2$ has been reported to occur by energy transfer from both the singlet and triplet excited states of the sensitizer, but that from the triplet excited state is highly preferred because singlet–triplet interaction is of very low probability. The lifetime of $^1O_2$ is highly dependent on the solvent medium, and the presence of scavengers or oxidizable acceptors. It was determined to be about 3.1×10^{-6} s in water (Rodgers and Snowden, 1982) and $50–100 \times 10^{-6}$ s in lipid (Henderson and Dougherty, 1992). A half-life in tissue was estimated to be less than 5×10^{-7} s (Patterson *et al.*, 1990). Singlet oxygen might diffuse about 1 mm in a cellular environment (Moan *et al.*, 1979). While the energy of $^1O_2$ is only 92 kJ/mol higher than that of ground state oxygen, its chemical reactivity is completely different because it is now spin-matched with ground state molecules susceptible to oxidation. Thus $^1O_2$ is capable of oxidizing a large variety of substances including biological cell components such as DNA, protein and lipids. Since many sensitizers are themselves in a reduced form, they also may act as substrates, giving fully oxidized products. As a consequence many preparative organic chemical processes are carried out photochemically, with $^1O_2$ being the mediator.

2.8.1 Quenchers of singlet oxygen

Singlet oxygen is deactivated by either physical or chemical quenching agents. The two physical mechanisms are energy-transfer and charge-transfer quenching. The carotenoid pigments play an important role in the protection of biological systems, apparently as they are particularly efficient energy-transfer quenchers. β-Carotene is the most studied member of this group. The extended conjugated π-system has triplet energies close to or below that of singlet oxygen, so that collisional energy transfer occurs. Subsequently the excited β-carotene decays itself by vibrational relaxation, so that no net chemical changes accrue (Gorman and Rodgers, 1981).

Amines generally are capable of quenching singlet oxygen via a charge-transfer process, but may react chemically as well. The primary process is envisaged as formation of a complex between the electron-donating quencher and the electron-deficient oxygen species; the quenching rate constants correlate with amine ionization potentials. The resulting triplet complex either dissociates with loss of energy by vibational relaxation, or forms oxidation products. Formation of products requires an abstractable hydrogen α to the nitrogen; N-methyl groups are particularly susceptible. Diazabicyclo-octane (DABCO) is unable to react chemically, presumably on steric grounds, but is an efficient physical quencher. Some phenols are also able to quench singlet oxygen by a mixture of physical and chemical processes, e.g., the 2,4,6-trisubstituted phenols used as antioxidants, BHT and α-tocopherol.

Other chemical reactions or quenching of singlet oxygen rely on the fact that singlet oxygen is more electrophilic than ground state oxygen and therefore can react selectively with electron-rich regions of many molecules, e.g., olefins and aromatics. Some examples of the addition of singlet oxygen are given in Fig. 2.9, including the *ene*-reaction in which an olefin possessing an allylic hydrogen form allylic hydroperoxides, and endoperoxide formation by 1,4-addition to π-systems such as furan and anthracene derivatives. As with other oxidation reactions the initial products are metastable and secondary reactions will occur but on a slower

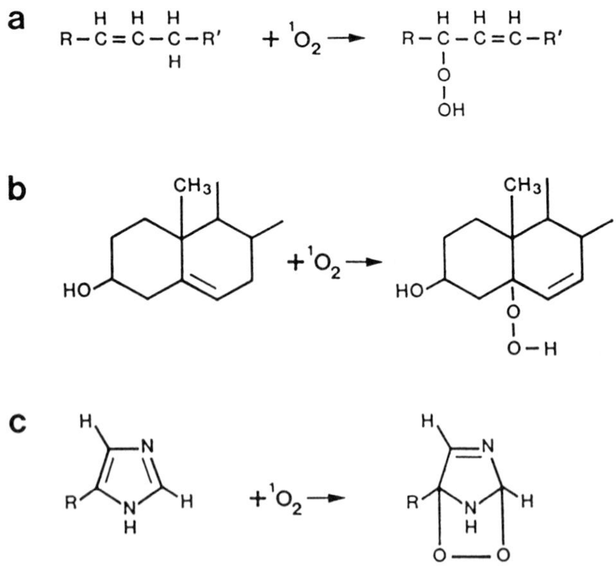

Figure 2.9 Chemical quenching of singlet oxygen: (a) the *ene* reaction – addition of singlet oxygen to an olefin with allylic hydrogen; (b) the *ene* reaction of cholesterol; (c) endoperoxide formation by singlet oxygen to imidazole residue as in histidine

time scale. Dioxetan formation occurs by singlet oxygen addition to olefins in which the double bond possesses an electron-donating heteroatom, generally N, O or S, ultimately leading to cleavage of the double bond, in a similar way to the reaction of superoxide in equation (2.17). The similarity leads to some controversy as to the mechanism of dioxetan formation (Gorman and Rodgers, 1981).

2.8.2 Detection of singlet oxygen

There are several methods for the detection of $^1O_2$ generated in an irradiated solution. A characteristic luminescence at 1270 nm, corresponding to the return to the ground state, can be detected with the appropriate equipment (Hall *et al.*, 1987). The alternative is to measure the rates of reaction in the presence of molecules which react readily with or quench singlet oxygen. Here the choice depends on the solvent being used, with sodium azide, 2,5-dimethylfuran and the amino acid histidine being suitably soluble for aqueous systems, while β-carotene, DABCO and diphenylisobenzofuran (DPBF) being more readily used in organic solvents. Analysis of the reaction rates is achieved in terms of oxygen uptake measured with an oxygen electrode (Moore, 1977), or by product separation and quantification. DPBF absorbs intensely at 415 nm and reacts rapidly with singlet oxygen to form a colourless intermediate endoperoxide. The DPBF reaction can be used as a benchmark against which the effect of an added quencher is compared. A note of caution must be applied. The use of inhibitors and quenchers alone is not unambiguous in its outcome and should strictly be supplemented with flash photolysis experiments. Thus, if a photosensitized reaction is quenched by millimolar concentrations of azide ion, it should also be established that azide does not quench the triplet state of the sensitizer directly, since that would also affect the reaction rate. It has also been reported that the furans and histidine can be oxidized to the same products by free radical processes. Nevertheless, these compounds have such a high reactivity with singlet oxygen that they are very rarely wrong as indicators of its generation by a photosensitizer. Cholesterol is regarded as an unambiguous trapping compound, since singlet oxygen reacts with it to form a single product, the 5-α-hydroperoxide, whereas reaction with radicals gives a mixture of other products (Spikes, 1989). The analytical procedure is more technically demanding than that employed with histidine or the furans.

Another kinetic technique is to compare the rates in heavy water (D_2O) with normal water, since the lifetime as noted above is about 10 times greater in D_2O. This will only achieve a meaningful result when singlet oxygen deactivation by the solvent is the rate-determining process. Frequently other species in the solution are capable of reacting with singlet oxygen and the effect of the longer lifetime is not manifest.

Typical photosensitizers which generate singlet oxygen include dyes such as methylene blue, Rose Bengal and rhodamine. Many drug molecules such as phenothiazines, quinine and other antimalarials, thiazides, naproxen and other anti-inflammatories, and psoralens have been demonstrated to generate singlet oxygen under the influence of UV-R or visible light. Environmental contaminants such as the polycyclic aromatic hydrocarbons also are very efficient $^1O_2$ generators.

2.9 Active forms of oxygen and oxidant species

As noted above, the formation of free radicals or singlet oxygen is very often accompanied by the generation of various other short-lived species (such as hydroxyl radicals, superoxide radicals and peroxyl radicals) which together with singlet oxygen are termed reactive oxygen species (Pryor, 1986). For example, superoxide radicals can be generated following photoionization (equation (2.4)), from singlet oxygen by electron transfer between $^1O_2$ and either the ground state sensitizer (equation (2.18)) or appropriate substrates (equation (2.19)). In some cases, the subsequent reactions may result in the formation of toxic hydrogen peroxide (equation (2.20)). This in turn decomposes to produce hydroxyl radicals (equation (2.21)) (Proctor and Reynolds, 1984).

$$^1O_2 + D_0 \rightarrow O_2^{\cdot -} + D^{\cdot +} \qquad (2.18)$$

$$^1O_2 + AH \rightarrow O_2^{\cdot -} + A^{\cdot} + H^+ \qquad (2.19)$$

$$O_2^{\cdot -} + H^+ \rightarrow HO_2 \rightarrow H_2O_2 + O_2 \qquad (2.20)$$

$$H_2O_2 \rightarrow 2\ ^{\cdot}OH \qquad (2.21)$$

Apart from the photodynamic reactions, a photosensitized reaction may proceed through the direct photoionization of the sensitizer in which oxygen is not required (equation (2.4)). Since photoionization is found to occur particularly from molecules containing one or more hetero-atoms, there are a significant number of drugs which undergo photoionization although, in general, higher energy radiation (<300 nm) is required. Drugs which are said to photoionize include chlorpromazine (Navaratnam *et al.*, 1978), 4-hydroxybenzothiazole (Chedekel *et al.*, 1980), sulfacetamide (Land *et al.*, 1982) and hydrochlorothiazide (Tamat and Moore, 1983). The hydrated electron is one of the strongest known reducing agents, reacting rapidly with oxygen to form the superoxide radical anion ($O_2^{\cdot -}$), the precursor of hydrogen peroxide (Buxton *et al.*, 1988). Photoionization and superoxide formation are supported to a greater extent by an aqueous medium because the electrons are readily stabilized by the solvent.

2.10 Consequences of the excited state processes to drug stability *in vitro*

The photodecomposition reactions of a large number of drug substances were recently reviewed by Greenhill and McLelland (1990). A catalogue therein of the most common reaction types shows that a drug might experience addition, cyclization, N-dealkylation, decarbonylation, decarboxylation, dehalogenation, dimerization, oxidation, reduction, isomerization, rearrangement, and/or hydrolysis following photon absorption. It is also possible that a drug sensitizes its own degradation. There is great difficulty in elucidating the mechanism for some photodegradation reactions because several pathways are reported for many drugs. Additionally, there are significant differences in the irradiation conditions used by different laboratories, in terms of wavelength range, time of exposure, and drug concentration. Another factor that should be considered is the effect of oxygen on the product distribution. In many cases oxidation may be responsible for products which are secondary or even tertiary in the overall sequence when long exposures are used. Because of the broad array of chemical structures of drug molecules and the variety of processes which can occur following the absorption of electromagnetic radiation, it is possible to predict the possible outcome for a new compound only if it is a close structural analogue of a previously studied compound.

Two case studies, naproxen (Moore and Chappuis, 1988) and sulfamethoxazole (Zhou and Moore, 1994), will be described briefly to illustrate how information concerning the excited state(s) assists in the elucidation of the mechanism of photodegradation. Their spectroscopic properties are detailed in Table 2.1 together with several other drugs given for comparative purposes. The state of ionization of the molecule can be a factor in the photochemistry, so it is relevant to show the information for both neutral and anionic sulfamethoxazole (pKa 5.6). Ionization of naproxen (pKa 4.2) has no significant effect on its spectroscopic properties as the carboxyl group is not directly attached to the naphthalene chromophore.

Both molecules have the capacity to emit fluorescence and phosphorescence, the latter suggesting an appreciable population and lifetime of their respective triplet states. The first photochemical tests involve analysing the reaction mixtures after irradiation for varying times, using spectrophotometry and liquid chromatography. It is important to follow the reaction from the very early stages to ensure that the products that are seen are those formed from the very beginning of the reaction. Some primary products may decompose in secondary thermal and/or photochemical reactions.

The data from flash photolysis experiments (266 nm excitation) on naproxen in aqueous solution at pH 7 are shown in Fig. 2.10. The transient spectra were identified as the solvated electron and a triplet state. Similar species were found

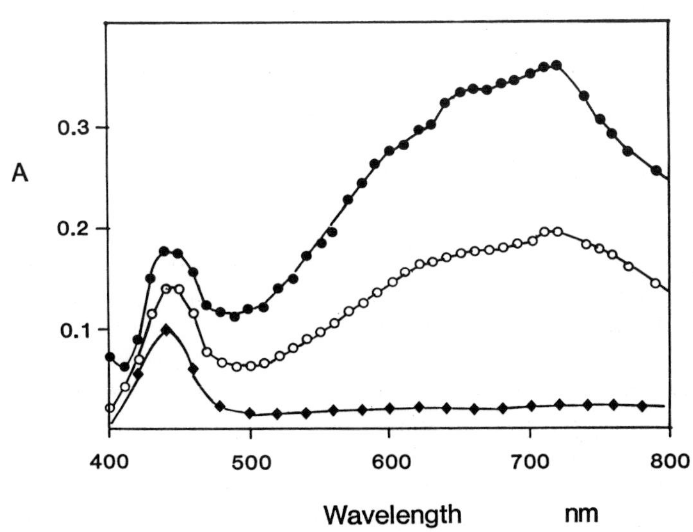

Figure 2.10 Transient absorption spectra obtained on laser flash photolysis at 266 nm of naproxen (0.1 mM, pH 7.0). The spectra were recorded 0.02 (●), 0.26 (○) and 1.29 μsec (♦) after the flash

Table 2.1 Spectroscopic data (absorption, fluorescence and phosphorescence) and photosensitization reaction rates (photooxidation and photopolymerization) for selected drugs in aqueous solution

Drug 5.0×10^{-5} M	Buffer	λ_{max} nm	ε_{max} $M^{-1}\,cm^{-1}$	λ_F nm	ϕ_F	$\lambda_P{}^1$ nm	Relative photoabsorption[b]	Oxygen uptake rate $mmol\,l^{-1}\,min^{-1}$		Polymerization rate $mmol\,l^{-1}\,min^{-1}$
								Drug alone	+DF	
Sulfamethoxazole	pH 9.0	258	17 000	342	0.0020	—	25.8	<0.3	8.2	0.7
Sulfamethoxazole	pH 3.0	268	17 360	342	0.0019	410	12.1	<0.4	9.6	1.3
Diclofenac[c]	pH 7.0	285	8 220	—	0	460	100	<0.2	3.9	1.79
Chlorpromazine[d]	pH 7.0	315	4 550	455	0.13	nd	860	3.1	32.8	3.42
Frusemide[d]	pH 7.0	330	5 800	400	0.06	nd	630	0.7	2.8	0.59
Naproxen[e]	pH 7.0	316, 331	1972, 1988	355	0.40	510	1 500	5.3	25.1	0.24

nd means 'not determined'. DF is 2,5-dimethylfuran.

[a]Phosphorescence measured in an ethanol glass at 77 K (drug in neutral form).

[b]Relative photoabsorption is the overlap integral of the drug's absorption spectrum and the mercury arc output spectrum, normalized to diclofenac set to 100.

[c]Data from Moore *et al.* (1990).

[d]Data from Moore and Burt (1981).

[e]Data from Moore and Chappuis (1988).

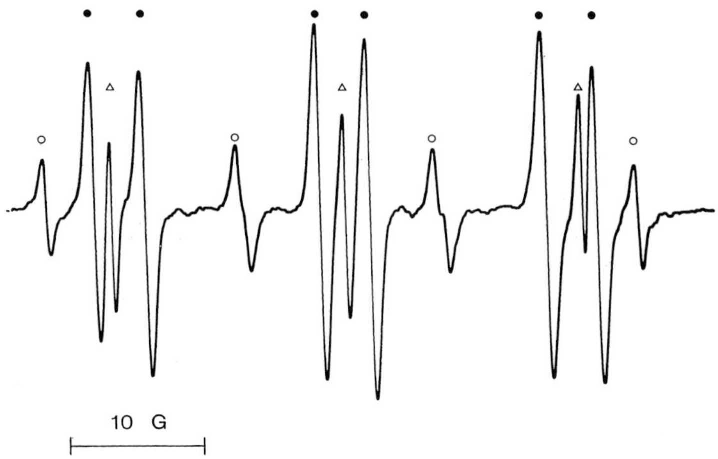

Figure 2.11 EPR spectrum obtained on photolysis of naproxen at 330 nm in the presence of 2-methyl-2-nitrosopropane (MNP). The spectrum contains contributions from DTBN (△), MNP-H (○) and decarboxy-naproxen (●)

for sulfamethoxazole. Under this pH condition, both molecules are present as the anion, so photoionization is a likely process with the high energy excitation. The solvated electron generation implied a radical production so EPR with spin trapping was performed. Sulfamethoxazole failed to produce any trappable radicals with an array of different spin traps, but naproxen afforded the EPR spectrum shown in Fig. 2.11 when irradiated with 330 nm UV-R in the instrument cavity in the presence of 2-methyl-2-nitroso-propane (MNP). The spectrum contains contributions from di-t-butyl nitroxide, a known photoproduct of MNP. The H-atom adduct MNP-H also evident can arise by several different mechanisms, including the trapping of an H atom by MNP, the reaction of MNP with an electron followed by protonation, and the direct reduction of MNP by an excited state species. In view of the flash photolysis results, it was concluded that photoionization was the major precursor of MNP-H. The third radical corresponded to a C-centred radical carrying a single H atom, leading to the postulate of a decarboxylation reaction as the primary photochemical step. Confirmation of the participation of free radical intermediates came from the initiation of the free radical polymerization of acrylamide with rates as shown in Table 2.1.

While sulfamethoxazole did not yield trappable radicals, there was evidence of free radical intermediates from polymerization experiments (Table 2.1) together with results from oxygen uptake experiments using 2,5-dimethylfuran as substrate. The latter measurements are an indication of the relative capability of the drugs to generate singlet oxygen, in confirmation of the population of the triplet state. All of these data, in combination with the identification of the isolated photoproducts, play a role in postulating intermediates and fully elucidating the mechanism of the photodegradation, shown in Fig. 2.12 for naproxen and Fig. 2.13 for sulfamethoxazole.

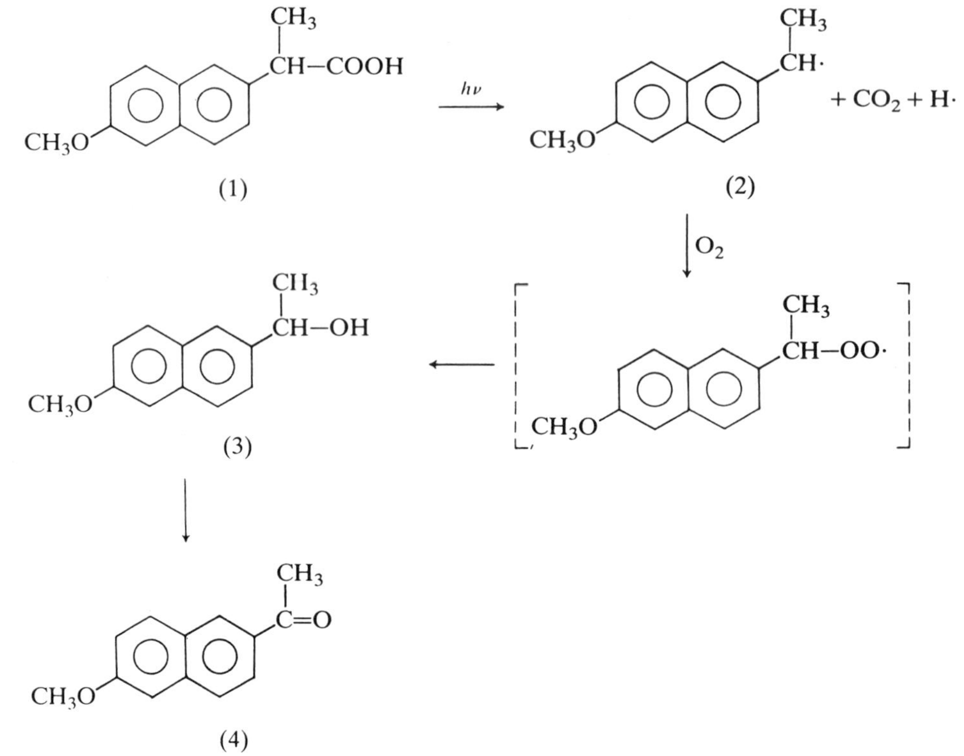

Figure 2.12 Reaction mechanism for the photodegradation of naproxen in air

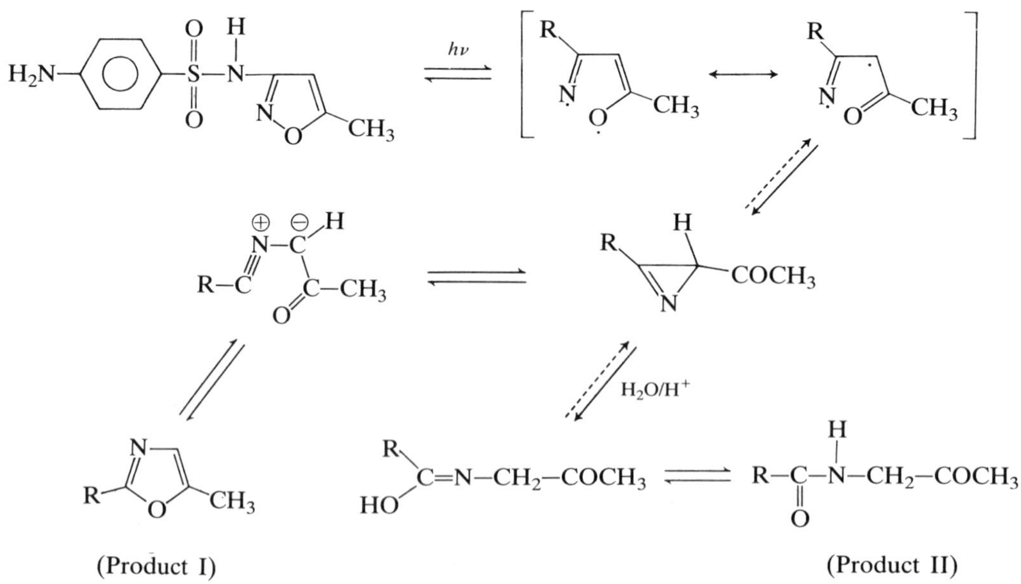

($R = H_2NC_6H_4SO_2NH$)

Figure 2.13 Reaction mechanism for the photodegradation of sulfamethoxazole

2.11 Consequences of the excited state processes to adverse effects*in vivo*

A drug which displays photochemical reactivity *in vitro* may give rise to adverse photosensitivity effects in patients, manifest in responses which have been labelled as *phototoxicity* and *photoallergy*. Phototoxicity is defined as an alteration of cell function by an interaction between a chemical and non-ionizing radiation, the response being likened to an exaggerated sunburn. The reaction occurs upon simultaneous exposure to a phototoxic chemical and radiation of the appropriate wavelength. The wavelength of radiation implicated in most phototoxic reactions ranges from 300 to 400 nm. Most drug-induced phototoxic reactions are acute, occurring within a few minutes to several hours after exposure. They reach a peak from several hours to several days later, and usually disappear within a short time period after stopping either the drug or the exposure to radiation (Epstein and Wintroub, 1985). Photoallergy is an acquired altered capacity of the skin to react to sunlight. Thus it is immune-mediated wherein a drug molecule is stimulated by radiation to combine with a protein or other biomacromolecule in the skin to form an antigen. Photoallergy is believed to require several exposures and an induction period before the response is observed, and it has a different histology compared with phototoxicity (Epstein, 1989).

It is believed that phototoxicity causes damage of cells by direct modification of certain targets, such as DNA, lipids and/or amino acids and proteins. In principle, this can occur by the photosensitization reactions described above, and there is a reasonable correlation between the capacity of a drug to participate in Type I and II processes *in vitro* and the number of adverse photosensitivity reports registered against them (Moore, 1990). It is, of course, difficult to compare different drugs in this way because a number of extra variables enter, for example, the dose and pharmacokinetics of the drug's biodistribution.

Phototoxic reactions may be oxygen dependent (photodynamic) or oxygen independent (non-photodynamic). In general, it is the capacity of the drug to generate free radicals that has been regarded as the most potentially damaging characteristic, because of the possibility of chain reactions subsequently occurring (Trush *et al.*, 1982). Chlorpromazine is the classic example of this, with a record as the most photosensitive drug (Magnus, 1976) and a high yield of a long-lived free radical on irradiation (Table 2.1, and Kochevar, 1981). Nonetheless, the importance of singlet oxygen can be seen by the fact that the polycyclic aromatic hydrocarbons are both efficient producers of singlet oxygen and acute phototoxicity responses (Kochevar *et al.*, 1982; Burt and Moore, 1987).

In addition, toxic photoproducts may also be formed by the action of sunlight on the drug in the epidermal layers of the skin of patients and may thereby cause adverse photosensitivity effects either because they possess undesirable physiological properties, or because they can easily transfer energy to body compounds (Rahn *et al.*, 1974). For example, a stable photoproduct of chlorpromazine was cytotoxic to the plasma membrane such that it could result in hemolysis of erythrocytes in the dark (Kochevar and Lamola, 1979). Another example is the non-steroidal anti-inflammatory drug, benoxaprofen, which was the subject of many adverse reports before its withdrawal in 1982. A two-step mechanism was proposed for benoxaprofen photosensitization. First, photoexcited benoxaprofen decarboxylates and thereby loses its ability to dissolve in aqueous media. The lipophilic photoproduct partitions into the lipid bilayer of the red blood cell membrane where it acts as a singlet oxygen generator to lead to the oxidation of membrane lipids, ultimately disrupting the membrane structure (Reszka and Chignell, 1983; Kochevar *et al.*, 1984).

In certain cases, a direct reaction between the excited state of a phototoxin and a biological target may result in the formation of a covalent photoaddition product which is comprised of the compound itself and the biological target. Such a mechanism was demonstrated in the phototoxicity of psoralen in which a direct photochemical reaction of psoralen with DNA occurred (Song and Tapley, 1979). For some photosensitizers, such as amiodarone and chlorpromazine, metabolites may play a significant role in their phototoxic reactions (Llunggren and Möller, 1977; Ferguson *et al.*, 1985).

The mechanism of photoallergy is postulated in a similar fashion to the non-photodynamic processes with the drug becoming attached to a protein by a free radical process, thereby creating a hapten which can lead to an immune response. It is presumed that the reason for the low incidence of photoallergic reactions is the body's immune response mechanisms generally deactivate the hapten, or the oxygen present in the blood is able to scavenge the free radical intermediates.

2.12 Approaches to the stabilization of formulations against photodegradation

In principle, formulations containing drugs susceptible to photoreactions should be clearly marked and stored appropriately. However, there are situations where the ideals are not maintained and it is worth considering whether special procedures or additives should be included.

The use of an inert atmosphere in the container is an important first approach. Clearly this will limit reactions involving oxygen, but many examples have been found in which oxygen does not take part, e.g., sulfamethoxazole. The procedure would have to be considered on an individual basis following a full study of the role of oxygen in the photoreactions of a particular drug. On the other hand, it would not be advisable for intravenous fluids to be depleted of their oxygen prior to injection, so the application to which the formulation will be put is also a factor for consideration.

Where photoreactions proceed through a Type I or II photosensitization mechanism, the possibility of adding quenchers to the formulation can be considered. If such quenchers can be clearly shown to be non-toxic, and to not affect the therapeutic action, this is perhaps an option. The major possibilities would be substances such as ascorbic acid, α-tocopherol and BHT, which are capable of acting as free radical scavengers as well as weak singlet oxygen quenchers. They are already in use as food additive antioxidants. β-Carotene is the only major singlet oxygen quencher which may be regarded as a possible food additive, but the trace concentration which can be used may not be effective. L-Histidine reacts rapidly with singlet oxygen but addition of this to the formulation may disturb a patient's amino acid balance. In view of the present regulatory environment, the use of these and other quenchers would involve a considerable amount of development work.

2.13 Influence of excipients on drug stability

The compatibility of the drug with the excipients in a given formulation must be established early in the development programme. For solutions and suspensions, this is usually a matter of choice of buffer and pH, determined from solution kinetic studies. For parenterals, compatibility with packaging components (plugs, plastics), and the possibility of metal ion contamination is also investigated. For solid products, the investigation is more complex because of the greater range of excipients. The choice of excipient is company dictated and may include any of the following: lactose, dicalcium phosphate anhydrous, dicalcium phosphate dihydrate, corn starch, mannitol, terra alba, sugar, magnesium and calcium stearates, talc, stearic acid, polyvinylpyrrolidone (PVP), crosspovidone and modified starches (Carstensen, 1990).

Photostability testing adds another dimension to the testing with excipients, and what is not mentioned in the preceding list are the colouring agents frequently used in elixirs, ointments, tablets and capsules. To consider those first, it is obvious that such colouring agents should be of negligible reactivity when irradiated so that photosensitized decomposition of the therapeutic agent is not an issue. Many vegetable dyes are regarded as safe although they may themselves gradually fade. It is not always clear whether any drug decomposition is occurring under those circumstances, because the product would be discarded if the colour changes.

In relation to other excipients, one can see some components liable to participate in photochemical processes. For example, lactose, mannitol, sugar, starches and PVP are all susceptible to free radical attack in that they have abstractable hydrogens. In that sense they may be regarded as 'good' excipients, since they would be the free radical transfer reagents to inhibit the degradation of the drug substance. Obviously their effectiveness will depend on the proportions of all components in the formulation, making the testing of the photostability of the final product a mandatory exercise in the product development.

References

BENSASSON, R. V., LAND, E. J. and TRUSCOTT, T. G., 1983, *Flash Photolysis and Pulse Radiolysis*. Oxford: Pergamon Press, pp. 1–19.

BRULS, W. A. G., SLAPER, H., VAN DER LEUN, J. C. and BERENS, L., 1984, Transmission of human epidermis and stratum cornea as a function of thickness in the ultraviolet and visible wavelengths, *Photochem. Photobiol.*, **40**, 485–94.

BURT, C. D. and MOORE, D. E., 1987, Photochemical sensitization by 7-methylbenz[c]acridine and related compounds, *Photochem. Photobiol.*, **45**, 729–39.

BUXTON, G. V., GREENSTOCK, C. L., HELMAN, W. P. and ROSS, A. B., 1988, Critical review of rate constants for reactions of hydrated electrons, hydrogen atoms and hydroxyl radicals ($^{\cdot}OH/^{\cdot}O^{-}$) in aqueous solution, *J. Phys. Chem. Ref. Data*, **17**, 513–886.

CADET, J., ANSELMINO, C., DOUKI, T. and VOITURIEZ, L., 1992, Photochemistry of nucleic acids in cells, *J. Photochem. Photobiol. B: Biol.*, **15**, 277–98.

CALVERT, J. G. and PITTS, J. N., 1966, Experimental methods in photochemistry. *Photochemistry*. New York: John Wiley & Sons, pp. 783–804.

CARSTENSEN, J. T., 1990, *Drug Stability*. New York: Marcel Dekker, Inc., pp. 288–9.

CHEDEKEL, M. R., LAND, E. J., SINCLAIR, R. S., TAIT, D. and TRUSCOTT, G., 1980,

Photochemistry of 4-hydroxybenzothiazole: a model for pheomelanin degradation, *J. Am. Chem. Soc.*, **102**, 6587–90.

CHIGNELL, C. F., MOTTEN, A. G. and BUETTNER, G. R., 1985, Photoinduced free radicals from chlorpromazine and related phenothiazines: relationship to phenothiazine-induced photosensitization, *Environmental Health Perspectives*, **64**, 103–10.

CORNELISSEN, P. J. G., BEIJERSBERGEN VAN HENEGOUWEN, G. M. J. and GERRITSMA, K. W., 1979, Photochemical decomposition of 1,4-benzodiazepines. Chlordiazepoxide, *Int. J. Pharm.*, **3**, 205–20.

DAVIES, A. K., HILAL, N. S., MCKELLAR, J. F. and PHILLIPS, G. O., 1975, Photodegradation of salicylanilides, *Brit. J. Derm.*, **92**, 143–7.

DAVIES, A. K., NAVARATNAM, S. and PHILLIPS, G. O., 1976, Photochemistry of chlorpromazine (2-chloro-N-(3-dimethyl-aminopropyl)phenothiazine) in propan-2-ol solution, *J. Chem. Soc. Perkin Trans.* **2**, 25–9.

DIFFEY, B. L., 1983, A mathematical model for ultraviolet optics in skin, *Phys. Med. Biol.*, **28**, 647–57.

EPSTEIN, J. H., 1989, Photomedicine, in Smith, K. C. (ed.), *The Science of Photobiology*, pp. 155–92. New York: Plenum Press.

EPSTEIN, J. H. and WINTROUB, B. U., 1985, Photosensitivity due to drugs, *Drugs*, **30**, 42–57.

FERGUSON, J., ADDO, H. A. and JONES, S., 1985, A study of cutaneous of photosensitivity induced by aminodarone, *Br. J. Dermatol.*, **113**, 537–49.

FISHER, G. J. and LAND, E. L., 1983, Photosensitization of pyrimidines by 2-methylnaphtoquinone in water: a laser flash photolysis study, *Photochem. Photobiol.*, **37**, 27–32.

FOOTE, C. S., 1968, Mechanisms of photosensitized oxidation, *Science*, **162**, 963–70.

FREDERICK, J. E., 1993, Ultraviolet sunlight reaching the earth's surface: a review of recent research, *Photochem. Photobiol.*, **57**, 175–8.

FREDERICK, J. E., SNELL, H. E. and HAYWOOD, E. K., 1989, Solar ultraviolet radiation at the earth's surface, *Photochem. Photobiol.*, **50**, 443–50.

GATES, D. M., 1966, Spectral distribution of solar radiation at earth's surface, *Science*, **151**, 523–9.

GILBERT, A. and BAGGOTT, J., 1991, *Essentials of Molecular Photochemistry*. Oxford: Blackwell, pp. 145–228.

GORMAN, A. A. and RODGERS, M. A. J., 1981, Singlet molecular oxygen, *Chem. Soc. Rev.*, **10**, 205–31.

GREENHILL, J. V. and MCLELLAND, M. A., 1990, Photodecomposition of drugs, *Progr. Med. Chem.*, **27**, 51–121.

GRIMSHAW, J. and DE SILVA, A. P., 1981, Photochemistry and photocyclization of aryl halides, *Chem. Soc. Rev.*, **10**, 181–203.

GROSSWEINER, L. I., 1989, Photophysics, in Smith, K. C. (ed.), *The Science of Photobiology*, pp. 1–45. New York: Plenum Press.

GROSSWEINER, L. I. and SMITH, K. C., 1989, Photochemistry, in Smith, K. C. (ed.), *The Science of Photobiology*. pp. 47–78. New York: Plenum Press.

HALL, R. D., BUETTNER, G. R., MOTTEN, A. G. and CHIGNELL, C. F., 1987, Near-infrared detection of singlet molecular oxygen produced by photosensitization with promazine and chlorpromazine, *Photochem. Photobiol.*, **46**, 295–300.

HENDERSON, B. W. and DOUGHERTY, T. J., 1992, How does photodynamic therapy work?, *Photochem. Photobiol.*, **55**, 145–57.

JAGGER, J., 1985, *Solar-UV Actions on Living Cells*. New York: Praeger Scientific, p. 119.

KOCHEVAR, I. E., 1981, Phototoxicity mechanisms: chlorpromazine photosensitized damage to DNA and cell membranes, *J. Invest. Dermatol.*, **77**, 59–64.

KOCHEVAR, I. E., ARMSTRONG, R. B., EINBINDER, J., WALTHER, R. R. and HARBER,

L. C., 1982, Coal tar phototoxicity: active compounds and action spectra, *Photochem. Photobiol.*, **36**, 65–9.

KOCHEVAR, I. E., HOOVER, K. W. and GAWIENOWSKI, M., 1984, Benoxaprofen photosensitization of cell membrane disruption, *J. Invest. Dermatol.*, **82**, 214–18.

KOCHEVAR, I. E. and LAMOLA, A. A., 1979, Chlorpromazine and protriptyline phototoxicity: photosensitized oxygen-independent red cell hemolysis, *Photochem. Photobiol.*, **29**, 1177–97.

KRAMER, H. E. A. and MAUTE, A., 1973, Sensitized photooxygenation: change from Type I (radical) to Type II (singlet oxygen) mechanisms, *Photochem. Photobiol.*, **17**, 413–23.

LAND, E. J., NAVARATNAM, S., PARSONS, B. J. and PHILLIPS, G. O., 1982, Primary processes in the photochemistry of aqueous sulfacetamide: a laser flash photolysis and radiolysis study, *Photochem. Photobiol.*, **35**, 637–42.

LEE, P. C. C. and RODGERS, M. A. J., 1987, Laser flash photokinetic studies of rose bengal sensitized photodynamic interactions of nucleotides and DNA, *Photochem. Photobiol.*, **45**, 79–86.

LLUNGGREN, B. and MÖLLER, H., 1977, Phenothiazine phototoxicity: an experimental study on chlorpromazine and its metabolites, *J. Invest. Dermatol.*, **68**, 313–17.

LOVELL, W. W., 1993, A scheme for *in vitro* screening of substances for photoallergic potential, *Toxic. in Vitro*, **7**, 95–102.

MAGNUS, I. A., 1976, *Dermatological Photobiology*, London: Blackwell, pp. 213–16.

MASON, R. P. and CHIGNELL, C. F., 1982, Free radicals in pharmacology and toxicology – selected topics, *Pharmacol. Rev.*, **33**, 189–211.

MOAN, J., PETTERSON, E. O. and CHRISTENSEN, T., 1979, The mechanism of photodynamic inactivation of human cells *in vitro* in the presence of haematoporphyrin, *Br. J. Cancer*, **39**, 398–407.

MOORE, D. E., 1976, Antioxidant efficiency of polyhydric phenols in photooxidation of benzaldehyde, *J. Pharm. Sci.*, **65**, 1447–51.

1977, Photosensitization by drugs, *J. Pharm. Sci.*, **66**, 1282–4.

1990, Photochemistry of photosensitizing drugs, *Trends Photochem. Photobiol.*, **1**, 13–23.

MOORE, D. E. and BURT, C. D., 1981, Photosensitization by drugs in surfactant solutions, *Photochem. Photobiol.*, **34**, 431–9.

MOORE, D. E. and CHAPPUIS, P. P., 1988, A comparative study of the photochemistry of the non-steroidal anti-inflammatory drugs, naproxen, benoxaprofen and indomethacin, *Photochem. Photobiol.*, **47**, 173–81.

MOORE, D. E. and HEMMENS, V. J., 1982, Photosensitization by antimalarial drugs, *Photochem. Photobiol.*, **36**, 71–7.

MOORE, D. E., ROBERTS-THOMSON, S., DONG, Z. and DUKE, C. C., 1990, Photochemical studies on the antiinflammatory drug diclofenac, *Photochem. Photobiol.*, **52**, 685–90.

MOORE, D. E., and SITHIPITAKS, V., 1983, Photolytic degredation of frusemide, *J. Pharm. Pharmacol.*, **35**, 489–93.

MOORE, D. E. and ZHOU, W., 1994, Photodegradation of sulfamethoxazole: a chemical system capable of monitoring seasonal changes in UVB intensity, *Photochem. Photobiol.*, **59**, 497–502.

NAVARATNAM, S., PARSONS, B. J., PHILLIPS, G. O. and DAVIES, A. K., 1978, Flash photolysis study of the photoionization of chlorpromazine and promazine in solution, *J. Chem. Soc. Faraday Trans. I*, **74**, 1811–19.

PARRISH, J. A., ANDERSON, R. R., URBACH, F. and PITTS, D., 1978, *UV-A: Biological Effects of Ultraviolet Radiation with emphasis on Human Responses to Long-wave Ultraviolet*. New York: Plenum Press, p. 121.

PATTERSON, M. S., MADSEN, S. J. and WILSON, B. C., 1990, Experimental tests of the feasibility of singlet oxygen luminescence monitoring *in vivo* during photodynamic therapy, *J. Photochem. Photobiol. B: Biol.*, **5**, 69–84.

PROCTOR, P. H. and REYNOLDS, E. S., 1984, Free radicals and disease in man, *Physiol. Chem. Phys. Med. NMR*, **16**, 175–95.

PRYOR, W. A., 1986, Oxy-radicals and related species: their formation, lifetimes, and reactions, *Annu. Rev. Physiol.*, **48**, 657–67.

RAHN, R. O., LANDRY, L. C. and CARRIER, W. L., 1974, Formation of chain breaks and thymine dimers in DNA upon photosensitization at 313 nm with acetophenone, acetone or benzophenone, *Photochem. Photobiol.*, **19**, 75–8.

RESZKA, K. and CHIGNELL, C. F., 1983, Spectroscopic studies of cutaneous photosensitizing agents – IV. The photolysis of benoxaprofen, an anti-inflammatory drug with phototoxic properties, *Photochem. Photobiol.*, **38**, 281–91.

RODGERS, M. A. J. and SNOWDEN, P. T., 1982, Lifetime of $O_2(^1D_g)$ in liquid water as determined by time-resolved infra-red luminescence measurements, *J. Am. Chem. Soc.*, **104**, 5541–3.

SONG, P. and TAPLEY, Jr, K. J., 1979, Photochemistry and photobiology of psoralens, *Photochem. Photobiol.*, **29**, 1177–97.

SPIKES, J. D., 1989, Photosensitization, in Smith, K. C. (ed.), *The Science of Photobiology*, pp. 79–110. New York: Plenum Press.

TAMAT, S. R. and MOORE, D. E., 1983, Photolytic decomposition of hydrochlorothiazide, *J. Pharm. Sci.*, **72**, 180–4.

THORINGTON, L., 1985, Spectral, irradiance and temporal aspects of natural and artificial light, *Ann. N.Y. Acad. Sci.*, **453**, 28–54.

TRUSH, M. A., MIMNAUGH, E. S. and GRAM, T. E., 1982, Activation of pharmacological agents to radical intermediates. Implications for the role of free radicals in drug action and toxicity, *Biochem. Pharmacol.*, **31**, 3335–46.

ZHOU, W. and MOORE, D. E., 1994, Photochemical decomposition of sulfamethoxazole, *Int. J. Pharm.*, **110**, 55–63.

3

Technical Requirements and Equipment for Photostability Testing

J. BOXHAMMER

3.1 Introduction

Photostability tests as discussed in the pharmaceutical industry have been carried out on technical materials and products for many years in the most varied fields of industrial application. The stage of discussion on accelerated photostability tests of drugs with artificial light sources clearly reflects the long years of research and permanent revision and redefinition of standardized test methods for technical materials.

Precise knowledge about the spectral energy distribution of the radiation that shall be simulated and intensified is a basic prerequisite for designing test methods under simulated conditions. Those data are available today for the global solar radiation and may serve as a concrete basis for simulation purposes.

Conducting tests under accelerated more controlled conditions compared with natural exposure and good correlation as well as repeatability and reproducibility of test results are the main objectives of tests with laboratory light sources. In the equipment relating technical requirements, the most important stress conditions have to be defined precisely in standardized test methods to meet the mentioned objectives.

The present state of knowledge and findings and their conversion into testing procedures and techniques on an international level might also provide impetus in the near future for a standardized test of the photostability of drugs.

Specification of the spectral distribution of radiation, level of irradiance, surface temperature and description of necessary measuring devices are focal points of the newest international standards. Various improvements in equipment technology are necessary to meet the specified requirements.

Comparison of the laboratory light sources used in general material testing with those for photostability testing of drugs as well as the contents of international standards to the present state of discussion in pharmaceutical industry provides a basic concept for a well-defined guideline for performance tests of drugs concerning technical requirements for the equipment.

3.2 Spectral energy distribution of solar radiation

Most drug compounds and drug formulations will be subjected to light during production, storage or distribution and use by the patient. It is obvious that various kinds of artificial light sources as well as sunlight present in rooms behind window glass may cause material degradation processes. Considering real room conditions, sunlight directly behind the window may represent the 'worst case' concerning stress on drugs.

As a basis for interpretation of test results under daylight itself as well as for establishing test methods for simulated conditions with artificial light sources, a very thorough knowledge about sunlight is necessary.

It is well known that the spectral distribution and irradiance of the solar radiation at the earth surface depends on the location and is subjected to seasonal and diurnal variations. There is a need for a reference spectrum as a basis for comparison with the spectral distribution of artificial light sources. Data from CIE No. 15: 1971 (colorimetry, official recommendations of the international commission on illumination) recommending a standard illuminant D 65 with a scheduled relative spectral distribution between 300 and 830 nm in 5 nm steps for a correlated colour temperature of approximately 6500 K have been used as a basis over the years.

The first recommendations for the integrated irradiance and the spectral distribution of simulated solar radiation for testing purposes have been published in 1972 as publication Commission Internationale de l'Eclairage (CIE) No. 20 (recommendations for the integrated irradiance and the spectral distribution of simulated solar radiation for testing purposes). The scheduled spectral energy distribution is given in 40 nm steps up to 800 nm and contains the infrared part of radiation up to 3000 nm. For testing the resistance of technical objects to deterioration the maximum value of total irradiance has been specified. As an international criterion for comparison of artificial sources with natural daylight, specific spectral data which are more precise as compared with CIE No. 20 have been published in CIE No. 85: 1989 (Technical Report: Solar Spectral Irradiance) and are used today.

The sum of direct and diffuse solar radiation on a horizontal plane at the earth's surface, called global solar radiation, in the emission range between 300 nm and 2450 nm is given as 1090 W/m^2 (relative air mass = 1). The spectral energy distribution is shown in Fig. 3.1.

Compared with outdoor measurements the curve is smoothed since the spectral data (5 nm steps in UV range) are centred spectral irradiance values averaged over 20 nm. The increase of the spectral data in UV is rather equivalent to mean values of the UV spectral irradiance as measured in Arizona (Fig. 3.2).

Based on these data the spectral energy distribution of global solar radiation behind a window depends on the type and thickness of the window glass used (Fig. 3.3). With an increasing thickness of the window glass the total irradiance is reduced and the cut-off in the UV range of radiation shifts towards longer wavelengths.

The total radiation on a horizontal measurement plane is the sum of direct radiation and the part that is scattered by the earth's atmosphere. Depending on the altitude of the sun the sky radiation may be between 30 and 90 per cent of total radiation. At very low altitudes of the sun the total energy level decreases

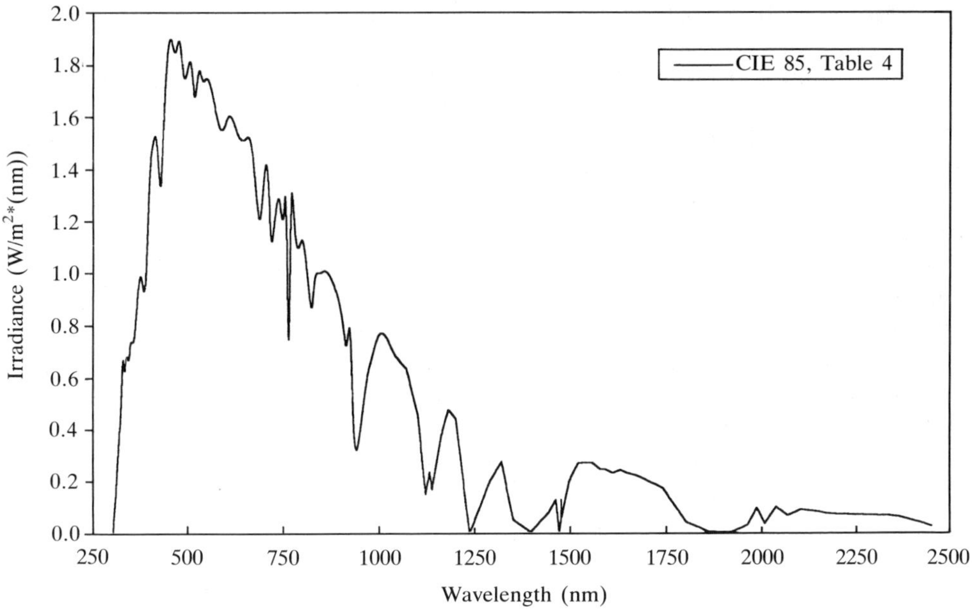

Figure 3.1 Spectral energy distribution of global solar radiation according to publication CIE No. 85, Table 4 (189)

and the spectral distribution is shifting towards the red. In sky radiation, the shorter wavelengths are scattered more than the longer ones. Therefore, the sky radiation shows a spectral distribution shifted towards the blue. The irradiance of the total radiation may be reduced to 20 per cent on a totally cloudy day as compared with cloudless days.

3.3 Absorption of solar radiation in materials

Based on the incident spectral irradiance of the sunlight only the radiation which is absorbed by a material according to its specific degree of absorption $\varepsilon(\lambda)$, which depends in general on the wavelength, may produce an effect in the material.

A very small part of the absorbed radiation will result in primary photochemical processes. The UV range and partially the visible wavelength range of solar radiation have a major influence. The determination of the 'spectral sensitivity' (Trubiroha, 1987) and the 'activation spectrum' may give valuable information on the wavelength bands of incident radiation which cause photodegradation. As those data are not available in most cases there is a need to simulate the spectral distribution of solar radiation in UV and visible wavelength range as closely as possible.

The major part of incident radiation absorbed by a material leads to a heating up of the surfaces especially of solid samples above ambient temperature depending on the irradiance level as well as on other climatic environmental conditions (wind velocity, humidity) and the thickness as well as the optical and thermal properties of the sample material. A specific situation is given for

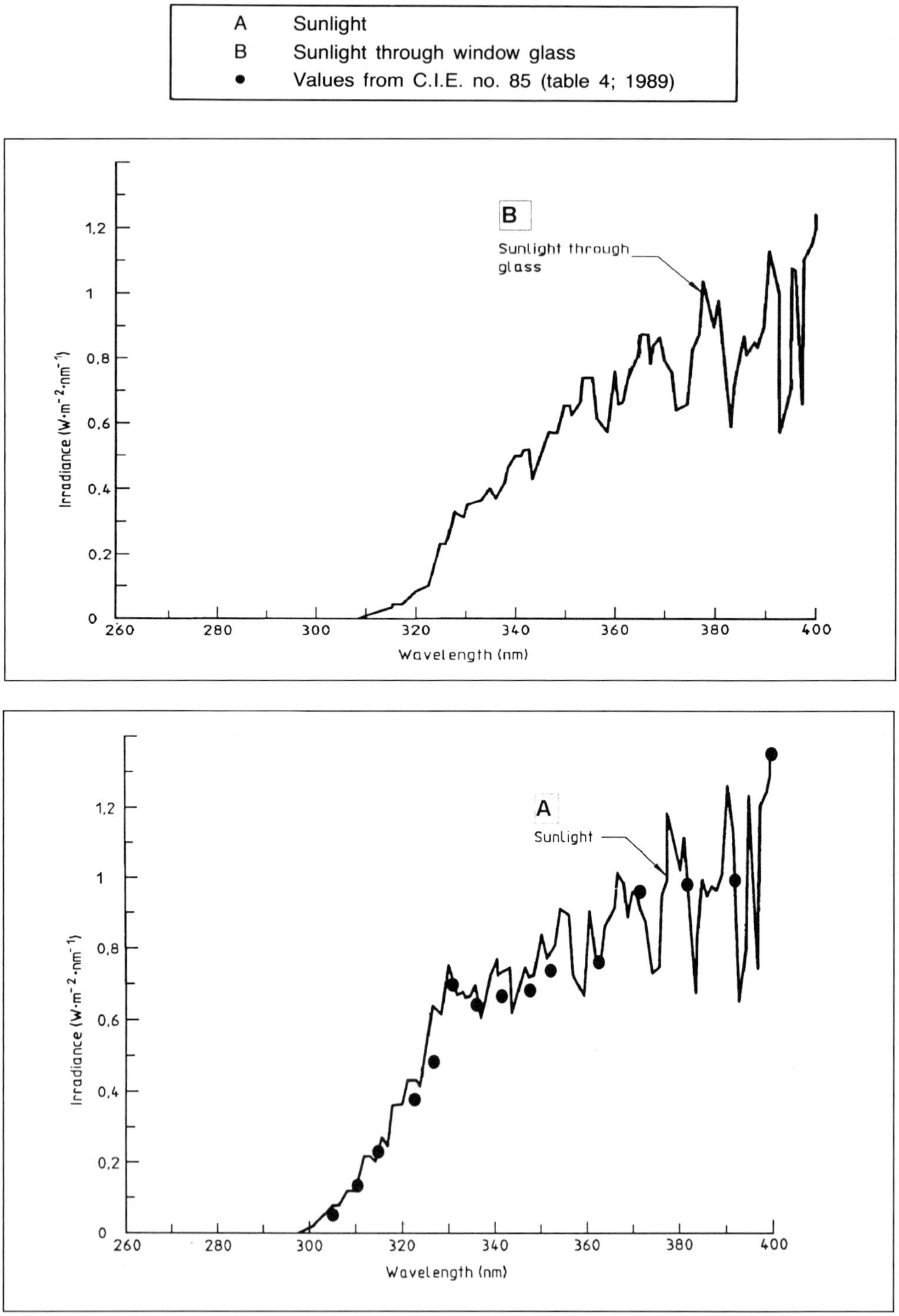

Figure 3.2 Sunlight spectral energy distribution in UV (ISO 4892, Part 3 – Information Annex); Phoenix, Arizona (USA) at solar noon at the summer solstice with a clear sky; spectroradiometer on an equitorial follow-the-sun mount

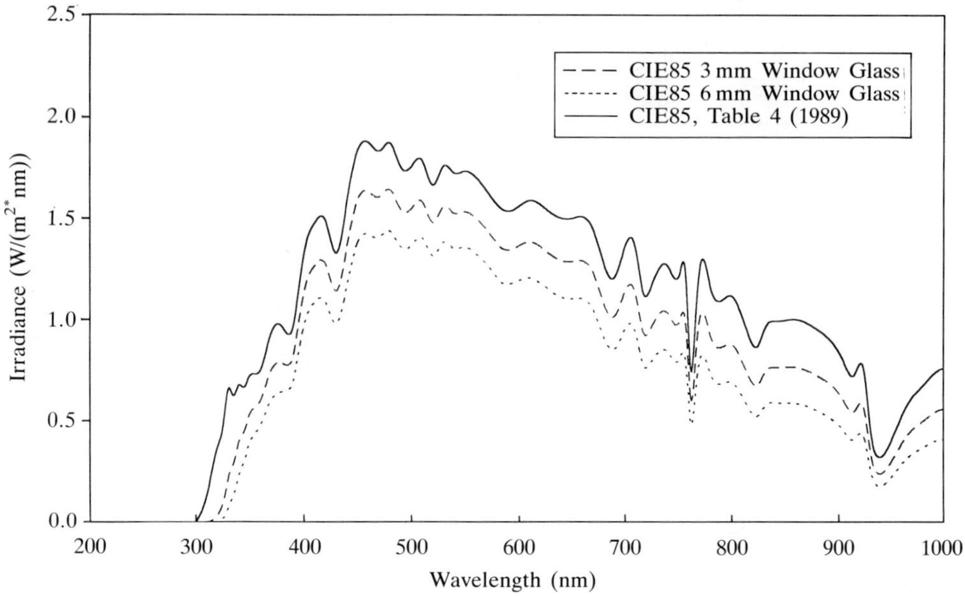

Figure 3.3 Spectral energy distribution of global solar radiation behind window glass

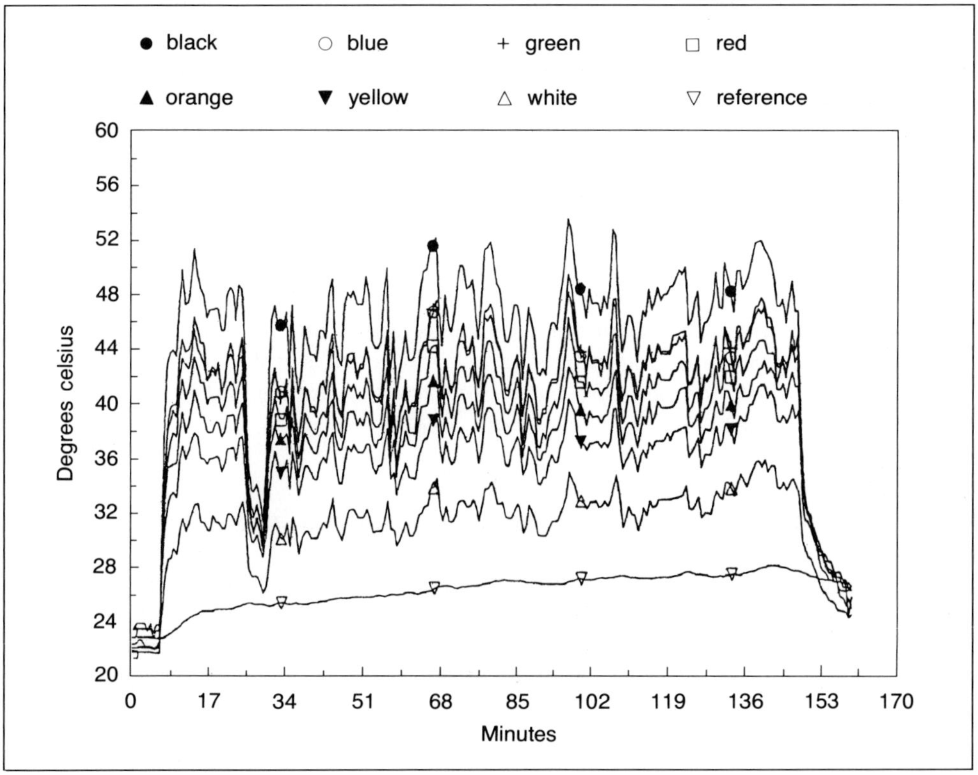

Figure 3.4 Temperature profile for coloured samples as collected from the multichannel recorder (Minnesota 30° backed, July 23 1991) (according to Fischer, 1994)

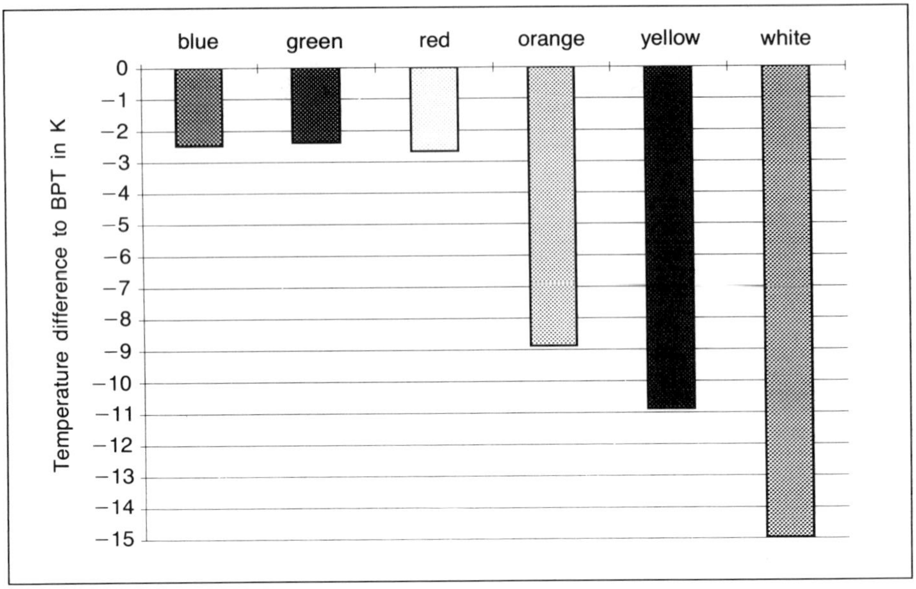

Figure 3.5 'Typical' temperature profile for coloured samples exposed 45° open (BPT = 53°C) exposure (from Minnesota 45° open temperature model) according to Fischer, 1994.

translucent samples where very complex relationships between heat absorption by radiation and heat emission can occur (Wendisch, 1966).

Nearly all photochemical ageing processes are influenced by the temperature. Considering that a temperature increase of $\Delta T = 10\,K$ may double the reaction rate, the increased temperature of the exposed surface is a far more important parameter than the temperature of the ambient air.

Measurements on differently coloured PVC films in outdoor exposure have shown temperature differences between black and white samples of 15 to 23 K, the coloured samples ranging in between (Fischer and Ketola, 1994). An example is shown in Fig. 3.4 (temperature measurements – example) and Fig. 3.5 ('typical' temperature profile).

3.4 Accelerated photostability tests with artificial light sources

By studying the literature on photostability testing of drugs under simulated conditions and the present status of a guideline (Anderson, 1994; Sandeep Nema *et al.*, 1995) it can be seen that various different artificial light sources have been used in the past and are still in use. Due to this and the lack of an official guideline it seems that there are as many testing procedures being performed for light stability testing of pharmaceuticals as there are pharmaceutical companies.

The present stage of discussions on photostability testing of drugs with artificial light sources clearly reflects the long years of research and permanent revision and redefinition of standardized test methods for photostability testing on materials and products in non-pharmaceutical industries.

Considering the stress conditions which lead to reactions in materials caused by daylight there are only a few very complex factors that have to be defined as precisely as possible in a standardized test method. These parameters have to be specified for the area or volume where the substances or products are really exposed.

In the exposed samples themselves there are still other factors concerning the whole test procedure which may influence the reactions and/or the reaction rate and have to be described and possibly specified in detail. In the following only the technical requirements for the equipment for photostability tests will be discussed.

The present state of experiences and their conversion into testing procedures and techniques on an international level might also provide impetus for defining acceptable performance criteria in combination with the current approach of the drug industry.

Some most important thoroughly revised international standards in the field of testing under simulated conditions on plastics, paints and textiles are listed in Table 3.1.

The 'General Guidance' – Part 1 of ISO 4892 provides general information on testing procedures, especially for plastics, under simulated natural conditions that may also be useful in other industries.

The introduction of this standard states 'accelerated exposure in artificial light devices is conducted under more controlled conditions that are designed to accelerate degradation and product failures'.

The introduction also states that generally valid correlations between the ageing processes which occur during artificial and natural exposure cannot be expected because of the large number of factors involved. The prerequisites for running tests considering correlation to tests under natural conditions as well as repeatability and reproducibility of tests under accelerated conditions are described.

The focal points for accelerated tests are as follows:

- Use of CIE Publication No. 85 as a basis for comparison of artificial sources with natural daylight.
- Statements concerning factors tending to decrease the degree of correlation as there are:
  - use of ultraviolet radiation of wavelength shorter than those occurring in natural exposure;
  - use of a spectral distribution that differs widely from that of daylight;
  - use of a very high light-flux; and
  - use of high specimen temperatures, particularly with materials which readily undergo changes from thermal effects alone.
- Statement that all testing conditions as there are
  - spectral distribution of radiation;
  - level of irradiance; and
  - surface temperature level.

These have to be ensured on specimen area where the materials are really exposed.

Table 3.1 Materials–methods of exposure to laboratory light sources – International Standards

ISO/TC	Standard	Title	Status
61	ISO 4892	*Plastics:* Methods of exposure to laboratory light sources	
	Part 1	General guidance	published 1994
	Part 2	Xenon-arc sources	published 1994
	Part 3	Fluorescent UV-lamps	published 1994
	Part 4	Open flame carbon-arc lamps	published 1994
35	ISO 11 341	*Paints and varnishes:* Artificial weathering of coatings and exposure of coatings to artificial radiation – Exposure to filtered xenon-arc radiation	published 1994
	ISO/DIS 11 507	*Paints and varnishes:* Exposure of coatings to artificial weathering in apparatus – Exposure to fluorescent ultra-violet and condensation conditions	in progress
38	ISO 105	*Textiles:* Tests for colour fastness	
	Part B06[a]	Colour fastness and ageing to artificial light at high temperatures: xenon-arc fading lamp test	German proposal for revision: 1995
61	ISO/DIS 9370	*Plastics:* Instrumental determination of radiant exposure in weathering tests – General guidance and basic test method	in progress
	ISO/TR 9673	Solar radiation and its measurement for determining outdoor exposure levels	in press

Key
ISO – International Standardization Organization
TC – Technical Committee
DIS – Draft Standard

Note:
[a]Automotive applications.

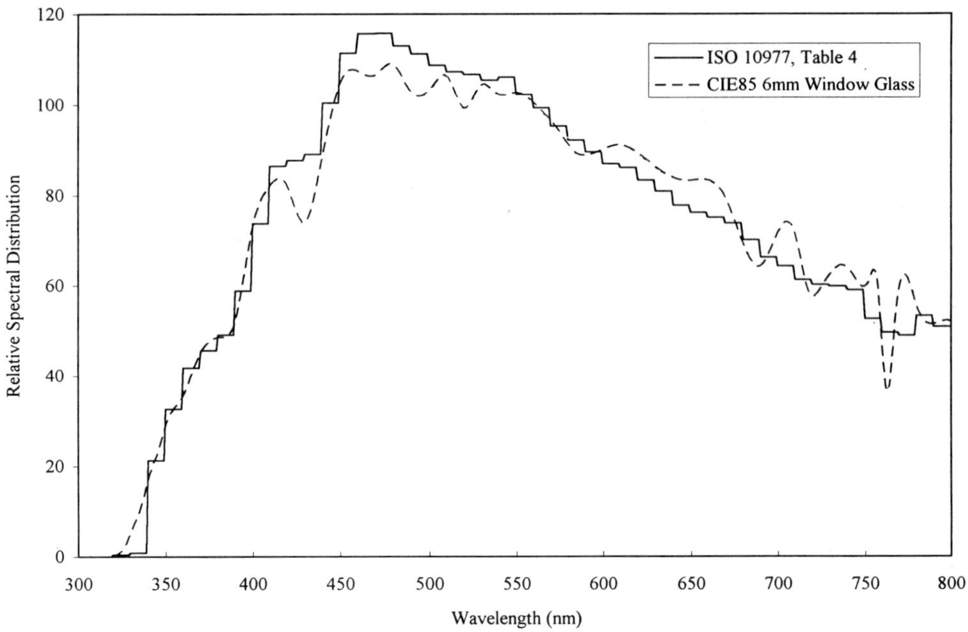

Figure 3.6 Relative spectral distribution of CIE No. 85 with 6 mm window glass compared with the data of ISO 10977 simulated indoor indirect daylight

- Description and recommendation of methods and devices for determining irradiance and radiant exposure as well as measuring of temperature.

From the various artificial light sources that were used in the past in the general field of material light stability testing, specially filtered xenon arc sources and specific fluorescent UV lamps were found to be useful and are therefore specified in the basic standards.

There are still other international standards with photostability tests as part of the contents. The previously mentioned ISO 10977: 1993 (Photography – Processed photographic colour films and paper prints – Methods for measuring image stability) may be an example. In this standard relative spectral distributions for simulated indoor indirect daylight (Fig. 3.6, compared with CIE No. 85 and 6 mm window glass) as well as different artificial light sources have also been scheduled, especially under the aspect of radiation behind window glass.

Only in some national standards are specific metal halide lamps currently described as the basis for the simulation of solar radiation – DIN 75220 (1992) 'Ageing of automotive components in solar simulation units' is an example.

Some basic aspects concerning the requirements for simulation of solar radiation are best discussed on the basis of the technical contents of the ISO – standards.

Independent of the type of radiation system used, the spectral energy distribution of global solar radiation according to CIE No. 85, Table 4 represents the basis for an evaluation of the quality of simulation.

For sunlight behind window glass the CIE data are integrated with the spectral transmittance of a window glass with a thickness of 3 or 4 mm respectively.

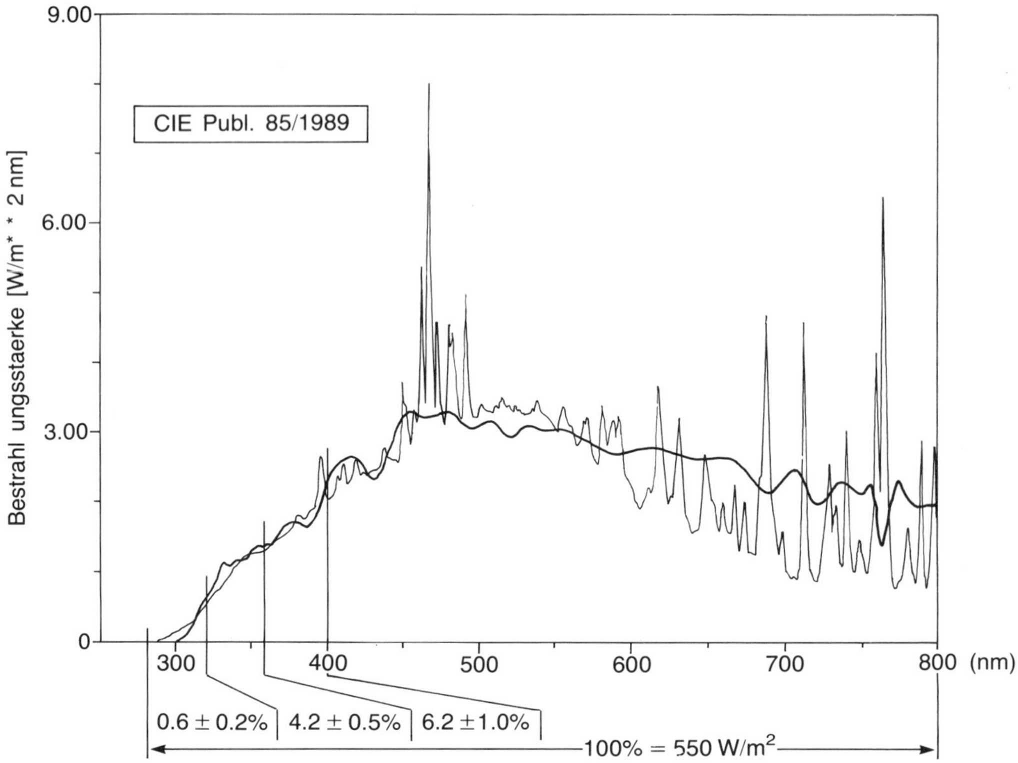

Figure 3.7 Radiation function – ISO 4892, Part 2 (1994); filtered xenon arc radiation for comparison

Especially for filtered xenon arc sources as artificial radiation systems for simulating natural conditions outdoors as well as behind window glass, relative spectral functions have been specified using the values of CIE No. 85 in 40 nm steps in the UV range and the wavelength range up to 800 nm set to 100 per cent (Fig. 3.7). In an additional step the spectral function is extended to separated wavelength steps in the visible wavelength range. The data as specified in the ISO standards are shown in Table 3.2.

These spectral functions have to be realized within the given tolerances on the sample area of test cabinets depending on effects such as changes of laboratory voltage or lamp/filter age of the used light sources. For fluorescent UV lamps the situation is not as clear as for filtered xenon arc sources. At the present time in the standard itself only different lamp types are described and preliminary spectral data are listed and compared with global radiation in an annex for information (Table 3.3). It is already clearly stated that those spectral data should be part of the standard in the near future as already realized in a previously published German standard (DIN 53 384, 1989).

The spectral data being discussed for fluorescent UV lamps are already irradiance values (20 nm steps in UV). For filtered xenon radiation the irradiance level has been set to 550 W/m^2 in the wavelength range up to 800 nm for the purpose of reference. Other values may be used but have to be reported.

Table 3.2 Radiation functions (basis for simulation of global radiation with filtered xenon arc lamps) as specified in International Standards

Standard	ISO 4892, Part 2 ISO 11 341		ISO 105-B06[a]
Method resp. set of exposure conditions	A (artificial weathering)	B (daylight behind window glass)	1, 2 and 3
Wavelength (nm)		Relative irradiance (%)[b]	
<290	0[c]	–	0
<300	–	0	<0.05
280–320	0.6 ± 0.2	<0.1	<0.1
320–360	4.2 ± 0.5	3.0 ± 0.5	3.0 ± 0.85
360–400	6.2 ± 1.0	6.0 ± 1.0	5.7 + 2.0/−1.3
400–520	–	–	32.2 +3.0/−5.0
520–640	–	–	30.0 ± 3.0
640–800	–	–	29.1 ± 6.0
<800	100	100	100

[a]Proposal for revision (1995). There are several sets of explosive conditions in ISO-B06 but the specific spectral function is only specified for sets 1 to 3.
[b]As a percentage of the total irradiance in the wavelength range up to 800 nm.
[c]Xenon arcs operating as specified in method A emit a small amount of radiation below 290 nm. In some cases, this can cause degradation reactions which do not occur in outdoor exposures.

Table 3.3 Spectral data for fluorescent UV-lamps as described in ISO 4892, part 3 (informative annex)

Spectral band nm	Type I (340 nm peak)	Type I (351 nm peak)	Type II (313 nm peak)	Combination
<270	0	0	0	0
270 to 300	0.1	0	5.2	0.3
301 to 320	3.0	0.8	13.1	3.0
321 to 360	25.1	22.6	12.1	22
361 to 400	11.0	12.7	1.1	18

Note: The data shown in this table are preliminary only. Additional work is under way in ASTM Committeee G03, Durability of Non-metallic Materials, to develop more complete and technically valid data on irradiance specifications for fluorescent UV lamps.

It is most important that the used irradiance shall not vary by more than ± 10 per cent comparing any two points in the sample plane parallel to the lamp axis.

Using fluorescent lamps, especially those with emission of radiation only in the UV part of radiation, differences in temperature between specimen surfaces and the ambient air as experienced in natural exposure are very small. For laboratory light sources with an amount of radiation at longer wavelengths, especially in the infrared region as for example in the emission of xenon arcs and reduced by specific filter elements, the temperature level has to be thoroughly considered and is specified as the surface temperature of a described specific measuring element that will be discussed later.

The specified requirements for artificial tests are experienced as really necessary in the general field of material testing for an improved correlation to natural exposure as well as an improved repeatability and reproducibility of accelerated tests.

The characteristics of artificial light sources and most of the used filter elements are subject to change during use due to ageing and other influencing factors with the effect of changing spectral distributions, irradiance levels and surface temperatures on the sample area.

To meet the discussed requirements, improved equipment technology was necessary in already existing devices and is realized in a range of today's commercially available testing equipment, especially with xenon arc sources from different manufacturers.

With partially different solutions some focal points are:

- improvements in lamp and filter technology;
- variable irradiances on specimen area;
- measuring and controlling of preset values for irradiance and surface temperature at sample level; and
- development of specific measuring devices for irradiance/radiant exposure and surface temperature.

These essentials should always be considered when designing test systems for photostability tests on the mere basis of any artificial light sources and measuring devices available in the market.

The bench top light exposure cabinet Suntest CPS+ (Heraeus Company) which is already widely used in Europe by many pharmaceutical and cosmetic companies (Thoma and Kübler, 1994) may serve as an example for the high equipment technology available today. This unit completes a range of equipments with filtered xenon arc radiation and contains most of the features previously mentioned. The focal points are as follows:

- The irradiance level on the sample area can be varied in a wide range without changing the spectral energy distribution (Fig. 3.8).
- Specific UV cut-off filters are used for realizing specific requirements on the simulation of solar radiation (Fig. 3.9).
- The preset irradiance is continuously measured and controlled constant by a filter radiometer calibrated in the wavelength range 300 to 800 nm.
- Tests may be run to preset values of radiant exposure (300 to 800 nm).
- The installed filter radiometer is calibrated/recalibrated with instruments that are traceable to national standards.
- The temperature on the sample area can be varied in a wide range and is measured and controlled constantly by a black standard thermometer.
- The measured values are continuously monitored and can be documented on a chart recorder or data longer.

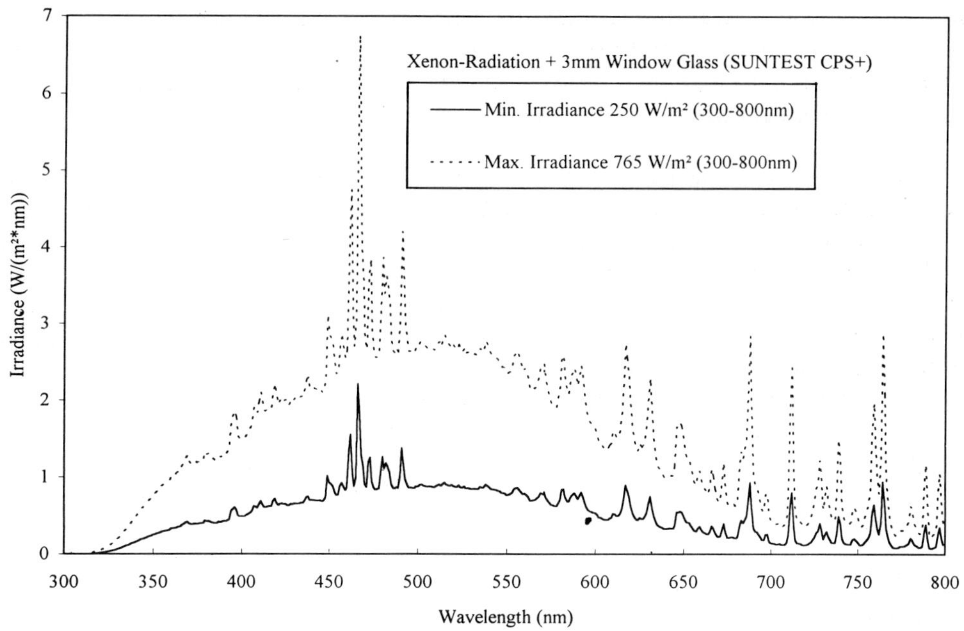

Figure 3.8 Spectral distribution (min./max.) on sample level in a Suntest CPS+

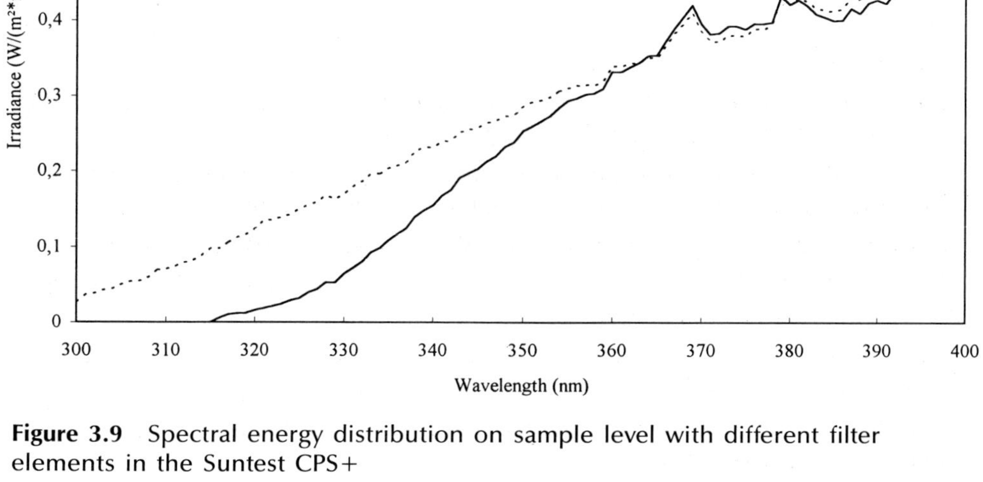

Figure 3.9 Spectral energy distribution on sample level with different filter elements in the Suntest CPS+

3.5 Measurement of radiation and temperature at sample level

3.5.1 Irradiance/radiant exposure

In the available technical data for the various artificial radiation sources in use for photostability testing there is some confusion concerning the use of photometric and radiometric units. Both have their proper place in science and engineering. We must be aware of the differences in terminologies and uses and determine which is most appropriate for each type of study. Based on a given relative spectral distribution of light sources' radiometric units for a specific wavelength, intervals may be calculated from photometric units by the use of determined conversion factors. Since radiometric units are not limited to visual response and the UV part of solar radiation is most important for photoreactions, radiometric units are widely used in photostability tests.

Two approaches to the measurement of radiation (with emphasis on the ultraviolet wavelength range) are commonly used. The first is the use of a physical standard (chemical actinometer as well), for example, to expose a reference material (substance) which shows a change in property proportional to the dose of incident radiation (UV or other specific wavelength bands). The preferred approach is to use a radiometer which responds to the defined wavelength band (ISO/DIS 9370).

Precise information on the spectral irradiance of incident radiation on a sample surface is only available using spectroradiometers. A schematic view of a measuring system which can be used for measurements on sample area in any equipment is shown in Fig. 3.10 as an example. Those systems are very expensive and not usable for the daily practice in stability tests.

Therefore, UV and visible radiation-measuring instruments (filter radiometer) described in the basic ISO standards are used in practice. A schematic view of a measuring system is shown in Fig. 3.11.

ISO/DIS 9370 (Table 3.1) contains valuable information and recommendations on the important characteristics for the instruments used and provides a guide for the selection and use as well as calibration procedures of these radiometers including both natural and simulated exposure testing. Instrumental techniques include the continuous measurement of irradiance in specific wavelength bands and the accumulation (or integration) of instantaneous data to provide a total radiant exposure (dosage).

Modern commercially available equipment for photostability testing where the level of irradiance can be varied in a wide range, especially devices with filtered xenon arc radiation, are equipped with those filter radiometers to control the preset level of irradiance as well as to accumulate the current data. Depending on the type of the various existing apparatus, the preferred wavelength bands are either 300 to 400 nm or 300 to 800 nm or measurements at a specific wavelength (e.g. 340 or 420 nm) (Boxhammer, 1994). There are still other filter radiometers in the market for different wavelength bands. Comparing measuring results those differences have to be considered thoroughly. Filter radiometers shall be calibrated to a given source. When used with sources having a different spectral distribution of radiation the radiometer must be recalibrated to that source.

Calibration of spectrally selective radiometers should be by substitution, using

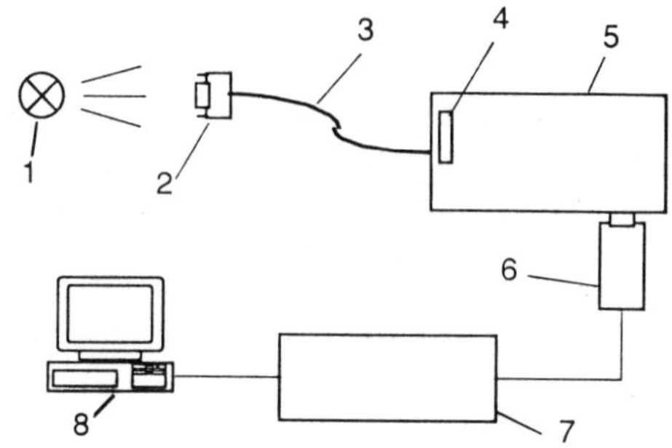

1: Lamp-Filter-System, 2: Entrance system with cosine adaption; 3: Optical fiber; 4: Filter set; 5: Polychromator; 6: Line camera; 7: Controller; 8: PC

Figure 3.10 Spectroradiometric measurements on sample area (principle)

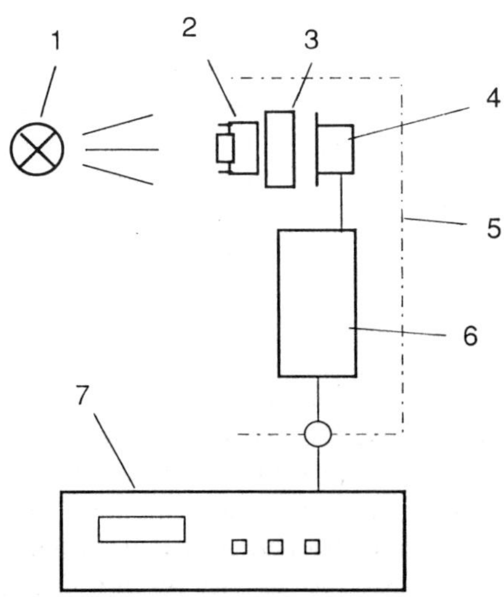

1: Lamp-Filter-System, 2: Entrance system with cosine adaption; 3: Band filter; 4: Photo diode; 5: Measuring head case; 6: Measuring amplifier with memory unit

Figure 3.11 Measurement of irradiance and radiant exposure (bandpass) on sample area (principle)

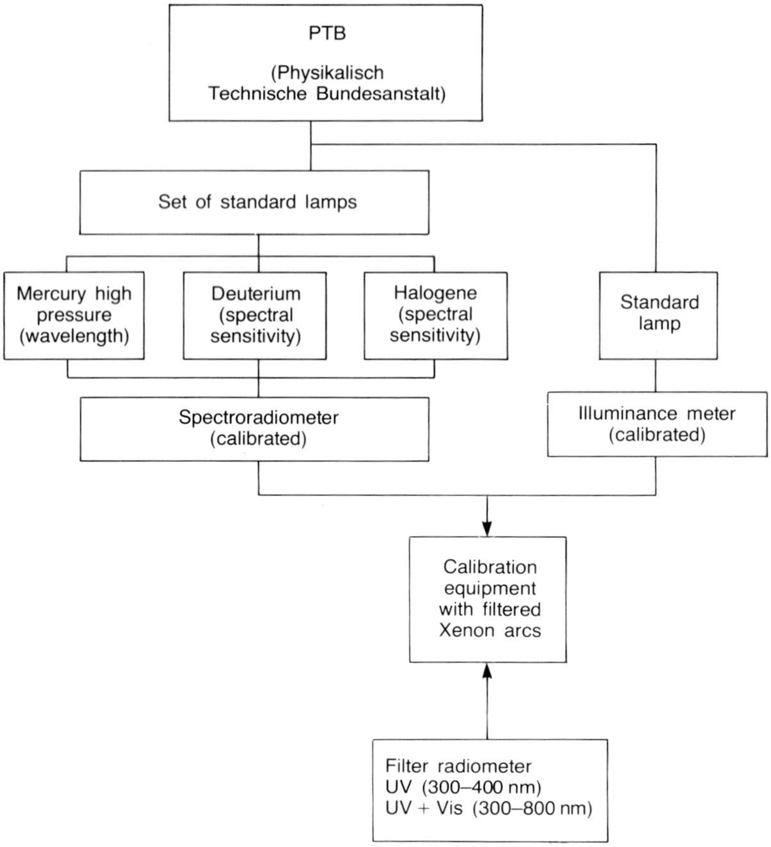

Figure 3.12 Calibration procedure – Xenon equipment (Heraeus Industrietechnik)

a spectroradiometer calibrated to standard lamps in the appropriate wavelength range traceable to national primary standard lamp and a radiation source that is identical to the source that will be measured. One example for a calibration procedure is given in Fig. 3.12. Calibration standard lamps traceable to national primary standard lamp may also be used.

3.5.2 Surface temperature

Since it is not practicable to monitor the surface temperature of individual exposed samples, a specified black-coated flat plate sensor has been used for decades to measure and control temperature on the sample surface in material testing.

At the present time a so-called black standard thermometer is described and specified in the basic ISO standards for characterization of the highest possible surface temperature of a dark solid sample with poor thermal conductivity under a given incident radiation and other surrounding conditions (air temperature, air velocity and humidity) (Boxhammer, 1994).

Samples with different coloured or even white surfaces show different deviations to the black standard temperature. A so-called white standard thermometer, also

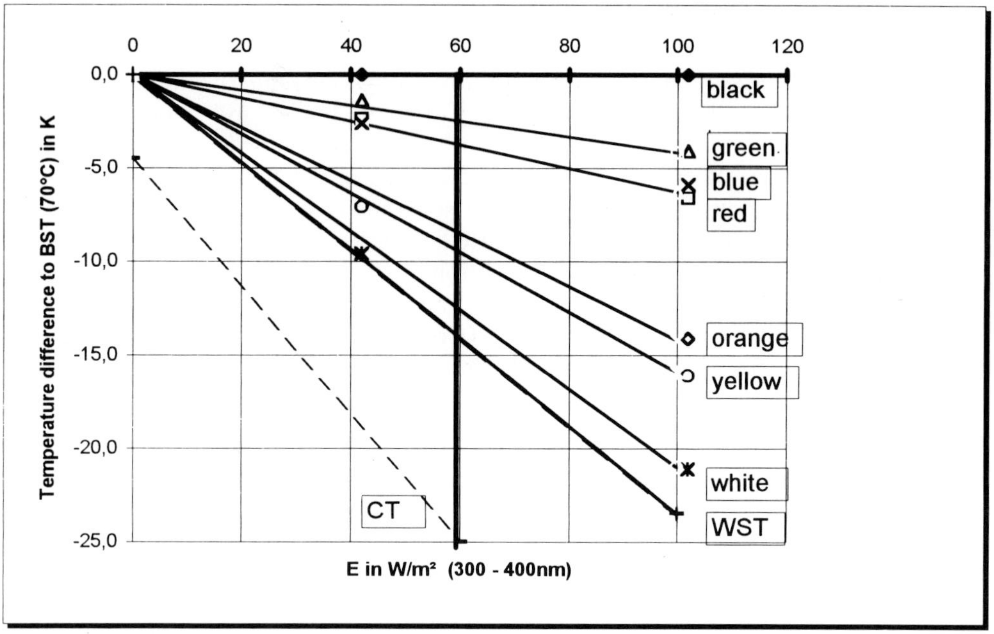

Figure 3.13 Surface temperatures of coloured samples in Xenon equipment Xenotest 1200; FS: 3 Suprax dishes) (according to Fischer, 1994)

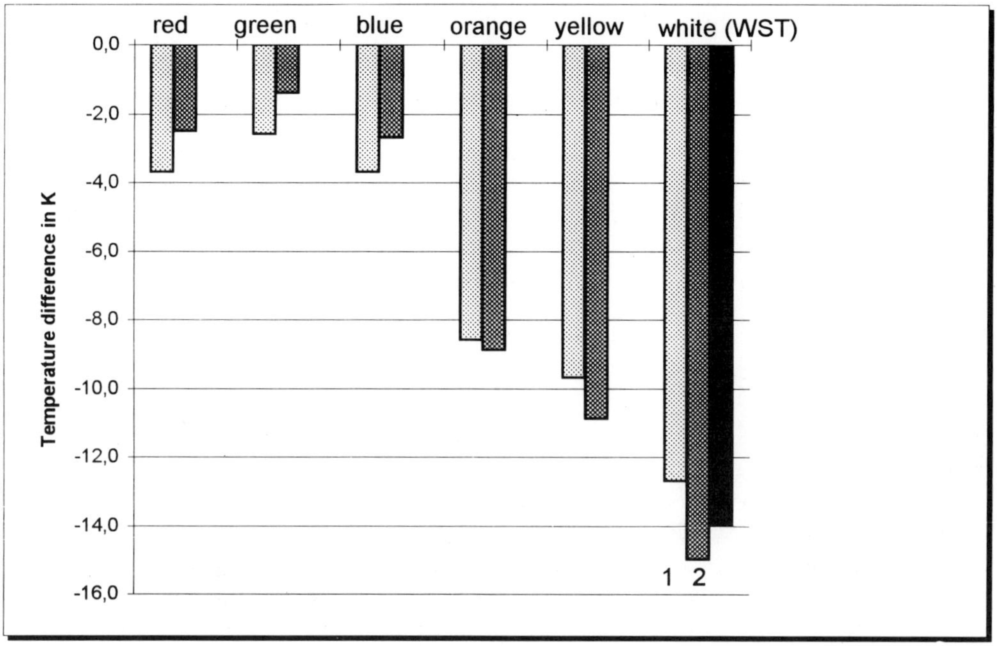

Figure 3.14 Temperature differences in Xenon equipment ($E_{(uv)}$ = 60 W/lm^2) compared with outdoor exposure (45° open): 1 – Xenon equipment; 2 – outdoor exposure (according to Fischer, 1994)

described in the basic ISO standards can be used, indicating the lowest surface temperature of a sample whose surface is directly absorbing and reflecting.

The temperatures of samples with coloured surfaces range between the black and the white standard temperatures. Constant black and white standard temperatures under given exposure conditions during a test indicate constant surface temperature of any exposed sample even if in the field of pharmaceutical substances the real surface temperature of specifically prepared samples may be above or below the characterizing standard temperatures.

Various measurements at varied conditions in equipment with filtered xenon arc radiation have shown that there is a linear function between the standard temperatures and the level of irradiance (Boxhammer *et.al.*, 1993) and that the surface temperatures of coloured samples measured in equipment can be correlated to those under natural conditions (Figs 3.13 and 3.14) (Boxhammer, 1995).

3.6 Spectral distributions of artificial light sources as proposed for photostability tests of drugs compared with global radiation behind window glass

Curves of the relative spectral distribution of artificial light sources (only examples) as used for photostability tests on drugs in the past and recommended in the fourth draft tripartite guideline – The photostability testing of new drug substances and products (1994) – compared with global radiation behind window glass are shown in Figs 3.15 and 3.16. The graphs are normalized to 100 per cent in the wavelength region 300 to 800 nm (Fig. 3.15) and 300 to 400 nm (Fig. 3.16) respectively.

The spectral distribution over the visible wavelength range may be approxi-

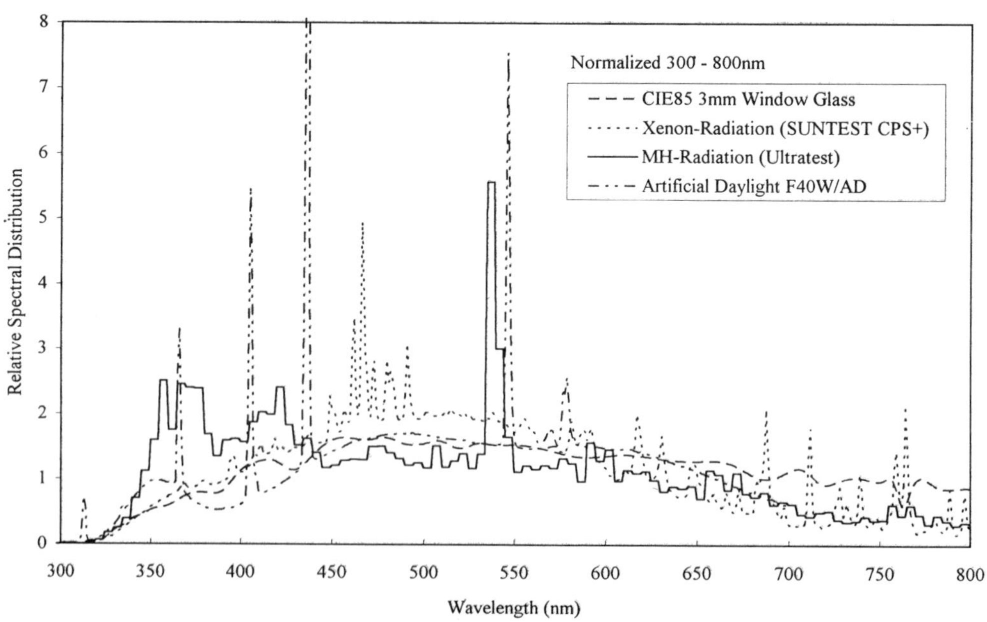

Figure 3.15 Relative spectral distribution of different laboratory light sources compared with CIE No. 85 (UV + visible range)

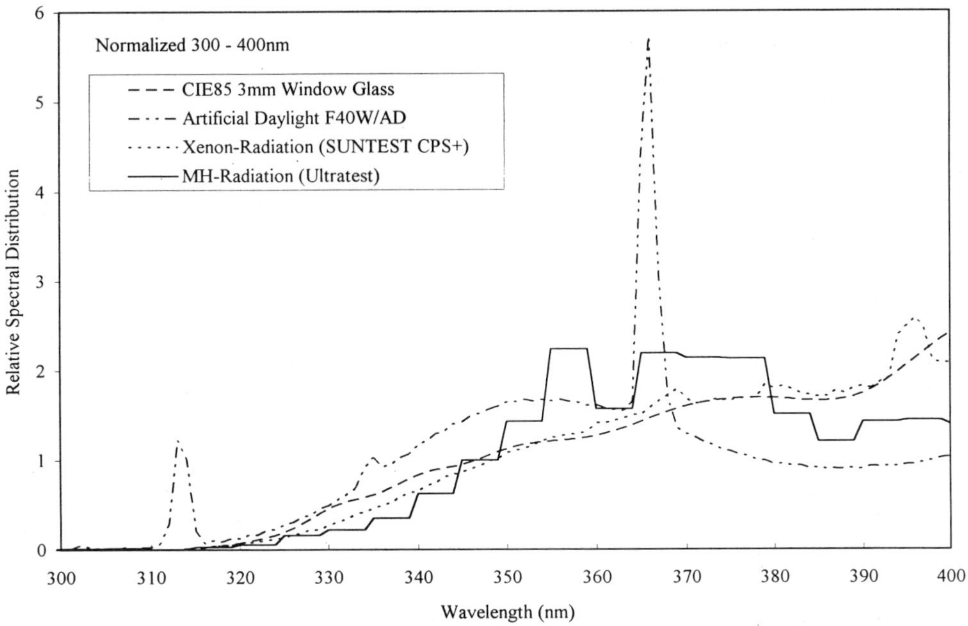

Figure 3.16 Relative spectral distribution of different laboratory light sources compared with CIE No. 85 (UV-wavelength range)

mately met by those different light sources, but there are significantly different spectral distributions especially in the UV range of radiation. This picture may be completed by the measured spectral energy distribution in the UV range of radiation of a so-called full spectrum fluorescent lamp (Fig. 3.17) and a specific

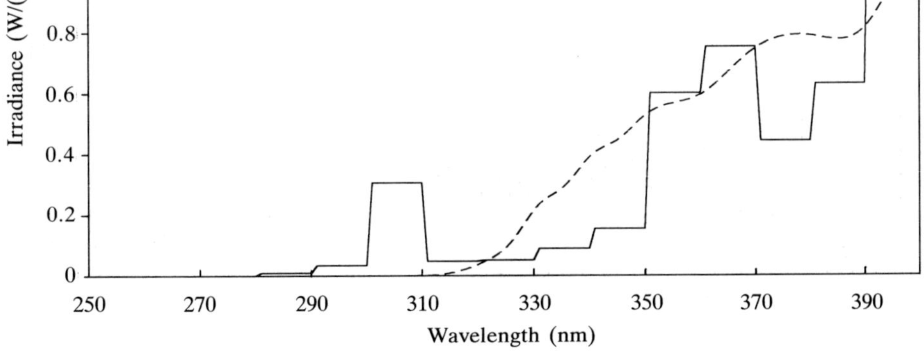

Figure 3.17 Spectral energy distribution of a 'full spectrum' fluorescent lamp compared with CIE No. 85 (UV-region)

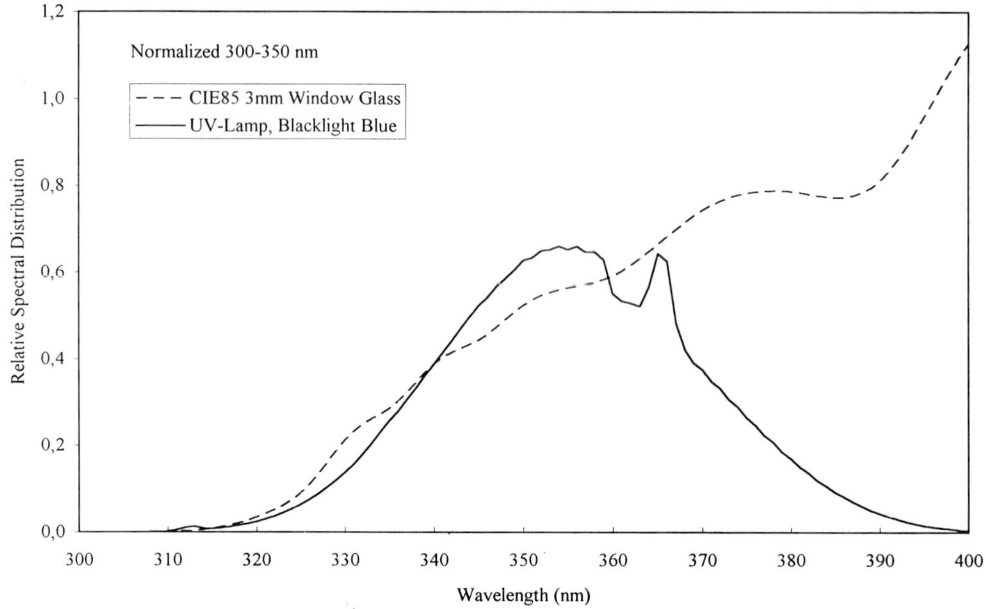

Figure 3.18 Relative spectral power distribution of a fluorescent UV-lamp (example) compared with CIE No. 85 normalized to wavelength region 300 – 350 nm)

fluorescent UV lamp (Fig. 3.18). Furthermore, it has to be considered that only a general description of wavelength ranges of emission allows for using various types of the respective light sources with quite different spectral distributions in specific wavelength ranges. An example is given in Fig. 3.19 for two fluorescent lamps both used as 'daylight lamps' for colour-matching purposes.

There is a similar situation concerning metal halide lamps in general and those types used for photostability testing of materials (Boxhammer, 1983) as well as of drugs (Thoma and Kerker, 1992) in the past with an enhanced amount of radiation in the UV wavelength range relative to the visible region (Fig. 3.15). This does not correspond to the requirements on simulation of sunlight as specified today for solar simulation units (DIN 75220) and covered by some types of available specific metal halide lamps.

A rather uniform basic spectrum is given for xenon arc radiation at least as used in commercially available equipment. Differences in detail in the spectral distribution are mostly determined by the various used filter elements for the infrared part (surface temperature) and the UV cut-off as well as the increase of radiation in the UV wavelength range in a different kind of equipment.

The different spectral distributions of used laboratory light sources especially in the UV range of radiation (Fig. 3.16) and partially in the visible wavelength range (Fig. 3.15) may lead to test results which are not comparable since interactions between reactions caused by different wavelength bands of radiation may happen. This is especially true for using light sources with an emission only either in the UV range of radiation or in the visible wavelength range and running separate tests. This procedure seems to be very questionable under the interactions between reactions caused by different wavelength bands of radiation.

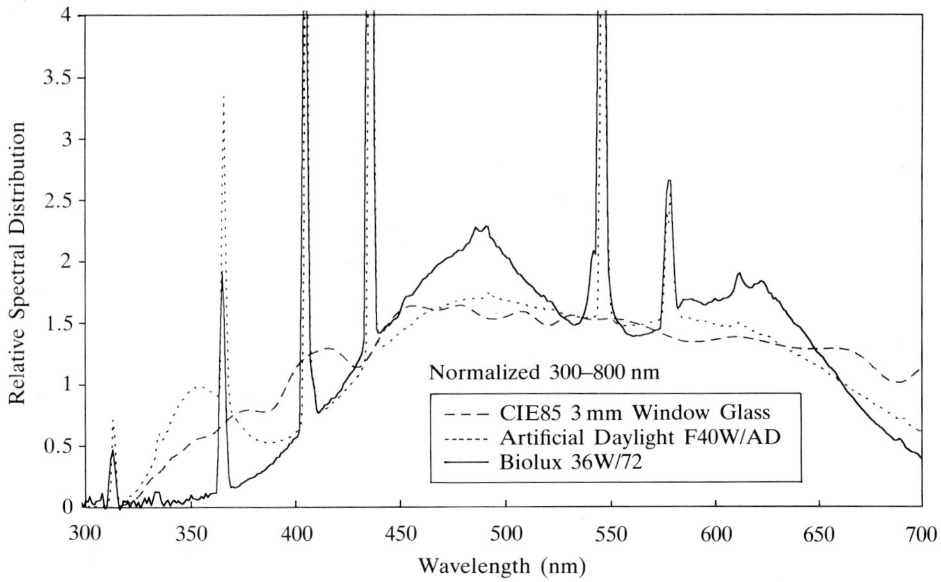

Figure 3.19 Relative spectral distribution of fluorescent lamps (examples) compared with CIE No. 85

3.7 Level of irradiance on sample area and radiant exposure in photostability testing

The levels of irradiance and radiant exposure are two very important parameters for reproducible photostability tests and they should be measured on sample level (where the drug f.ex. is exposed to the light source) and not somewhere else.

Assuming that the spectral distribution of an artificial light source does not contain radiation below the cut-off of natural solar radiation behind window glass, the acceleration of degradation processes is primarily determined by the irradiance level. The reaction rates may be influenced by other factors such as temperature and humidity.

The level of irradiance for accelerated photostability testing of technical materials is based on the highest value of global solar radiation of about 550 W/m^2 between 300 and 800 nm using filtered xenon arc radiation with only some exceptions to lower or higher levels.

The very low illuminance level of 6 Klx, equivalent to about 30 W/m^2 between 300 and 800 nm using filtered xenon arc radiation for light stability tests of colour photographs (ISO 10977) as an example shows the wide range of recommended respectively specified irradiances dependent on the type of materials and application.

The levels of illuminance or irradiance proposed for photostability testing of drugs vary considerably; for fluorescent lamps between 10^3 and 10^6 lux and for metal halide or xenon lamps between 200 and 750 w/m^2 in the wavelength range 300 to 800 nm.

The conversion between photometric and radiometric units may be done

Table 3.4 Relative irradiance values and conversion of illuminance into irradiance

	Relative irradiance for different wavelength bands in %			Conversion factor Illuminance in lux/ irradiance
	300–400	400–800	300–800	(300–800 nm) in W/m^2
CIE No. 85, Table 4	11.0	89.0	100	181
CIE No. 85 + 3 mm window glass	9.0	91.0	100	191
CIE No. 85 + 6 mm window glass	7.6	92.4	100	197
ISO 10977, Table 4	7.2	92.8	100	199
Fluorescent lamp artificial daylight	10.3	89.7	100	216
Metal halide lamp (Ultratest[a]) + 3 mm window glass	19.5	80.5	100	195
Xenon arc lamp (Suntest, CPS +[a]) + 3 mm window glass	9.2	90.8	100	220

[a]Heraeus Company

by using the values for some of the discussed radiation sources as listed in Table 3.4.

Using quite different levels of irradiance as previously mentioned, the mechanism of material degradation may change compared with natural exposure even if the spectral energy distribution of radiation remains stable.

This has to be considered thoroughly under the aspect that depending on the different lamp technologies and the required size of exposure area with a satisfactory uniformity of irradiance, a wide range of irradiances may be realized as already mentioned. In addition, it has to be considered that the total irradiance determines the surface temperature of exposed samples at a given ambient temperature and velocity of air.

Currently, the proposals for the photostability testing of drugs only specify end-point criteria as luminant exposure (luxh) based on artificial daylight and radiant exposure (Wh/m^2) for the UV part of radiation independent of the used level of irradiance. This may cause different results even if the tests have been run to the same defined end-point.

3.8 Performance test of drugs

From discussion of the already specified technical requirements for photostability tests on materials other than pharmaceuticals and comparison with the present state of the proposed performance test of drugs, it can be concluded that there is a need for defining the important factors determining correlation and variability of test results more precisely.

This concerns in a first step the definition of the spectral energy distribution of the radiation during natural exposure as a real basis for simulation. Assuming that different laboratory light sources should be allowed for use in accelerated photostability testing, the spectral distribution of these lamps as well as levels of irradiance should be defined in separated test methods as precisely as possible. This should be done for that location where the samples are really exposed. Preference should be given to those laboratory systems which closely meet the basic requirements for simulation. Requirements for measuring irradiance/radiant exposure on the sample area should be described.

The influence of temperature should be considered by specifying a maximum surface temperature level characterized by a black standard/white standard temperature rather than an ambient temperature.

In designing such a test method the discussed contents of the existing international standards may be helpful. In general it should clearly be stated that test results on samples from different test methods should not be compared unless correlation has already been established.

3.9 Conclusion

Conducting tests under accelerated controlled conditions compared with natural exposure having good correlation as well as repeatability and reproducibility of test results are the main objectives of tests with laboratory light sources in the general field of material testing.

Assuming that sunlight directly behind window glass represents the 'worst case' concerning stress on drugs today, available data for the global solar radiation have been discussed.

Those data are an essential basis for designing methods for testing under simulated conditions using artificial light sources. Well-defined test methods prepared on international level already exist. Focal points have been described to provide impetus for a standardized test method of the photostability of drugs.

Comparing those standards to the present state of discussion in the pharmaceutical industry, a basic concept for a well-defined guideline concerning the technical requirements for the equipment is presented as a basis for discussion.

References

ANDERSON, N. H., 1994, Photostability testing design and interpretation of tests of drug substances and dosage forms, *Conferences of Stability Testing*, London 27/28 April.

BOXHAMMER, J., 1983, Einfluß unterschiedlicher Strahlungssysteme auf das Alterungsverhalten von Hochpolymeren, exemplarisch dargestell anhand der Änderung des photochemischen Aufbaues einiger Polyäthylen- Werkstoffe, *Die Angewandte Makromolekulare Chemie*, **114**, 59–67.

1994, Current status of light and weatherfastness standards – new equipment technologies, operating procedures and application of standard reference materials, 2nd International Symposium on Weatherability, Tokyo.

1995, Oberflächentemperaturen und Temperaturmeßtechnik in der Ebene exponierter

Proben bei der zeitraffenden Bestrahlung/Bewitterung in Geräten, Seminar 'Natürliches und Künstliches Bewittern polymerer Werkstoffe', TA-Wuppertal; 28/29 March.

BOXHAMMER, J., KOCKOTT, D. and TRUBIROHA, P., 1993, Black standard thermometer, *Materialprüfung* **35**, 143–7.

FISCHER, R. M. and KETOLA, W. D., 1994, Surface temperatures of materials in exterior exposures and artificial accelerated tests; accelerated and outdoor durability. Testing of Organic Materials, *ASTM, STP* 1202.

SANDEEP NEMA, WASHKUHN, R. J. and BEUSSINK D. R., 1995, Photostability testing: An overview, *Pharmaceutical Technology*, March, 170–85.

SEARLE, D., 1985, Spectral factors in photodegradation: activation spectra using the spectrograph and sharp cut filter techniques, 7th Int. Conference on Advances in the Stabilisation and Controlled Degradation of Polymers; Luzern 22 to 24 May.

THOMA, K. and KERKER, R., 1992, Standardized photostability tests – a neglected field of drug safety. Personal communication.

THOMA, K. and KÜBLER, N., 1994, Photostabilitätsprüfungen und andere Stabilitätsprüfungen, *Grundlagen der Körperpflegemittel, Kap.* 10; DKG; October.

TRUBIROHA, P., 1987, Ermittlung der spektralen Empfindlichkeit der photochemischen Alterung durch monochromatisches Bestrahlen, International Symposium on Weathering; Essen; 28/29 September.

WENDISCH, P., 1966, Zum Temperaturverlauf in einer Platte bei Erwärmung durch intermittierende Infrarotbestrahlung; *Plaste und Kautschuk*, **13**, 344–6.

4

Standardization of Photodegradation Studies and Kinetic Treatment of Photochemical Reactions

D. E. MOORE

4.1 The need for uniformity in photodegradation studies

Thermal stability studies on pharmaceutical formulations have been formalized for many years now. Specific protocols have been developed to provide data from which a shelf-life determination can be made (Carstensen, 1990). Thus, the procedures followed in one laboratory are reproduced in another and the shelf life is a consistent estimation for the product, independent of the laboratory where the data were gathered. In thermal stability studies, the principal consideration is how long the drug substance or formulation is exposed to a particular temperature. The nature of the apparatus to be used is not important as long as the temperature of the sample is uniform. Thus, the sample may be contained in a flask, a bottle or a tube, held in a water thermostat or an air incubator, whatever is most convenient for the study. Also, the concentration of the drug being studied is not crucial as a thermal degradation usually proceeds by first-order kinetics for which the half-life is independent of the starting concentration.

The same is not true in respect of the investigation of degradation caused by light. The problem is that the exposure of a pharmaceutical to light is a more difficult situation to control and quantify. It is not simply a matter of how long a drug or its formulation is exposed to light, but what wavelength range is involved and how intense the light is at the surface of the sample being tested. Also important is the concentration of the sample, because the percentage change does depend on the starting concentration, even though first-order kinetics may apply to the reaction.

As a consequence of these factors, there has been great variability in the manner in which drug photostability has been determined in pharmaceutical laboratories (Anderson *et al.*, 1991), so that only general statements such as 'Protect from Light' are to be found in the monographs of drugs which are decomposed in some way following exposure to light.

In order to bring uniformity to the studies, as well as gaining more useful information which is transferable from one environment to another, standardized protocols must be introduced covering all the variables that can enter into photochemical reactions. To this end, the apparatus must be defined in detail and the procedures to be followed fully explained. We can approach the discussion of these issues under three headings, namely, the light or photon source, the measurement of the amount of light absorbed by the sample by physical and chemical methods, and the treatment of experimental data.

4.2 Light sources for photodegradation studies

The first question to be answered relates to the nature of the irradiation source to which the drug should be exposed. Some may argue that, when handled and stored correctly, a pharmaceutical should never be exposed to direct sunlight and therefore sunlight should not be contemplated as an irradiation source. While in the best of worlds that may be possible, it is an erroneous view, as studies have shown that numerous situations occur when various formulations are exposed to direct or filtered sunlight in hospital pharmacies while they are held in clear glass or plastic containers (Tønnesen and Karlsen, 1995). There is also the possibility of inadvertent exposure to sunlight during manufacture and transport, or by the consumer in the course of a treatment. The bathroom windowsill is a well-known storage place for medicines – it is considered more important to be out of the reach of children than protected from sunlight exposure. Because of these possibilities, photostability studies should be undertaken to include exposure to sunlight-simulating conditions, both unfiltered and filtered by glass and/or plastic.

In principle, photodegradation studies could be performed by exposing samples to natural sunlight and analysing after varying times. However, the intensity of sunlight, particularly the UV component, varies according to the weather, the latitude, the time of day and the season of the year. Nonetheless, it would be a realistic situation to set samples in a window where direct sunlight could fall on them to varying extents in the course of the day. If the study was continued for a period of at least a year, the conditions would average over all seasons, but would only apply to a particular region. Thus, the quantitative data from experiments on the same formulation performed in different laboratories are unlikely to be in agreement in most cases. The use of natural sunlight is not a viable option given the variability of conditions involved.

For consistency one needs to use an artificial light source which has an output with a spectral power distribution as near as possible to that of sunlight. The output from the source may then be filtered with, for example, glass or plastic, as required to simulate that type of packaging. The spectral power distribution of sunlight and the transmission characteristics of certain glass and plastic filters have been given in Chapter 2. There are two main types of photon source, namely, arc lamps and fluorescent tubes, having specific applications in photochemistry and photobiology because of their resemblance to the sunlight spectrum. The third major type, incandescent (filament) lamps have a spectral output with relatively high infrared and low ultraviolet components compared with sunlight (see Chapter 2) and are not used for this purpose. Arc lamps generate a high intensity from a relatively small size, so that the light can be readily focused in optical systems for accelerated studies. On the other hand, fluorescent tubes are simpler and cheaper to use, with the contrasting performance of being able to irradiate a large area at a lower intensity. A brief description of these sources is given here.

4.2.1 ***Arc lamps***

Mercury arc lamps can be constructed in three ways, with the mercury vapour at low, medium or high pressure, each variant having specific characteristics in emphasizing certain aspects of the mercury emission spectrum. The low-pressure arc emits 90 per cent of its energy as a line at 254 nm, so is of no direct use in a sunlight-simulating experiment. The medium-pressure arc is also a line source, producing greater intensities at the other characteristic mercury emission wavelengths, 302, 313, 334, 366 and 405 nm. Because this arc lamp is moderate in cost, has a long life and gives a good representation of emission in the UV region, it has been widely used in drug photostability studies with a glass filter to shield the sample from the 254 nm radiation (Moore, 1987). The principal application is as an irradiation source to determine degradation pathways. The high-pressure mercury arc emits the same lines, as well as a continuous background radiation right across the solar UV and visible regions. The emission spectrum of the high-pressure form is shown in Fig. 4.1.

The arc lamp having the best resemblance to sunlight is the xenon arc lamp, although the development of new metal-halide lamps has led to competition for that claim. The spectral outputs of these two types are shown in Figs 4.2 and 4.3.

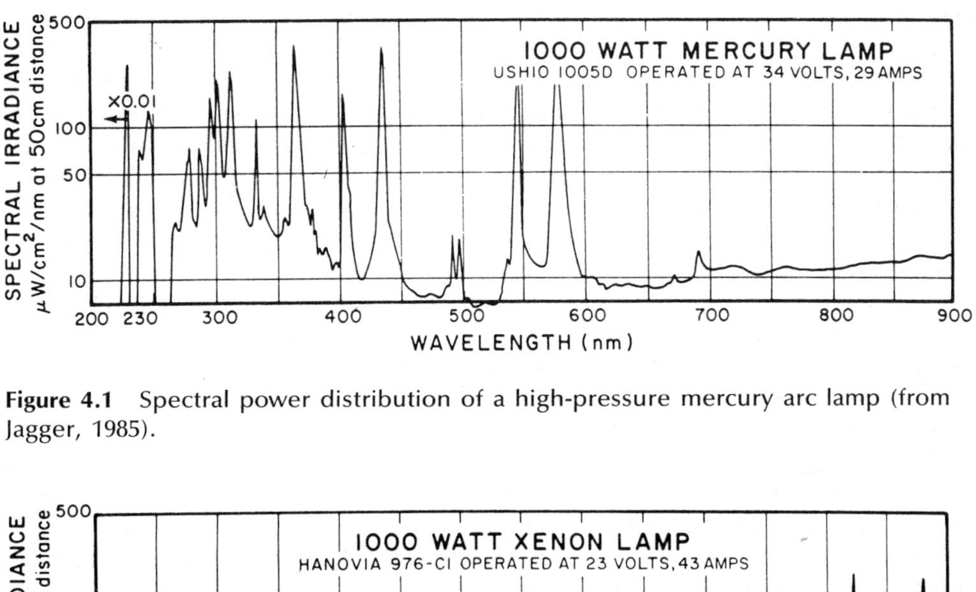

Figure 4.1 Spectral power distribution of a high-pressure mercury arc lamp (from Jagger, 1985).

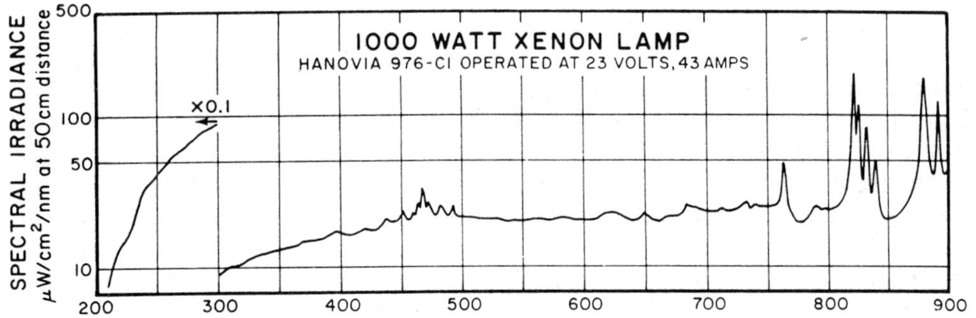

Figure 4.2 Spectral power distribution of a xenon arc lamp (from Jagger, 1985).

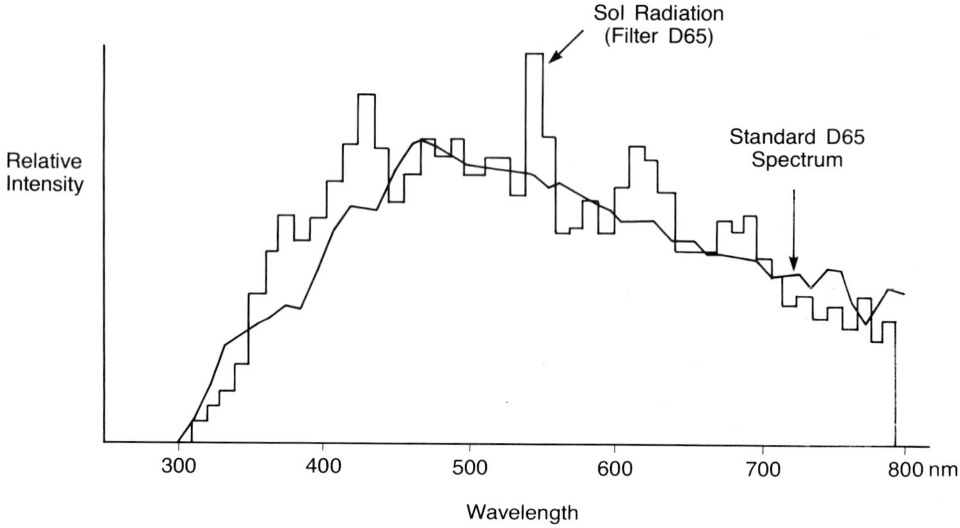

Figure 4.3 Spectral power distribution of a metal halide lamp compared to sunlight.

The xenon arc has a relatively smooth continuous output spectrum with some line emissions superimposed in the region 450–500 nm, whereas the metal-halide lamp is more uniform across the 350 to 550 nm region. One disadvantage of all the arc lamps is the high heat output, as seen by the continued output above 500 nm, but this can be dissipated by the use of a heat filter, usually containing water. Since photochemical reactions are generally initiated by UV radiation, adjustment of the intensity above 500 nm is most unlikely to lead to erroneous photostability data. On the other hand, overheating of the sample by the lamp may lead to thermal decomposition processes to complicate the issue.

The other principal disadvantage of both xenon and metal-halide sources is their comparatively short life span – 750 h for the metal-halide and 1500 to 2000 h for the xenon arc. Their initial cost is high and the focus of their irradiation is such that only a relatively small number of samples can be irradiated at one time.

4.2.2 *Fluorescent lamps*

Fluorescent lamps have been used in photostability testing by a number of laboratories. The operating principle of fluorescent lamps is based on mercury vapour discharge at very low pressure, producing the 254 nm emission which is converted to higher wavelengths by the phosphor coating on the inside surface of the tube. The emission characteristics are determined by the particular phosphor used in the manufacture. Some examples have been given in Chapter 2. For possible application to drug photostability studies are the 'daylight' and 'cool-white' and near-UV fluorescent tubes, all of which have the advantage that they can be set up in large banks at relatively low cost to irradiate large numbers of samples at one time. It is not possible to achieve a sunlight-simulating spectrum with just one type of fluorescent lamp; a combination must be used to get the

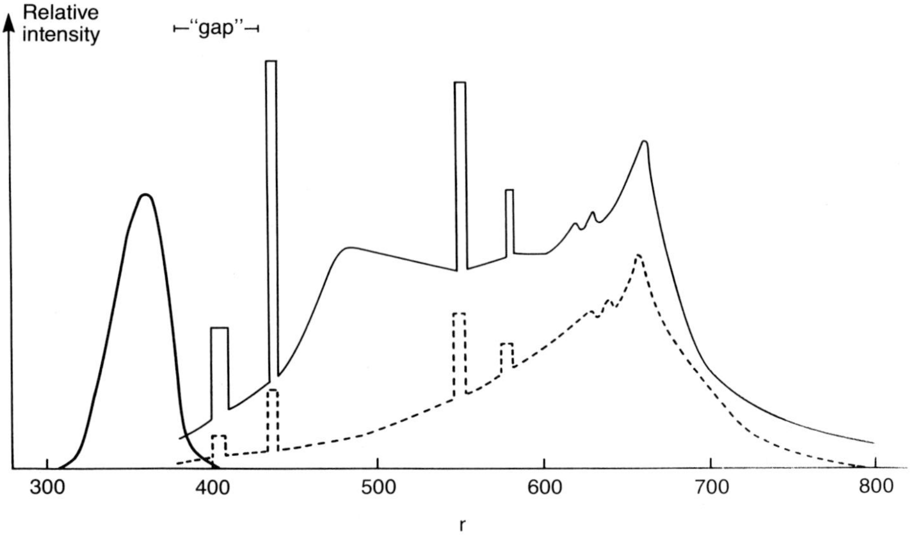

Figure 4.4 Relative spectral distribution of a UVA 'black lamp' and two representative 'daylight' fluorescent lamps. (Redrawn from the Osram Product Catalogue)

appropriate amounts of UV-A, UV-B and visible components. Other combinations which have been suggested involve the use of a 'black-light' UV-A source with daylight fluorescent tubes as shown in Fig. 4.4, but these are unsatisfactory because the correct balance between the UV and visible region intensities is not achieved for all the irradiated samples without diffusers which reduce the overall intensity (Tønnesen and Moore, 1993). While the fluorescent lamps have long lifetimes in excess of 20 000 h and do not cause a problem with respect to heat output, the intensity developed is low compared with the xenon arc and exposure times of the order of several weeks may be needed to gain a change that occurs in a few hours with the xenon arc.

In a number of laboratories at the present time, qualitative photostability testing of drugs involves a combination of a UV-A fluorescent lamp with a daylight fluorescent lamp. The sample is exposed to light for a fixed period of time and if any changes in the physical or chemical properties of the sample are observed compared with a reference sample, it is recommended that the drug should be protected from light. Such a test is inexpensive and easy to perform, but must be regarded as a minimum level of exposure, and more stringent testing should be performed in appropriate cases when significant changes are observed. Reproducible results will not be obtained nor practical conclusions reached unless the output spectra of the recommended light sources are standardized.

4.3 Measurement of the light intensity

The intensity of the photon source is a major variable which depends very much on the nature of the lamp (factors such as its age and design). Its position with respect to the sample will determine the incident dose that the sample receives.

Any attempt to standardize photostability testing must define the type of light source to be used, and include in the experimental arrangement a means of monitoring the amount of light falling on the sample so that the total exposure reaches a certain predetermined value.

Light measurement is termed radiometry and can be achieved by physical instruments such as a radiometer, a lux meter or a thermopile or by chemical actinometry using a reaction of known photochemical efficiency (quantum yield). First it is necessary to define briefly the terms involved. Fuller accounts are given by Jagger (1985) and Thorington (1985).

4.3.1 *Light intensity – irradiance, fluence and dose*

Radiant intensity is the power emitted per unit solid angle of the source, whereas the radiance is the intensity per unit area of the source. Thus a fluorescent lamp has an intensity similar to a filament lamp, but comparatively a low radiance.

The measurement of the light incident upon a sample can be expressed in terms of the number of photons of a particular wavelength crossing a unit area in a unit time. Using the Planck equation,

$$E = \mathrm{h}\nu = \mathrm{hc} / \lambda$$

this number can be converted into an energy, so that the energy fluence, or irradiance, is obtained, expressed in units of Joules $\mathrm{m}^{-2}\,\mathrm{sec}^{-1}$, or watts m^{-2}. That definition is complicated when the irradiating source covers a range of wavelengths, i.e., a range of energies. In this case, the response is integrated over the wavelength range, but it is difficult to obtain an absolute value for the irradiance without reference to a calibrated system of measurement.

In the visible region, the terminology is somewhat different, being called photometry and based on the human response to radiation. The unit of source intensity is the candle, and the radiance of the source becomes the brightness, while the irradiance is called the illuminance (measured in lux) or the illumination (measured in foot-candles). The lumen is the unit of power (in watts) while lux is the power per unit area, expressed as lumen m^{-2} or watts m^{-2}. Note that the units of photometry relate to standardized human perception. Thus a monochromatic UV source may have a high irradiance but a zero illuminance, since none of the energy can be perceived by the human eye. However, many sources of UV-R are also powerful visible light sources, and are rated by manufacturers in photometric terms.

The total dose is determined from the length of time the sample is irradiated, expressed in watt h m^{-2}. There is a fundamental difference between irradiance and dose, in that the former describes a beam of light, whereas the latter relates to the irradiated sample. In other words, the irradiance of a light beam may be constant, but the dose will vary according to how an irregularly shaped sample is orientated with respect to the beam. In most experimental work, it is the irradiance that is measured, when strictly it is the dose incident on the sample which is more important. The radiation may be absorbed, transmitted, scattered or reflected, but in terms of photochemical reaction, that which is absorbed is the critical quantity. Measuring the quantity of light absorbed can only be achieved for liquid samples which transmit the unabsorbed light to be measured by a detector behind the sample.

4.3.2 Physical instrumentation for light intensity measurement

Radiometers are devices based on various types of photocells which generate a current when light falls upon them. The bimetallic photovoltaic cell, the basic component of solar energy converters, is the first and simplest example but these converters have been replaced by photodiodes in recent times. The photomultiplier tube can also be used, but is regarded as too cumbersome for incorporation into a portable meter. All these devices have a response which varies with the wavelength of the incident light, but their use can be designed for a particular spectral region by a specific filter. Each filter-photocell combination requires a separate calibration, as does each different source that is used. Thus UV filter-radiometers are designed to measure incident radiant power in the UV region transmitted by the particular filter. By appropriate choice of filter it is possible to get a reading of the UV-B or the UV-A intensity. Examples of these detectors are shown in Fig. 4.5, with their relative spectral response in Fig. 4.6.

While this provides a convenient means of measuring the output of a photon source, their response is dictated by the characteristics of the filter, which differ between manufacturers and also suffer from change with time. Due to the lack of standardized filters, different meters may measure different fractions of the radiant energy. They must be calibrated regularly to allow for changing responses due to alterations in performance of the filter and photodiode. *A* considerable amount of study of this type of instrument has been promoted by measuring the changes in UV-B intensity associated with the depletion of ozone in the stratosphere (De Luisi *et al.*, 1992; Smith *et al.*, 1993). The quantity measured is an instantaneous measurement of the irradiance in watts per square metre. From the viewpoint of drug photostability testing, the radiometers do not give the desired output of an integrated exposure reading over the testing period.

The lux meter is used in photometry and is simply a radiometer which has a

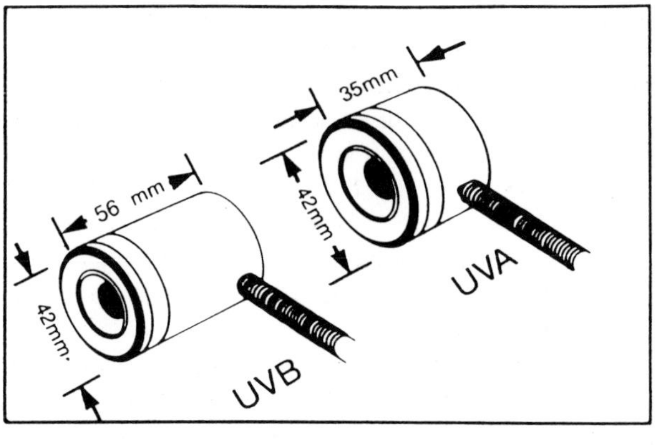

Figure 4.5 Schematic representation of radiometers designed for the UV-A and UV-B regions (International Light Inc.)

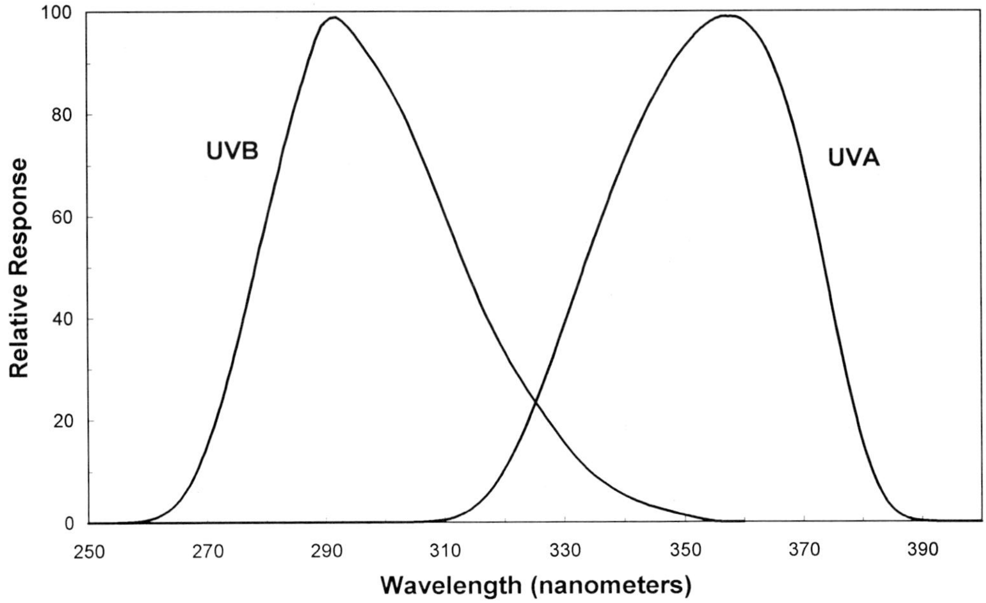

Figure 4.6 Relative spectral sensitivity of radiometers designed for the UV-A and UV-B regions (International Light Inc.)

spectral responsiveness that closely matches the visual response of the human eye, thus measuring incident radiant power in the visible region of the electromagnetic spectrum. In this case the unit of measurement is the illuminance in lux, and is calibrated against a specific tungsten lamp. The lux meter should be provided with a set of correction factors to enable compensation for differences in spectral response for lamps with emission spectra different from the calibration lamp.

The spectroradiometer combines a monochromator with a photomultiplier detector as in a spectrophotometer, to gain a detailed estimate of the light intensity as a function of wavelength. It is used for checking radiometer calibration and monitoring sunlight intensity variations with season, particularly in the UV-B region (De Luisi *et al.*, 1992).

A thermopile is a thermal detector capable of measuring the total incident radiant flux (in W/m^2) in the UV, visible and IR regions of the spectrum, with a response that is essentially independent of wavelength. It is the reference against which the other devices can be calibrated, but is not recommended for direct use itself because it is expensive and not a straightforward instrument to use correctly (Jagger, 1985).

4.4 Chemical actinometry

While physical instrumentation is more convenient, the lack of an integrated output and the need for regular calibration are hindrances to its widespread application to measurement of the number of photons absorbed by a sample. The alternative is the use of a chemical actinometer system in which a photochemical reaction of known characteristics is monitored when subjected to the same irradiation conditions as the test sample. When used in the fashion described below, the chemical actinometer measures the amount of light actually absorbed by the sample, rather than that which is incident. The most important property of an actinometer system is that it should give a response over the full range of wavelength to which the various samples are tested. In the case of drug photostability studies, this can mean from the UV-B through the visible regions, although in practice it would be sufficient if only the UV-B and UV-A regions were covered, since very few drug substances absorb in the visible. The various systems applicable to all types of photoprocesses have been summarized by Kuhn *et al.* (1989) for the International Union of Pure and Applied Chemistry Commission on Photochemistry.

4.4.1 Quantum yield of a photochemical reaction

Chemical actinometer systems have been widely used in basic photochemical studies to enable the determination of the quantum yield of a photochemical reaction. The quantum yield is defined as follows:

$$\varphi = \frac{\text{Number of molecules reacted per unit volume per unit time}}{\text{Number of photons absorbed per unit volume per unit time}} \tag{4.1}$$

The reason for the interest of photochemists in the quantum yield of a photochemical reaction is that it is the measure of the amount of reaction corresponding to the light actually absorbed by the sample, and therefore is the true constant describing the rate or efficiency of that reaction independent of the experimental arrangement used. When correctly measured, the quantum yield of a photochemical reaction should be the same, irrespective of whether it is determined in Oslo, Sydney or Kalamazoo, so that when reporting the photochemical reactivity of a drug in a quantitative sense, the quantum yield is what should be quoted for each reaction (Calvert and Pitts, 1966; Moore, 1987). The fundamental problems are that actinometry and the measurement of the quantum yield are technically demanding, and it is then a complicated matter to convert the quantum yield to a reaction rate applicable to a particular experimental setting. First, the procedure will be briefly described for actinometry and quantum yield determination, then it will be explained how the rate of a photochemical reaction can be predicted from a knowledge of the quantum yield and the incident light intensity.

4.4.2 The actinometer experimental design

In order to determine the amount of light absorbed by a drug when irradiated in solution an experiment is set up with a two-cell arrangement of the drug solution and the actinometer solution, as shown in Fig. 4.7. Cell B is filled with the actinometer solution, and two sets of irradiations are performed. In Set 1, cell A contains the solution of drug to be tested at a concentration such that its absorbance is between 0.5 and 0.8, while in Set 2 cell A is filled with the solvent. The contents of the cells should be efficiently stirred throughout. The samples are analysed after a series of different times of irradiations and the response of the actinometer

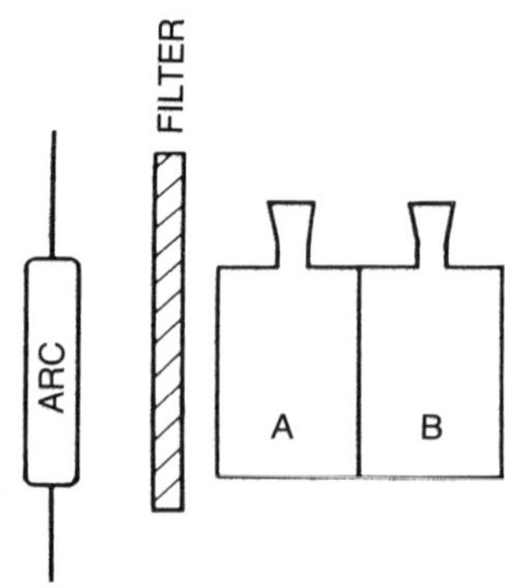

Figure 4.7 Arrangement of reaction vessels for actinometry experiments

determined for the two sets. A plot such as that given in Fig. 4.8 should be obtained and the difference between the slopes for Sets 1 and 2 represents the amount of light absorbed by the drug sample. Thus, the number of photons absorbed per unit time is calculated by:

$$N_{drug} = R \, [\text{slope(Set 1)} - \text{slope(Set 2)}] \, / \, \varphi_{act} \tag{4.2}$$

where φ_{act} is the quantum yield of the actinometer system, and R is the factor necessary to convert the analytical response of the actinometer to a number of molecules reacted. For example, if the actinometer is analysed in terms of a product formation detected by a change in absorbance, then the factor R would include the molar absorptivity of the actinometer product, and Avogadro's Number (and any dilution factors arising from the analysis procedures).

In parallel, the extent of reaction of the drug sample in cell A is determined by the appropriate method of analysis, so that the number of molecules reacted per unit time can be calculated for determination of the quantum yield of the drug degradation, according to equation (4.1). This experiment has been routinely

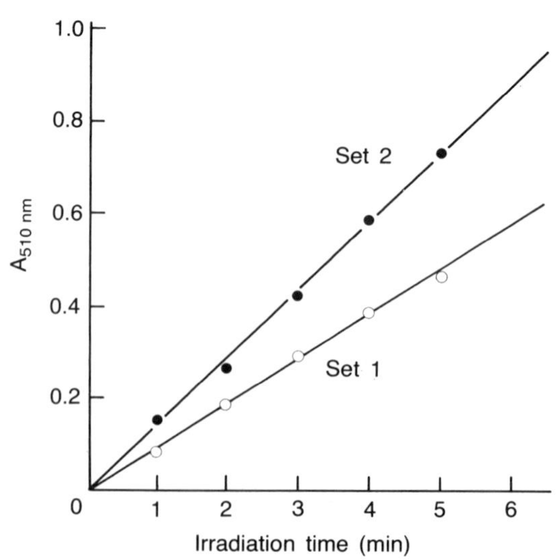

Figure 4.8 Typical plot of results of an actinometry experiment

performed by photochemists with chemical actinometers, but, in principle, there is no reason why a reliable integrating physical instrument could not be used in the same way.

The experimental arrangement described here applies to samples which transmit some of the incident light, implying a homogeneous solution. For many pharmaceutical formulations such as solids, suspensions and ointments, the incident light is likely to be reflected or scattered, in which case the actinometer solution or physical instrument would need to be placed beside the sample. While that will not give a measure of the light absorbed by the sample, it is at least a measure of the incident light from which an estimate of the relative reactivity can be gained.

4.4.3 *The ferrioxalate actinometer*

The most widely used actinometer system is that based on potassium ferrioxalate, since it fulfils the condition of applicability over a broad wavelength range. Potassium ferrioxalate is readily prepared by reaction of ferric chloride with potassium oxalate. The details of the photochemistry were worked out by Hatchard and Parker (1956). When acidic solutions (6 mM) of potassium ferrioxalate are irradiated by light in the range 250 to 570 nm, 99 per cent of the light is absorbed; Fe(III) is reduced, while oxalate is oxidized. The progress of the reaction is monitored by detecting the Fe(II) as a complex with 1,10-phenanthroline at 510 nm. The product Fe(II) and its oxalate complex do not absorb the incident radiation to a measurable extent, so there is no back reaction. The quantum yield of Fe(II) formation varies with wavelength of the irradiation, but for the 300–400 nm region, the value can be taken as 1.2 ; the stoichiometry of the reaction indicates that two atoms of Fe(III) are reduced per photon absorbed, i.e., a theoretical maximum quantum yield of 2.

There are two disadvantages which affect its use in drug photostability studies with sunlight-simulating irradiation. One is minor, in that the procedure to be followed to measure the extent of the photoreaction of ferrioxalate is technically cumbersome. The major concern is that while the photoreaction occurs with a constant quantum yield, it is rapid in relation to most drug photodecompositions when used with a wide range of wavelengths. That is, there is sufficient reaction of the actinometer in 5 to 10 minutes to enable precise determination of the light falling on the sample, yet the exposure times required for the drug sample to show a suitable change for detection is often of the order of several hours. If it can be assumed that the photon source is uniform in its output this would not affect the result, but unfortunately that is not the case as xenon arc lamps may vary by as much as 20 per cent over a period of 5 to 6 h. Fundamentally, the ferrioxalate system remains the actinometer of choice when a narrow range of wavelength is being used for the irradiation.

4.4.4 *Other actinometer systems*

While the review of actinometers by Kuhn *et al.* (1989) covered a vast array of well-documented gaseous, liquid and solid phase systems, there are few which might be considered useful for the drug photostability situation, principally because they are not applicable to the sunlight wavelength range. There is considerable interest in the development of solid state 'sunburn dosimeters' for the measurement of UV-B, based on the photochemical change of a polysulfone film (Diffey 1987; Herlihy *et al.*, 1994) but the wavelength range is too narrow for the actinometry application in stability studies.

Uranyl oxalate is the only other chemical solution system which responds over a wide wavelength range including the UV-A and visible regions. Its reactivity is slower than ferrioxalate, but any advantage there is outweighed by its expense and toxicity.

Quinine hydrochloride solution has been investigated as an actinometer system with slower reactivity than ferrioxalate in sunlight-simulating conditions (Yoshioka *et al.*, 1994). It has been suggested that the increase in absorbance at 400 nm of a 4 per cent solution of quinine hydrochloride in water be used as a measure by which exposure times be determined for formulations. The quantum yield for quinine photodecomposition has not been determined, but the proposal is for a unit of exposure time expressed as $\Delta A(400\,nm)$. However, to measure incident light in terms of one species implies that its absorption spectrum covers the full range of incident wavelengths to be used. This is not the case for quinine whose absorption maximum is at 335 nm, rapidly falling off as wavelength increases. The relatively high concentration has been used as an attempt to extend the absorption as well as to ensure that all radiation in its absorption range is absorbed. While the quinine system might be satisfactory for many drugs whose absorption spectrum does not extend much beyond the UV-B region, many other examples can be given (e.g., tetracycline, nifedipine) for which the quinine spectrum does not overlap and is therefore unsatisfactory. At this time, no chemical actinometer appears ideal in all respects.

4.5 Kinetic treatment of photochemical reactions

Testing the photostability of a drug substance at the preformulation stage invariably involves a study of the rate of degradation of the drug in solution when exposed for a period of time to a source of irradiation. The experimenter wishes to know the kinetics of the process but will find that the value of the rate constant depends very much on the design of the experiment, for the same reasons as outlined above.

The factors that determine the rate of a photochemical reaction are simply the rate at which light is absorbed by the test sample (i.e., the number N of photons absorbed per second), and the efficiency of the photochemical process (i.e., the quantum yield of the reaction, φ). If one is using a monochromatic light source, the number of photons absorbed depends upon the intensity of the photon source and the absorbance at that wavelength of the absorbing species. As explained above, the quantum yield is the characteristic constant for the process in question. Thus the rate of a photochemical reaction is defined as:

$$\text{Rate} = \text{Number of molecules transformed per second}$$
$$= N\,\varphi$$

In the first instance, the rate can be determined for a homogeneous liquid sample in which the only light absorption is due to the drug molecule undergoing transformation, with the restriction that the concentration is low so that the drug does not absorb all of the light available in the wavelength range corresponding to its absorption spectrum.

The value of N can be derived at a particular wavelength λ and is given by:

$$N_\lambda = I_\lambda - I_t = I_\lambda \, (1 - 10^{-A}) \tag{4.1}$$

where I_λ and I_t are the incident and transmitted light intensities, respectively, and A is the absorbance of the sample at the wavelength of irradiation. This expression can be expanded as a power series:

$$N_\lambda = 2.303 \, I_\lambda \, (A + A^2/2 + A^3/6 + \ldots) \tag{4.2}$$

When the absorbance is low ($A < 0.02$), the second- and higher-order terms are negligible and the expression simplifies to the first term. Given the Beer's Law relation between absorbance and concentration, N can be seen to be directly proportional to concentration:

$$N_\lambda = 2.303 \, I_\lambda \, A = 2.303 \, I_\lambda \, \varepsilon_\lambda \, \mathrm{b} \, C \tag{4.3}$$

where ε_λ is the molar absorptivity at wavelength λ, and C the molar concentration of the absorbing species, and b is the optical path length of the reaction vessel. Now I_λ and ε_λ vary with wavelength so the expression has to be integrated over the relevant wavelength range where each has a non-zero value:

$$N = 2.303 \, \mathrm{b} \, C \int (I_\lambda \, \varepsilon_\lambda) \, d\lambda \text{ integrated from } \lambda_1 \text{ to } \lambda_2$$

$$\text{Thus Rate} = 2.303 \, \mathrm{b} \, C \, \varphi \int (I_\lambda \, \varepsilon_\lambda) \, d\lambda \tag{4.4}$$

Now the overlap integral ($\int I_\lambda \, \varepsilon_\lambda \, d\lambda$) is a constant for a particular combination of photon source and absorbing substance, b is determined by the reaction vessel chosen, and φ is a characteristic of the reaction, so by grouping the constant terms into an overall constant k, the expression is simplified to:

$$\text{Rate} = \mathrm{k} \, C$$

According to this equation, first-order kinetics apply, i.e., a plot of ln (concentration of drug remaining) should be linear with slope equal to (– k). Data were given in Fig. 4.9 showing the photodegradation of sulfamethoxazole in aqueous solutions of varying pH.

Strictly, it is the overlap integral which determines the rate and therefore the rate constant for the reaction, and when the substance being examined has only a relatively small absorption in the UV-B and UV-A regions, the overlap integral is small and first-order kinetics are observed even though a large amount of the absorbing substance is present in the system. Thus most photodegradation reactions do follow first-order kinetics, but the rate constant derived from a study performed in one laboratory will not be the same as that recorded in another because of the inherent difficulty in reproducing exactly the experimental arrangement of photon source and sample irradiation geometry. An obvious question that arises is whether there is any value in kinetic measurements in uncalibrated apparatus. The answer is that relative values are useful in a given experimental arrangement for making comparisons of degradation of the absorbing substance in different formulations, for example those which contain ingredients designed to inhibit the photoreaction.

The use of rate constants is helpful for comparative purposes when studying a number of different reaction mixtures under the same irradiation conditions. However, the reaction order and numerical values of the rate constants are all relative to those conditions. Quantitative expressions of photochemical rate should be given in terms of the quantum yield φ, which will normally have a value between 0 and 1, except for the rare case when a chain reaction is involved causing multiple degradation events for every photon absorbed. The quantum yield is a wavelength-independent quantity, and so can be determined at any wavelength that the substance absorbs, although clearly a greater precision can be gained near the absorption maximum.

4.5.1 Deviations from first-order kinetics

It should be recognized that the derivation in equation (4.4) is an approximation for low values of the absorbance A, and that at higher values, second- and third- order terms of the power series would come into play, leading to a deviation from the first-order kinetic equation. When the absorbance approaches 2, essentially all of the light is absorbed by the drug, so the rate-limiting factor is the intensity of the incident light and the reaction follows pseudo-zero-order kinetics. Experimentally, it is important that the irradiated solution is stirred effectively to ensure that a uniform concentration is maintained throughout the solution, and that products are being dispersed to minimize secondary degradation. In the high concentration situations it is rare to follow the reaction to the extent (>50 per cent conversion) necessary to clearly decide which kinetic order is applicable, so most experimenters treat their data according to first-order equations. Thus a typical plot of first-order rate constant versus concentration of absorbing species will have the appearance of a saturation kinetics system for a given experimental arrangement.

It is therefore recommended that preformulation studies of photodegradation be conducted with low solution concentrations so that first-order kinetics apply and the reaction rate is limited by the drug concentration rather than the light intensity. Note that treatment of concentration data in terms of percentage remaining of the initial concentration will produce rate constants which are *inversely proportional* to the initial concentration.

The derivation of equation (4.4) would need to be modified if other components of the system absorb light. Two possibilities arise depending upon whether the other component(s) participate in subsequent reactions. If the other absorbing components do not react, the only effect is a reduction or filtering of the incident light. The fraction of the light absorbed by the drug can be expressed as:

$$F_C = \varepsilon_\lambda \, C \, / \, (\alpha_\lambda + \varepsilon_\lambda)$$

where α_λ is the attenuation coefficient of the medium at wavelength λ (sum of the absorbances of components other than the drug of interest).

Where the other absorber(s) react and influence the degradation of the drug, for example, an impurity or additive which sensitizes the photoreaction, the overall kinetics will be a combination of both pathways. Thus the reaction rate may depend on both the drug and sensitizer concentrations. Photochemical stability of a drug in a formulation may not necessarily be predicted only from the absorption spectrum and stability studies in a pure solvent; the final form of the drug in the formulation should also be considered.

4.5.2 *Application of the derived rate constant expression to sunlight exposure*

In practice, the integration in equation (4.4) is replaced by a summation of the values measured for finite wavelength bands ($\Delta\lambda$) across the region of interest, so that the rate constant expression becomes:

$$k = 2.303\ \varphi\ b\ \Sigma\ (I_\lambda\ \varepsilon_\lambda\ \Delta\lambda)\ \text{summed from}\ \lambda_1\ \text{to}\ \lambda_2 \qquad (4.5)$$

Using equation (4.5) the reaction rate constant for the photodegradation of a particular compound can be predicted from the quantum yield and the molar absorptivities at specific wavelengths, together with the light source intensity data, for example, when the sample is exposed to direct sunlight. On the other hand, a determination of the rate constant of a well-documented photodegradation occurring in a particular experimental arrangement can provide information about the light intensity of the source used, in the region of absorption by the solute. In other words, it can act as an actinometer system.

An example of the use of equation (4.5) is that the rate constant can be calculated for outdoor exposure of a sample because the intensity of sunlight has been recorded for a full range of locations (Leifer, 1988). This can be demonstrated by a study of the degradation kinetics of sulfamethoxazole exposed to sunlight (Moore and Zhou, 1994).

The first step was the determination of the quantum yield of the photodegradation of sulfamethoxazole in aqueous solution using narrow wavelength range irradiation (xenon arc + monochromator). As indicated by the data in Fig. 4.9,

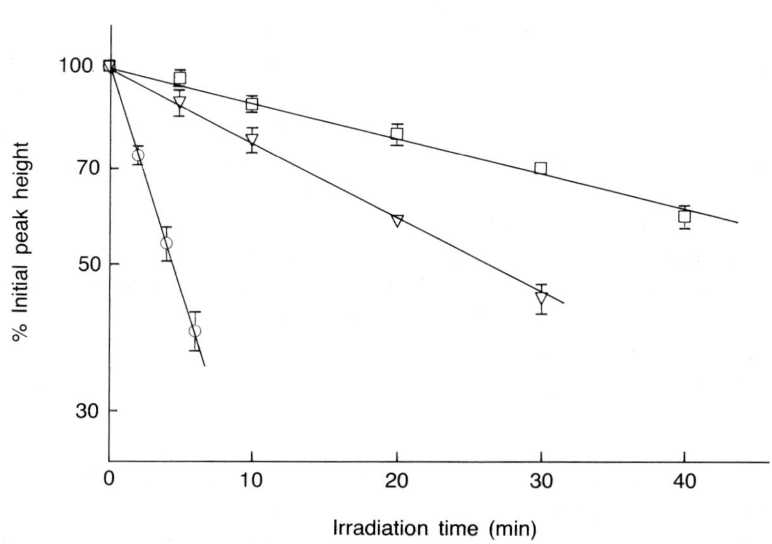

Figure 4.9 Photodegradation kinetics of sulfamethoxazole (5×10^{-5} M) in buffered aqueous solution analysed by HPLC (from Moore and Zhou, 1994). (○) pH 3.0, (▽) pH 7.0, (□) pH 9.0

the reaction is strongly dependent on the pH and the measured quantum yield was found to be 0.47 ± 0.05 at pH 3 (268 nm) and 0.084 ± 0.016 at pH 9 (257 nm) by ferrioxalate chemical actinometry. Relative to other drugs, this is a very high quantum yield in acid solution. The fact that sulfamethoxazole shows only slight absorptivity to UV-R above 290 nm might lead one to think that it should not be very susceptible to sunlight. However, the spectrum is broad and does extend into the UV-B region, so that direct photoreaction is observed when exposed to natural sunlight. As discussed above, the rate constant of the direct photoreaction can be calculated according to equation (4.5) using the quantum yield and the sunlight intensity. Values for the average solar irradiance for clear skies at midday Z_λ (photon $cm^{-2}s^{-1}(2.5\,nm)^{-1}$) at 2.5 nm bandwidths over the relevant wavelength range 297.5 to 320 nm as a function of season are available in the published literature only for 40 °N (Leifer, 1988). On the other hand, day-average solar irradiance data are available for all latitudes (Zepp and Cline, 1977). The ratio of the 40 °N values was used to estimate the midday irradiances for all other latitudes expressed for 2.5 nm bandwidths as U_λ in millieinstein.$cm^{-2}\,min^{-1}(2.5\,nm)^{-1}$. U_λ is therefore equivalent to $I_0\ \Delta\lambda$ in equation (4.5) which now becomes

$$k = 2.303\ \varphi\ b\ \Sigma\ (\varepsilon_\lambda\ U_\lambda) \qquad (4.6)$$

by which the photodegradation rate constant can be calculated using the molar absorptivity of sulfamethoxazole measured at the midpoint of the specific wavelength intervals corresponding to those for the solar irradiance. The summation term represents the area under the overlap integral of the sulfamethoxazole absorption spectrum and the midday sunlight intensity, shown in Fig. 4.10 for Sydney (33.5 °S) in the various seasons.

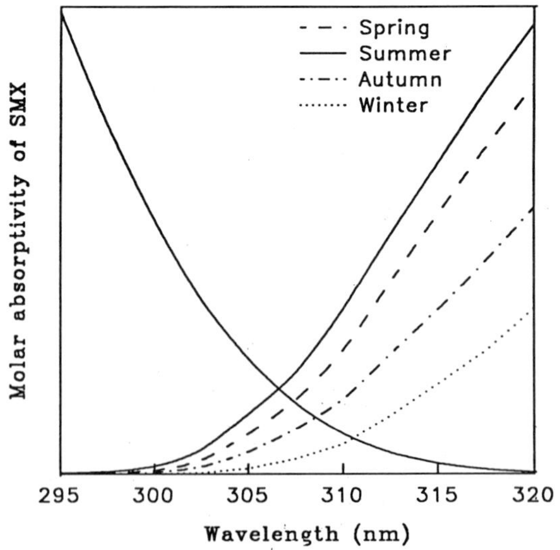

Figure 4.10 Molar absorption spectrum of sulfamethoxazole, and sunlight intensity as a function of season, at midday (latitude 33.5°) calculated by Moore and Zhou (1994) from data in Leifer (1988) and Zepp and Cline (1977)

Table 4.1 Predicted and experimental photochemical shelf-life of sulfamethoxazole in acidic aqueous solution at midday in direct sunlight as a function of season in Sydney (latitude 33.5 °S)

Season 1992–1993	Measured midday UV-B[a] 290–320 nm ($W.m^{-2}$)	Shelf-life (mins)	
		Predicted[b]	Measured[c]
Summer	6.08 ± 0.31	9.02	8.87 ± 0.2
Autumn	3.81 ± 0.20	16.0	16.3 ± 0.3
Winter	2.09 ± 0.12	42.7	87.9 ± 8.1
Spring	4.39 ± 0.22	11.3	12.7 ± 0.3

[a]Instantaneous UV-B intensity recorded with an International Light radiometer.
[b]Calculated using equation (4.6).
[c]By HPLC analysis of sunlight-exposed solutions.

Thus, the photochemical shelf-life t_{10} (time for 10 per cent decomposition) of dilute acidic solutions of sulfamethoxazole in direct sunlight can be predicted as a function of latitude and season of year. These are shown for Sydney (33.5 °S) in Table 4.1 together with the results from degradation experiments performed at midday on clear sunny days at the summer and winter solstices and the autumn and spring equinoxes. The agreement between the predicted and experimental values is within experimental error (except for the winter measurement when the sunlight UV-B was relatively low and the extent of reaction small). This is a significant result considering the variation of weather condition, uncertainty of atmospheric pollution and the possible difference in atmospheric aerosols and ozone levels between the northern and southern hemispheres.

Another component of this study is shown in Fig. 4.11. The rate constant was measured at various times of the day, and a close correlation can be seen with

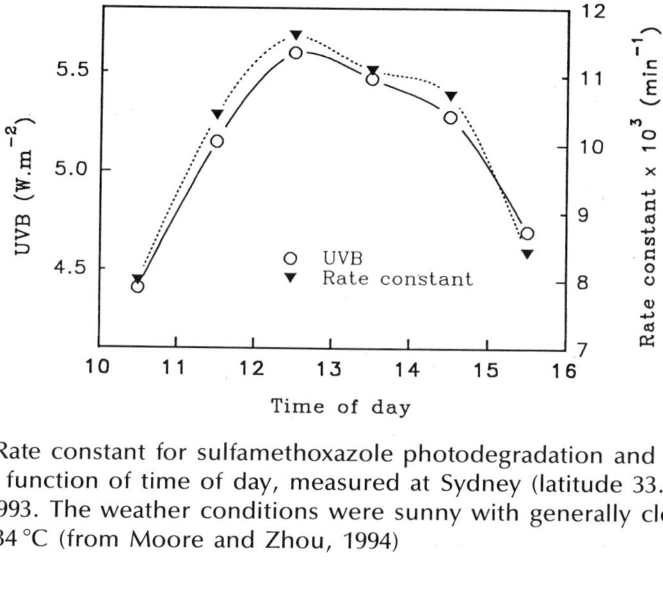

Figure 4.11 Rate constant for sulfamethoxazole photodegradation and UV-B intensity as a function of time of day, measured at Sydney (latitude 33.5 °S) on February 4, 1993. The weather conditions were sunny with generally clear sky, air temperature 34 °C (from Moore and Zhou, 1994)

the UV-B intensity measured with the UV-B radiometer. In other words, the rate constant for this particular reaction provides a means of determining the component of the irradiating light that sulfamethoxazole is able to absorb. Sulfamethoxazole can therefore serve as an actinometer for the UV-B region.

This successful demonstration that laboratory testing can correlate with outdoor exposure points the way in which the determination of the quantum yield for the drug degradation can be used to predict the shelf-life of the product when exposed to sunlight. It is applicable to all solution preparations and all lighting conditions for which the irradiance has been measured. An assumption which is implicit in the calculation is that all of the incident radiation within the overlap integral is absorbed by the substance. Therefore the principle can be extended to other formulations for which the absorption characteristics are known and the quantum yield can be determined with sufficient precision.

4.5.3 Activation energy of photochemical reactions

The Arrhenius equation expresses the variation with temperature of the rate constant of a chemical reaction in the form:

$$k = A \exp(-E_a / R T)$$

where E_a is the activation energy of the reaction, interpreted simplistically as the energy barrier over which the molecule has to be raised to bring about the reaction. For a thermal reaction, such as ester hydrolysis, E_a has a value between 50 and 150 kJ mol^{-1} with the higher values indicating a slower reaction rate. When more complex mechanisms apply, the E_a value applies to the rate-determining step. On the other hand, in a chain reaction, such as a peroxide-initiated oxidation, the high energy initiation step is compensated by the much lower energy propagation step. Once the reaction reaches steady state, kinetic analysis shows the overall activation energy is composed of the E_p for the propagation step plus half the E_i for the initiation step. Values of the order of 20–50 kJ mol^{-1} are typical for chain reactions (Dainton, 1966).

In a photochemical reaction, the photon which is absorbed by a molecule provides the energy to raise the molecule to its excited or reactive state from which the initial products are formed. The energy corresponding to a 300 nm photon is 400 kJ mol^{-1} well in excess of a thermal activation energy. Subsequent processes from the excited state represent the dissipation of energy. Thus a true photochemical process will be independent of temperature. However, the initial products which may be peroxides in the case of photooxidation, are likely to undergo secondary reactions of a thermal nature, so that a temperature dependence would then be observed. In a chain reaction, initiated photochemically, the overall energy of activation would be that corresponding to the propagation step.

4.5.4 The use of accelerated tests

In the study of thermal stability, accelerated testing in the form of elevated temperatures has been used by many pharmaceutical companies to minimize the time involved in the testing process. This procedure is only valid for simple formulations in which the single major ingredient is broken down by a thermal reaction. In practice regulatory authorities demand that a shelf-life determined by extrapolation of accelerated test data should be supported by actual stability data obtained by normal temperature storage (Carstensen, 1990). This is because degradation of a product by microbial contamination may well be inhibited at elevated temperatures.

For photostability testing, there appears to be no similar reason why accelerated testing in the form of exposure of samples to high intensity sources should not be effective in providing data for direct consideration in shelf-life determination. If the study is correctly set up, the extent of degradation should be directly proportional to the number of photons absorbed. The principal experimental factors would be to ensure (1) adequate mixing of liquid formulations, and turning of solids, and (2) the samples are maintained at a uniform temperature.

Certainly, irradiation with a high intensity photon source is a necessary form of stress testing in the preformulation stage, designed to evaluate the overall sensitivity of the material for purposes of analytical method development aimed at determining the presence of potential degradation products.

References

ANDERSON, N. H., JOHNSTON, D., MCLELLAND, M. A. and MUNDEN, P., 1991, Photostability testing of drug substances and drug products in UK pharmaceutical laboratories, *J. Pharm. Biomed. Anal.*, **9**, 443–9.

CALVERT, J. G. and PITTS, J. N., 1966, Experimental methods in photochemistry, in *Photochemistry*, pp. 783–804. New York: John Wiley & Sons.

CARSTENSEN, J. T., 1990, *Drug Stability*. New York: Marcel Dekker, Inc., pp. 288–9.

DAINTON, F. S., 1966, *Chain Reactions*. London: Methuen, pp. 52–3.

DELUISI, J., WENDELL, J. and KREINER, F., 1992, An examination of the spectral response characteristics of seven Robertson-Berger meters after long-term field use, *Photochem. Photobiol.*, **56**, 115–22.

DIFFEY, B. L., 1987, A comparison of dosimeters used for solar ultraviolet radiometry, *Photochem. Photobiol.*, **46**, 55–60.

HATCHARD, C. G. and PARKER, C. A., 1956, The photochemistry of the ferrioxalate actinometer, *Proc. R. Soc. London*, **A235**, 518–30.

HERLIHY, E., GIES, P. H., ROY, C. R. and JONES, M., 1994, Personal dosimetry of solar UV radiation for different outdoor activities, *Photochem. Photobiol.*, **60**, 288–94.

JAGGER, J., 1985, *Solar-UV Actions on Living Cells*. New York: Praeger Scientific, pp. 159–76.

KUHN, H. J., BRASLAVSKY, S. E. and SCHMIDT, R., 1989, Chemical actinometry, *Pure Appl. Chem.*, **61**, 187–210.

LEIFER, A., 1988, *The Kinetics of Environmental Aquatic Chemistry – Theory and Practice*. New York: American Chemical Society, pp. 255–64.

MOORE, D. E., 1987, Principles and practice of drug photodegradation studies, *J. Pharm. Biomed. Anal.*, **5**, 441–53.

MOORE, D. E. and ZHOU, W., 1994, Photodegradation of sulfamethoxazole: a chemical system capable of monitoring seasonal changes in UV-B intensity, *Photochem. Photobiol.*, **59**, 497–502.

SMITH, G. J., WHITE, M. G. and RYAN, K. G., 1993, Seasonal trends in erythemal and carcinogenic ultraviolet radiation at mid-southern latitudes 1989–1991, *Photochem. Photobiol.*, **57**, 513–17.

THORINGTON, L., 1985, Spectral, irradiance and temporal aspects of natural and artificial light, *Ann. N.Y. Acad. Sci.*, **453**, 28–54.

TØNNESEN, H. H. and KARLSEN, J., 1995, Photochemical degradation of components in drug formulations, *Pharmeuropa*, **7**, 137–41.

TØNNESEN, H. H. and MOORE, D. E., 1993, Photochemical degradation of components in drug formulations, *Pharm. Technol. Int.*, **5**, 27–33.

YOSHIOKA, S., ISHIHARA, Y., TERAZONO, T., TSUNAKAWA, N., MURAI, M., YASUDA, T., KITAMURA, KUNIHIRO, Y., SAKAI, K., HIROSE, Y., TONOOKA, K., TAKAYAMA, K., IMAI, F., GODO, M., MATSUO, M., NAKAMURA, K., ASO, Y., KOJIMA, S., TAKEDA, Y. and TERAO, T., 1994, Quinine actinometry as a method for calibrating ultraviolet radiation intensity in light stability testing of pharmaceuticals, *Drug Dev. Ind. Pharm.*, **20**, 2049–62.

ZEPP, R. G. and CLINE, D. M., 1977, Rates of direct photolysis in aquatic environment. *Environ, Sci. Technol.*, **11**, 359–66.

5

Is the Photodecomposition of Drugs Predictable?

J. V. GREENHILL

5.1 Introduction

A body of data relating to the light-induced decomposition of well over 100 medicinal compounds has been accumulated and added to the wider knowledge of photochemistry in the literature (Greenhill and McLelland, 1990; Beijersbergen van Henegouwen, 1991). Little effort has so far been made to rationalize these data so that the photoactivity of new drugs may be predicted. Photochemistry is such a diverse subject and rates of photoreaction are so variable that it may never be possible to predict this property with complete confidence. However, it is already clear that certain molecular structures are vulnerable to photodecomposition. This chapter attempts to rationalize these reactions and to alert the formulation pharmacist to possible storage problems and the doctor to watch for signs of phototoxicity. The list of photodegradation products given for each example is not intended to be comprehensive, but rather those important products which illustrate the particular theme are emphasized.

In surveying and assessing the literature reports the reader is immediately impressed by the wide variation in light-emitting devices employed in drug photochemical studies. Fluorescent or tungsten lights, direct or indirect sunlight and the whole range of mercury and xenon lamps have been employed by different investigators, usually without reference to quantum yield. Not only the intensity, but also the frequency spectra of these sources varies greatly. While the need for international agreement on the standards for testing is urgent, our ability to predict photochemical activity in several molecular types should already alert us to possible dangers and pitfalls.

5.2 Olefines

Olefinic carbon–carbon double bonds **1** are photoactive, particularly when conjugated with another olefine, a carbonyl group or an aromatic ring. When UV energy is absorbed, one of the π-electrons is promoted to a non-bonding orbital to give a form simply shown as **2**. The diradical **2** is free to rotate about the olefinic σ-bond and *cis–trans* isomerism occurs. From a pure isomer **1** or **3** a dynamic mixture results, Scheme 5.1. For example, preparations containing retinoic acid **4**, sometimes applied topically in the treatment of acne and other skin disorders, have been shown to decompose rapidly in sunlight (Brisaert *et al.*, 1992). The compound has four *trans* double bonds of which those at C–9, C–11 and C–13 are vulnerable to photoisomerization when irradiated at 366 nm. The photostationary state consists of a mixture of all possible geometric combinations including the original all-*trans* form (Curley and Fowble, 1988). Presumably steric hindrance by the cyclic methyl groups prevents significant amounts of the 7-*cis* derivative forming. Nitrofurazone (**5**) is a topical antiseptic, often exposed to artificial light after application. In solution the imine group is isomerized to the *syn* form (**6**) by tungsten or fluorescent light. It reverts to the *anti* form on standing in the dark (Quilliam *et al.*, 1987), Scheme 5.1.

There is an interesting example of an advantageous use of photodecomposition in medicine in the treatment of hyperbilirubinemia. Neonates, particularly if premature, are often jaundiced due to their inability to eliminate the toxic and insoluble bilirubin **7**. Incubators are equipped with fluorescent lights which induce *cis–trans* isomerism at both exocyclic olefinic bonds (C–4 and C–15). The result is a mixture of the four possible geometric isomers, at least one of which is more readily eliminated, presumably because of higher solubility. Infants soon develop the ability to conjugate bilirubin as the glucuronide for normal renal excretion (Dobbs and Cremer, 1975; Ennever, 1988).

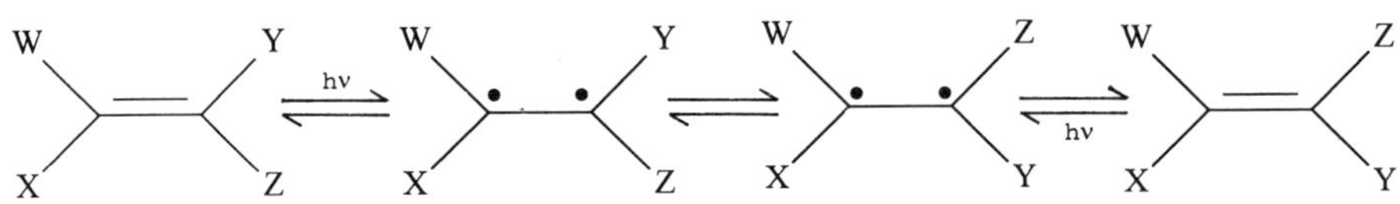

1 **2** **3**

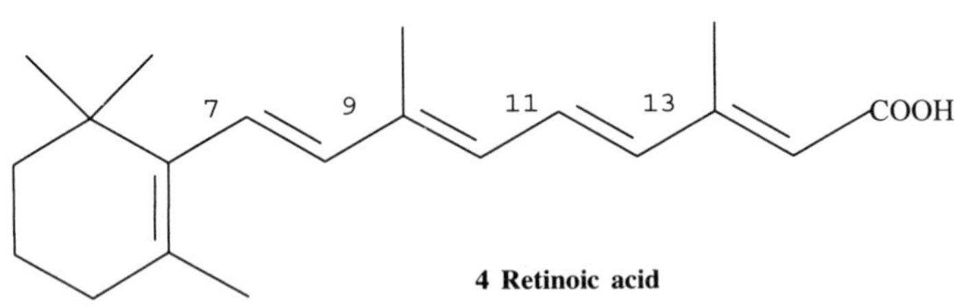

4 Retinoic acid

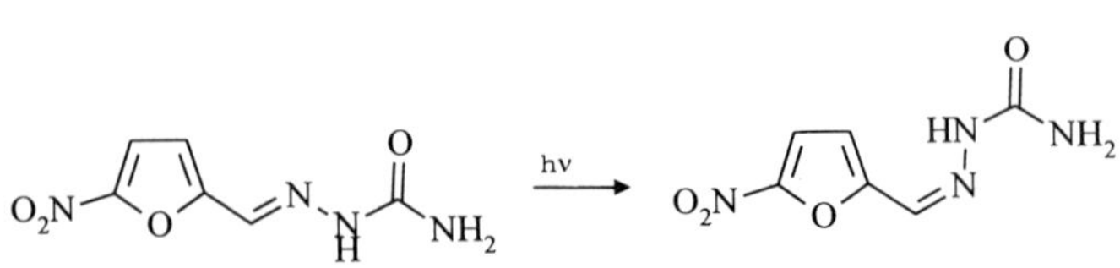

5 Nitrofurazone **6**

Scheme 5.1

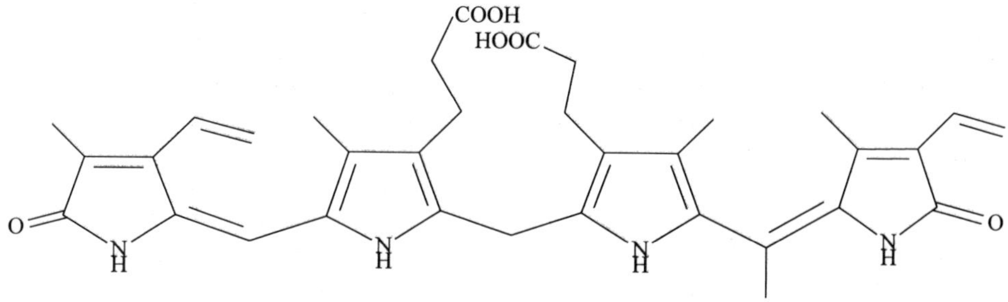

7
4Z,15Z-Bilirubin

In several drugs, geometric isomerization of an olefine brings two rings close together and allows photocyclization to give a third ring. Stilboestrol (**8**) is a pure *trans* compound, but on exposure of an aqueous methanol solution to light of 254 nm photoisomerization gives some of the *cis* form **9**. A symmetry allowed conrotatory photocyclization followed by spontaneous *enol–keto* isomerization then gives the tricyclic dione **10**. As this locks up the *cis* form further isomerization of stilboestrol occurs and the usual photoequilibrium of the geometric isomers does not appear (Scheme 5.2); (Hugelstofer *et al.*, 1960; Doyle *et al.*, 1976).

Dienoestrol is a conjugated diene; although the central C–C bond is traditionally shown single, orbital overlap restricts rotation. The drug in the pure *transoid* form is shown (**11**) (Koch, 1948). It was suggested that photoisomerism gives **12** which is changed via a symmetry allowed [1,5] sigmatropic rearrangement to the 3-*cis* olefine **13**. Cyclization as for stilboestrol then gives the dione **14** as the only isolated product, Scheme 5.3 (Doyle *et al.*, 1978).

Clomiphene is used as a mixture of *cis* (**15**) and *trans* (**16**) isomers. Photolysis of a chloroform solution with a high-pressure mercury lamp gives the expected ring closure to a mixture of the dihydrophenanthrenes which readily lose H_2 to give the phenanthrenes **17** and **18**. If the separate isomers from clomiphene are irradiated they both give mixtures of phenanthrenes showing that the photoisomerization is more rapid than the photocyclization, Scheme 5.4 (Frith and Phillipou, 1986).

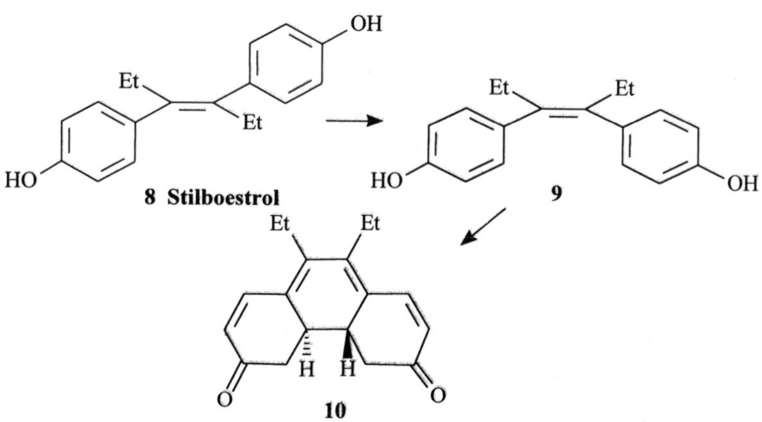

Scheme 5.2

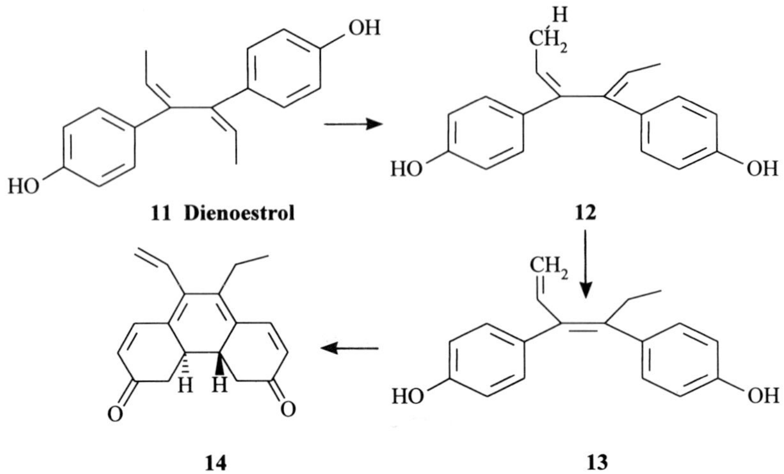

Scheme 5.3

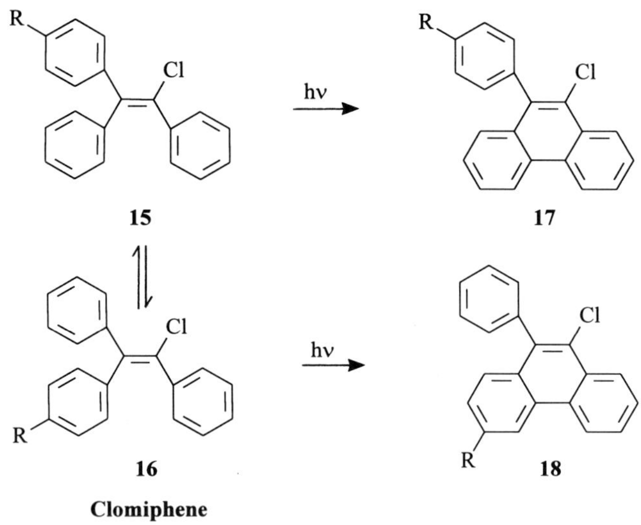

$R = OCH_2CH_2NEt_2$

Scheme 5.4

Photochemical reactions between singlet oxygen and olefines **19** are well known. Intermediate formation of a perepoxide **20** is followed by a further transformation, the direction of which is not always predictable (Kagan, 1993). Hindered olefines often give epoxides **21**, while a common reaction of an olefine next to a methylene group is to give a rearranged olefine peroxide **22**, Scheme 5.5.

An open solution of menadione (**23**) left in the sun for 10 minutes gives the epoxide **24** (Mee *et al.*, 1975). Similarly, norethisterone (**25**) exposed to 300 nm light for 30 minutes in aqueous buffer pH 7.4 gives the β-epoxide **26**. All epoxides are reactive, but this is stable long enough to be transported to the inner organs.

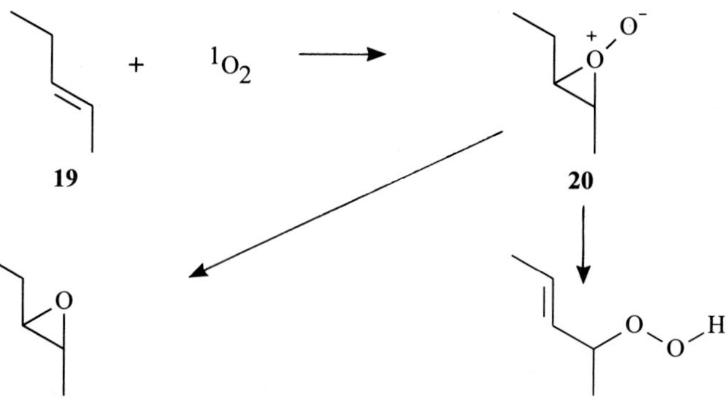

19 20 21 22

Scheme 5.5

It has been suggested that it is the cause of the non-dermatological side-effects associated with norethisterone containing oral contraceptives (Sedee *et al.*, 1983, 1985).

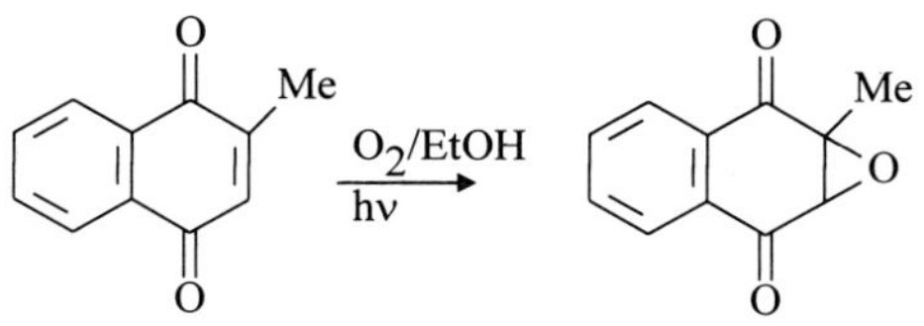

23 Menadione **24**

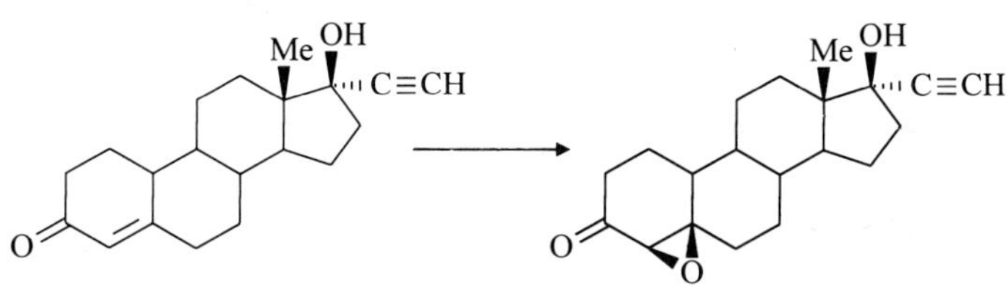

25 Norethisterone **26**

Cyclobarbitone **27** is photo-oxidized at 254 nm to the cyclohexenone **28** presumably via the olefine peroxide. It is noteworthy that this is the sole product of photo-oxidation and the alternative product **29** is the sole product of chemical or enzymatic oxidation (Bouche *et al.*, 1978).

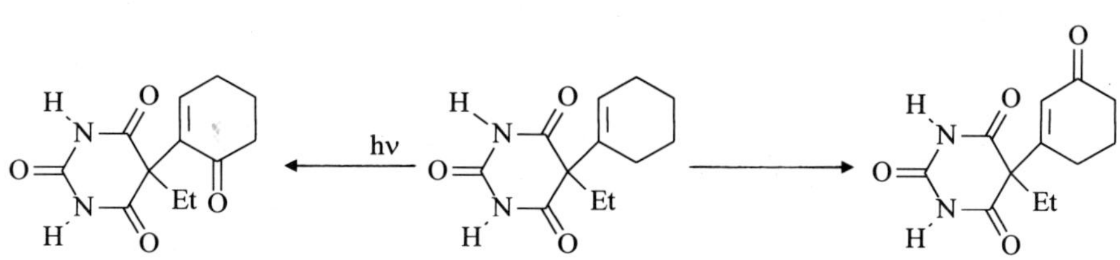

28 **27 Cyclobarbitone** **29**

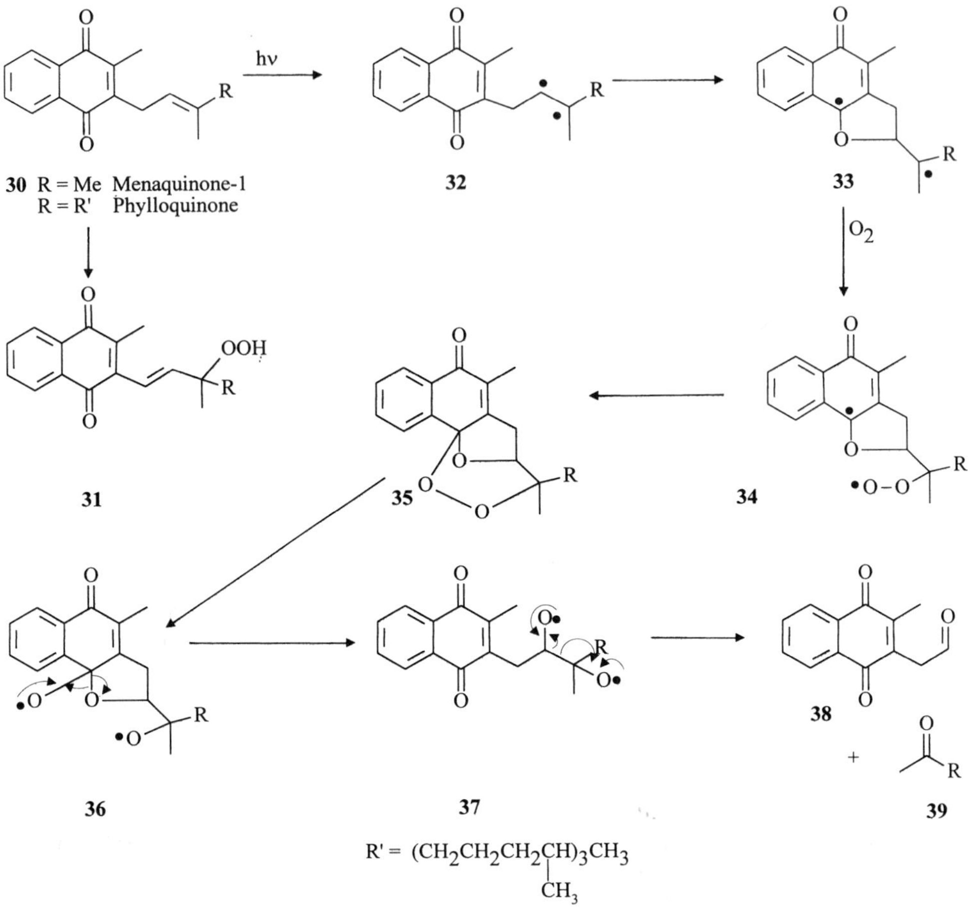

Scheme 5.6

Menaquinone-1 (**30**, R=Me) irradiated at 458 nm gives a number of oxidized products including the olefine hydroperoxide **31** (R=Me) and the trioxane **35** (R=Me). The mechanism **32–34** was suggested to explain the formation of **35** which does not absorb at 458 nm, but at 334–364 nm it gives the aldehyde **38** and acetone (**39**, R=Me) (Wilson *et al.*, 1980). The peroxide link is weak and readily homolysed by UV light so the reaction presumably follows the route **36–37** shown, Scheme 5.6.

A third reaction between an olefine and singlet oxygen commonly seen in photochemistry is $2\pi + 2\pi$ cycloaddition to give a dioxetane **40** which usually decomposes to a mixture of carbonyl compounds, **41** and **42**. Phylloquinone (**30**,

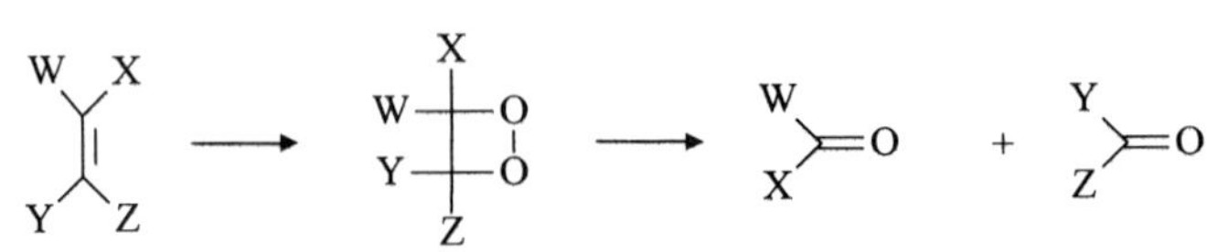

1 **40** **41** **42**

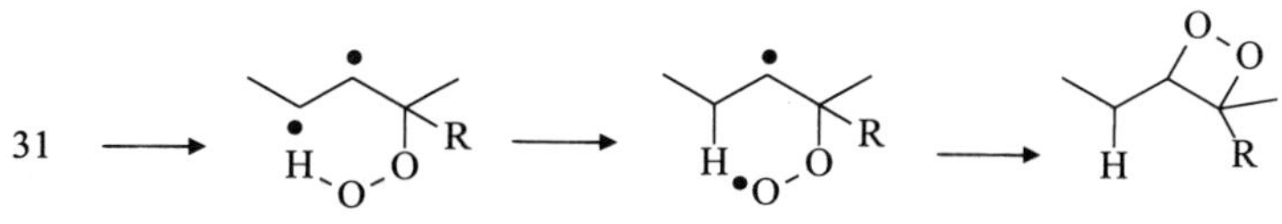

Scheme 5.7

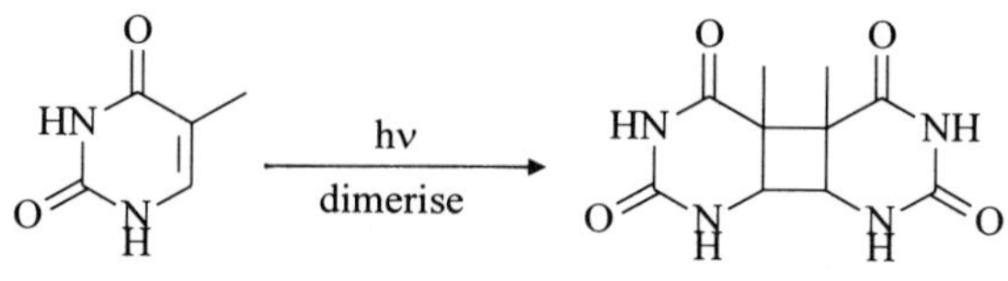

43 Thymine **44**

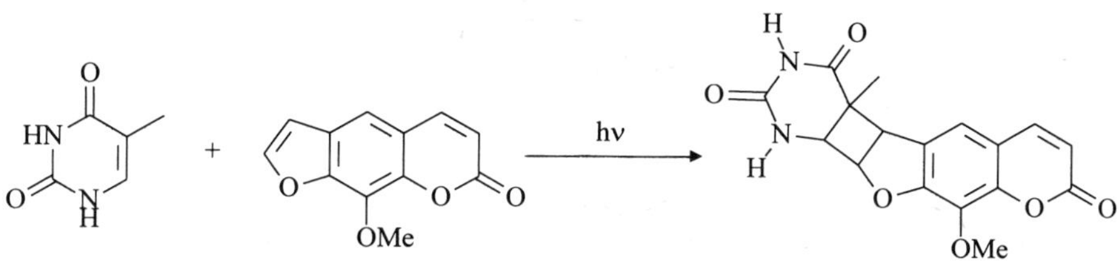

43 **45 Methoxsalen** **46**

Scheme 5.8

R=R′) photolysed in a solution exposed to the atmosphere gave 10 per cent of the enehydroperoxide **31** (R=R′) and 50 per cent of the ketone **39** (R=R′). Direct singlet oxygen addition to phylloquinone may account for some of this ketone, but isolation and anaerobic photolysis of the enehydroperoxide gave the same ketone in 91 per cent yield (Snyder and Rapoport, 1969). This suggests the alternative mechanism for dioxetane formation shown in Scheme 5.7. The hydrogen shift from oxygen to carbon gives the more stable form with oxygen carrying the radical electron. It should be noted that there are examples in organic chemistry of the photochemical formation of dioxetanes from hindered olefines which could not give enehydroperoxides. Any olefine may be expected to react with singlet oxygen by one or more of these three routes.

Cycloaddition ($2\pi + 2\pi$) of two olefine groups to give a cyclobutane is a common photochemical reaction. A well-known biological example is the dimerization of thymine (**43**) to give **44** and its isomers. This reaction may occur when methoxsalen (**45**) is used in conjunction with UV to treat psoriasis, where the drug would act as a photosensitizer. However, the olefine bonds of the two outer rings of methoxsalen also undergo ($2\pi + 2\pi$) cycloadditions and derivatives such as **46** have been isolated, Scheme 5.8.

5.3 Benzyl radicals

Any radical centred on a benzylic atom – one alpha to an aromatic ring – is stabilized and therefore more likely to form. This is usually seen in the propionic-

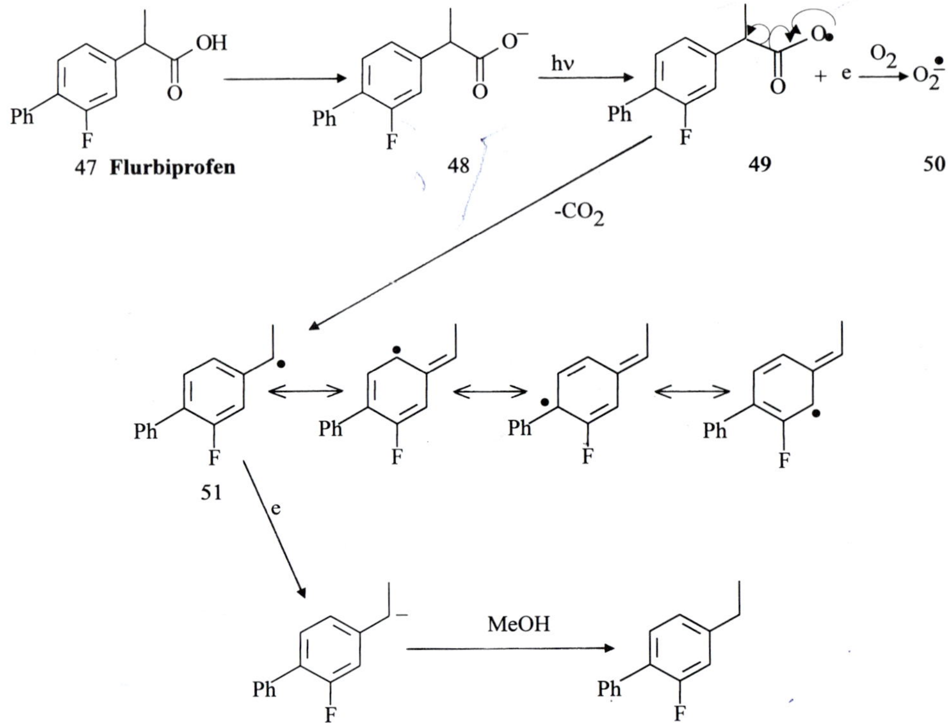

47 **Flurbiprofen** 48 **49** **50** 51 **52** **53**

Scheme 5.9

acid-derived non-steroidal anti-inflammatory drugs. From a photochemical point of view, these are better described as arylacetic acids since they all carry α-aromatic groups. For example flurbiprofen (**47**) in methanol photodecomposes by decarboxylation to 1-ethyl-3-fluoro-4-phenylbenzene **53**, along with a secondary alcohol and a ketone as for naproxen, below. Since phenylacetic acids are substantially ionized at physiological pH, it is probable that the anion **48** loses an electron and after loss of CO_2 from radical **49** gives the new radical **51** which is stabilized by relay into the ring as shown. The stability of this radical is the driving force of the reaction. When **51** picks up the electron, probably carried as a superoxide radical anion **50**, it gives the anion **52** which is also stabilized by relay of the charge into the ring. This anion would be protonated by solvent to give the observed product, Scheme 5.9 (Boscá *et al.*, 1990; Fox, 1990; Castell *et al.*, 1992; Budac and Wan, 1992, but see also Epling and Lopes, 1977).

Similar reactions are reported for benzoxaprofen (Navaratham *et al.*, 1985) and (recently) for butibufen (Castell *et al.*, 1992). In the case of butibufen, a radical dimer and a primary alcohol from reduction of the acid group were also positively identified. The methanolic solvent is capable of delivering a hydrogen atom (Schemes 5.25, 5.26 and the Conclusion) which may account for this reduction. Naproxen (**54**) in aqueous solutions purged with oxygen was photolysed to a mixture of the alcohol and the ketone. Presumably the initially formed radical **55** reacted with O_2 to give the peroxide **59** which decomposed via homolysis to the

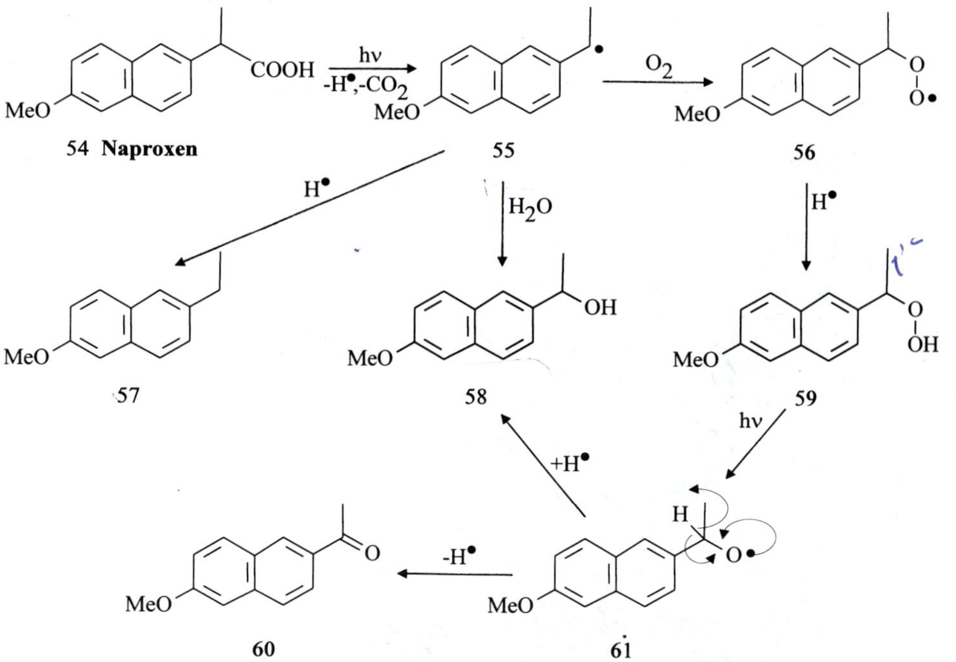

Scheme 5.10

alcohol **58** or ketone **60**. However, deaerated solutions gave mixtures of the alcohol and 2-ethyl-6-methoxynaphthalene (**57**). Here, it was suggested, the radical **55** abstracted either H˙ or OH˙ from the water, Scheme 5.10 (Moore and Chappuis, 1988). The participation of water in photochemical changes is unusual; perhaps a complex between the released hydrogen atom and a water molecule is responsible for these reactions.

Tiaprofenic acid reacts similarly, although significant amounts of the decarboxylated derivative are detected even under aerobic conditions (Castell *et al.*, 1993). A careful recent study of ketoprofen (**158**) in sunlight under oxygen demonstrated the rapid production of a hydroperoxide, an alcohol, a ketone and the decarboxylated derivative; samples of all of them were isolated (Boscá *et al.*, 1994). This work is also notable for the photoproducts of the benzophenone moiety; all the products are shown in Scheme 5.22. Ketoprofen is frequently administered topically in an ointment base and it is clear that significant photodecomposition occurs during a few minutes' exposure to the sun. There are a number of reports of photocontact dermatitis and it has been shown that the photodecarboxylation products all induce haemolysis *in vitro*, but the hydroperoxide and the alcohol are the most potent haemolytic agents (Costanzo *et al.*, 1989).

Aqueous solutions of chloramphenicol (**62**) give several degradation products in sunlight, tungsten light or UV light (Shih, 1971; Reisch and Weidmann, 1971). All probably derive from an initial homolysis to the radicals **63** and **64**. The benzylic radical **63** readily loses a hydrogen atom to give 4-nitrobenzaldehyde **65**. The aliphatic radical **64**, mesomerically stabilized by the neighbouring nitrogen, expels a hydrogen atom to give the imine **66** which hydrolyses to dichloroacetamide **67**

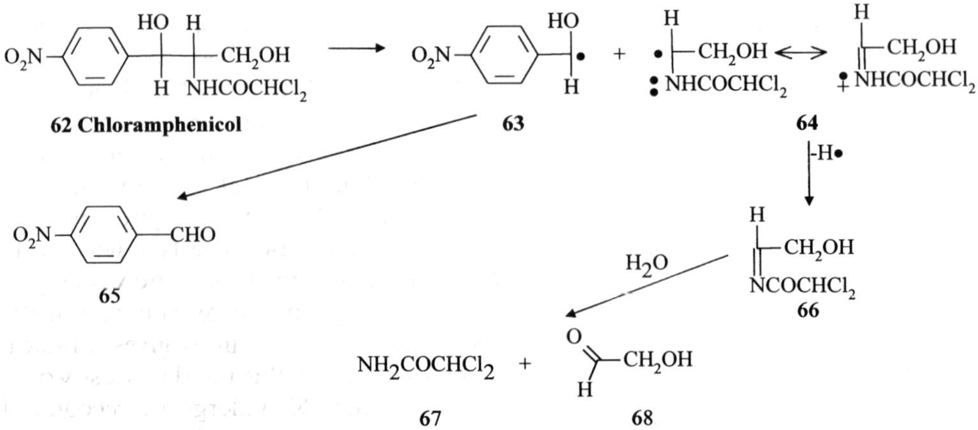

Scheme 5.11

and glycol aldehyde **68**, Scheme 5.11. Further degradations of 4-nitrobenzaldehyde are discussed under nitro compounds, below.

5.4 N-dealkylations

In the presence of oxygen, N-dealkylation is a common photodecomposition. A tertiary amine **69** undergoes homolysis of the α-C–H bond to give a radical **70**, stabilized by form **71**. The radical is attacked by ground state oxygen **73** leading to a peroxide **76** which decomposes in the usual way to give a secondary amine **74** and an aldehyde **78**, Scheme 5.12. The product **74** may undergo further N-dealkylations (Lewis and Ho, 1980). Alternatively, the peroxide may homolyse to **75** and lose a hydrogen atom to give an amide **72**.

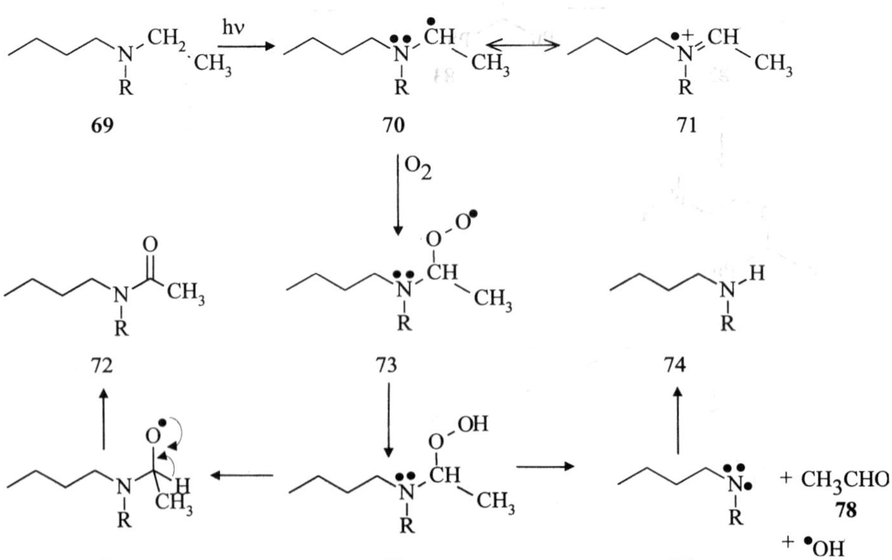

Scheme 5.12

Diphenhydramine (**79**) gives a complex mixture on photolysis which includes not only the simple N-demethylated derivatives **82** and **85**, but rearrangement products **80** and **83** and their dealkylated derivatives **81**, **84** and **86**, Scheme 5.13. Irradiation of a pure sample of **82** gave a mixture of **83**, **84** and **86**, but irradiation of **85** did not give any rearranged product, which is evidence for the intramolecular nature of the rearrangements of **79** and **82**. In addition, irradiation of **79**, **82** or **85** gives diphenylmethane (**87**), diphenylmethanol (**88**) and benzophenone (**89**) (Beijersbergen van Henegouwen *et al.*, 1987). Scheme 5.14 suggests a mechanism for the interesting rearrangement **79** to **93**. Homolysis of the C–O bond would give a benzylic radical **90** stabilized by two benzene rings and an oxygen radical **91**. A hydrogen shift through a six-membered ring intermediate in **91** gives a radical **92** stabilized by the adjacent nitrogen. Ion-pairing during this rapid process would allow the new C–C bond to form to give **93**. Compound **93** undergoes a recognized photofragmentation to eliminate formaldehyde, shown in **94**, and give the tertiary amine **95**. The sequence **96** to **101** explains the dealkylation.

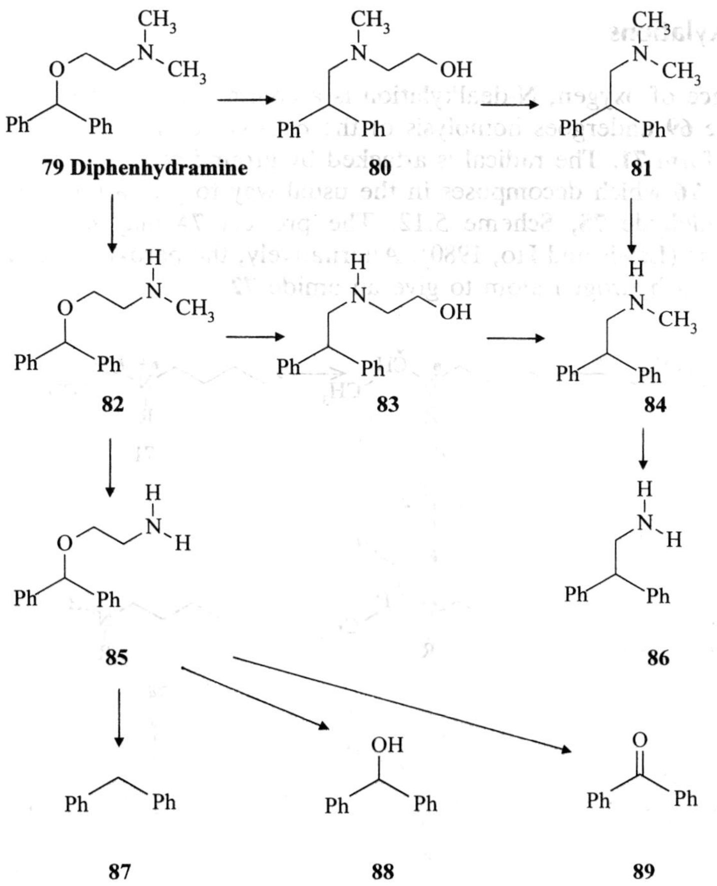

Scheme 5.13

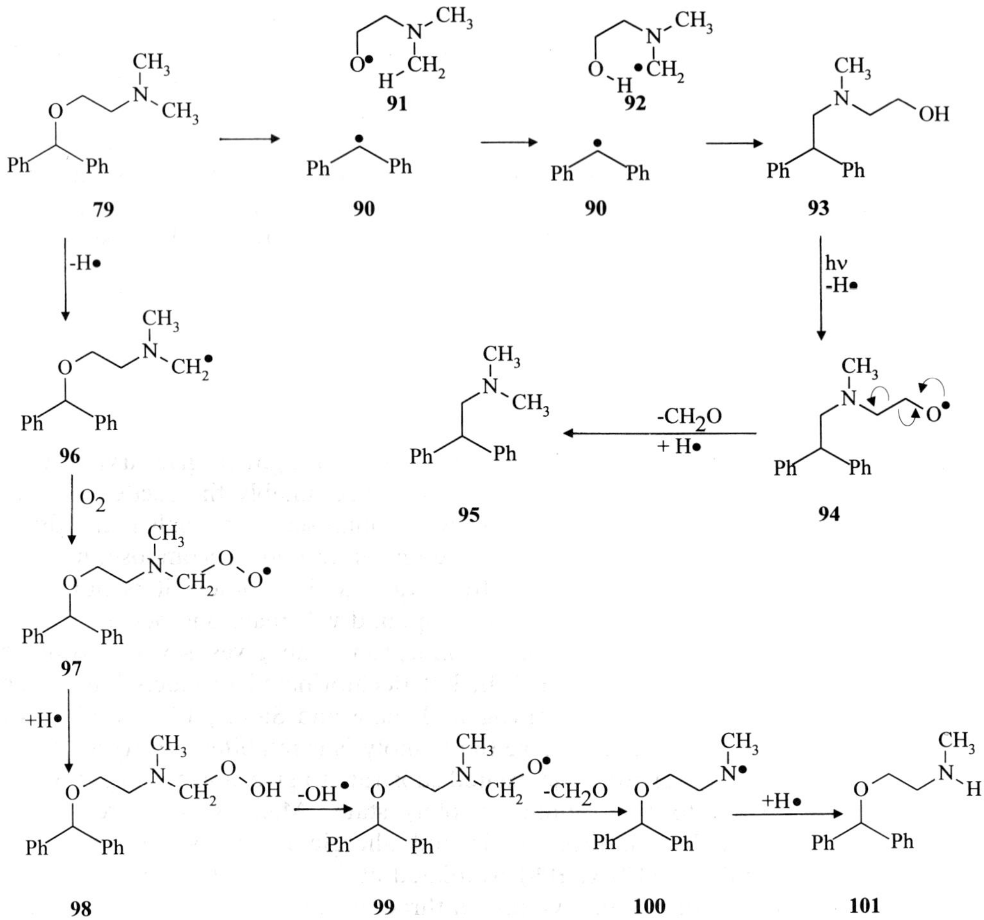

Scheme 5.14

Chloroquine and hydroxychloroquine also give mixtures of N-dealkylated products when irradiated in aqueous solution in the presence of air (Tønnesen *et al.*, 1988; Nord *et al.*, 1991).

Methotrexate (**102**) under fluorescent light N-dealkylates at either position shown with loss of a methyl group or a pteridine aldehyde unit (Chatterji and Gallelli, 1978). Folic acid (**103**) reacts similarly to give pteridine aldehyde (Lowry *et al.*, 1949). When methotrexate (**102**) is irradiated in neutral aqueous solution

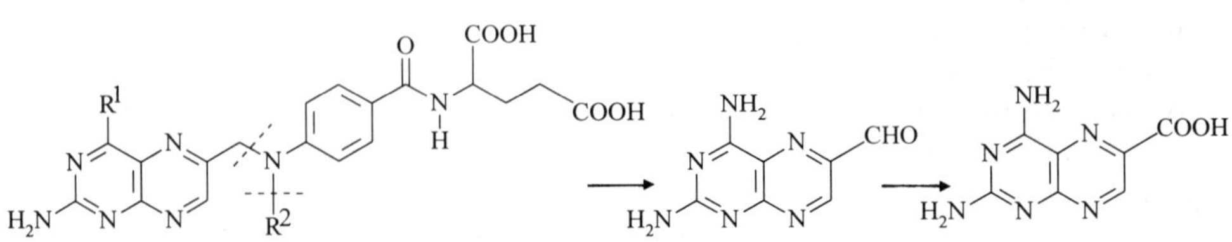

102 $R^1 = NH_2, R^2 = CH_3$ **Methotrexate** **104** **105**
103 $R^1 = OH, R^2 = H$ **Folic acid**

with oxygen present both the aldehyde **104** and the derived carboxylic acid **105** are obtained. It has been demonstrated that **104** is photo-oxidized to **105** (Chatterji and Gallelli, 1978; Chahidi *et al.*, 1986). The products sensitize the oxidation of histidine and tryptophane by a singlet oxygen mechanism. Patients undergoing high-dose methotrexate chemotherapy show photosensitization. When methotrexate is used in conjunction with UV-A to treat severe psoriasis a much lower UV dose is needed compared with conventional treatment using UVA alone. The photosensitization by the aldehyde **104** and the acid **105** has been suggested as an explanation of this (Chahidi *et al.*, 1983).

5.5 Photodehalogenations

Aryl-halogen bonds **106** are readily homolysed by UV light to give aryl radicals **107** which usually pick up a hydrogen atom **108**. Presumably the excited state of the substituted aromatic ring spontaneously decomposes. Although many drugs with halogen-substituted aromatic groups are known to photodecompose, in some cases the dehalogenation is quenched by oxygen and in others it is not. It is impossible to predict how any particular compound will react photochemically in the presence of oxygen. For example, chlorpromazine gives several oxidized products on irradiation in an open dish, but dechlorinated products have been found only after photolysis under nitrogen (Huang and Sands, 1964, 1967). For some drug/solvent combinations where the homolysis is inhibited by oxygen there must be an energy transfer from the excited substrate to give singlet oxygen which subsequently returns to the ground (triplet) state. Many halogenated drugs are known to cause light-induced toxic and allergic reactions. For example, tetrachlorosalicylanilide (TCSA, **109**) irradiated at 365 nm suffered rapid loss of the chlorine from position 3 and within 1 h three of the four chlorine atoms were lost. In the past TCSA was used as a bacteriostat in soaps until it was shown to induce a long-lasting photosensitivity (sometimes >1 year) in some users. The aryl radical from the dechlorination at C–3 was shown to combine readily with γ-globulin and protein. The bound form was assumed to remain in the skin for several months and to be the photosensitizer which caused the long-term effects, since the isolated dechlorinated products did not cause photoallergy, Scheme 5.15 (Davies *et al.*, 1975; Kochevar, 1979).

A dose-related photosensitivity to amiodarone **110** is seen in many patients (Zachary *et al.*, 1984; Paillous and Verrier, 1988). Irradiation of an oxygen-free ethanol solution at 300 nm gave first the mono-deiodinated derivative, but within 10 h the drug was completely deiodinated. Spin-trapping experiments on both amiodarone and its major metabolite desethylamiodarone indicated the formation of aryl radicals and superoxide was detected. The aryl radicals could extract hydrogen atoms from several donors including linoleic acid. It was suggested that this could lead to peroxy radical formation and explain the deposition of lipofuscin, a product of lipid peroxidation, in the skin of sensitive patients (Li and Chignell, 1987). Other similar work demonstrated that amiodarone and desethylamiodarone sensitized singlet oxygen formation under aerobic conditions, but that photodehalogenation proceeds via a triplet state under anaerobic conditions, Scheme 5.15 (Paillous and Verrier, 1988).

106 **107** **108**

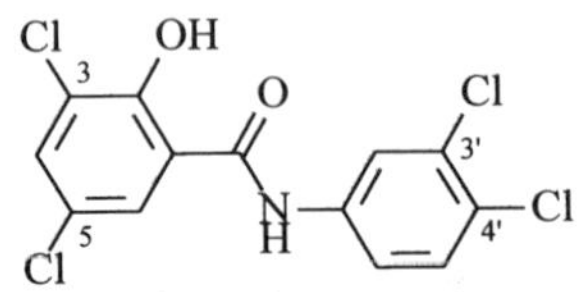

109 Tetrachlorosalicylanilide

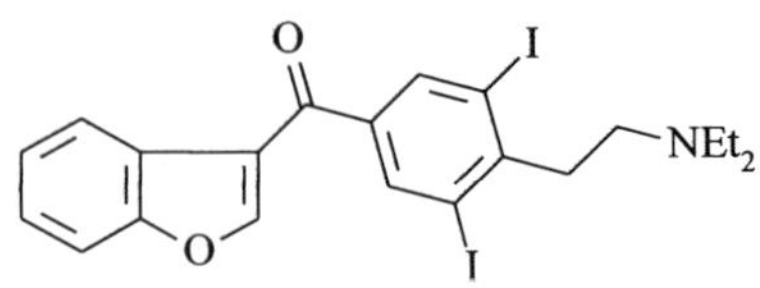

110 Amiodarone

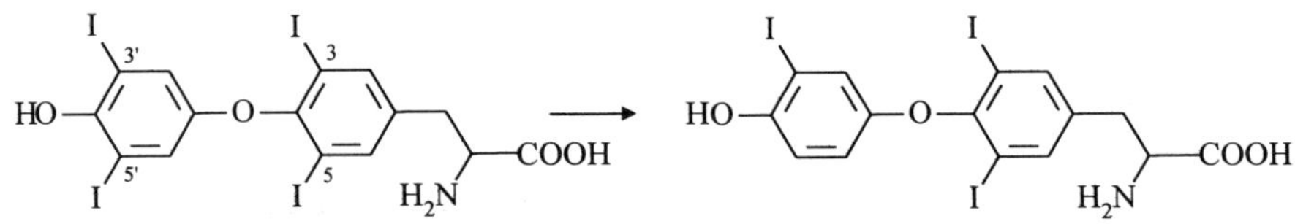

111 Thyroxine **112**

Scheme 5.15

Irradiation of thyroxine (**111**) in a mixture of ammonium hydroxide and methanol with UV wavelengths above 340 nm showed preferential loss of iodine from the phenol ring to give 3,3′,5-triiodothyronine **112**. 3,3′,5′-Triiodothyronine similarly gave 3,3′-diiodothyronine. The use of wavelengths >300 nm rapidly gave mixtures containing 3,5-diiodothyronine and 3-iodothyronine in major amounts. With each loss of halogen the UV spectrum experiences a hypsochromic shift so that the UV absorption by monoiodothyronine is very low at 300 nm and loss of the last iodine atom is slow. These techniques gave good yields of the thyroxine derivatives and were suggested as the preferable synthetic routes to the compounds, Scheme 5.15 (van der Walt and Cahnmann, 1982).

Chlorpromazine (**113**), prochlorperazine (**114**) and perphenazine (**115**) are three chlorosubstituted phenothiazines all associated with a high incidence of phototoxicity (Greenhill and McLelland, 1990). Chlorpromazine has long been known to cause serious side effects such as excessive sunburn, pigmentation of the skin and ocular opacity; it accumulates in the skin and cornea (Rosenthal *et al.*, 1978). Prochlorperazine and perphenazine are also associated with high phototoxicity (Ljunggren and Möller, 1977). Exposure of chlorpromazine solutions to the sun under nitrogen gives mixtures of dechlorinated products including promazine (**116**) and 2-hydroxypromazine (**117**), Scheme 5.16 (Huang and Sands, 1964, 1967). Solutions of chlorpromazine and prochlorperazine bases in methanol, irradiated at wavelengths >300 nm under nitrogen, are rapidly dechlorinated, but the reaction is quenched by oxygen. The same photodegradation occurs in water under nitrogen, but this time is only slightly slowed by oxygen (Moore and Tamat, 1980). Solutions of DNA and chlorpromazine under filtered xenon light give radicals via photodechlorinations which cause breaks in the DNA strands.

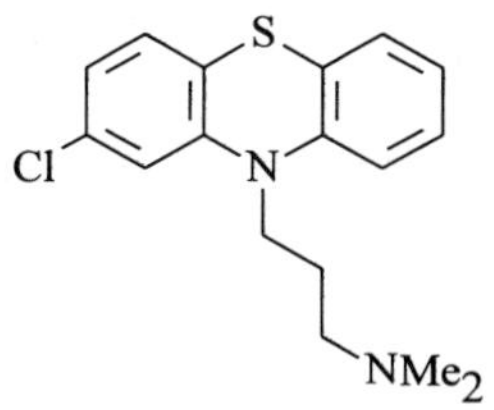

113 Chlorpromazine

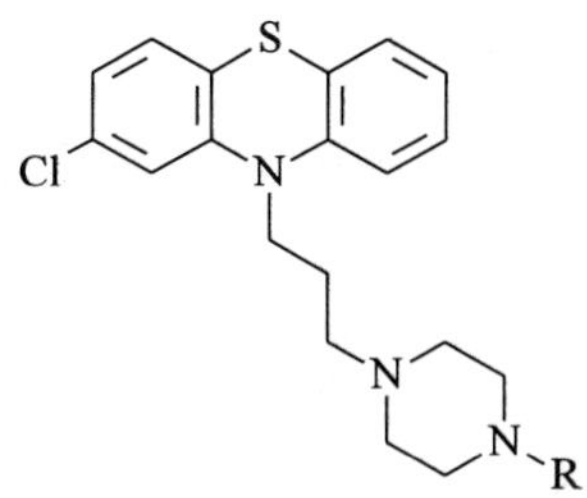

114 R = Me, Prochlorperazine
115 R = CH_2CH_2OH, Perphenazine

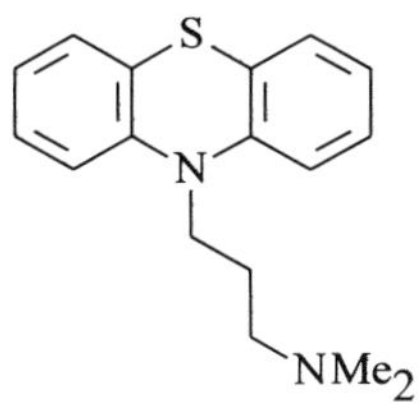

116 Promazine

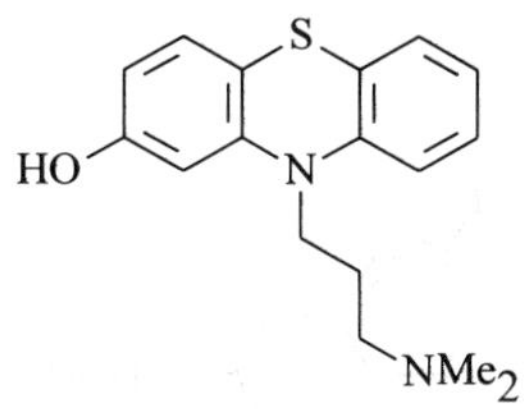

117 2-Hydroxypromazine

Scheme 5.16

Promazine, trifluopromazine and methoxypromazine (none of which have aryl-ring halogen substituents) are less vigorous strand-breakers and do so via direct reaction of the cation radicals formed by near-UV irradiation and superoxide radicals. An excited promazine molecule transfers an electron to oxygen to give a superoxide radical and if DNA is present this gives an $OH^{\cdot}$ radical, probably via a Haber-Weiss reaction catalysed by a DNA-iron complex (Decuyper *et al.*, 1984).

Frusemide (**118**) shows rapid photodechlorination in water under oxygen or nitrogen (Moore and Tamart, 1980). An oxygen-free methanol solution contained detectable amounts of the dechlorinated derivatives **119** and **120** after 10 minutes' irradiation at 365 nm (Moore and Sithipitaks, 1983). The methoxy compound **120** is presumed to form by reaction of the aryl radical with solvent. The neutral form of frusemide dispersed in micelles with non-ionic surfactants has been shown to be the most active in producing free radicals (Moore and Burt, 1981). Hydrochlorothiazide (**121**) has been shown to cause skin photosensitization by both free-radical and singlet oxygen mechanisms. There is also evidence for a light-activated

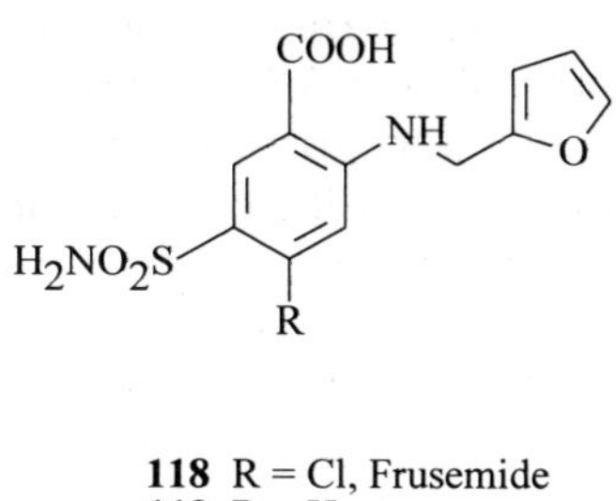

118 R = Cl, Frusemide
119 R = H
120 R = OMe

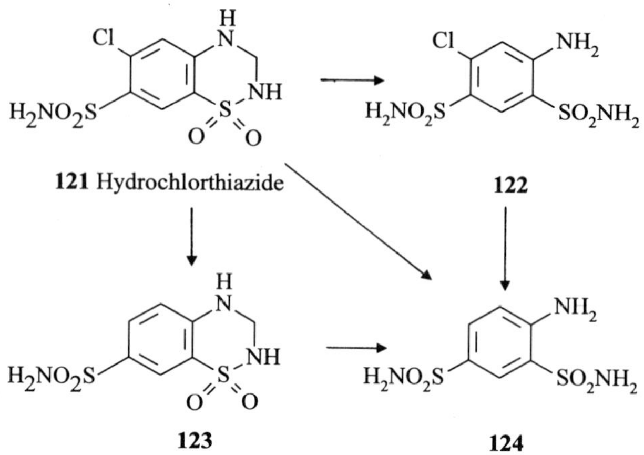

Scheme 5.17

mutagenicity. Exposure of methanol solutions to light of wavelengths > 310 nm gives about equal amounts of the ring opened derivative **122** and the dechlorinated compound **123**. Prolonged irradiation gives **124**. Under oxygen these reactions run at one-tenth of the rate under nitrogen. In water the main product is **124**, but again the reaction is slower under aerobic conditions, Scheme 5.17 (Moore and Tamat, 1980; Tamat and Moore, 1983).

Photodechlorination of diclofenac (**125**) gives an aryl radical **126** which attacks the other ring to give the carbazine derivative **128**. Subsequently, the second chlorine atom is lost. In water a mixture of the hydrocarbon **129** and the phenol **130** is obtained and in methanol the aryl radical gives **129** and the ether **131**, Scheme 5.18 (Moore *et al.*, 1990). It is interesting that there is no report of decarboxylation of the aryl acetic acid group. Such a photodecarboxylation would have to involve a primary radical – a much less stable species than the secondary radicals formed from the propionic acid-based NSAIDs.

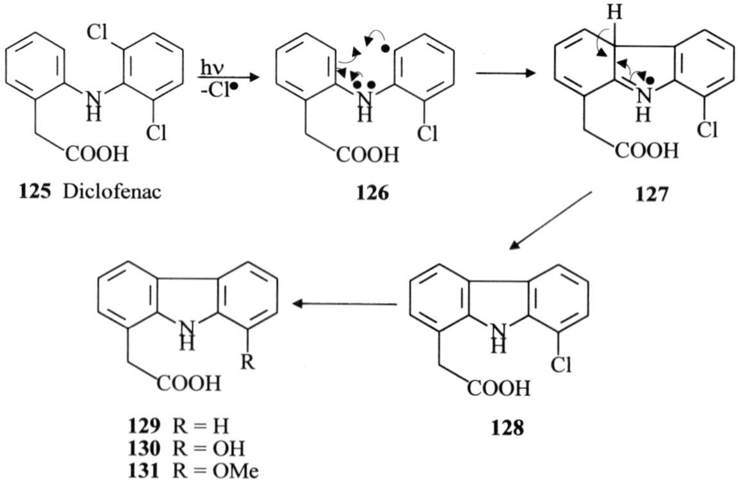

Scheme 5.18

5.6 Aromatic nitro compounds

Aromatic nitro groups are often photoactive. Rapid degradations in sunlight are seen with the 4-(nitrophenyl)-1,4-dihydropyridines typified by nifedipine (**132**). Here the nitro group is reduced to nitroso while the ring is oxidized. Overall the compound loses a molecule of water. Compound **133** is the product of exposure to sunlight, but under UV irradiation the nitroso group is reoxidized to give **134** (Testa *et al.*, 1979; Pietta *et al.*, 1981; Sadana and Ghogare, 1991). Other drugs for which the reaction has been demonstrated are furnidipine (Núñez-Vergara *et al.*, 1994) which has an *o*-nitro group and nitrendipine (Squella *et al.*, 1990), nimodipine (Zanocco *et al.*, 1992) and nicardipine (Bonferoni *et al.*, 1992), all of which carry *m*-nitro groups. It has been shown that similar compounds lacking nitro groups are photostable (Al-Turk *et al.*, 1988).

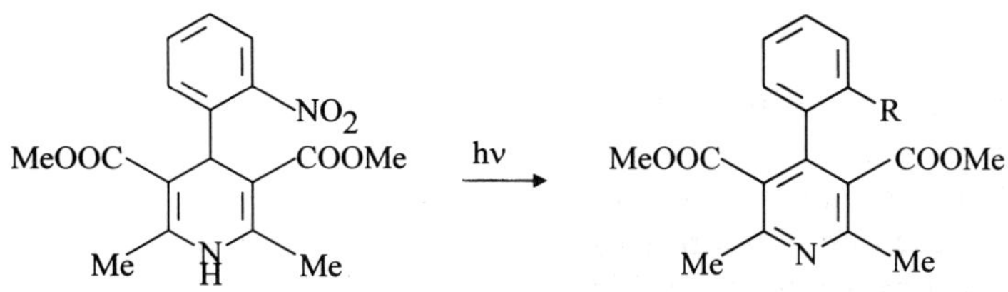

132 Nifedipine

133 R = NO
134 R = NO_2

The 4-nitrobenzaldehyde (**65**) shown in Scheme 5.11 as a photodegradation product of chloramphenicol gives some important secondary products. A photorearrangement gives 4-nitrosobenzoic acid (**135**), believed to be the cause of the aplastic anaemia associated with this drug. Photolysis of 4-nitrobenzaldehyde in water gives **135** in high yield. Chloramphenicol is little used now, except in eye

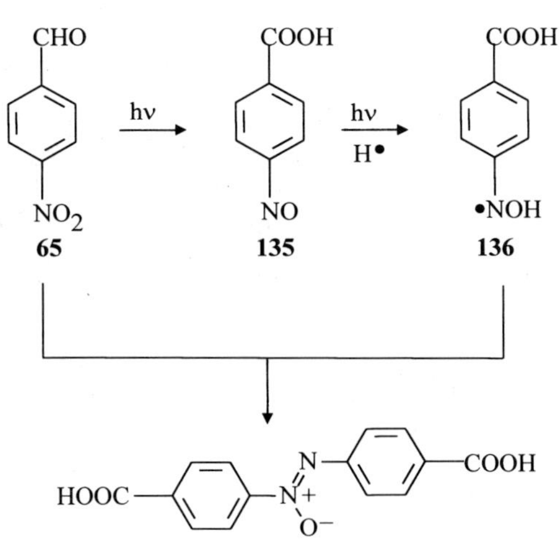

65 **135** **136**

137

Scheme 5.19

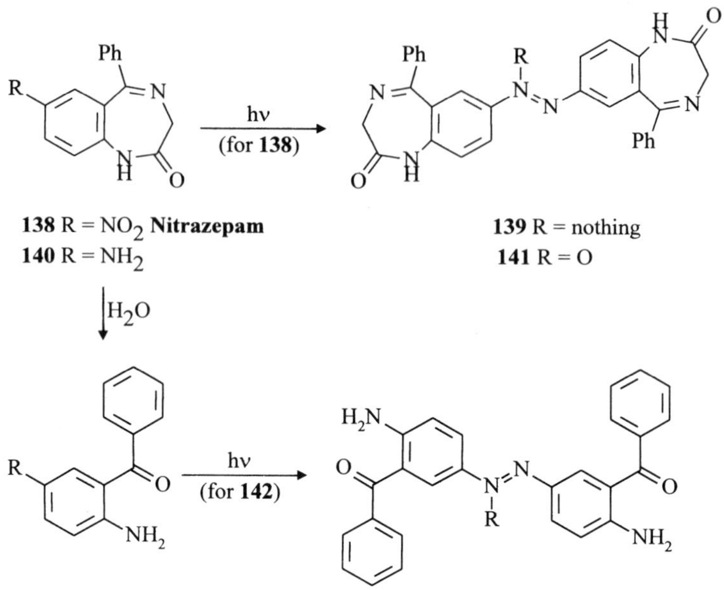

138 R = NO_2 **Nitrazepam**
140 R = NH_2

139 R = nothing
141 R = O

142 R = NO_2
144 R = NH_2

143 R = nothing
145 R = O

Scheme 5.20

drops, but *in vivo* as much as 45 per cent of the drug in the eyes is converted to 4-nitrosobenzoic acid and aplastic anaemia has been observed even after this use. Partial reduction of 4-nitrosobenzoic acid probably mediated by a hydrogen atom gives an intermediate **136** which attacks a molecule of 4-nitrobenzaldehyde to give the azoxybenzoic acid **137**, a typical reduction product of an aromatic nitro group (Mubarak *et al.*, 1982).

Nitrazepam (**138**) is photoreduced in the presence of transferable hydrogen to a mixture of the amino derivative **140** and the azo **139** and azoxy **141** dimers. The azoxy compound **141** has been separately photoreduced to **139**. Hydrolysis or metabolism of nitrazepam gives 2-amino-5-nitrobenzophenone (**142**) which undergoes a similar series of photoreductions (Roth and Adomeit, 1973).

Sulphanilamide (**146**) shows both oxidation and reduction when photolysed in ethanol. The first product is the nitro derivative **147** which is reduced to the diazobenzene **148** (Reisch and Niemeyer, 1972).

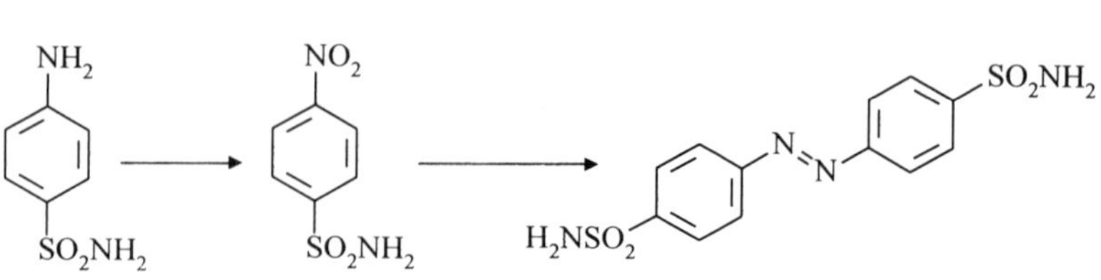

146
Sulphanilamide

147

148

5.7 Benzophenones

It has long been known that irradiation of benzophenone gives long-lived (approximately 10^{-4} s) triplet excited states which efficiently transfer energy to substrate molecules. It is likely, therefore, that at least some of the photosensitizing effects observed in patients taking benzophenone-based drugs result from triplet energy transfer to biomolecules. Fenofibrate (**149**) was recently (December 1993) approved by the FDA for prescription in the USA, but has been widely used worldwide for several years as a lipid-lowering agent. Light-induced cutaneous reactions have been reported during its use. Fenofibrate in methanol rapidly decomposes under irradiation from a medium-pressure mercury lamp to give 4-chloroperbenzoic acid (**155**). This powerful oxidizing agent was isolated and found to induce haemolysis in the absence of irradiation. A mechanism to account for **155** and the other main product methyl 4-chlorobenzoate (**153**) is suggested in Scheme 5.21. The minor product, 4-chlorobenzoic acid (**154**), would arise from decomposition of the peracid or hydrolysis of the ester, possibly during work-up. It is interesting that no products from radical **156** were reported. The ether group (R) would probably photolyse early in the proceedings and this may give small molecules which would escape detection (Vargas *et al.*, 1993).

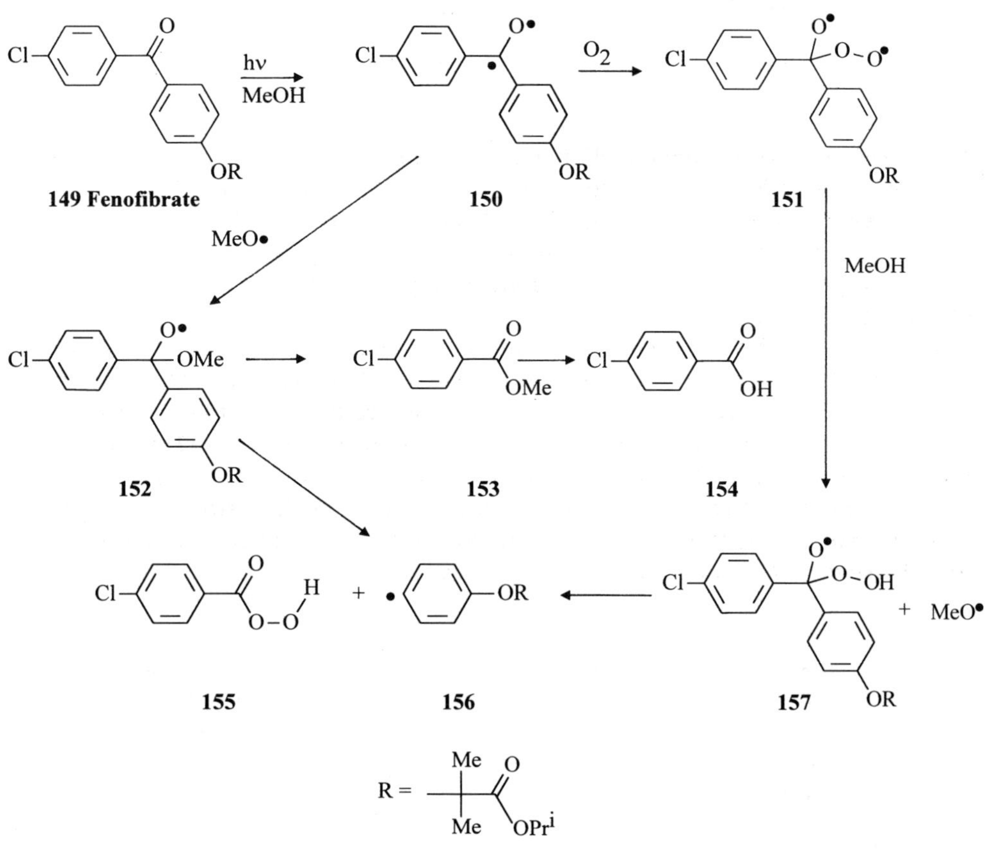

Scheme 5.21

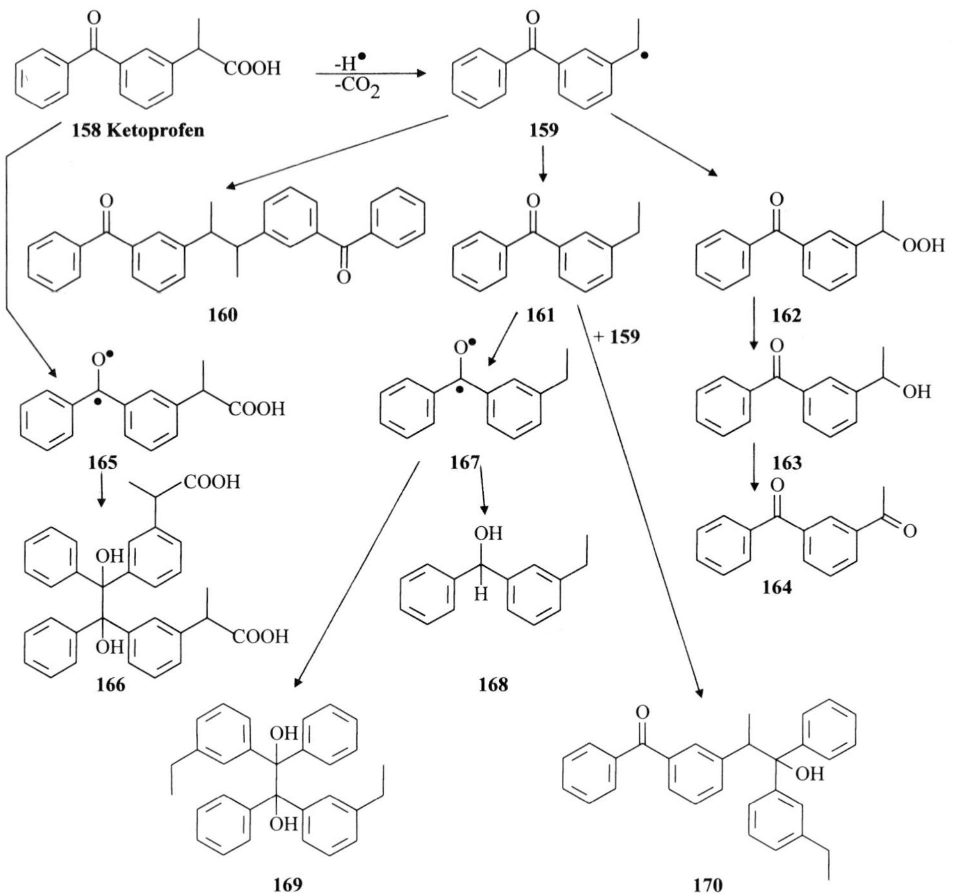

Scheme 5.22

Ketoprofen (**158**) gives several products of arylacetic acid photodecomposition, but it also gives typical radical derivatives of the benzophenone group, Scheme 5.22 (Boscá *et al.*, 1994). The drug is decarboxylated to the benzylic radical **159** which is the intermediate for the observed products **160** to **164** as described above for reactions of benzyl radicals. In addition, the benzophenone moiety of ketoprofen absorbs irradiation to give the excited state **165** which can dimerize, or more likely attack an unexcited molecule, to give **166**. Similar reactions of the main photoproduct **161** account for the formation of compound **169** via **167**. The reduction product **168** may have formed by reaction with the solvent or with one of the components of the phosphate buffer. Reaction between radical **159** and **161** would account for the observed product **170**, but attack of **159** on unchanged ketoprofen (**158**) followed by decarboxylation is the most likely route to **170**.

The antipsoriatic action of dithranol (**171**) is believed to result from its ability to photosensitize the conversion of ground state $^3O_2$ to $^1O_2$. The singlet oxygen is known to oxidize polyunsaturated fatty acids (as in Scheme 5.5) of the type present in psoriatic skin in greatly increased concentrations. Drug decomposition results from addition of $^1O_2$ to the stabilized anion **172** (present in polar, aprotic

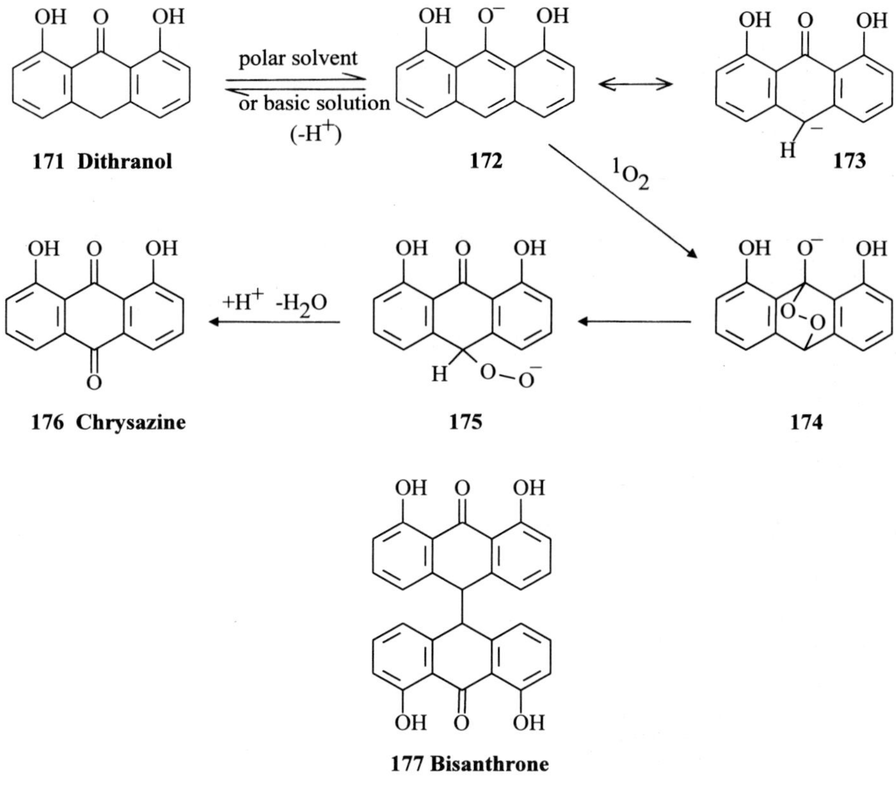

Scheme 5.23

solvents or in mildly basic solution) to give the *endo*-peroxide **174** which rearranges to **175** and gives chrysazine (**176**) as shown. Triplet oxygen ($^3O_2$) in base converts dithranol to bisanthrone (**177**) which also gives chrysazine with $^1O_2$ (Müller *et al.*, 1986a and b). Dithranol also gives superoxide under UV irradiation, which may contribute to the medicinal action (Müller *et al.*, 1987).

5.8 N-oxides

Chlordiazepoxide (**178**) and two of its metabolites, demethylchlordiazepoxide (**183**) and demoxepam (**184**), all show similar phototoxicity. When irradiated at 350 nm for a few minutes, all three compounds gave oxaziridines, for example, **179** (Bakri *et al.*, 1985). All three N-oxides were phototoxic to a bacterial cell preparation, but the reduced derivatives were not. There was a close correlation between the phototoxicities of the N-oxides and the toxicities in the dark of their derived oxaziridines (Cornelissen *et al.*, 1980). The half-life of the oxaziridine **179** in the presence of plasma proteins is about 30 minutes, so it has time to be transported from the skin to internal organs. It has been shown to be covalently bound to biomolecules not only in the irradiated skin, but also in organs such as liver and kidney, where it probably is the cause of damage observed in the rat (Bakri *et al.*, 1986).

The oxaziridine **179** has very low absorption at 350 nm and is only slowly photodegraded to its secondary products. Solutions of chlordiazepoxide in methanol or methanol/water irradiated at 254 or 300 nm for 50 minutes, give mainly the quinoxaline **181** and the benzoxadiazocine **182** along with some oxaziridine **179** (Cornelissen *et al.*, 1979). Homolysis of the strained 3-membered ring would give the diradical **180**. Transannular attack by the nitrogen radical (route a) would produce ring contraction to **181** while similar attack by the oxygen radical (route b) would cause ring expansion to **182**. Methaqualone (**185**) does not photodecompose, apparently, but its metabolite methaqualone N-oxide (**186**) gives the isolatable oxaziridine **187** which photodegrades further to compounds **189**, **192**, **194** and **195** (Stevens and Gunn, 1972; Theil *et al.*, 1985). A reasonable intermediate **188** from homolysis of the aziridine accounts for the observed products. A transannular reaction by the nitrogen radical would give the benzopyrazole **189**. A 1,2-methyl shift gives the rearranged system **192**. Addition of solvent methanol gives **191** which would spontaneously rearrange to produce **195**. An alternative route in which the methanol acts as a reducing agent with the elimination of formaldehyde proceeds via **190** and **193** to the diamide **194**. A minor second metabolite **196** is also photoactive and gives the diazo compound **197**, Scheme 5.25.

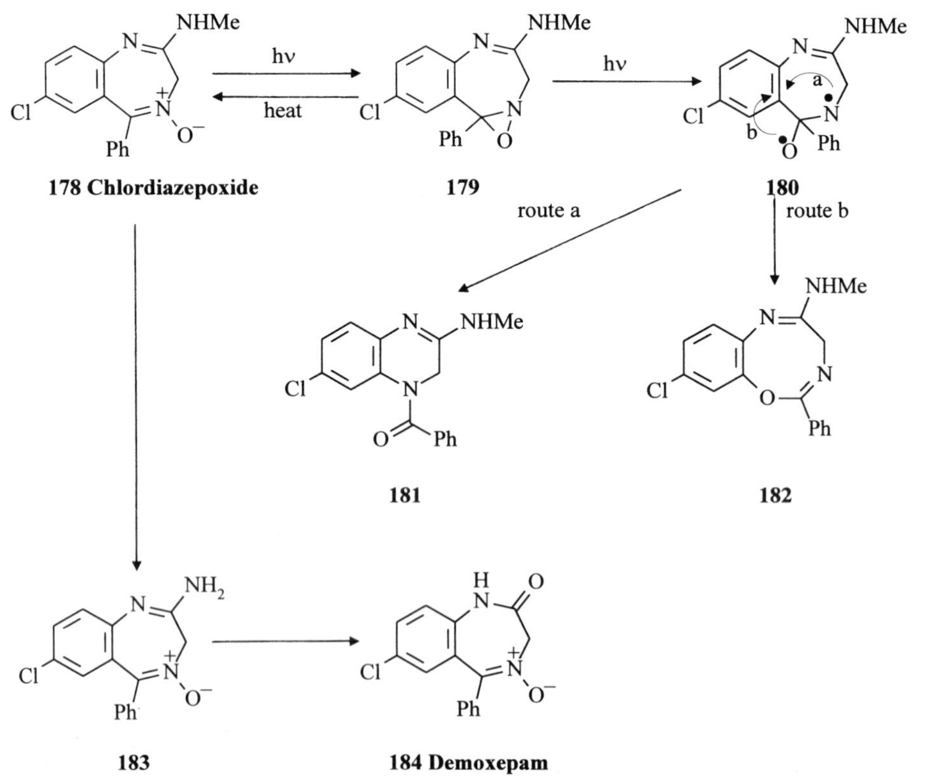

Scheme 5.24

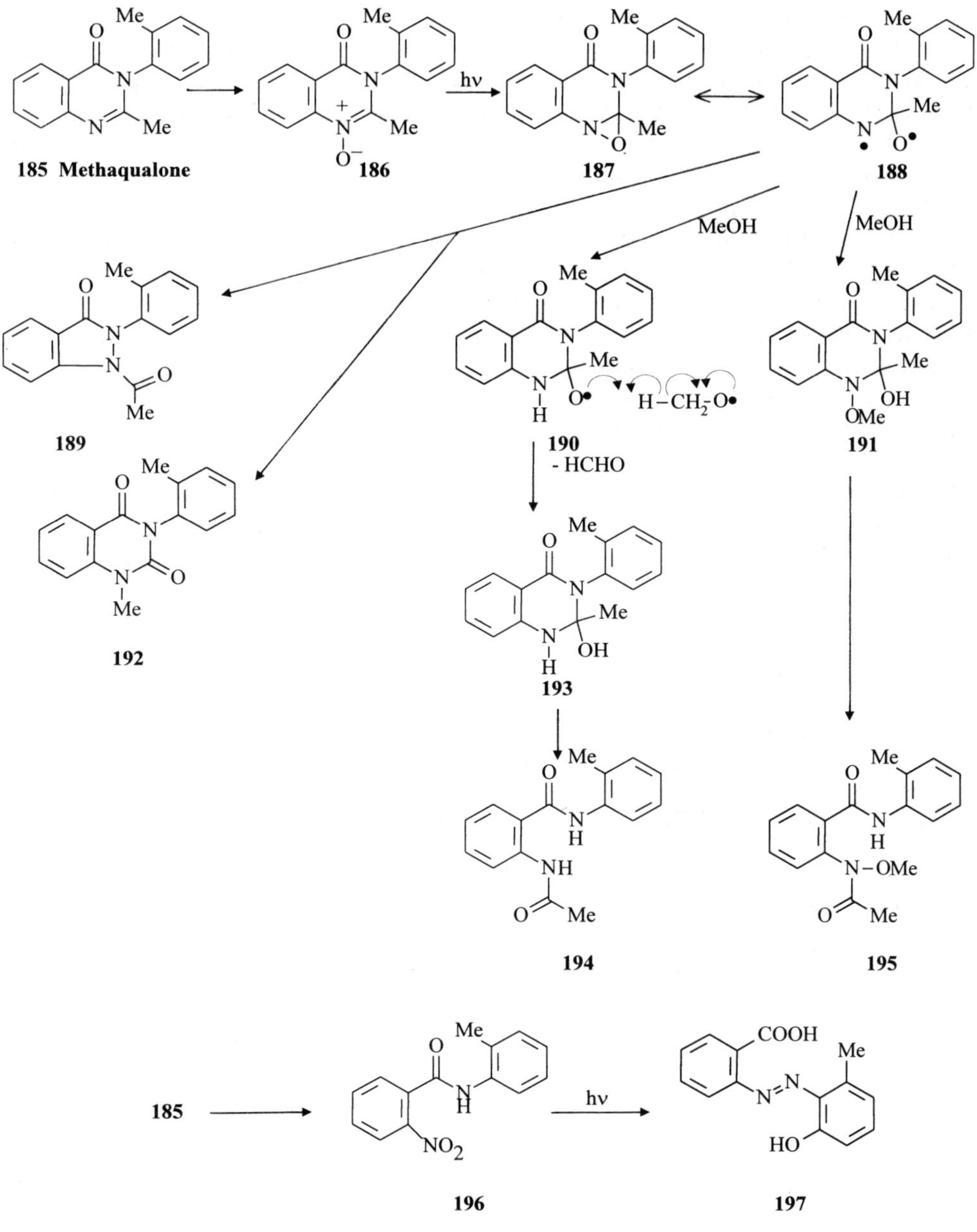

Scheme 5.25

5.9 Conclusions

The various chemical groups discussed in this chapter highlight chemical features known to be vulnerable to photodecomposition. Indeed, these groups encompass a substantial proportion of all the groups found in medicinal chemicals. In most cases it would not be wise for the drug designer to try to avoid the use of light-sensitive groups in new molecules, rather it is the job of the pharmacist to protect his products from light and of the other health care professionals to inform the patient of the risks. When the responsible scientists are able to agree on standard methods of testing for photodecomposition it is possible that some of the reactions discussed here will be shown to be too slow to cause phototoxicity, but until that time it is essential that all precautions are taken with any drug known to be photosensitive.

All aspects of experimental design may affect photochemical changes. For example, it is best to conduct studies in buffered aqueous solutions, but this is not always possible. Because of its excellent solvent power, methanol is sometimes used as solvent or co-solvent. However, under conditions which induce radical formation, methanol is known to act as a reducing agent. For example, the isolation of the diamide **194** as a photodecomposition product of methaqualone may result from methanolic reduction as shown in Scheme 5.25. Carboxylic acids photolysed in any alcoholic solvent usually give a proportion of the appropriate ester. This can easily be discounted, although in a recent report of flurbiprofen no less than 47 per cent conversion to the methyl ester was observed along with only low

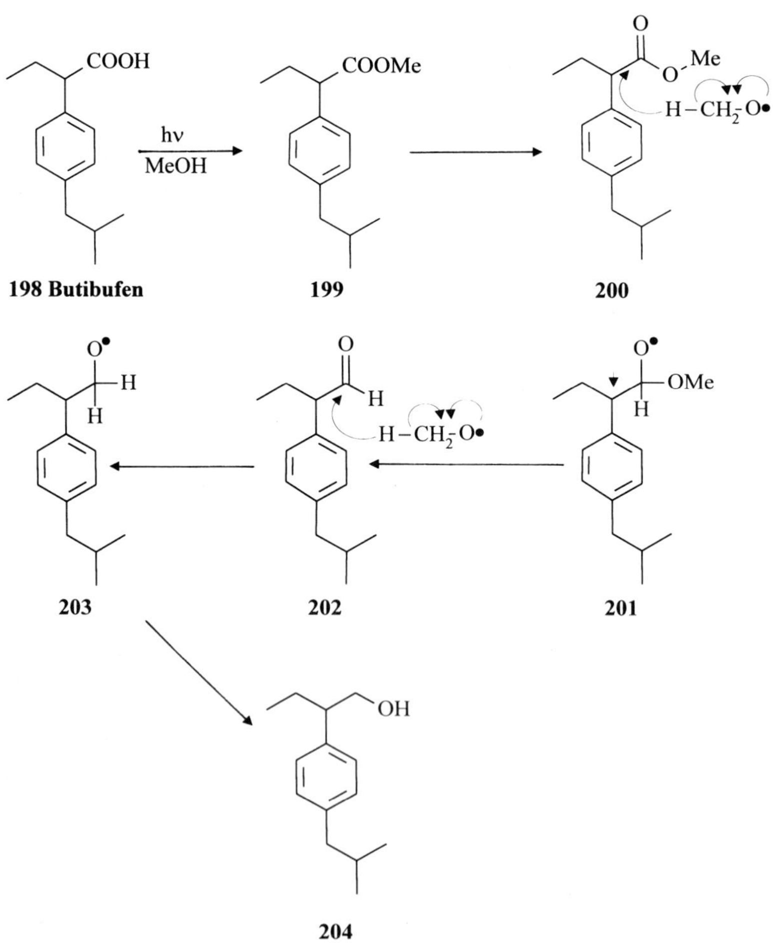

Scheme 5.26

amounts of the expected decarboxylation product (1 per cent), alcohol (4 per cent) and ketone (4 per cent) compare other NSAIDs, above (Castell *et al.*, 1992). The same report includes similar products from butibufen (**198**) plus a decarboxylated dimer and a primary alcohol **204** in a very significant yield. A similar primary alcohol is given by ibuprofen after 7 h irradiation with a medium-pressure mercury lamp (Castell *et al.*, 1987). This was assumed to come from attack by the benzylic radical on a solvent molecule. However, that mechanism is difficult to imagine; more likely reduction of the ester or acid gave **204**, as shown in Scheme 5.26. Here again, homolysis of methanol gives an active species to provide a hydrogen atom with the loss of a formaldehyde molecule. The ester group is reduced first to give the aldehyde **202**, then a second reduction cycle gives the primary alcohol.

References

AL-TURK, W. A., MAJEED, I. A., MURRAY, W. J., NEWTON, D. W. and OTHMAN, S., 1988, Some factors affecting the photodecomposition of nifedipine. *Int. J. Pharm.*, **41**, 227–30.

BAKRI, A., BEIJERSBERGEN VAN HENEGOUWEN, G. M. J. and CHANAL, J. L., 1985, Involvement of the N-4 oxide group in the phototoxicity of chlordiazepoxide in the rat, *Photodermatology*, **2**, 205–12.

BAKRI, A., BEIJERSBERGEN VAN HENEGOUWEN, G. M. J. and SEDEE, A. G. J., 1986, Irreversible binding of chlordiazepoxide to human plasma protein induced by UV-A radiation, *Photochem. Photobiol.*, **44**, 181–5.

BEIJERSBERGEN VAN HENEGOUWEN, G. M. J., 1991, New trends in photobiology, *J. Photochem. Photobiol. B:Biol.*, **10**, 183–210.

BEIJERSBERGEN VAN HENEGOUWEN, G. M. J., VAN DE ZIJDE, H. J., VAN DE GRIEND, J. and DE VRIES, H., 1987, Photochemical decomposition of diphenhydramine in water, *Int. J. Pharm.*, **35**, 259–62.

BONFERONI, M. C., MELLERIO, G., GIUNCHEDI, P., CARAMELLA, C. and CONTE, U., 1992, Photostability evaluations of nicardipine HCl solutions, *Int. J. Pharm.*, **80**, 109–17.

BOSCÁ, F., MIRANDA, M. A., CARGANICO, G. and MAULEÓN, D., 1994, Photochemical and photobiological properties of ketoprofen associated with the benzophenone chromophore, *Photochem. Photobiol.*, **60**, 96–101.

BOSCÁ F., MIRANDA, M. A., VAÑÓ, L. and VARGAS, F., 1990, New photodegradation pathways for naproxen, a phototoxic non-steroidal anti-inflammatory drug, *J. Photochem. Photobiol. A: Chem.*, **54**, 131.

BOUCHE, R., DRAGUET-BRUGHMANS, M., FLANDRE, J. P., MOREAUX, C. and VAN MEESCHE, M., 1978, Ketonic oxidation products of cyclobarbital, *J. Pharm. Sci.*, **67**, 1019–22.

BRISAERT, M. G., EVERAERTS, I., BULTINCK, G. R. and PLAIZIER-VERCAMMEN, J. A., 1992, Onderzoek naar de chemische stabiliteit van tretinoine in dermatologische preparaten, *Farmaceutisk Tijdschrift voor Belgie*, **69**, 194–9.

BUDAC, D. and WAN, P., 1992, Photodecarboxylation: mechanism and synthetic utility, *J. Photochem. Photobiol. A: Chem.*, **67**, 135–66.

CASTELL, J. V., GÓMEZ-LECHÓN, M. J., GRASSA, C., MARTINEZ, L. A., MIRANDA, M. A. and TARREGA, P., 1993, Involvement of drug-derived peroxides in the phototoxicity of naproxen and tiaprofenic acid, *Photochem. Photobiol.*, **57**, 486–90.

CASTELL, J. V., GÓMEZ-LECHÓN, M. J., MIRANDA, M. A. and MOREA, I. M., 1987, Photolytic degradation of ibuprofen, *Photochem. Photobiol.*, **46**, 991–6.

1992, Phototoxicity of non-steroidal anti-inflammatory drugs: *in vitro* testing of the photoproducts of butibufen and flurbiprofen, *J. Photochem. Photobiol. B:Biol.*, **13**, 71–81.

CHAHIDI, C., GIRAUD, M., AUBAILLY, M., VALLA, A. and SANTUS, R., 1986, 2,4-diamino-6-pteridinocarboxaldehyde and an azobenzene derivative are produced by UV photodegradation of methotrexate, *Photochem. Photobiol.*, **44**, 231–3.

CHAHIDI, C., MORLIERE, P., AUBAILLY, M., DUBERTRET, L. and SANTUS, R., 1983, Photosensitization by methotrexate photoproducts, *Photochem. Photobiol.*, **38**, 317–22.

CHATTERJI, D. C. and GALLELLI, J. F., 1978, Thermal and photolytic decomposition of methotrexate aqueous solutions, *J. Pharm. Sci.*, **67**, 526–31.

CORNELISSEN, P. J. G., BEIJERSBERGEN VAN HENEGOUWEN, G. M. J. and GERRITSMA, K. W., 1979, Photochemical decomposition of 1,4-benzodiazepines. Chlordiazepoxide, *Int. J. Pharm.*, **3**, 205–20.

CORNELISSEN, P. J. G., BEIJERSBERGEN VAN HENEGOUWEN, G. M. J. and MOHN, G. R., 1980, Structure and photobiological activity of 7-chloro-1,4-benzodiazepines. Studies on the phototoxic effects of chlordiazepoxide, desmethylchlordiazepoxide and demoxepam using a bacterial indicator system, *Photochem. Photobiol.*, **32**, 653–9.

COSTANZO, L. L., DE GUIDI, G., CONDORELLI, G., CAMBRIA, A. and FAMA, M., 1989, Molecular mechanism of drug photosensitization – Photohemolysis sensitized by ketoprofen, *Photochem. Photobiol.*, **50**, 359–65.

CURLEY, Jr, R. W. and FOWBLE, J. W., 1988, Photoisomerization of retinoic acid and its photoprotection in physiologic-like solutions, *Photochem. Photobiol.*, **47**, 831–5.

DAVIES, A. K., HILAL, N. S., MCKELLAR, J. F. and PHILLIPS, G. O., 1975, Photochemistry of tetrachlorosalicylanilide and its relevance to the persistent light reactor, *Br. J. Dermatol.*, **92**, 143–7.

DECUYPER, J., PIETTE, J., LOPEZ, M., MERVILLE, M.-P. and VAN DE VORST, A., 1984, Induction of breaks in deoxyribonucleic acid by photoexcited promazine derivatives, *Biochem. Pharmacol.*, **33**, 4025–31.

DOBBS, R. H. and CREMER, R. J., 1975, Phototherapy, *Arch. Dis. Child.*, **50**, 833–5.

DOYLE, T. D., BENSON, W. R. and FILIPESCU, N., 1976, Photocyclization of diethylstilbestrol. Isolation of a stable, self trapping dihydrophenanthrene intermediate, *J. Amer. Chem. Soc.*, **98**, 3262–7.

1978, Spectrophotometric study of dienestrol photoisomerization, *Photochem. Photobiol.*, **27**, 3–8.

ENNEVER, J. F., 1988, Phototherapy for neonatal jaundice, *Photochem. Photobiol.*, **47**, 871–6.

EPLING, G. A. and LOPES, A., 1977, Fragmentation pathways in the photolysis of phenylacetic acid, *J. Amer. Chem. Soc.*, **99**, 2700–4.

FOX, M. A., 1990, Photoinduced electron transfer, *Photochem. Photobiol.*, **52**, 617–27.

FRITH, R. G. and PHILLIPOU, G., 1986, Applications of clomiphene photolysis to assays based on analysis of the derived phenanthrenes, *J. Chromatog.*, **367**, 260–6.

GREENHILL, J. V., 1995, Photodecomposition of drugs, in Swarbrick, J. and Boylan, J. C. (eds), *Encyclopedia of Pharmaceutical Technology*, 105–35. New York: Marcel Dekker, Vol. 12.

GREENHILL, J. V. and MCLELLAND, M. A., 1990, Photodecomposition of drugs, in Ellis, G. P. and West, G. B. (eds), *Progress in Medicinal Chemistry*, 51–121. Amsterdam: Elsevier, Vol. 27.

HUANG, C. L. and SANDS, F. L., 1964, The effect of UV irradiation on chlorpromazine, *J. Chromatogr.*, **13**, 246–9.

1967, Effect of ultra-violet irradiation on chlorpromazine. II Anaerobic condition, *J. Pharm. Sci.*, **56**, 259–64.

HUGELSTOFER, P., KALVODA, J. and SCHAFFNER, K., 1960, Lichtcatalysierte cyclodehydrierung von 1,2-diaryläthylenen und azobenzol, *Helv. Chim. Acta*, **43**, 1322–32.

KAGAN, J., 1993, in *Organic Photochemistry, Principles and Applications*. London: Academic Press, p. 87.

KOCH, H. P., 1948, Configuration of synthetic oestrogens, *Nature*, **161**, 309.

KOCHEVAR, I. E., 1979, Photoallergic responses to chemicals, *Photochem. Photobiol.*, **30**, 437–42.

LEWIS, F. D. and HO, T.-I., 1980, On the selectivity of tertiary amine oxidations, *J. Amer. Chem. Soc.*, **102**, 1752–3.

LI, A. S. W. and CHIGNELL, C. F., 1987, A spin trapping study of the photolysis of amiodarone and desethylamiodarone, *Photochem. Photobiol.*, **45**, 191–7.

LJUNGGREN, B. and MÖLLER, H., 1977, Phenothiazine phototoxicity: an experimental study on chlorpromazine and related tricyclic drugs, *Acta Derm.-Venereol.*, **57**, 325–9.

LOWRY, O. H., BESSEY, O. A. and CRAWFORD, E. J., 1949, Photolytic and enzymatic transformations of pteroglutamic acid, *J. Biol. Chem.*, **180**, 389–98.

MEE, J. M. L., BROOKS, C. C. and YANAGIHARA, K. H., 1975, Gas chromatographic-mass spectrometric investigation of the photo-epoxidation of vitamin K3, *J. Chromatog.*, **110**, 178–81.

MOORE, D. E. and BURT, C. D., 1981, Photosensitization by drugs in surfactant solutions, *Photochem. Photobiol.*, **34**, 431–9.

MOORE, D. E. and CHAPPUIS, P. P., 1988, A comparative study of the photochemistry of the non-steroidal anti-inflammatory drugs naproxen, benzoxaprofen and indomethacin, *Photochem. Photobiol.*, **47**, 173–80.

MOORE, D. E., ROBERTS-THOMPSON, S., ZHEN, D. and DUKE, C. C., 1990, Photochemical studies on the anti-inflammatory drug diclofenac, *Photochem. Photobiol.*, **52**, 685–90.

MOORE, D. E. and SITHIPITAKS, V., 1983, Photolytic degradation of frusemide, *J. Pharm. Pharmacol.*, **35**, 489–93.

MOORE, D. E. and TAMAT, S. R., 1980, Photosensitization by drugs: photolysis of some chlorine containing drugs, *J. Pharm. Pharmacol.*, **32**, 172–7.

MUBARAK, S. I. M., STANDFORD, J. B. and SUDGEN, J. K., 1982, Some aspects of the photochemical degradation of chloramphenicol, *Pharm. Acta Helv.*, **57**, 226–30.

MÜLLER, K., EIBLER, E., MAYER, K. K., WIEGREBE, W. and KLUG, G., 1986a, Dithranol, singlet oxygen and unsaturated fatty acids, *Arch. Pharm.* (Weinheim), **319**, 2–9.

MÜLLER, K., MAYER, K. K. and WIEGREBE, W., 1986b, $^1O_2$ oxidation of dithranol to chrysazin, *Arch. Pharm.* (Weinheim), **319**, 1009–18.

MÜLLER, K., WIEGREBE, W. and YOUNES, M., 1987, Dithranol, active oxygen species and lipid peroxidation *in vivo*, *Arch. Pharm.* (Weinheim), **320**, 59–66.

NAVARATHAM, S., HUGHES, J. L., PARSONS, B. J. and PHILLIPS, G. O., 1985, Laser flash and steady state photolysis of benzoxaprofen in aqueous solution, *Photochem. Photobiol.*, **41**, 375–80.

NORD, K., KARLSEN, J. and TØNNESEN, H. H., 1991, Photochemical degradation of chloroquine, *Int. J. Pharm.*, **72**, 11–18.

NÚÑEZ-VERGARA, L. J., SUNKEL, C. and SQUELLA, J. A., 1994, Photodecomposition of a new 1,4-dihydropyridine: furnidipine, *J. Pharm. Sci.*, **83**, 502–7.

PAILLOUS, N. and VERRIER, M., 1988, Photolysis of amiodarone, an antiarrhythmic drug, *Photochem. Photobiol.*, **47**, 337–43.

PIETTA, P., RAVA, A. and BIONDI, P., 1981, HPLC of nifedipine, its metabolites and photochemical degradation products, *J. Chromatogr.*, **210**, 516–21.

QUILLIAM, M. A., MCCARRY, B. E., HOO, K. H., MCCALLA, D. R. and VIATEKUNAS, S., 1987, Identification of the photolysis products of nitrofurazone irradiated with laboratory illumination, *Cand. J. Chem.*, **65**, 1128–32.

REISCH, J. and WEIDMANN, K. G., 1971, Photo- und Radiolyse des chloramphenicols,

Arch. Pharm. (Weinheim), **304**, 911–19.

REISCH, J. and NIEMEYER, D. H., 1972, Photochemische studien am sulfanilamid, sulfacetamid und 4-aminobenzoesäure-äthylester, *Arch. Pharm.* (Weinheim), **305**, 135–40.

ROSHENTHAL, I., BEN-HUR, E., PRAGER, A. and RIKLIS, E., 1978, Photochemical reactions of chlorpromazine; chemical and biochemical implications, *Photochem. Photobiol.*, **28**, 591–4.

ROTH, H. J. and ADOMEIT, M., 1973, Photochemie des nitrazepams, *Arch. Pharm.* (Weinheim), **306**, 889–97.

SADANA, G. S. and GHOGARE, A. B., 1991, Mechanistic studies on photolytic degradation of nifedipine by use of ^{1}H NMR and ^{13}C NMR spectroscopy, *Int. J. Pharm.*, **70**, 195–9.

SEDEE, A. G. J., BEIJERSBERGEN VAN HENEGOUWEN, G. M. J. and BLAAUWGEERS, H. J. A, 1983, Isolation, identification and densitometric determination of norethisterone-4β,5β-epoxide after photochemical decomposition of norethisterone, *Int. J. Pharm.*, **15**, 149–58.

SEDEE, A. G. J., BEIJERSBERGEN VAN HENEGOUWEN, G. M. J., DE VRIES, H., GUIJT, W. and HAASNOOT, C. A. G., 1985, The photochemical decomposition of the progestrogenic 19-nor-steroid, norethisterone, in aqueous medium, *Pharm. Weekbl. Sci. Ed.*, **7**, 194–201.

SHIH, I. K., 1971, Photochemical products of chloramphenicol in aqueous solution, *J. Pharm. Sci.*, **60**, 1889–90.

SNYDER, C. D. and RAPOPORT, H., 1969, Photooxygenation of phylloquinone and menaquinones, *J. Amer. Chem. Soc.*, **91**, 731–7.

SQUELLA, J. A., ZANOCCO, A., PERNA, S. and NUÑEZ-VERGARA, L. J., 1990, A polarographic study of the photodegradation of nitrendipine, *J. Pharm. Biomed. Anal.*, **8**, 43–7.

STEVENS, M. F. G. and GUNN, B. C., 1972, Photolysis of a methaqualone metabolite, *J. Pharm. Pharmacol.*, **24**, 141P.

TAMAT, S. R. and MOORE, D. E., 1983, Photolytic decomposition of hydrochlorthiazide, *J. Pharm. Sci.*, **72**, 180–3.

TESTA, R., DOLFINI, E., RESCHIOTTO, C., SECCHI, C. and BIONDI, P. A., 1979, GLC determination of nifedipine, a light sensitive drug, in plasma, *Farmaco, Ed.Prat.*, **34**, 463–73.

THEIL, F.-P., PÖHLMANN, H., PFEIFER, S. and FRANKE, P., 1985, Fotochemische reaktivität von methaqualon-1-oxid, *Pharmazie*, **40**, 328–31.

TØNNESEN, H. H., GRISLINGAAS, A.-L., WOO, S. O. and KARLSEN, J., 1988, Photochemical stability of antimalarials, *Int. J. Pharm.*, **43**, 215–19.

VAN DER WALT, B. and CAHNMANN, H. J., 1982, Synthesis of thyroid hormone metabolites by photolysis of thyroxine and thyroxine analogs in the near UV, *Proc. Nat. Acad. Sci. USA*, **79**, 1492–6.

VARGAS, F., RIVAS, C. and CANUDAS, N., 1993, Formation of a perbenzoic acid derivative in the photodegradation of fenofibrate: phototoxicity studies on erythrocytes, *J. Pharm. Sci.*, **82**, 590–1.

WILSON, R. M., WALSH, T. F. and GEE, S. K., 1980, Laser photochemistry: the wavelength dependent oxidative photodegradation of vitamin K analogs, *Tetrahedron Lett.*, **21**, 3459–63.

ZACHARY, C. B., SLATER, D. N., HOLT, D. W., STORY, G. C. A. and MCDONALD, D. M., 1984, The pathogenesis of amiodarone-induced pigmentation and photosensitivity, *Br. J. Dermatol.*, **110**, 451–6.

ZANOCCO, A. L., DIAZ, L., LÓPEZ, M., NUÑEZ-VERGARA, L. J. and SQUELLA, J. A., 1992, Polarographic study of the photodecomposition of nimodipine, *J. Pharm. Sci.*, **81**, 920–4.

6

Photodecomposition and Stabilization of Compounds in Dosage Forms

K. THORMA

6.1 Actual importance

Photodecomposition and stabilization of compounds in dosage forms are increasingly gaining significance. This results from the required protection of the drug not only at storage, but also at production and application. The number of photoinstable substances with therapeutic importance has considerably grown in recent years. Character and extent of photodecomposition can strongly be influenced by dosage forms.

Photoinstable drugs are found among different therapeutic groups (Table 6.1).

6.2 Particular characteristics of photodecomposition in the solid state

In contrast to drug solutions, photodegradation in the solid state only takes place at the surface. The depth of light penetration is markedly restricted by absorption and reflection. Various factors influence the rate of decomposition:

- size and surface of particles;
- colour and crystalline structure and
- layer thickness and excipients (Carstensen, 1974; Reisch, 1979; Hüttenrach *et al.*, 1986; Takács *et al.*, 1990).

6.2.1 *Crystalline state*

Photodegradation of solid drugs depends on whether the state is crystalline or amorphous, as is shown here by the example of ergocalciferol (Fig. 6.1).

The liver drug cianidanol appears in five various crystalline forms. Only the monohydrate II has been shown to be photostable. The other crystalline forms were decomposed by air humidity during irradiation (Akimoto *et al.*, 1985).

Table 6.1 Examples of photoinstable drugs

Antibiotics	Tetracyclines, doxycycline, rifampicin, rifamycin, cephalosporins, chloramphenicol
Cytotoxic agents	Daunorubicin, dacarbazine, vincristine, cisplatin, methotrexate
Chemotherapeutics	Mefloquine, sulfonamids, nitrofurazone, ciprofloxacin
Calcium antagonists	Nifedipine, nitrendipine, nicardipine, verapamil
Vasodilators	Molsidomine
Antihypertensive agents/diuretics	Nitroprussid, furosemide
Psychopharmacological agents	Phenothiazines, haloperidol, tricyclic antidepressants, benzodiazepines
Dermatics	Corticosteroids, antimycotics
Other drugs	Dihydroergotaminemethanesulfonate, colchicine, emetin, ubidecarenone, vitamins

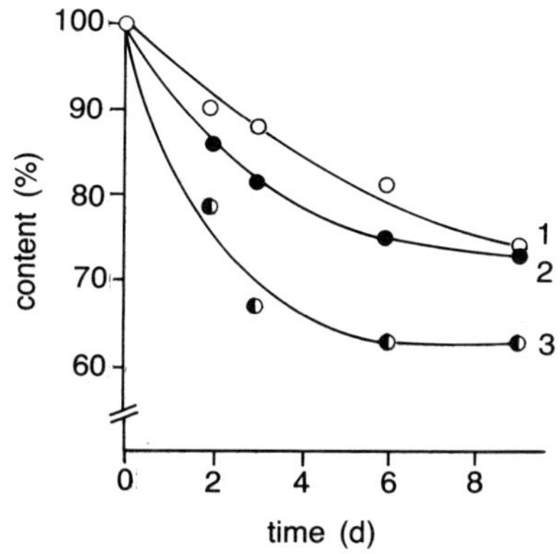

Figure 6.1 Photodegradation of ergocalciferol by daylight (9d) depending on state of order (Hüttenrauch *et al.*, 1986); (1) crystalline; (2) amorphous; (3) amorphous and ground

6.2.2 Modifications

Another factor influencing photodegradation is polymorphism. Furosemide and carbamazepin, for example, showed different rates of photodecomposition depending upon their polymorphous modification (De Villiers *et al.*, 1992; Matsuda *et al.*, 1994; Nyqvist and Wadsten, 1986) (Fig. 6.2). Figure 6.2 shows that only the least discoloured modification (I) is sufficiently photostable.

6.2.3 Dimerization and isomerization

Under certain conditions concerning structure and crystallinity, solid drugs can dimerize under irradiation. Photodimerization has been reported for naphthoquinone, digitoxin and levonorgestrel (Dekker *et al.*, 1968; Reisch *et al.*, 1994).

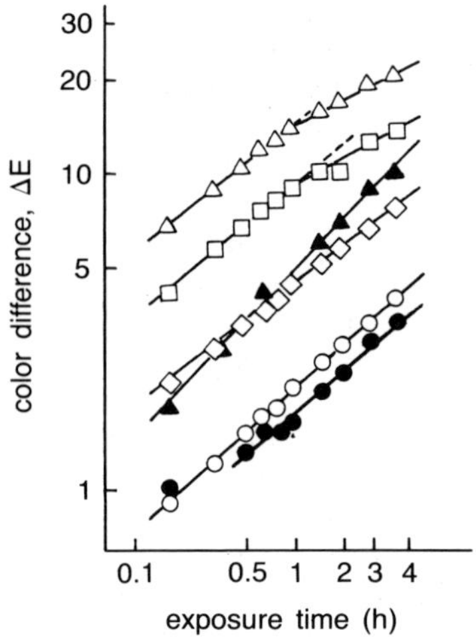

Figure 6.2 Colouration process of furosemide modifications under irradiation (Matsuda and Tatsumi, 1990) ○ Mod. I, tablet; □ Mod. III, tablet; △ DMF solvate; ◇ dioxan solvate; ● Mod. I, powder; ▲ Mod. III powder

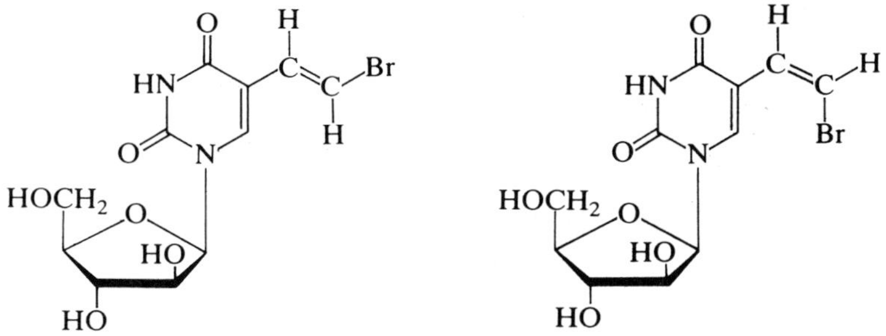

Figure 6.3 Structural formulae of sorivudine and its Z-isomer

The antiviral agent sorivudine decomposes when irradiated with light. The main photodegradation product is the less active Z-isomer (Fig. 6.3) (Desai *et al.*, 1994).

6.2.4 *Decomposition products in the solid state*

The number and structure of photodegradation products can be influenced by the reaction medium. Different results can be obtained by irradiating the drug in the solid state or in solution (Fig. 6.4). Figure 6.4 shows that crystalline aminophenazone is oxidized in the presence of light. In solution, however, other degradation products are yielded under the influence of the solvent. These products are partly photoinstable and can themselves decompose again (Reisch and Abdel-Khalek, 1979).

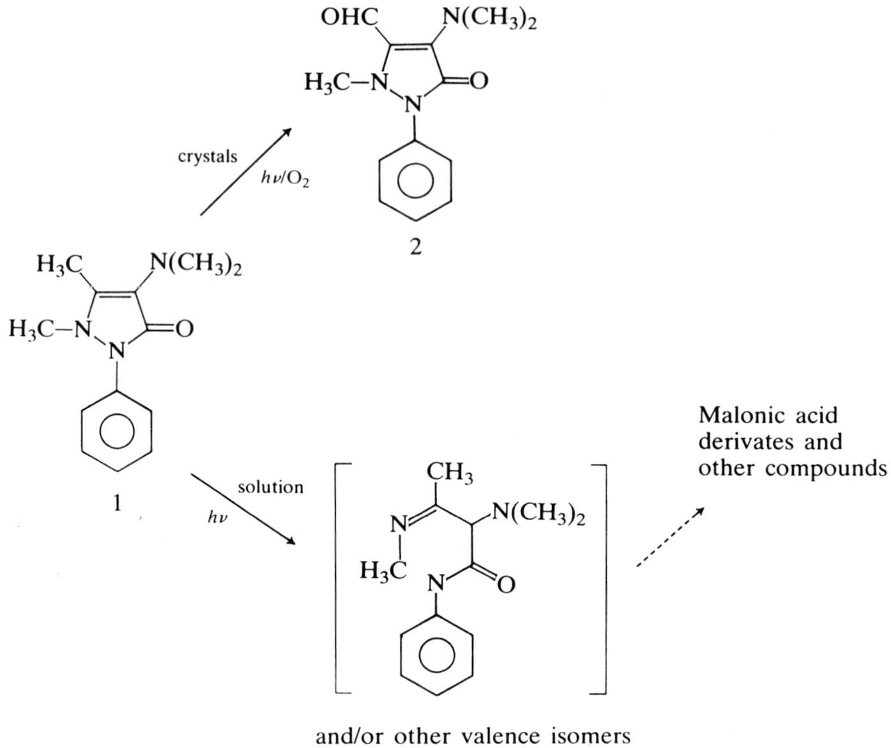

Figure 6.4 Photodegradation of aminophenazone

6.2.5 *Reaction order of photodegradation*

Although interest in chemical degradation processes of photoinstability has markedly grown in recent years, knowledge of photodecomposition of crystalline drugs is poor.

Clarified reaction paths indicate that photodegradation chiefly follows zero, first order or apparent first order kinetics (Reisch and Reisch, 1980; Marciniec, 1983). Investigations of the photostability of ubidecarenone in the solid state show the degree of degradation as a function of the light-absorption properties of the yellow-coloured substrate. The photolytic degradation followed apparent first-order kinetics, promoted by temperature or by irradiation wavelengths (Fig. 6.5). The Arrhenius plot shows activation energy to be different in the solid from the liquid state (Matsuda and Masahara, 1983).

6.3 Photodecomposition of drugs in solid dosage forms and effects on drug safety

In the last 25 years methods have also been developed for the stabilization of drugs in the solid state. The effect of photoinstability on drug safety and the resulting necessity for light protection will be examined through the example of two highly instable drugs, nifedipine and molsidomine.

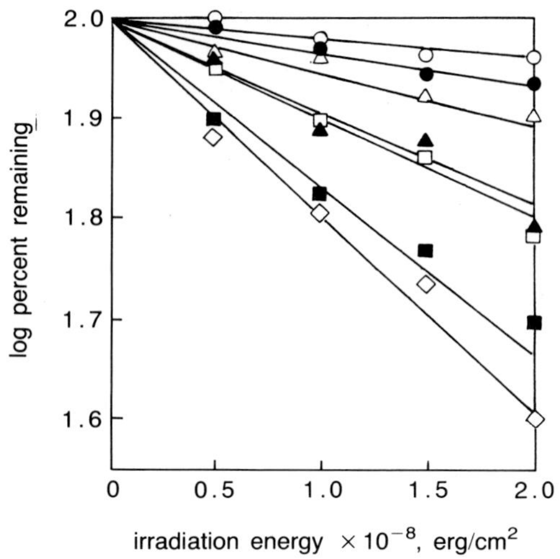

Figure 6.5 Semilogarithmic plot for photolytic degradation profiles of ubidecarenone at various wavelengths: ○ 480 nm; ● 464 nm; △ 426 nm; ▲ 400 nm; □ 373 nm; ■ 347 nm; ◇ 290 nm

6.3.1 ***Photoinstability of dosage forms containing nifedipine***

6.3.1.1 Solutions

Nifedipine in solution shows high photosensitivity in the presence of daylight. Depending on light intensity $t_{90\%}$ is attained within one minute during the month of May (Fig. 6.6) (Thoma and Klimek, 1985a and b). On exposure of nifedipine solutions to daylight the pharmacologically inactive nitrosophenylpyridine derivative and the nitrophenylpyridine derivative result from photodegradation.

Two other decomposition products have been detected in small amounts after irradiation in the solid state, among them the azoxy derivative (Fig. 6.7).

The European Pharmacopoeia and USP XXIII proscribe the presence of nitrosophenylpyridine and nitrophenylpyridine derivative as analysed by HPLC.

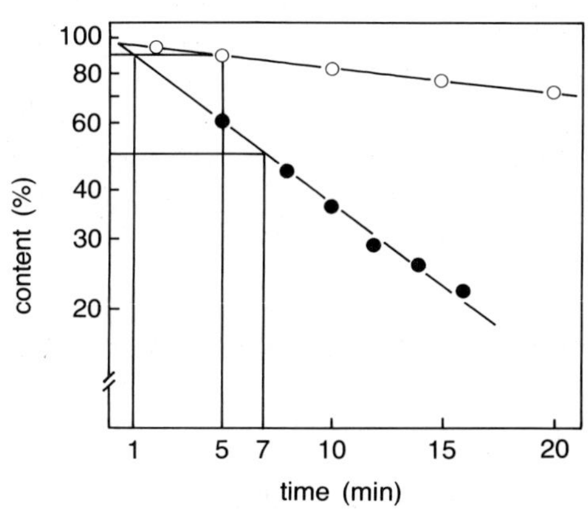

Figure 6.6 Photoinstability of nifedipine solution depending on light intensity: ○ November $t_{90 \text{ per cent}} = 5$ min; ● May $t_{90 \text{per cent}} = 1$ min

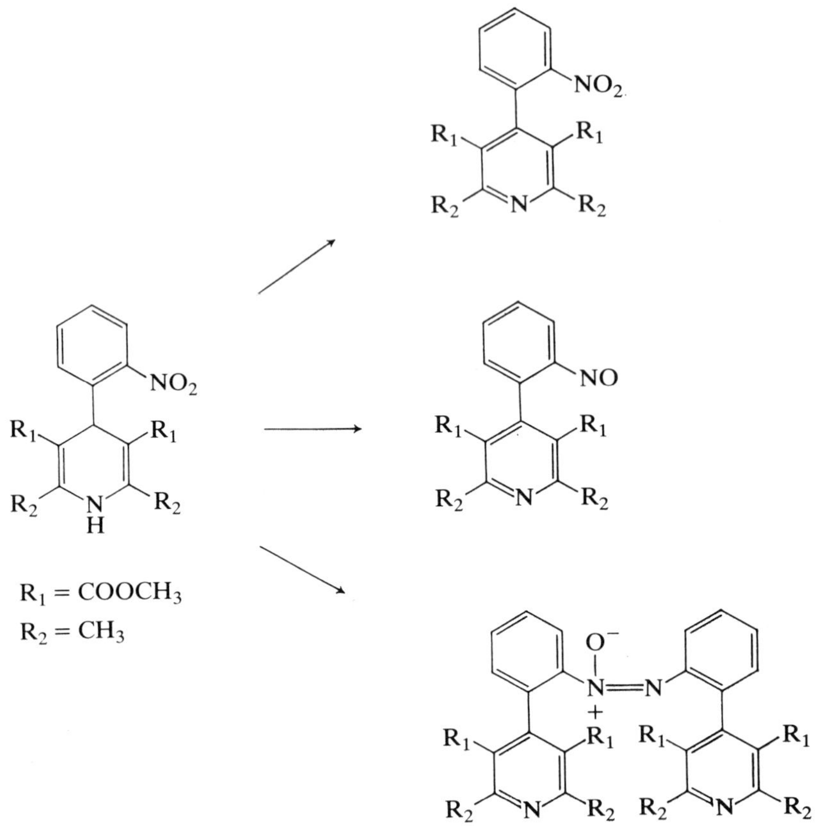

Figure 6.7 Photochemical decomposition products of nifedipine: (1) nitrophenylpyridine derivative; (2) nitrosophenylpyridine derivative; (3) azoxy derivative

6.3.1.2 *Crystalline form*

The photodegradation of nifedipine in the crystalline state and in solution were compared. Within 40 minutes, 20 per cent of the crystalline nifedipine decomposed. Further degradation did not occur during the next 80 minutes. A nifedipine solution decomposed completely during the period (Fig. 6.8) (Thoma and Klimek, 1985). Although nifedipine shows high photosensitivity even in the solid state only few studies have been conducted in regard to its photostability in drug products. Investigations of 60 commercial nifedipine preparations in 1989 lead to the following results (Thoma and Kerker, 1992a):

6.3.1.3 *Protective effect of blisters on coated tablets*

The tablets were irradiated both in and out of the blister. This procedure simulated storage as it is probably performed by the patient. Irradiation in the light cabinet corresponded to exposure to daylight through a window for a period of about 4 to 6 weeks. Outside the blister, the maximum loss of nifedipine content was 5.6 per cent in total. In the blister degradation was limited to a 2 per cent maximum. The strong variations of the content of nitrosophenylpyridine derivative (Fig. 6.9)

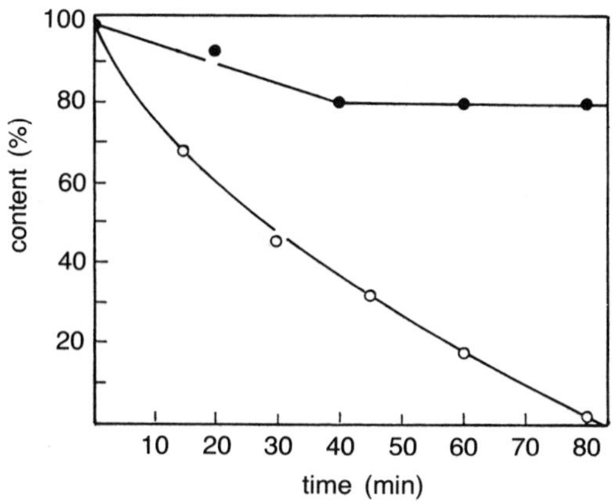

Figure 6.8 Photoinstability of nifedipine crystals (≤5 μm) compared with nifedipine solution :– •– nifedipine crystals; –○– nifedipine solution, c_0 = 3 mg/50 ml

indicate the considerable differences in quality between the various preparations, for example, between T4 and T12.

6.3.1.4 *Protective effects of blisters on soft gelatine capsules*

The layer thickness of the coloured or pigmented coating varied mainly between 20 and 32 μm; the blister foils were all red with the one exception which was yellow. Irradiation of one preparation of nifedipine soft capsules resulted in a 31 per cent loss of content out of the blister and a 26 per cent loss in the blister. The other

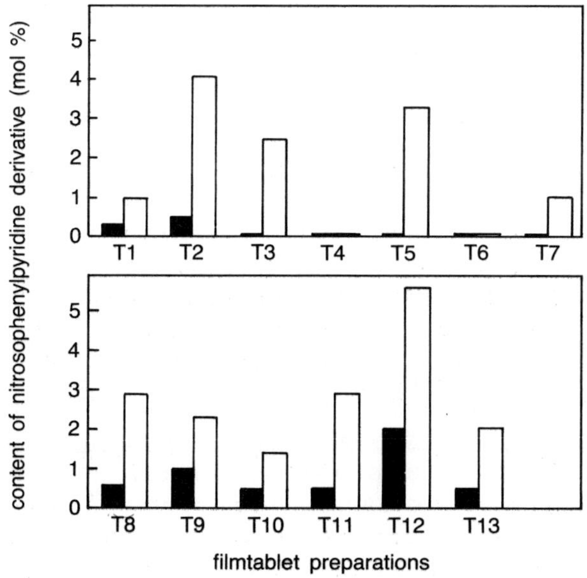

Figure 6.9 Contents of nitrosophenylpyridine derivative in 13 coated nifedipine tablet preparations after 72 h of irradiation in the Novasoltest: ■ in blister; □ out of blister

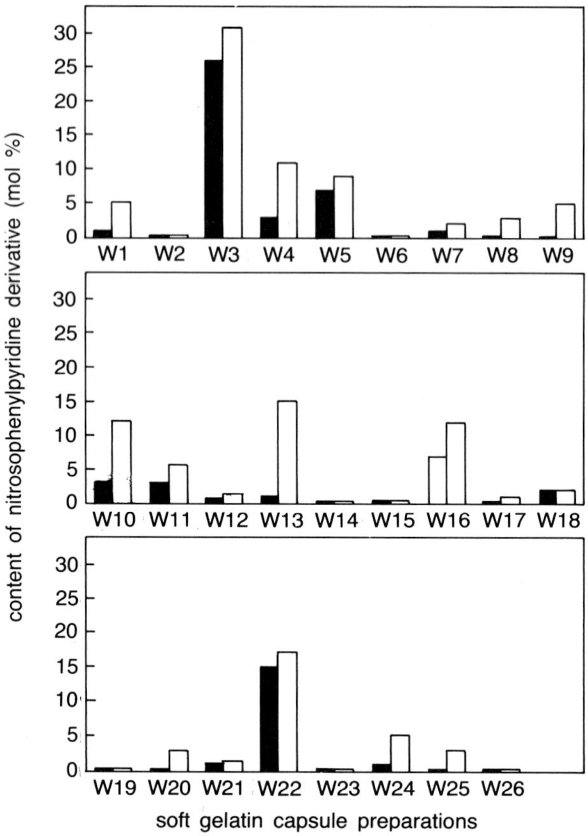

Figure 6.10 Content of nitrosophenylpyridine derivative of 26 nifedipine soft gelatine capsule preparations after 72 h of irradiation in the Novasoltest: ■ in blister; □ out of blister

preparations showed differing stability, but all were of a noticeable quality (Fig. 6.10). Gelatine shells of 13 preparations were either brown or orange coloured, but were stored in clear blisters. A strong correlation of the thickness of the gelatine shell and the photoprotective effect was shown in soft gelatine capsules.

6.3.1.5 *Protective effect of blisters on hard gelatine capsules*

Stabilization of hard gelatine capsules can be obtained by similar methods to those used for film tablets and soft gelatine capsules. Alteration of the colouring/pigmentation and thickness of the gelatine shell and colouring/coating of the capsule's filling all lead to improvement. In nifedipine preparations, hard gelatine capsules are chiefly used in sustained release formulations. Nine of the 13 tested preparations showed controlled drug release; eight of those were filled with pellets (Thoma and Kerker, 1992a).

Before the stability testing, the content of related compounds met the requirements of the USP monograph. The hard gelatine capsules were irradiated under the same conditions by a Novasoltest with Supraxfilter as soft gelatine capsules and film tablets (Thoma and Kerker, 1992a).

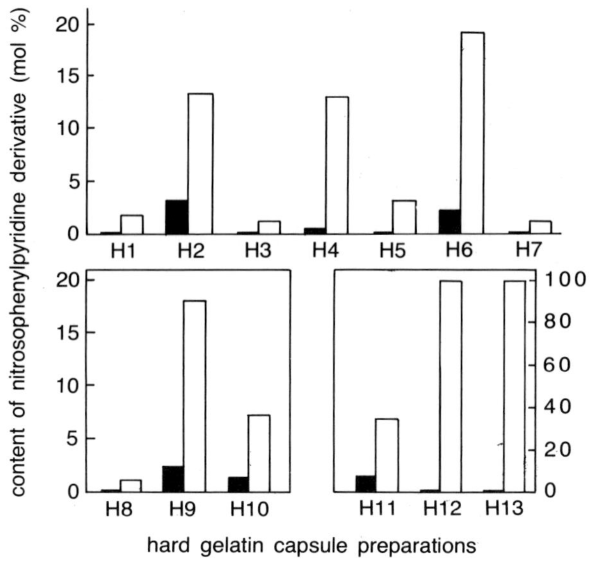

Figure 6.11 Content of nitrosophenylpyridine derivative in 13 hard gelatine capsule preparations after 72 h of irradiation in the Novasoltest with Supraxfilter: ■ with blister; □ without blister

Investigation of the photostability of hard gelatine capsules revealed two preparations irradiated without a blister as having a high degradation rate. These two preparations were produced by the same manufacturer but contained different doses: their shells are faintly coloured and pigmented. Nifedipine was filled to the capsule as a semi-solid solution. The preparations are protected from light by aluminium blisters (Thoma and Kerker, 1992a) (Fig. 6.11). To prove the comparability of the results of artificial irradiation to daylight conditions, the samples were also exposed to daylight. As illustrated in Fig. 6.12, rapid

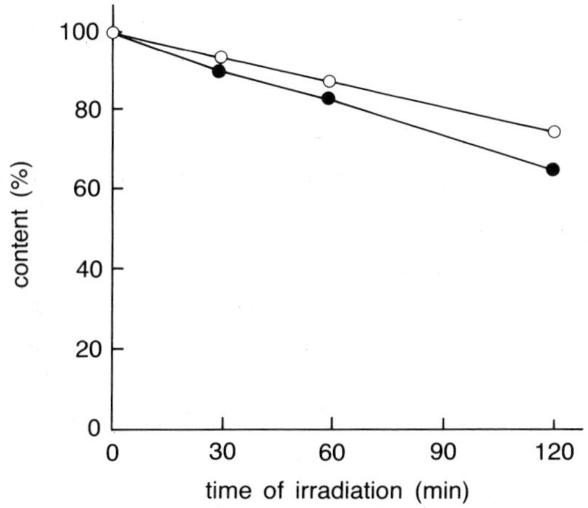

Figure 6.12 Residual Nifedipine content in hard gelatine capsule preparation H12 after irradiating in the Novasoltest and in direct daylight behind window glass: • daylight; ○ Novasoltest

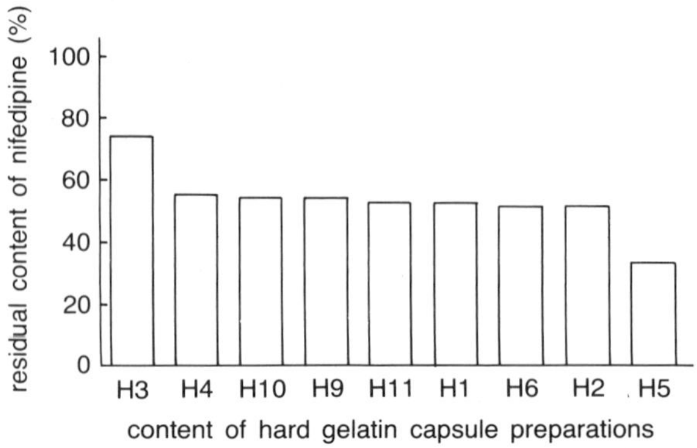

Figure 6.13 Residual nifedipine content of the capsule fillings of hard gelatine capsule preparations after one hour of irradiation in the Novasoltest (filling: pellets, except of preparation H5 (granule))

decomposition occurs under bright daylight. The rate of degradation is higher than in the Novasoltest with Supraxfilter. After one hour, less than 90 per cent nifedipine is detectable (Thoma and Kerker, 1992a). Comparison of the photostability of the other hard gelatine capsule preparations also showed noticeable differences (Fig. 6.11). In three preparations the amount of nitrosophenylpyridine derivative was small: about 1 per cent of the derivative could be detected. In contrast one preparation contained 34 per cent derivative.

To investigate whether the different photostability results depended not only on varying transmission of the gelatine shell but also on the differing stability of the filling's formulation, irradiation tests of the fillings outside the capsules were also conducted (Thoma and Kerker, 1992a) (Fig. 6.13). The granule was more sensitive to light than the pellets. In the case of the pellets, only preparation H3 showed obviously higher photostability. An explanation would be the red colouring of the pellet coating. Preparation H3 is the only preparation with a coating.

6.3.2 *Photoinstability of dosage forms containing molsidomine*

6.3.2.1 Solutions

In 1989 two guttae preparations of molsidomine were taken from the market. Photodegradation leading to morpholine during their use could not be conclusively excluded. Morpholine may react with nitrite in the acidic medium of the stomach, causing toxicologic risks (Fig. 6.14). The photostability of solutions depends on

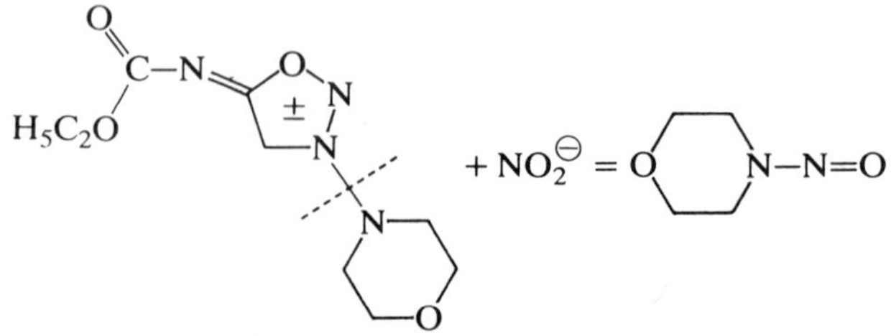

Figure 6.14 Scheme of the photodegradation of molsidomine to morpholine and potential transformation to nitrosomorpholine in the stomach

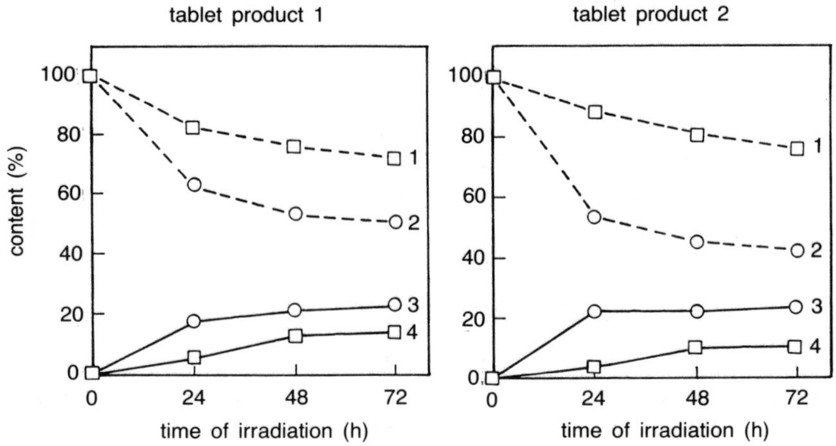

Figure 6.15 Increase of morpholine and decrease of molsidomine content of two tablet preparations in relation to the time of irradiation: –□– morpholine (irradiation in blister); ––□–– molsidomine (irradiation in blister); –○– morpholine (irradiation out of blister); ––○–– molsidomine (irradiation out of blister)

the transmission of the packaging material and the type of the solvent. The extent of photodegradation increased from water to propylene glycol < ethanol < macrogol (Thoma and Kerker, 1992c).

6.3.2.2 Tablets

Molsidomine tablet preparations were irradiated for 72 h (Novasoltest). No morpholine could be detected when the tablets were stored in both blisters and an outer container. These results illustrate the importance of packaging for the stability of molsidomine. In many cases the outer container must be included among the stabilizing precautions.

Figure 6.15 shows the importance of a suitable blister material for the photostability of molsidomine tablets (Thoma and Kerker, 1992c).

6.4 Stabilization of photoinstable drugs in dosage forms

In addition to improving light protection through packaging, photostabilization by other means has become increasingly important. Dosage forms that are exposed to light either during the manufacturing process or during application by the patient, need additional protection. Table 6.2 shows the possible precautions.

Table 6.2 Photostabilization of film tablets and gelatine capsules

Colouring/pigmentation of coatings
Film thickness of coatings
Colouring/pigmentation of the core
Size of granules/powder particles
Size/height of the core
Colouring/pigmentation of the gelatine shell
Thickness of the gelatine shell

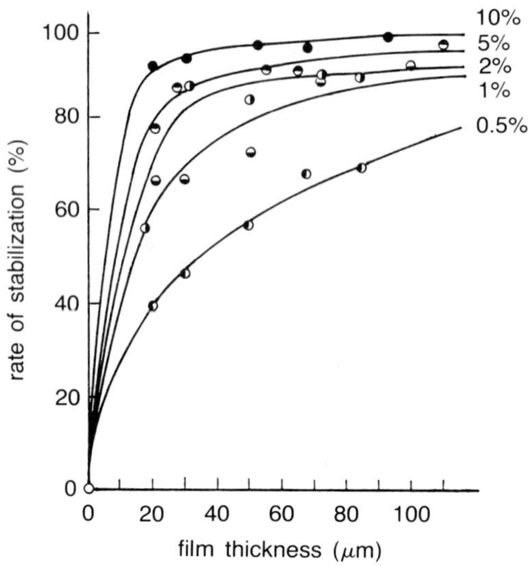

Figure 6.16 Effect of film thickness and concentration of oxybenzone on stabilization rate of coated sulfisomidine tablets (Matsuda *et al.*, 1978)

6.4.1 *Stabilization of film tablets*

6.4.1.1 *Photostabilization of the coating film*

Addition of pigments to coatings, for example, titanium dioxide and iron oxide, as part of light protection has been investigated in previous papers (Nyqvist and Nicklasson, 1982). Iron oxide in particular showed good light-protection qualities.

In studies of the opacity of tablet film coatings that have been conducted (Rowe, 1984a and b), titanium dioxide and iron oxide as well as coloured pigments showed a special improvement in the opacity of tablet film coatings (Matsuda *et al.*, 1978). Accordingly, among film tablets whose coating contained the UV-absorber oxybenzone, coloration and photolytic degradation of sulfisomidine can be prevented. Film thickness and the concentration of the UV-absorber were correspondingly important for the stabilizing effect (Fig. 6.16).

The particular influence of film thickness and the concentration of titanium dioxide was also examined for nifedipine (Béchard *et al.*, 1992).

Nifedipine in film tablets could be completely stabilized by a 60-μm-thick coating if the film suspension contained both 0.7 per cent titanium dioxide and 0.7 per cent tartrazine. The degradation rate constant showed a linear correlation with the per cent transmittance in the wavelength range relating to the photolytic degradation of the drug (Fig. 6.17) (Teraoka *et al.*, 1988).

6.4.1.2 *Photostabilization of the tablet core*

Granulation. Photostability of nifedipine is influenced by the characteristics and the granulation of the drug powder (Ogawa *et al.*, 1990).

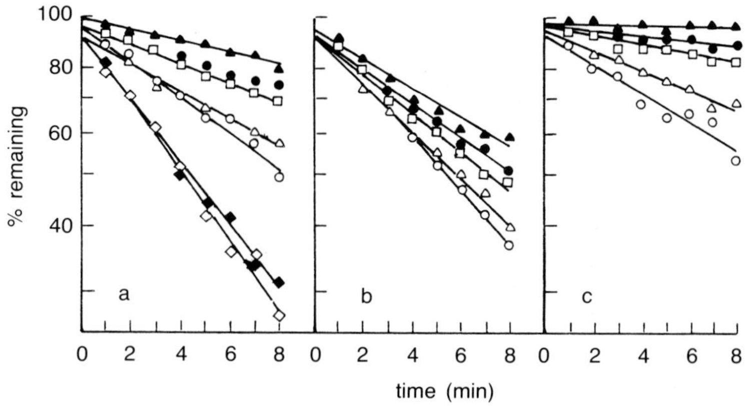

Figure 6.17 Photodegradation of nifedipine coated with film coating 0.5 per cent of tartrazine (a), titanium dioxide (b) and 0.5 per cent of each of these colourants (c) film thickness (◇) 0%/20 μm; (◆) 0%/100 μm; (○) 20 μm; (△) 40 μm; (□) 60 μm; (●) 80 μm; (▲) 100 μm (Teraoka *et al.*, 1988)

Iron oxides. Synthetic iron oxides are potent absorbers of wavelengths below 400 nm with a strong protection capacity. These characteristics of iron oxides were used for photostabilization of sorivudine and nifedipine by incorporation in the tablet core.

A combination of 0.05 per cent red and 0.05 per cent yellow iron oxides into uncoated tablets leads to a better level of light protection than higher concentrations (0.2 per cent) of one single oxide alone (Desai *et al.*, 1994).

UV-absorbers. The UV-absorber 2,4-dihydroxy-benzophenone was among the first compounds investigated as a photostabilizer for tablets. The photostabilizing qualities of the compound were assessed through application to certified dyes (FD&C Yellow No. 5, Blue No. 1 and Red No. 3) (Lachman *et al.*, 1962).

Food Colourants. Addition of light-protecting stabilizers, for example, the food colorant acid yellow (E105), during the granulation process of nifedipine tablets can cause a considerable improvement of stability. As can be seen from Fig. 6.18, unprotected nifedipine tablets show a loss of content of 30 per cent after 10 days in the daylight. Further exposure to light only leads to a minor loss of content. After addition of the light-absorbing dye acid yellow (E105), the loss of content was only about 4 per cent within the first three days (Fig. 6.18) (Thoma and Klimek, 1991a, b and c).

6.4.2 *Stabilization of gelatine capsules*

As mentioned above, photosensitive drugs in gelatine capsules need special protection since light penetrates unprotected capsule shells. In this context the photostability of indomethacin was investigated in relation to the thickness of the capsule shell (50–150 μm) and the concentration of titanium dioxide (0–1.5 per cent) used as opacificient.

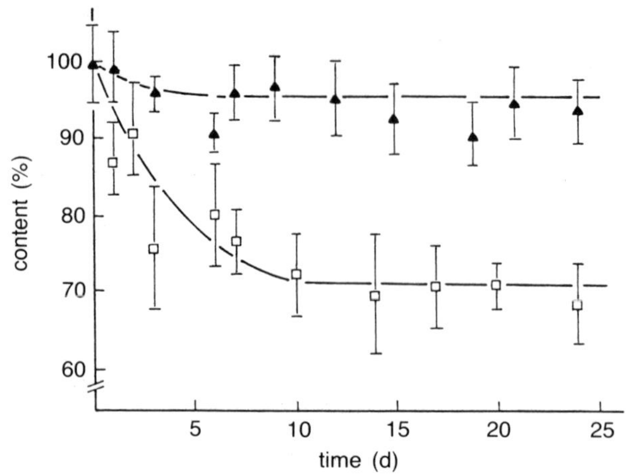

Figure 6.18 Photostabilization of nifedipine (4 mg) in tablets by Acid Yellow (E105) 4 mg/tablet (exposed to daylight, n ≥ 5): □ without dye; ▲ dye in the granule

The colouration rate of the instable indomethacin was directly dependent on the average transmittance through the capsule shell over the wavelength range relating to the photostability of indomethacin (Matsuda *et al.*, 1980). Stabilization experiments of nifedipine in soft gelatine capsules showed that the drug could be stabilized by dyeing the capsule shell with the photostabilizer acid yellow (E105) (Fig. 6.19). Adding the stabilizer to the capsule's filling leads to similar results. As illustrated in Fig. 6.19, the best results are obtained by twofold stabilization of capsule shell and filling (Thoma and Klimek, 1991a, b and c).

Influence of the thickness of the capsule shell. Soft gelatine capsules containing pigments as photo-stabilizers need a shell that is of both sufficient and homogeneous thickness. Lower-shell thickness may occur especially in the seal area.

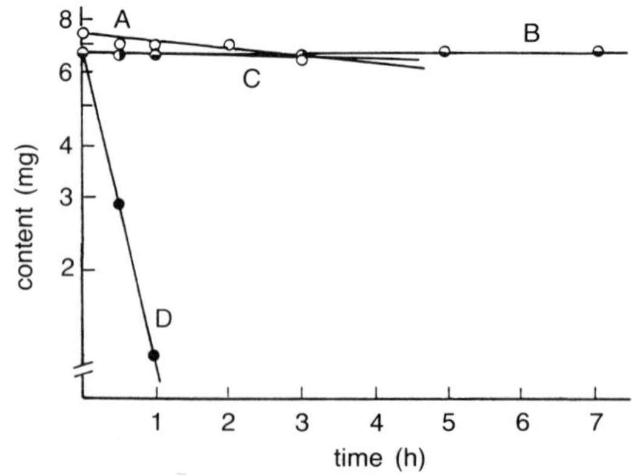

Figure 6.19 Stabilization of nifedipine in soft gelatine capsules by use of Acid Yellow (E105): A (○) Dye in the capsule shell; B (◒) Dye in capsule shell and filling; C (◑) Dye in capsule filling; D (●) without dye

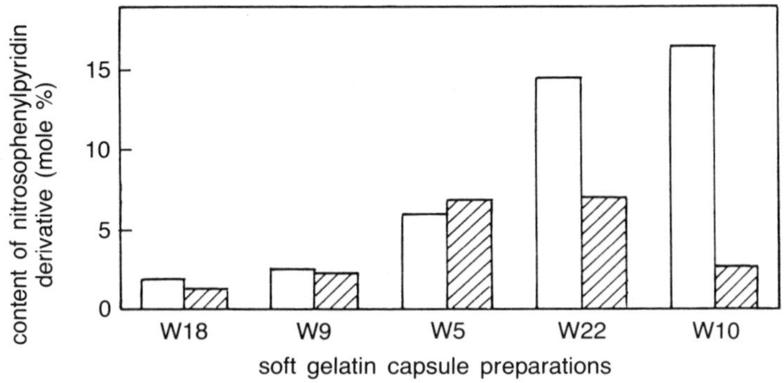

Figure 6.20 Content of nitrosophenylpyridine derivative in soft gelatine capsules after irradiation of variable capsule shell areas: □ in the seal area; (▨) out of the seal area

Figure 6.20 shows that inhomogeneous shell thickness can lead to varying contents of nitrosophenylpyridine derivative. Preparation W10 contains a tenfold amount of degradation product because the shell thickness drops from 260 to 85 μm around the seal (Thoma and Kerker, 1992a).

Stabilization of nifedipine, chloramphenicol, furosemide and clonazepam in soft gelatine capsules. Soft gelatine capsule shells containing 0.4 per cent curcumin have a photoprotective effect for the photolabile drugs nifedipine, chloramphenicol, furosemide and clonazepam. These drugs were dissolved in polyethylene glycol 400. A curcumin content of 0.4 per cent in the capsule shell resulted in a threefold or higher increase in the half-life of the test compounds (Fig. 6.21) (Tønnesen and Karlsen, 1987).

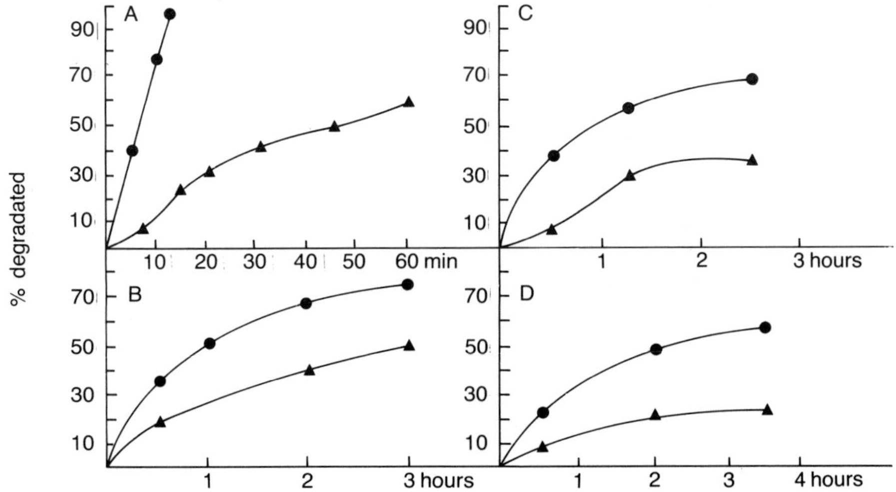

Figure 6.21 Increase of the photostability of nifedipine (A), chloramphenicol (B), furosemide (c), clonazepam (D): ● capsule shell without curcumin; ▲ capsule shell coloured with curcumin

Stabilization of ubidecarenone. Stabilization tests of ubidecarenone were carried out by microencapsulation with each of the three fat-soluble vitamins tocopherol, tocopherol acetate and phytonadione (Matsuda and Teraoka, 1985). Photostability of a redispersible dry emulsion could be improved by the dyes Oil Yellow OB (Y–3) and Oil Red XO (R–5). In contrast to these compounds, riboflavin butyrate accelerated the photodecomposition of the drug. The cause of this phenomenon was ascribed to the excitation of molecules of riboflavin butyrate to higher energy levels by UV irradiation, thereby promoting the decomposition of the drug via the process of fluorescence emission (Takeuchi *et al.*, 1992).

6.4.3 Stabilization of solutions and topical preparations

Protection by spectral overlay with suitable excipients can stabilize different photosensitive drugs and their dosage forms, for example, solutions and hydrogels (Thoma, 1985; Thoma and Klimek, 1991a).

Molsidomine. According to this principle solutions of molsidomine can be stabilized by suitable excipients, for example, yellow orange S, vanillin or flavonoids (Fig. 6.22) (Voegele and Laudenbach, 1985; Thoma and Kerker, 1992c).

Dihydroergotamine. Dihydroergotamine degenerates when irradiated with light. Photooxidation of the indole ring and other parts of the molecule results in an oxo derivative. As this substance shows an absorption maximum between 250 and 320 nm, a protective effect should be achievable by using substances which absorb in this region. This has been established for vanillin and methylgallate (Thoma and Strittmatter, publication in preparation; Thoma and Klimek, 1991c) (Fig. 6.23).

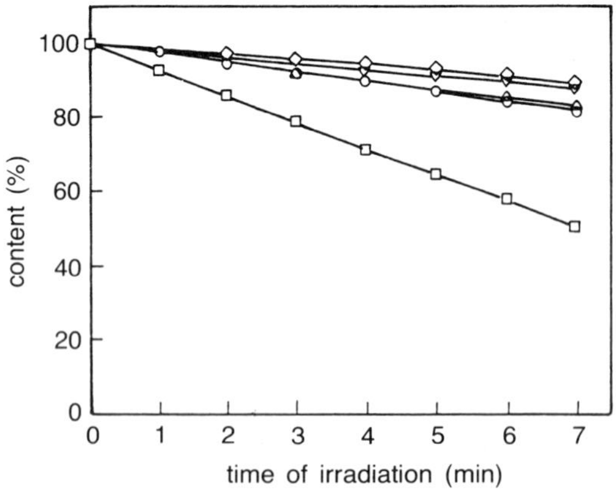

Figure 6.22 Photodegradation of molsidomine after adding flavonoids: □ without additive; ○ with hesperidin 0.2 per cent; ◇ with luteolin 0.2 per cent; ▽ with morin 0.2 per cent; △ with rutin 0.2 per cent

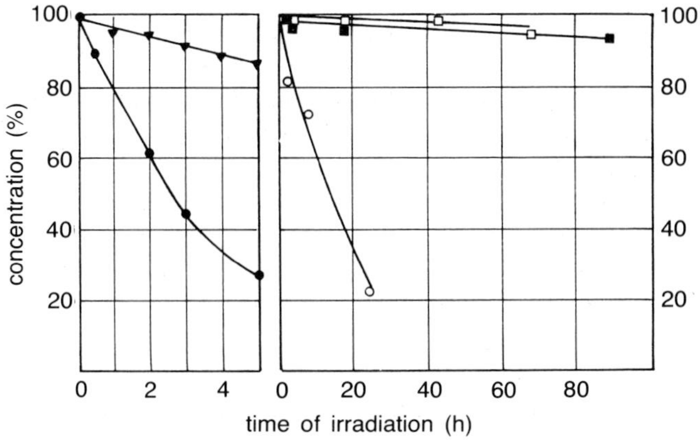

Figure 6.23 Rate of degradation of stabilized dihydroergotamine solutions (light testing cupboard, filtered mercury high-pressure lamp): ● dihydroergotamine methanesulfonate 1 mg/ml in methanol (Stock solution); ▼ stock solution with vanillin added (1 mg/ml); ■ stock solution with methylgallate added (1 mg/ml; ○ dihydergot injection solution (1 mg/ml dihydroergotamine methanesulfonate)); □ dihydergot injection solution with methylgallate added (1 mg/ml)

As is clear from the left-hand section of Fig. 6.23, the concentration of a non-stabilized solution falls within 5 h by about 75 per cent. Following stabilization with vanillin, only about 10 per cent degrade over the same period of time. Stabilization with methylgallate prolongs the $t_{90\%}$ from 5 h to over 100 h.

Addition of stabilizer to a commercial preparation increases the $t_{90\%}$ from less than 5 to more than 70 h. The stabilizing effect of methylgallate is not only due to its photo-protective action but also to its properties as a redoxstabilizer.

Nitrofurazone, furosemide, haloperidol and thiothixene. Further investigations have shown that mixtures of stabilizers can produce even greater photoprotection, if they thereby extend the absorption profile.

The examples given in Table 6.3 concern proprietary medicinal products containing nitrofurazone, furosemide and haloperidol. The $t_{90\%}$ of these agents is increased by a factor of between 4 and 15. For haloperidol and thiothixen, combinations of stabilizers gave the best results (Thoma and Strittmatter, in preparation; Thoma and Klimek, 1991c). These methods can not only be applied for solutions, but also for other dosage forms, tablets and topical preparations.

Stabilization of photolabile drugs in gels. As shown by the following model experiment, it is also possible to stabilize photosensitive gels with acceptable results (polyacrylate gel with curcumin (E100)): The $t_{90\%}$ of nifedipine is only 12 min when exposed without protection to light. When stabilized with curcumin the compound can be handled within safe bounds for up to five times longer (Thoma and Klimek, 1991c).

Table 6.3 Effects of photostabilizers and mixtures on the $t_{90\%}$ of drugs and proprietary medicinal products under photostability testing (mercury high-pressure lamp)

		$t_{90\%}$	
Drug	Photostabilizer	Unstabilized	Stabilized
Nitrofurazone 0.2 per cent (ear drop preparation)	Curcumin 0.05 per cent	6 h	22 h
Furosemide 1 per cent (injection)	Vanillin 1.0 per cent	1.5 h	~7 h
Haloperidol 0.5 per cent (injection)	Benzyl alcohol and vanillin (each 0.5 per cent)	25 h	>120 h
Thiothixene 0.2 per cent	Quinosol and vanillin (each 0.3 per cent)	3 min	45 min

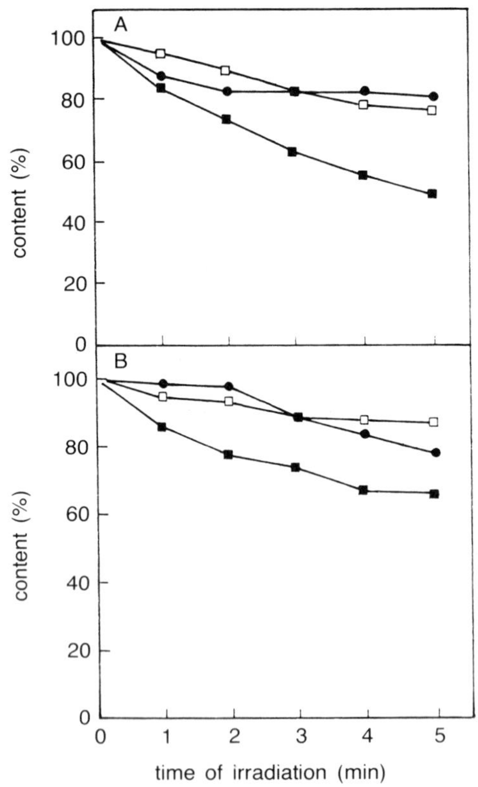

Figure 6.24 Degree of retinoic acid in 0.05 per cent methanolic solution (A) and liposomes (B) (liposomes were made with SLH/Ch 1:1) after irradiation in a light testing box (Novasoltest with Suprax-filter): ■ retinoic acid without stabilizer; □ adding of curcumin (12 mg/100 ml); ● adding of curcumin (24 mg/100 ml)

Stabilization of photolabile drugs in serum samples. Addition of curcumin can protect serum samples with nifedipine so that there is a sixfold increase in half-life. For furosemide and clonazepam in serum curcumin had a slight stabilizing effect (Tønnesen and Karlsen, 1988).

Stabilization of retinoic acid in liposomes. In the presence of sunlight retinoic acid liposomes showed a loss of 23 per cent of drug content after 5 min (Thoma and Lauchert, in preparation). It is also known that a substantial part of topically applied retinoic acid decomposes through photoisomerization on the surface of the skin (Schaefer and Zesch, 1975). In order to protect liposomes containing retinoic acid, curcumin can be added to the drug-loaded vesicle. Investigations revealed an improvement in stability of 17 per cent (Fig. 6.24) (Thoma and Lauchert, in preparation).

Photodegradation and stabilization of a corticosteroid hydrogel on the skin. After irradiation of solutions a decrease in the content of some corticosteroids could be determined. Investigations into possible photodegradation of hydrogels containing corticosteroids were carried out. Measurements were taken after application of an isopropanolic polyacrylate hydrogel containing 0.05 per cent betamethasone-17-valerate to the skin. After 20 minutes of irradiation with Novasoltest and in sunlight the corticosteroid gels showed a 17 per cent greater loss of drug content than tests carried out in darkness. Measurements were taken of the amount of undecomposed betamethasone-17-valerate still detectable after removal from the skin. After addition of 4 per cent of the stabilizing agent 2-phenylbenz-imidazole-5-sulfonic acid to the gel, irradiation was shown to have no effect on the drug content (Thoma and Kerker, 1992b) (Fig. 6.25).

6.5 Principle of photostabilization by spectral overlay with suitable excipients

The degradation kinetics of a photolabile drug show that they depend on both the intensity and spectral distribution of the light source used. It is very often

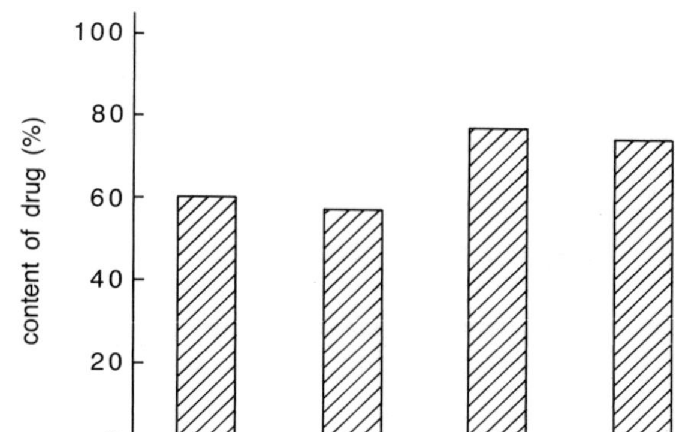

Figure 6.25 Comparison of photodegradation of betamethasone-17-valerate in an isopropanolic polyacrylate hydrogel, residual amounts after 20 minutes of irradiation

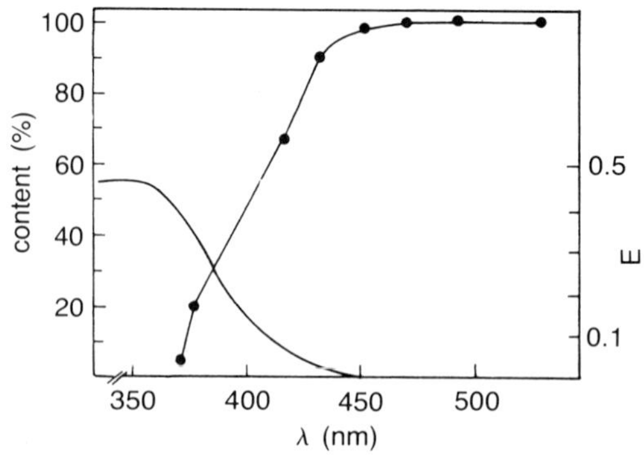

Figure 6.26 Influence of the wavelength of the irradiation light on the photoability of nifedipine: (–•–) dependence of the residual concentration on the wavelength of xenon radiation (left ordinate); (–) long-wavelength section of the nifedipine adsorption spectrum (right ordinate)

imagined that UV light in the energy-rich short-wavelength region is the cause of the degradation of drugs. However, a distinct spectral region of visible light is responsible for the photolysis of nifedipine (Thoma and Klimek, 1991c).

The left-hand side of Fig. 6.26 shows the absorption spectrum of nifedipine in the long-wavelength region between about 350 and 450 nm. The right-hand curve compares the range of stability of the dissolved nifedipine with the irradiation wavelength. In each case, the points represent the lower limit of the wavelengths used. The graphs show that the solution is stable down to a wavelength of 475 nm. Photolysis starts exactly at the point where nifedipine absorption begins at 450 nm. Photolysis increases considerably up to about 400 nm. Nifedipine is thus completely degraded by light in the rather long-wavelength region within 10 minutes. If this wavelength region corresponds to the region of intrinsic instability, it should be possible to protect specifically the problem compound.

Figure 6.27 demonstrates that by using the natural food colorant curcumin, the relevant long-wavelength region of nifedipine's spectrum between 300 and 450 nm is well covered. This does not apply to the short-wavelength region below 300 nm. Addition of curcumin in roughly equimolar proportions leads to remarkably good photostabilization by a factor of 60, relative to the half-life in daylight (Thoma and Klimek, 1991c). Other yellow food colorants can be used to produce similar stabilizing effects. The differences in the effects of stabilization are demonstrated in Fig. 6.28. If one plots the decrease in content versus exposure time to daylight, then Fig. 6.28 shows that three groups can be distinguished, with different stabilization effects: The upper group demonstrates the influence of the yellow colorants, which covers the long-wavelength nifedipine peak very well or reasonably well, the middle group consists of yellow and orange colorants which absorb daylight in this critical region only to a limited extent. The line marked 9 is that obtained with the red colorant Ponceau 4R and Cochineal Red A, which, because it does not absorb at all, shows similar instability values as those for line 10, that is, the unprotected solution of nifedipine (Thoma and Klimek, 1991c). To what extent can such a protection principle be applied to other drugs?

It is feasible to stabilize the highly light-sensitive, red-coloured cytostatic daunorubicin using red food colorants or other appropriate colorants (Thoma and

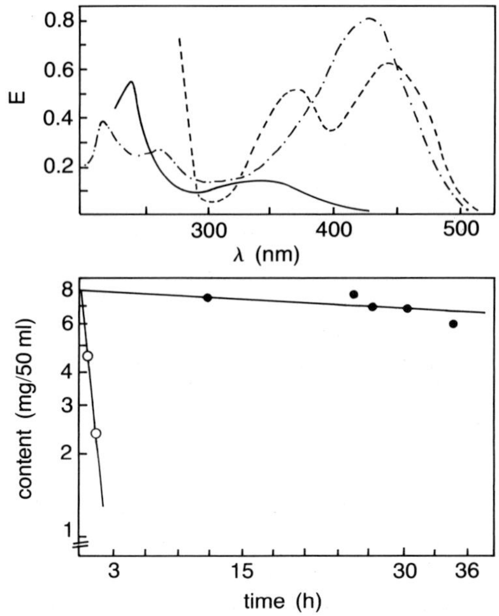

Figure 6.27 Light absorption of curcumin (●–●–●) 0.7 mg per cent and riboflavine-5′-phosphate Na (---) 2 mg per cent; nifedipine (–) 1 mg per cent (upper panel). Stabilization of nifedipine solutions with curcumin (E100): (-●-) curcumin (3.2×10^{-4} mol/litre), molar proportion 1:0.7; $t'_{50} = 5.5$ days; (–○-) without curcumin (lower panel)

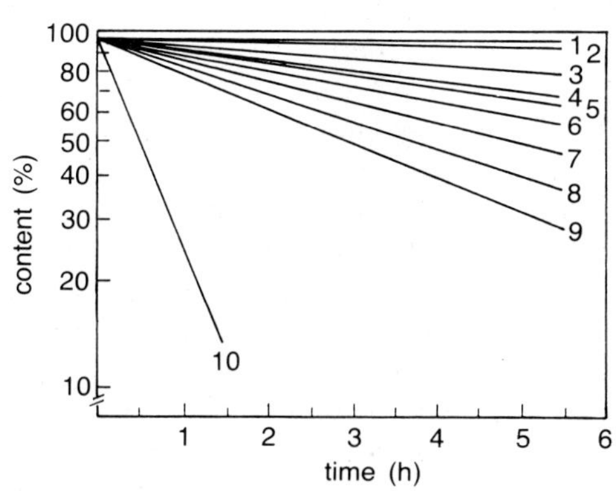

Figure 6.28 Comparative investigation of the stabilization of nifedipine solutions with food colourants (radiation source: Spectrotest, $c_0 = 4.6 \times 10^{-4}$ mol/litre): (1) Curcumin (3.2×10^{-4} mol/litre = 6 mg/50 ml); (2) Fast Yellow (2.9×10^{-4} mol/litre = 6 mg/50 ml; (3) Chrysoine (3.7×10^{-4} mol/litre = 6 mg/50 ml); (8) Apocarotinal (6 mg/50 ml); (9) Cochineal Red A (2.0×10^{-4} mol/litre = 6 mg/50 ml); (10) nifedipine solution without additive

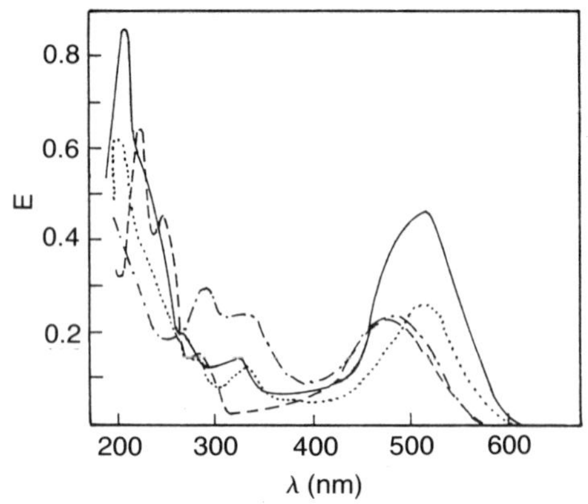

Figure 6.29 Absorption properties of the stabilizers of daunorubicin (c = 2 mg/100 ml): —— amaranth (E123); . . . . . ponceau 6R (E126; ·–·– scarlet GN (E125); –––– daunorubicin

Klimek, 1991b). These absorption spectra clearly show that the intense red colour of daunorubicin is based on its light absorption in the region of 370–570 nm (Fig. 6.29). Figure 6.29 also demonstrates that this absorption peak in the long-wavelength region is almost exactly overlapped by the food colorant Scarlet GN. The compounds Amaranth and particularly Ponceau leave the left flank of the peak partially unprotected. Thus Scarlet GN should provide very good stabilization. With the other two food colorants, stabilization will be less reliable. Tartrazine, a yellow colorant that absorbs only in the short-wavelength region, was also included just for comparison. Under the given irradiation conditions, the unprotected daunorubicin solution is quickly degraded. The yellow colorant tartrazine prolongs the $t_{90\%}$ to an insignificant degree. Ponceau and Amaranth produce better stabilization (Fig. 6.30).

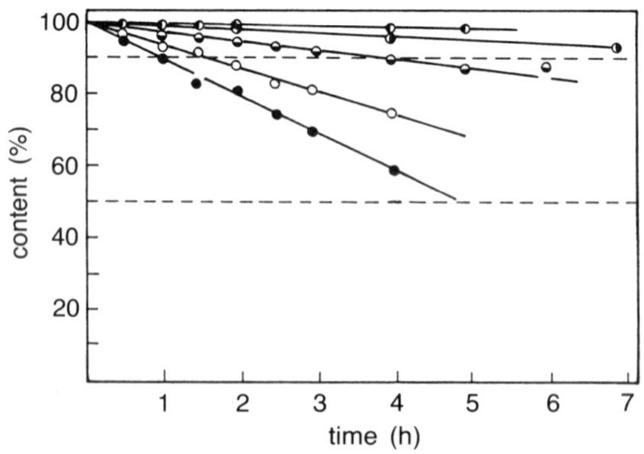

Figure 6.30 Stabilization of daunorubicin by colourants (radiation source: Spectrotest without filter = 700 – 290 nm; c_0 = 2 mg/50 ml): (◐) Scarlet GN (E125), $t_{90\%}$ = 19.5 h; (◑) Amaranth (E123), $t_{90\%}$ = 10.5 h; (◒) Ponceau 6 R (E126), $t_{90\%}$ = 4 h; (○) Tartrazine (E102), $t_{90\%}$ = 1.5 h; (●) without colourant, $t_{90\%}$ = 1 h

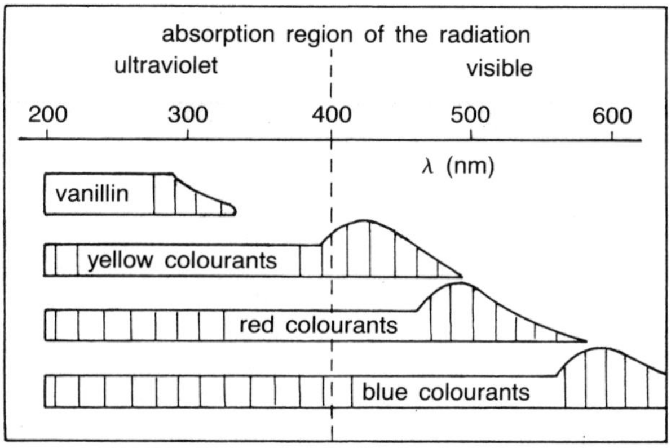

Figure 6.31 Principle of photostabilization through spectral overlay with absorbing excipients

However, the half-life is doubled again to 20 times the initial value with the colouring agent Scarlet GN. This is the only colorant in this test that perfectly covers the crucial region of the daunorubicin spectrum. The technique described above of photoprotection for light-sensitive drugs (Thoma and Klimek, 1991c) is based on finding suitable stabilizers with absorption spectra that overlap that of the respective drug.

In the ultraviolet region, these may be substances with benzene rings and suitable ligands; with yellow, red or blue drugs, suitable food colorants have proved effective (Fig. 6.31) (Thoma and Klimek, 1981). If necessary, photoprotection of highly light-sensitive drugs can actually be incorporated into the dosage form before further protection is given by packaging materials.

6.6 Improvement of photochemical stability of drugs by cyclodextrin complexation

Cyclodextrins (CyD) have aroused interest because of their ability to modify the physical and chemical properties of drugs by inclusion complexation. The extent of photodegradation of benzaldehyde (BA) has been significantly reduced by inclusion complexation with molar ratios BA/α-CyD 1:1, BA/β-CyD 3:2 and BA/γ-CyD 2:1 in water and in the solid phase (Uekama *et al.*, 1983b). The photostability of emetine and cephaeline in aqueous solution was improved by the complexation with γ- and 2,6-dimethyl-β-cyclodextrin. However photodegradation was intensified by complexation with β-cyclodextrin (Teshima *et al.*, 1989).

Photosensitivity of clofibrate was markedly reduced by solid complexation 1:1 with β- and γ-cyclodextrins (Fig. 6.32) (Uekama *et al.*, 1983b). Photodecomposition of nifedipine, hydrochlorothiazide, pyridoxine hydrochloride, clofibrate and retinol acetate in aqueous solution and in solid state decreased by adding β-cyclodextrin or other β-cyclodextrin derivatives (Tomono *et al.*, 1988). Table 6.4 shows that there is a slightly higher protecting effect for pyridoxine HCl and furosemide.

In solution as well as in the solid state there are significant differences of

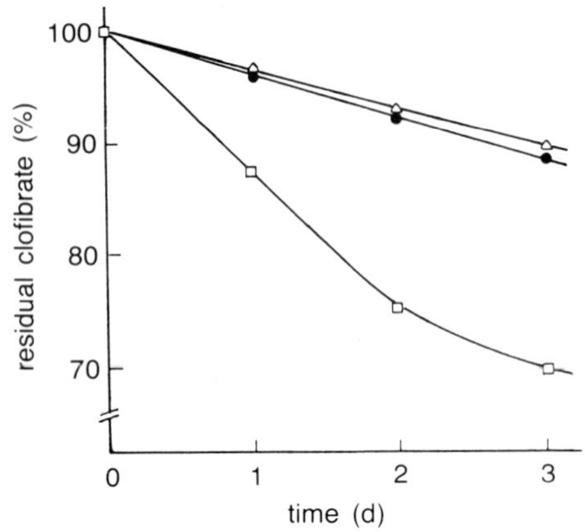

Figure 6.32 Photodegradation curves of clofibrate and its cyclodextrin complexes at 30 °C: □ clofibrate; ● clofibrate-β-cyclodextrin complex; △ clofibrate-γ-cyclodextrin complex

Table 6.4 Effect of various β-cyclodextrins on photodecomposition of drugs in the solid state: per cent of drug remaining after irradiation given below

CyD	Nifedipine 6 h	Furosemide 2 h	Pyridoxine HCl 6 h	Nitroglycerin 8 h
pure drug	71.1	68.1	37.4	99.0
β-CyD	56.0	87.2	74.3	99.0
CDPS	63.8	79.0	57.5	98.7
DM-β-CyD	74.2	98.0	79.2	99.2
TM-β-CyD	59.0	99.2	88.0	99.0

stabilizing effect between β-cyclodextrin, dimethyl-β-cyclodextrin, trimethyl-β-cyclodextrin and cyclodextrin polymer. In order to analyse the photostabilizing effect of cyclodextrin for nifedipine in the solid state, spray-dried complexes (1:1) with α-cyclodextrin, β-cyclodextrin, and γ-cyclodextrin were produced. A physical mixture of nifedipine and HP-β-cyclodextrin (1:1) was used as a comparison. The results show that the photostabilizing effect of complexation is low: the physical mixture contained 48 per cent nifedipine after irradiation lasting 30 minutes (Suntest CPS, UV-filter), while the spray-dried drug–cyclodextrin–complexes contained between 52 and 55 per cent (Thoma and Kübler in press). Complexation of molsidomine with α-cyclodextrin caused no improvement in photostability (Thoma and Kübler, in press).

6.7 Photostability of excipients

Compared with photosensitive drugs most excipients are less sensitive to light. However, far fewer studies concerning excipients have been conducted. There has also been little research into the possible catalytic influence of excipients on the photodegradation of drugs.

Dyes in coatings of tablets. A large number of dyes for tablets and tablet coatings was tested for the light-induced rate of fading (Hajratwala, 1985). In these investigations zero- and first-order reactions were found to have taken place. These investigations have provided information about the phenomenon of accelerated degradation processes under the influence of other excipients, e.g. tablet fillings and non-ionic tensids (Kuramoto *et al.*, 1958; Scott *et al.*, 1960). According to these results dyes may catalyse the degradation of other dyes (Evenson, 1939).

Dyes as photostabilizers in dosage forms. The photostability of the dyes themselves must be considered when they are used as spectral stabilizers. Synthetic azo-dyes are usually the more stable. Natural dyes, for example, riboflavin, curcumin or carotinoids, may be photosensitive depending on their medium (Kläui *et al.*, 1963). Riboflavin in solution degrades rapidly in daylight ($t_{50\%}$ = 170 min). This limits the general use as a photostabilizer, for example, for nifedipine (Klimek, 1978).

Curcumin shows photodecomposition under UV/visible radiation, both in solution and in the solid state. The main degradation component is a cyclization product, formed by loss of two hydrogen atoms. In combination with sensitizers like methylene blue, curcumin shows catalytic fading (Tønnesen *et al.*, 1986).

Other excipients. These are mainly excipients whose oxidizability is catalytically caused, or accelerated by light influence. Among these are all lipophilic compounds incorporating unsaturated fatty acids, for example, fat oils and liquid waxes. These must be stored under light protection. Another group is that of certain preservatives, for example, sorbic acid, thiomersal and phenoxyethanol (Thoma, 1986).

6.8 Present situation of photostability testing in the solid state

After completion of the manuscript the following topical requirements were announced at the ICH 3 in November 1995 and have been signed off at Step 2 of the ICH Process

Announcement of the ICH 3:
Stability Testing: Photostability Testing

Light Sources:	Spectral output similar to the D65/ID65 emission standard of CIE e.g. xenon or metal halide lamps, a cool white fluorescent lamp in combination with a UV lamp.
Radiant Exposure:	overall illumination $\geqslant$ 1.2 mlux-h near UV energy $\geqslant$ 200 W·h/m^2

Addition of dark controls (e.g. wrapped in aluminium foil)
Cooling of the samples

Drug Substance:	Liquid drug substances in chemically inert and transparent containers Solid drug substances in suitable glass or plastic dishes, layer thickness $\leqslant$ 3 mm
Drug Product:	Irradiation beginning with the fully exposed drug product, then in the immediate pack and finally in the marketing pack, if necessary.

In contrast to testing methods for thermal stability, the testing of photostability has still not been standardized. Several authors have called for standardization of photostability testing (Tønnesen, 1991; Anderson *et al.*, 1991; Tønnesen and Karlsen, 1993; Tønnesen and Moore, 1993; Thoma and Kerker, in preparation). The International Conference of Harmonization (ICH) (Yoshioka, 1993) and the European PA/SG Working Party 'Light Stability Testing' have also turned their attention towards this matter.

At this point it might be useful to give a survey of methods used for the testing of the photostability of drugs and dosage forms in the solid state, which are common in industrial practice. In 1995, 20 German manufacturers in the pharmaceutical industry were asked for details about the methods they use for testing of light stability. An evaluation of the answers gave the following results:

Sources of light. In 73 per cent of the testing methods xenon lamps are used (often adjusted with window glass filters). Fluorescent tubes (13 per cent) and daylight (9 per cent) are both becoming less important. Metal halide burners (Novasoltest) are used in 4.5 per cent of the cases. The specified light intensity varies. 585 W/m^2 (90 kLx) are usual in the case of xenon lamps.

Distance between sample and lamp. The distance between sample and lamp is determined by the apparatus. It varies between 20 and 90 cm. It was about 20 cm in the Suntest, the most commonly used light cabinet.

Duration of irradiation. How long the sample is irradiated differs considerably depending on the source of light, from between 1 and 3 months in the case of daylight to 2–72 hours in the experiments conducted with xenon lamps. Most commonly testing with xenon lamps lasts between 24 and 72 h.

Sample vessels. Examinations are chiefly carried out in petri dishes or other glass vessels with a diameter of 5–10 cm. Sometimes they are covered with foils or glass plates. Quartz vessels or the primary packaging are less frequently used.

Treatment of samples. With the exception of one case, all firms irradiate tablets on one side only. Powders are examined with a layer thickness varying from 1 to 5 mm. In about 40 per cent of the cases layer thickness is 1–2 mm. During the testing the surface is not changed.

Temperature in the testing apparatus and at the sample tray. Temperatures in the testing apparatus can vary considerably from 20–30 °C to 40–50 °C. Temperature on the surface of the samples was not determined at these photostability tests.

Reported influence of excipients on photostability. Although normally absorbing excipients in film coatings have a photoprotective action, in some cases negative effects have been reported, for example, the formation of peroxides, influence through buffer substances and discoloration of aromatic ingredients. Polymorphism of drugs or sensitization through excipients has not been observed during the practical tests.

6.9 Conclusion

As the present situation of photostability testing in the solid state shows, a considerable convergence among testing methods has taken place in recent years. Nevertheless, the information received from the manufacturers reveals also the necessity for establishing guidelines for the testing of light stability.

Acknowledgements

The author thanks Mr H. Spilgies, Mr W. Aman and Mrs I. Daxeder, Munich, for very good cooperation during the preparation of this chapter.

References

AKIMOTO, K., INOUE, K. and SUGIMOTO, I., 1985, Photostability of several crystal forms of cianidanol, *Chem. Pharm. Bull.*, **33**, 4050–3.

ANDERSON, N. H., JOHNSTON, D., MCLELLAND, M. A. and MUNDEN, P., 1991, Photostability testing of drug substances and drug products in UK pharmaceutical laboratories, *J. Pharm. Biomed. Anal.*, **9**, 443–9.

BÉCHARD, S. R., QURAISHI, O. and KWONG, E. 1992, Film coating: effect of titanium dioxide concentration and film thickness on the photostability of nifedipine, *Int. J. Pharm.*, **87**, 133–9.

CARSTENSEN, J.T., 1974, Stability of solids and solid dosage forms, *J. Pharm. Sci.*, **63**, 1–14.

DEKKER, J., VAN VUUREN, P. J. and VENTER, D. P., 1968, Photodimerization: I. The syn and anti Photodimers of 1,4-Naphthoquinone, *J. Org. Chem.*, **33**, 464–6.

DESAI, D. S., ABDELNASSER, M. A., RUBITSKI, B. A. and VARIA, S. A., 1994, Photostabilization of uncoated tablets of sorivudine and nifedipine by incorporation of synthetic iron oxides, *Int. J. Pharm.*, **103**, 69–76.

EVENSON, O. L., 1939, Effect of light on erythrosin and bromoacid tetrabromofluorescein, *J. Assoc. Official Agr. Chemists*, **22**, 773–5.

HAJRATWALA, B. R., 1985, Stability of colours, *STP Pharma*, **1** (6), 539–44.

HÜTTENRAUCH, R., FRICKE, S. and KNOP, M., 1986, Bedeutung des molekularen Ordnungszustands für die Lichtempfindlichkeit fester Wirkstoffe, *Pharmazie*, **41**, 664–5.

KERKER, R., 1991, Untersuchungen zur Photostabilität von Nifedipin, Glucocorticoiden, Molsidomin und ihren Zubereitungen, Dissertation, München.

KLÄUI, H., CAPEDER, A. and MÜNZEL, K. 1963, Wasserlösliche Karotinoidpräparate als Drageefarbstoffe, *Pharm. Ind.*, **25**, 173–8.

KLIMEK, R., 1978, Untersuchungen zur Stabilitätskinetik und Stabilisierung photoinstabiler Arzneistoffe, Dissertation, Frankfurt a.M.

KURAMOTO, R., LACHMAN, L. and COOPER, J., 1958, A study of the effect of certain pharmaceutical materials on color stability, *J. Am. Pharm. Assoc. Sci. Ed.*, **47**, 175–80.

LACHMAN, L., URBANYI, T., WEINSTEIN, S., COOPER, J. and SWARTZ, C. J., 1962, Color Stability of Tablet Formulations V Effect of Ultraviolet Absorbers on the Photostability of Colored Tablets, *J. Pharm. Sci.*, **51**, 321–6.

MARCINIEC, B., 1983, Photochemical Decomposition of Phenazone Derivatives Part 2: Kinetics of Decomposition in the Solid State, *Pharmazie*, **38**, 848–50.

MATSUDA, Y., AKAZAWA, R., TERAOKA, R. and OTSUKA, M., 1994, Pharmaceutical Evaluation of Carbamazepine Modifications: Comparative Study for Photostability of Carbamazepine Polymorphs by using Fourier-transformed Reflection-absorption Infrared Spectroscopy and Colorimetric Measurement, *J. Pharm. Pharmacol.*, **46**, 162–7.

MATSUDA, Y., INOUYE, H. and NAKANISHI, R., 1978, Stabilization of Sulfisomidine Tablets by Use of Film Coating Containing UV Absorber: Protection of Coloration and Photolytic Degradation from Exaggerated Light, *J. Pharm. Sci.*, **67**, 196–201.

MATSUDA, Y., ITOOKA, T. and MITSUHASHI, Y., 1980, Photostability of Indomethacin in Model Gelatin Capsules: Effects of Film Thickness and Concentration of Titanium Dioxide on the Coloration and Photolytic Degradation, *Chem. Pharm. Bull.*, **28**, 2665–2671.

MATSUDA, Y. and MASAHARA, R., 1983, Photostability of solid-state ubidecarenone at ordinary and elevated temperatures under exaggregated UV irradiation, *J. Pharm. Sci.*, **72**, 1198–203.

MATSUDA, Y. and TATSUMI, E., 1990, Physicochemical characterization of furosemide modifications, *Int. J. Pharm.*, **60**, 11–26.

MATSUDA, Y. and TERAOKA, R., 1985, Improvement of the photostability of ubidecarenone microcapsules by incorporating fat-soluble vitamins, *Int. J. Pharm.*, **26**, 289–301.

NYQVIST, H. and NICKLASSON, M., 1982, Studies on the physical properties of tablets and tablet excipients. V. Film coating for protection of a light-sensitive tablet-formulation, *Acta Pharm. Suec.*, **19**, 223–8.

NYQVIST, H. and WADSTEN, T., 1986, Preformulation of Solid Dosage Forms: Light Stability Testing of Polymorphs as a Part of a Preformulation Program, *Acta Pharm. Technol.*, **32**, 130–2.

OGAWA, S., ITAGAKI, Y., HAYASE, N., TAKEMOTO, I., KASAHARA, N., AKUTSU S. and INAGAKI, S., 1990, Photostability of Nifedipine in Powder, Obtained by Crushing Tablet, Granule of Fine-Granule, *Jpn. J. Hosp. Pharm.*, **16**, 189–97.

REISCH, G. und REISCH, J., 1980, Gitterkontrollierte Norrish III-Spaltung des kristallinen Methadons, *Pharmazie*, **35**, 402–4.

REISCH, J., 1979, Topochemische Lichtreaktionen an Arzneistoffen und Arzneizubereitungen, *Dtsch. Apoth. Ztg.*, **119**, 1–4

REISCH, J. und ABDEL-KHALEK, M., 1979, Zur Fotooxidation von kristallinem Aminophenazon, *Pharmazie*, **34**, 408–10.

REISCH, J., ZAPPEL, J., RAO, A. R. R. and HENKEL, G., 1994, Dimerisation of levonorgestrel in solid state ultraviolet light irradiation, *Pharm. Acta Helv.*, **69**, 97–100.

ROWE, R. C., 1984, The measurement of the opacity of tablet film coatings *in-situ*, *Acta Pharm. Suec.*, **21**, 201–4.

1984, The opacity of tablet film coatings, *J. Pharm. Pharmacol.*, **36**, 569–72.

SCHAEFER, H. and ZESCH, A., 1975, Penetration of vitamin A acid into human skin, *Acta Derm. Venereol.* (Stockholm), Suppl. 74, 50–5.

SCOTT, M. W., GOUDIE A, J. and HUETTEMAN, A. J., 1960, Accelerated color loss of certified dyes in the presence of nonionic surfactants, *J. Am. Pharm. Assoc. Sci. Ed.*, **49**, 467–72.

TAKÁCS, M., REISCH, J., GERGELY-ZOBIN, A. and GÜCER-EKIZ, N., 1990, Zur Untersuchung der Lichtempfindlichkeit von Festsubstanzen, *Sci. Pharm.*, **58**, 289–97.

TAKEUCHI, H., SASAKI, H., NIWA, T., HINO, T., KAWASHIMA, Y., UESUGI, K. and OZAWA, H., 1992, Improvement of photostability of ubidecarenone in the formulation of a novel powdered dosage form termed redispersible dry emulsion, *Int. J. Pharm.*, **86**, 25–33.

Teraoka, R., Matsuda, Y. and Sugimoto, I., 1988, Quantitative design for photostabilization of nifedipine by using titanium dioxide and/or tartrazine as colourants in model film coating systems, *J. Pharm. Pharmacol.*, **41**, 293–7.

Teshima, D., Otsubo, K., Higuchi, S., Hirayama, F., Uekama, K. and Aoyama, T., 1989, Effects of cyclodextrins on degradations of emetine and cephaeline in aqueous solution, *Chem. Pharm. Bull.*, **37**, 1591–4.

Thoma, K., 1978, Arzneimittelstabilität *Werbe- und Vertriebsgesellschaft Deutscher Apotheker mbH*, Frankfurt a.M.

1985, Stability of Drugs – Current problems in pharmaceutical technology, Kongressband, 10th Conference on Pharmaceutical Technology, Shirakabako, Japan, pp. 1–20.

1986, Wechselwirkungen zwischen Konservierungsmitteln und Rezepturbestandteilen, in *Deutsche Gesellschaft der Kosmetik-Chemiker e.V.* (Hrsg.) *Kreuznacher Symposium 1986*, Konservierung kosmetischer Mittel, pp. 53–78, Verlag für chem. Industrie H. Ziolkowsky KG, Augsburg 1987.

Thoma, K. and Kerker, R. 1992a, Photoinstabilität von Arzneimitteln 3. Mitteilung: Photoinstabilität und Stabilisierung von Nifedipin in Arzneiformen, *Pharm. Ind.*, **54**, 359–65.

1992b, Photoinstabilität von Arzneimitteln 5. Mitteilung: Untersuchungen zur Photostabilisierung von Glukokortikoiden, *Pharm. Ind.*, **54**, 551–4.

1992c, Photostabilität von Arzneimitteln 6. Mitteilung: Untersuchungen zur Stabilitt von Molsidomin, *Pharm. Ind.*, **54**, 630–8.

Proposal of a monograph describing the photochemical stability testing of drugs and drug formulations publication in preparation.

Thoma, K. and Klimek, R., 1980, Stabilitätsspezifische polarographische Gehaltsbestimmung von Nifedipin in Arzneiformen, *Deutsch. Apoth. Ztg.*, **120**, 1967–72.

1981, OS DE 3136282 (German Patent Number).

1985a, Untersuchungen zur Photoinstabilität von Nifedipin 1. Mitteilung: Zersetzungskinetik und Reaktionsmechanismus, *Pharm. Ind.*, **47**, 207–15.

1985b, Untersuchungen zur Photoinstabilität von Nifedipin 2. Mitteilung: Einfluß von Milieubedingungen, *Pharm. Ind.*, **47**, 319–27.

1991a, Untersuchungen zur Photoinstabilität von Nifedipin 3. Mitteilung: Photoinstabilität und Stabilisierung von Nifedipin in Arzneizubereitungen, *Pharm. Ind.*, **53**, 388–96.

1991b, Photoinstabilität und Stabilisierung von Arzneistoffen Möglichkeiten eines allgemein anwendbaren Stabilisierungsprinzips, *Pharm. Ind.*, **53**, 504–7.

1991c, Photostabilization of drugs in dosage forms without protection from packaging materials, *Int. J. Pharm.*, **67**, 169–75.

Thoma, K. and Kübler, N., Influence of different cyclodextrins on photostability of nifedipine in press.

Thoma, K. and Lauchert, M. (publication in preparation).

Thoma, K. and Strittmatter, T. (publication in preparation).

Tomono, K., Gotoh, H., Okamura, M., Ueda, H., Saitoh, T. and Nagai, T., 1988, Effect of β-cyclodextrin and its derivates on the photostability of photosensitive drugs, *Yakzaigaku*, **48**, 322–5.

Tønnesen, H. H., 1991, Photochemical degradation of components in drug formulations. Part 1: An approach to the standardization of degradation studies, *Pharmazie*, **46**, 263–5.

Tønnesen, H. H. and Karlsen, J., 1987, Studies on curcumin and curcuminoids. X. The use of curcumin as a formulation aid to protect light-sensitive drugs in soft gelatin capsules, *Int. J. Pharm.*, **38**, 247–9.

1988, Studies on curcumin and curcuminoids. XI. Stabilization of photolabile drugs in serum samples by addition of curcumin, *Int. J. Pharm.*, **41**, 75–81.

1993, Standardisation of photochemical stability testing of drugs, *Pharmeuropa*, **5**, 5.

Tønnesen, H. H., Karlsen, J. and Beijersbergen van Henegouwen, G., 1986,

Studies on curcumin and curcuminoids. VIII. Photochemical stability of curcumin, *Z. Lebensm. Unters. Forsch.*, **183**, 116–22.

TØNNESEN, H. H. and MOORE, D. E., 1993, Photochemical degradation of components in drug formulations, *Pharm. Technol. Int.*, **5**, 27–33.

UEKAMA, K., NARISAWA, S., HIRAYAMA, F., OTAGIRI, M., KAWANO, K., OHTANI, T. and OGINO, H., 1983, Improvement of thermal and photochemical stability of benzaldehyde by cyclodextrin complexation, *Int. J. Pharm.*, **13**, 253–61.

UEKAMA, K., OH, K., OTAGIRI, M., SEO, H. and TSUROKA, M., 1983, Improvement of some pharmaceutical properties of clofibrate by cyclodextrin complexation, *Pharm. Acta Helv.*, **58**, 338–42.

DE VILLIERS, M. M., VAN DER WATT J. G. and LÖTTER, A. P., 1992, Kinetic study of the solid-state photolytic degradation of two polymorphic forms of furosemide, *Int. J. Pharm.*, **88**, 275–83.

VOEGELE, D. and LAUDENBACH, P. (Cassella AG), 1985, Photostabilisierung von Molsidomin, *Offenlegungsschrift* DE 3346638 A1.

YOSHIOKA, S., 1993, Extension of the guideline: Photostability; in: Proceedings of the second International Conference on Harmonisation, Orlando 1993.

7

Addressing the Problem of Light Instability during Formulation Development

D. R. MERRIFIELD, P. L. CARTER, D. CLAPHAM and F. D. SANDERSON

7.1 Introduction

A number of case histories are presented which illustrate the formulation problems raised by light-sensitive drug substances. These range from the study and formulation of parenteral presentations, where the product is in solution or suspension form and the instability most closely mirrors that of the drug substance; pharmaceutical ointments where the drug substance is incorporated as a suspended material in a semi-solid base to the physical effects on a tablet formulation where the active is present as solid component.

It is argued that there is a need for a good understanding of the nature and extent of the photoinstability, the mechanism of the light-induced degradation reaction and the wavelengths causing the instability. The potential problems raised by photounstable formulations are such that it is necessary to develop a strategy to quantify and understand them at an early stage of development. This will permit formulation development to be progressed in a timely and efficient fashion.

Whilst there is a reasonable measure of agreement between regulatory bodies as to the nature of testing conditions for the stresses of temperature and humidity, it is only in very recent years that this has been attempted for the testing of light-sensitive materials (Anderson *et al.*, 1991a and b). The discussions surrounding the testing of light-sensitive drugs has culminated in draft guidelines on light-stability testing, with recommendations as to the extent of light exposure; the way in which this is conducted; how the results are interpreted and the implications for product labelling. These draft guidelines and the recommendations for testing and interpretation arising from them will allow a common approach in drug product registration. However, they will not necessarily provide a complete strategy for assessing the light sensitivity of a drug product, and the restrictions that light sensitivity place on the usage conditions of a drug product.

The conditions that the product will experience in use will clearly vary as a result of presentation type (especially depending on whether the drug is present in the solid or solution phase); the extent to which the dosage form is protected by the packaging; the mode of administration; and the geographical location of the market and the local conditions of the clinic. Whilst the conditions recommended in the ICH (International Commission on Harmonization) draft guidelines provide a common standard by which light sensitivity may be judged and compared, they can only act as a guide to understanding the nature of light instability such that the product may be handled appropriately by the patient or medical team. The conditions of the suggested ICH tests must be supplemented by other tests specifically designed to help define conditions of use of the formulated product.

7.2 Photoinstability of the drug substance

It should be established at a fairly early stage that a development compound is susceptible to light-induced degradation. Depending on the nature of product development within an organization, this will probably be known at an early stage of chemical development and may be established as an issue for a family of chemically similar compounds. At any rate the photoinstability will be established very early in analytical method development, and this will probably provide the first opportunity to quantify the extent of photoinstability. A commonly experienced problem at this stage will relate to the degree of confidence in the purity of the compound; the validity of the analytical methods; the degree of knowledge regarding degradation products; availability of the New Chemical Entity (NCE) and the relevance of testing conditions. Merely knowing that a material is light sensitive, by any given criteria, does not guarantee a clear appreciation of the relevance of that light instability as it relates to product usage.

The guidelines regarding testing set out in the ICH draft recommend exposure periods under given light intensities which enable the categorization of a material as light sensitive, along with a guide as to the requirement for protective packaging and subsequent labelling. These guidelines also provide a range of options as to the nature of the light source, with the degree of challenge being determined as a cumulative light exposure. Given the intensity of the sources used, the extent of exposure recommended represents a fairly severe test, likely to considerably exceed the conditions experienced by the drug product. It is nevertheless very valuable in setting a common criterion by which a material may be categorized as light sensitive or not.

At this early stage of development the available information on photostability is likely to be anecdotal in nature, based on laboratory observations made by development chemists. It is important that where light sensitivity is suspected, it is confirmed and quantified at the earliest opportunity, either by testing under the draft ICH recommended conditions or by an accepted in-house testing protocol (Nema *et al.*, 1995). This is important so that any photoinstability may be taken into account when considering results from early evaluation of the compound in screening tests and that a toxicology programme can be designed appropriately. Furthermore, the nature and extent of photoinstability are likely to impact on the objectives and timings of the development programme for the NCE and on the aims for its ultimate use as a product.

Recognizing that a candidate drug substance has potential photoinstability will require the development function to consider several approaches to assessing and solving the problem. It will be necessary, knowing the intended indication and likely mode of administration, to establish whether the problem may be largely resolved by the use of suitable packaging or protection. If the problem cannot be resolved in this way, the extent of the instability will need to be determined and decisions will need to be made as to whether this may be overcome by modifications to the formulation, or the process by which it is prepared. In any event, both primary (chemical) and secondary (pharmaceutical) processes will need to be reviewed to establish the extent to which the compound's susceptibility is likely to compromise the quality of the product during manufacture. Special conditions for handling the material at various stages of the process may need to be defined to alleviate this aspect of the problem.

Analytical development work should rapidly identify the extent of problem caused by the presence of photodegradation products. Identification of the degradation products is a necessary aspect of elucidating the reaction mechanism and will be required in any submissions made to regulatory groups. Their early identification and quantification is important in deciding the appropriate strategy for the toxicological assessment of the drug substance. Here it must be remembered that the photodegradation reaction will have to be sufficiently characterized and understood so that any degradation in the product occurs by the same mechanism, allowing the toxicity assessment of degradation products to equally apply to the final formulated product.

An important part in further elucidation of the degradation reaction relates to the specific wavelengths causing the degradation (Allwood and Plane, 1986; Moore, 1987). Knowledge of this allows the formulation group to select appropriate packaging to protect the product (Sugden and Bhadresa, 1981), with the packaging components attenuating those wavelengths causing most significant degradation. Knowledge of the causative wavelengths also enable the formulator to consider the use of formulation excipients (Thoma and Klimek, 1991) where appropriate, to absorb the light responsible for degradation, thereby reducing the rate at which the degradation occurs to a manageable and acceptable degree.

Establishing the kinetics of the degradation is a further important aspect of characterizing the nature of the photoinstability (Tønnesen, 1991). This will take the form of a series of experiments determining the reaction rate with the objective of establishing its dependence on concentration. The information provided from this study will assist in deciding the development strategy for the formulation and identifying any limitations that the degradation rate places on the product use.

7.3 Photoinstability of formulations

The approach to dealing with a drug's instability will clearly depend on the nature of the presentation and its mode of use (Tønnesen and Moore, 1993). Three case histories are presented which exemplify an approach to identifying, quantifying and resolving the issues raised by a photolabile drug substance or formulation.

7.3.1 *Intravenous infusions*

An intravenous infusion presentation offers considerable challenge to the formulation group as the problems associated with light instability will most closely resemble those of the drug substance itself. The drug is likely to be presented as a dilute solution and used over a number of hours in a clinical situation (Colardyn *et al.*, 1993). Furthermore, it is likely that, given hospital practices, the presentation will be made up by the pharmacy team some time before administration to the patient. It will be necessary that the reconstituted presentation is capable of withstanding a variable light-intensity challenge depending on geographical location of the clinic, the level of ambient light within the pharmacy and ward, and the duration of use. Given the highly variable nature of all these factors it is important that the directions for use of the product are very clear and that the presentation is capable of withstanding the harshest challenge likely to be met in practice. An additional complication in the formulation of light-sensitive drugs in parenteral infusion presentations arises from the need to demonstrate the usual freedom from contamination by extractives of the giving set and that the drug substance remains compatible with the range of infusion fluids likely to be used in clinical practice. It must be considered that any restrictions applying to the mode of use of the infusion fluid, or any contra-indicated infusion fluid will add to complication in use and be perceived as a competitive disadvantage for the product.

Compound A was observed to be light sensitive at an early stage of its development, consistent with its having a highly conjugated functional ring group. The degradation products were identified during early analytical development work and were consistent with a likely and recognized light-degradation mechanism. It was possible to isolate these degradation products following chemical synthesis by column chromatography or they could be prepared *in-situ* in solution by following a standard light exposure regime.

A series of photodegradation studies were conducted to establish the extent of the light-sensitivity problem, the implications for the proposed formulation and the manufacturing process.

7.3.1.1 Light degradation of the drug substance as a powdered solid

An early assessment of the substance was conducted on the dry powder material. It was expected that the drug substance was likely to be prepared either as a precipitated crystalline solid or as a lyophile. The stability of the drug substance raw material was determined by reverse phase HPLC analysis to help judge the feasibility of the chemical and pharmaceutical production processes, and to assess the likely shelf-life prior to reconstitution. This was conducted both as accelerated testing using a xenon light source (Xenotest) and metal halide light source (Dr Hölne SOL2 light cabinet), presenting the sample as thin powder sample (about 2 mm thick), and in real time in glass vials using sources of both artificial (laboratory) and natural light. The results of this test are shown in Table 7.1.

Apart from the expected problems arising from the variability of natural light, it was evident that degradation was occurring by the same pathway in each case (forming the degradation products A1 and A2), except in the instance of storage in the dark (vials overwrapped in aluminium foil). In the dark, the major degradation product was an alkaline degradation product (A3), with no evidence of formation of the main light-induced degradation product (A1).

In summary, the solid stability profile was extremely good for the class of

Table 7.1 Stability assessment of compound A as a powdered solid

Condition exposure period	Purity	Degradation products A1	A2	A3
Initial	85.3	nd	nd	0.2
6 weeks sunlight	75.1 (87%)	6.9	4.9	0.2
4 months sunlight	55.2 (65%)	22.2	14.2	1.2
6 months artificial	82.3 (96%)	5.0	2.8	0.4
6 months dark	85.0 (100%)	nd	0.3	0.3

Notes:
The powder was exposed to the light source as a shallow bed of about 2 mm depth.
Each content is given as %w/w in the dry powder, with the drug substance purity also expressed as % initial given in parenthesis.
nd = not detected.
The content was determined in each case by HPLC analysis.

compound when stored protected from light, but the material degraded rapidly in the presence of any level of natural or artificial light. These findings confirmed that the extent of instability would require it to be protected during storage and that special precautions were likely to be needed during the handling, especially during freeze drying of the compound.

7.3.1.2 *Light degradation of the drug substance in solution*

The rate of degradation of the drug substance as a function of concentration in solution is a crucial factor when considering the suitability of the NCE for use as an infusion presentation. Although a number of experiments had been conducted attempting to quantify the rate of degradation of solution under in-use conditions, the most valuable assessment of the extent of instability was gained by a series of studies conducted at different concentrations under standard lighting conditions (Table 7.2).

The experiment was conducted in a light cabinet specially constructed in-house to provide a high intensity of artificial light (2× Philips white 35 fluorescent bulbs) at a constant temperature. Constant temperature was achieved by fitting the cabinet with a powerful fan and situating the light cabinet in a constant temperature storage facility. Samples were taken over the course of the reaction and assayed for Compound A content by reverse phase HPLC analysis. The reaction was allowed to proceed to a sufficient extent to allow first-order rate constants to be determined.

The results in Table 7.2 demonstrate an interesting paradox for assessing the stability of a pharmaceutical active. It is customary within the industry to assess the proposed product by reference to the time taken to lose a certain proportion of the active constituent, in this case 10 per cent, otherwise referred to as T90. Reference to this figure demonstrates a very unacceptable stability profile for the most dilute solutions (0.05 and 0.025 mg/ml), making handling of such a product impractical. However, assessment of the data in molecular terms clearly shows a faster rate of reaction at the higher concentrations; there is an approximate

Table 7.2 Rate of degradation of compound A in solution

Concentration mg/ml	Rate mol/h	T90
2.5	3.6×10^{-4}	2.4 h
0.5	2.0×10^{-4}	0.9 h
0.25	1.8×10^{-4}	28 min
0.05	0.9×10^{-4}	11 min
0.025	0.5×10^{-4}	7 min

Notes:
Determined in clear glass vials at 25 °C and 10 kLux. (Light-testing cabinet fitted with 2 × Philips white 35 fluorescent bulbs.)
T_{90} = time taken for the compound to degrade to 90 per cent of its initial purity value.

Table 7.3 Dependency of reaction rate on light intensity

Source	Intensity	Rate mol min^{-1}	T90 minutes
Artificial[a]	10 kLux	1.24 × 10 − 5	66
'North light'	7 kLux	0.99 × 10 − 5	83
'South light'	70 kLux	8.22 × 10 − 5	10
Metal halide[b]	120 kLux	4.11 × 10 − 4	2

The solutions, prepared at 2.5 mg/ml, were tested in clear glass injection vials.

[a]Light testing cabinet, 2 × Philips white 35 fluorescent bulbs.
[b]Dr Hönle SOL 2 simulated sunlight tester.
T_{90} = time taken for the active to degrade to 90 per cent of its initial concentration.

doubling of rate with a tenfold increase in concentration. In this instance the acceptability of the product will need to be judged on the nature of degradation products as well as the rate of disappearance of the active constituent, as these will clearly be generated at a higher rate and to a greater extent, with increasing concentration. Further experiments were conducted at a constant concentration of 2.5 mg/ml in clear glass vials to determine the effects of light intensity and of light type (Tables 7.3 and 7.4).

The experiments shown in Tables 7.3 and 7.4 demonstrate the effect of light intensity and type. Taken together it can be seen that when degradation is allowed to proceed to an approximately equal extent (intended to be about 85 per cent of initial purity) the pattern of degradation products formed is similar in each case. Compounds A1 and A2 are the main degradation products in each instance and are formed in the same proportions. No other significant peaks were observed in the chromatogram, and the other known degradation product, due to alkaline decomposition A3, was not detected to any significant extent.

Table 7.4 Degradation product formation as a function of light type, conducted on solutions of compound A

Source	Potency and (% initial)	Degradation products A1	A2	A3
[a]Artificial (30 kLux.h)	0.224 (82%)	13.6 (0.7)	4.9 (0.25)	0.5 (0.02)
South light (7 kLux.h)	0.204 (94%)	3.9 (0.71)	1.5 (0.27)	nd
North light (15 kLux.h)	0.196 (84%)	13.3 (0.72)	4.7 (0.25)	0.2 (0.01)
[b]Metal halide (6 kLux.h)	0.204 (86%)	10.2 (0.71)	3.7 (0.26)	0.02 (0.001)

These studies were conducted at 2.5 mg/ml initial concentration, stored in clear glass injection vials. It was intended that each solution should be degraded by 15 per cent of initial value.

[a]Light testing cabinet, 2 × Philips white 35 fluorescent bulbs.
[b]Dr Honle SOL 2 simulated sunlight tester. Note that the light exposure is given as a cumulative challenge, for example, in lux hours.
Contents of degradation products are expressed as a percentage of the main peak in the chromatogram (intact drug) and as a relative proportion of the total degradation products in the chromatogram, given in parenthesis.

Given this common pattern of degradation, consistent with that of the powdered solid and the expected degradation mechanism, it is reasonable to interpret the results shown in Table 7.3 in terms of the effect of light intensity. The expected pattern is observed with the low-intensity North light being a far less severe stress than the metal halide source of the SOL2 sunlight simulator.

These experiments, along with others, defined the in-use and processing limitations of the product. Given its nature as a clear solution for injectable and infusion use, many of the approaches used to overcome light instability in the formulation were not available in this instance. The approach used for this product was via protection afforded by secondary packaging. Further experiments enabled the selection of a suitable protective plastic sleeving which greatly reduced the extent of degradation, allowing a usable solution shelf-life. These experiments were conducted using the tuneable wavelength source of a spectrofluorimeter to provide light of known wavelength to a solution held in the sample cuvette of the instrument. A number of wavelength-specific experiments were carried out to determine the dependency of degradation rate on wavelength. It can be seen from Fig. 7.1 that the wavelength dependency of the measured degradation rate closely mirrors the absorption spectrum of the compound.

7.3.2 ***Topical ointments***

The second case study represents an interesting situation combining the problems of a light-sensitive drug substance incorporated into a presentation which will be

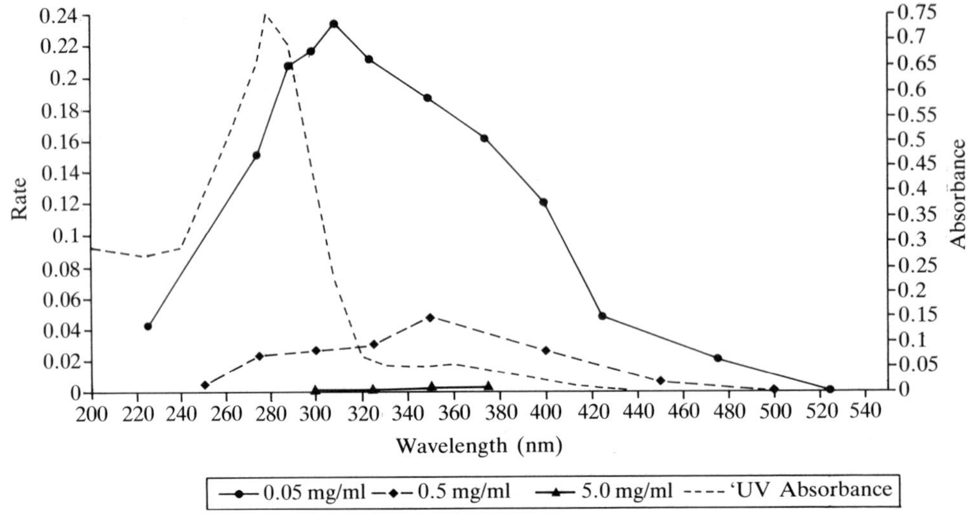

Figure 7.1 The degradation rate as a function of wavelength and the UV spectrum of compound A

exposed to sunlight during patient use. It also represents a situation part way between the solid stability of the drug substance and its photostability in the solution phase.

Compound B is intended for topical administration and is incorporated into an oily base. It was known to be photounstable from an early stage of development, and accordingly a number of salt forms of the drug substance were evaluated in preliminary screening tests. The parent compound (free acid form) was eventually selected as the appropriate form for development, and photostability of the ointment was studied incorporating the drug substance both in the precipitated form, and after micronization. As the clinical dose remained to be established, the drug was incorporated at a range of concentrations to assess whether this had any effect on the photoinstability.

The results in Table 7.5 show a number of interesting aspects. There is a clear dependence of the light stability of the compound on the nature of the salt form used. The effect is particularly striking for the disodium salt when compared with the free acid, given that the former is incorporated as an unmicronized powder.

For the free acid, the micronized material is less stable than the unmicronized form. This is indicative of a particle size effect where either the finer particle size receives a greater effective exposure to light because of the higher surface area available or that there is a higher proportion of the drug present actually in solution in this instance. From this limited data it does not appear that the more finely divided drug substance offers any self protection by physical obscuration of the light.

The results indicate that the extent of loss of the active is reduced as the concentration is increased. However, exactly as in the first case, when this is considered in molar terms it can be seen that the actual quantity of drug lost is greater at the higher concentrations. The complexity of the system must be

Table 7.5 Light stability of Compound B as the free acid and various salt forms as incorporated into a White Soft paraffin base

Salt/form	Per cent initial Compound B after 1 h exposure in SOL 2 0.1% conc.	0.5% conc.	2.0% conc.
Free acid (micronized)	51.0	80.0	85.9
Free acid (unmicronized)	69.4	nd	nd
Disodium salt (unmicronized)	9.9	3.6	nd
Ethylenediamine salt (unmicronized)	51.9	65.1	nd
Piperazine salt (unmicronized)	79.9	88.2	nd

nd = not done

The samples of ointment were exposed as thin films for 1 h in the Dr Honle SOL 2 simulated sunlight tester.
The concentration was determined by HPLC analysis.

Table 7.6 Light stability of Compound B free acid in various brands of paraffin ointment

Paraffin base	Colour	Per cent of initial (after 1 h exposure in SOL 2)
1	White	53.2
2	White	49.6
3	White	50.2
4	White	56.6
5	Yellow	70.0
6	White	73.2
7	Yellow	75.6
8	Yellow	77.4

The samples of ointment were exposed as thin films, for 1 h in the Dr Honle SOL 2 simulated sunlight tester.
The concentration was determined by HPLC analysis

considered. The solubility of the drug in the ointment base is approximately 0.05 per cent w/w. It is present in solution and in increasing amounts in suspension, in the higher concentration ointments. Various reactions will be occurring simultaneously, especially if the degradation mechanism differs between the solution and suspension phase.

The drug compound was incorporated at 0.1 per cent concentration, as a micronized powder, into a range of paraffin bases. Each were subjected to a 1 hour exposure in the SOL2 light simulation cabinet. The results of this screening test are shown in Table 7.6.

Table 7.7 Effect of alpha tocopherol on the light stability of Compound B

Formulation	Per cent initial after 1 h exposure in SOL 2
Brand 1	53.2
Brand 1 (+ 10 ppm)	48.3
Brand 6	74.2
Brand 6 (+ 250 ppm)	73.5
Brand 6 (+ 500 ppm)	72.5

The samples of ointment were exposed as thin films, for 1 h in the Dr Hönle SOL 2 simulated sunlight tester.
The concentration was determined by HPLC analysis.

The yellow paraffin bases can be observed to offer more protection to the drug substance than the white brands, with the exception of brand 6. It was known that this brand contained about 10 ppm of the antioxidant a-tocopherol, and it was considered that this constituent may be affording some protection. A further experiment was therefore conducted spiking the control blend, Brand 1, with a-tocopherol. This did not produce any noticeable effect on the stability of the drug, neither did the incorporation of varying levels of a-tocopherol in the best white paraffin brand, Brand 6. These results are shown in Table 7.7.

UV scans of the various paraffin bases dissolved in chloroform were obtained. This indicated a clear difference for the white paraffin base 6 compared with the other white bases. The UV spectrum of this brand was more similar to those of the yellow bases 5, 7 and 8. It is likely that the light-absorption properties of the paraffins account for their relative abilities to protect the active. Compound B has a UV spectrum with maxima at 215 nm and 330 nm, and it was inferred from the chemical structure that the maximum at 330 nm is most likely to be due to the part of the molecule susceptible to light. White soft paraffins generally have a minimal absorbance in this region of the UV spectrum, 290 nm being the longest wavelength absorbed compared with the yellow grades which showed a significant absorbance up to 340 nm. It is this broad absorbance which probably confers protection on the active Compound B.

These studies demonstrate an important approach to addressing the light instability of a drug substance within a formulation. If it is possible and acceptable to include a material that can block the passage (physical obscuration) (Béchard *et al.*, 1992) or compete with the drug substance for the available light energy (similar light absorbance spectrum) (Thoma and Klimek, 1991), then the photoinstability of the active compound should be reduced (Asker and Harris, 1988; Asker and Habib, 1989).

This approach was adopted for this compound, investigating the use of both light blockers and UV absorbers, resulting in an acceptable formulation with minimum photoinstability.

7.3.3 Solid dosage forms (tablets)

The third case study deals with a drug, not itself photolabile, incorporated into a solid dosage form, and demonstrates a different but equally important aspect of light stability, physical appearance (Jansson *et al.*, 1980; Nyqvist, 1984; Augustine *et al.*, 1986).

The material in this study is a marketed product, presented as a range of film-coated tablets of different active strength. They are actually prepared from a single formulation and are tabletted at a range of weights to provide the required dosages. It is known that the active compound itself is not light sensitive even under harsh conditions. In this study the effect of light on the physical appearance of the formulated product was examined.

The range of tablets, packaged in the intended market pack, were exposed to D65 white light radiation using the SOL2 sunlight simulator light cabinet. The packs (clear PVC/PVDC blister backed with aluminium foil) were exposed clear side uppermost to about 115 kLux radiation for 24 hours, using foil-wrapped packages as a control. After exposure the colour of the tablets was quantitatively determined using a Tristimulus colorimeter. The tablets were also analysed by reverse phase HPLC to determine the intact drug content and the two most significant degradation products.

The principle of operation of the Tristimulus colorimeter essentially mimics the red, green and blue detection of the human eye by measuring the amount of light reflected through wavelength-specific filters. These raw data are transformed mathematically to provide a uniform, reproducible numerical scale which can detect very slight differences in colour. The scale is depicted by a three-dimensional model, the L, a, b colour space. This represents colour on two scales, 'a' (red /green) and 'b' (yellow/blue), and colour brightness on a third, 'L'.

The technique is capable of being refined so that the light spot of the instrument can be tightly focused to a point on the tablet surface. This can allow comparisons between front (exposed) and back (unexposed) faces to be made.

The results in Table 7.8 illustrate an increased yellowing of the tablets, shown by an increasing yellow/decreasing blue 'b' component. It is also apparent from these data that the effect is more marked in the lower-strength tablets.

It was necessary to demonstrate that this appearance effect was not due to any degradation of the active material. To do this two additional experiments were conducted; these are shown in Tables 7.9 and 7.10.

These experiments demonstrate that the yellowing is not due to detectable degradation of the active component. The second experiment shows this most clearly; in each case an individual tablet from each of the three classifications yellowest, palest and mid-range were analysed for main peak and degradation products. This was done for tablets in each classification for each strength. Only the results for the 125 and 250 mg strengths are shown in Table 7.10.

Additionally the yellowest tablet seen was analysed. In this case the surface yellow layer was scraped off and assayed. Again, there was no observable increase in degradation products. No additional degradation products were observed in the HPLC chromatograms.

In this instance the nature of the effect of light could be overcome by appropriate packaging. In general this very sensitive technique allows a thorough quantification of the effect of light exposure on physical appearance of solid dosage forms.

Table 7.8 Measurement of change in colour of the tablets (as determined by Tristimulus colorimetry).

	Mean colour			Standard deviation		
Tablet strength	L	a	b	L	a	b
125 mg Control	99.5	2.60	5.32	2.4	0.6	0.6
Exposed	85.5	1.71	14.58	9.7	1.5	3.2
250 mg Control	99.0	2.60	4.92	2.7	0.6	0.2
Exposed	93.0	2.20	11.83	2.5	0.8	2.8
500 mg Control	95.5	3.02	5.36	3.1	1.0	0.3
Exposed	90.3	1.48	11.10	2.3	1.1	3.2
750 mg Control	95.8	2.58	5.18	1.5	0.3	0.2
Exposed	88.7	2.21	7.39	1.3	0.6	1.9

Notes:
The Tristimulus colour description is based on three axes. This allows any colour as perceived by the human eye to be expressed in terms of three numbers.
The tablets were exposed to 115 kLux intensity in the Dr Honle simulated sunlight tester for 24 h, the tablets being exposed on one side only.
L scale = colour brightness
a scale = red/green
b scale = yellow/blue, where a positive increase denotes a yellowing of tablets.

Table 7.9 HPLC analysis of tablets subjected to simulated sunlight testing

	Mean assay (%)		
Tablet strength	Main peak	Impurity 1	Impurity 2
125 mg Control	72.6	0.25	nd
125 mg Exposed	73.0	0.23	nd
250 mg Control	71.3	0.31	nd
250 mg Exposed	72.1	0.31	nd
500 mg Control	73.4	0.33	nd
500 mg Exposed	73.2	0.42	nd
750 mg Control	72.6	0.61	nd
750 mg Exposed	73.1	0.54	nd

Notes:
nd = not detected.
Contents of the degradation products 1 and 2 are given as % W/W.
Exposed tablets were subjected to 115 kLux for 24 h (Dr Hönle SOL 2), exposed on one side only.
Control tablets were overwrapped with foil.

Table 7.10 HPLC analysis on selected individual tablets

Tablet strength	Mean assay (%) Main peak	Impurity 1	Impurity 2
125 mg yellowest	73.4	0.26	nd
125 mg mid-range	73.0	0.21	nd
125 mg palest	72.0	0.22	nd
250 mg yellowest	71.7	0.37	nd
250 mg mid-range	72.7	0.28	nd
250 mg palest	71.8	0.29	nd
250 mg surface	72.1	0.28	nd

Notes:
nd = not detected.
Contents of the degradation products 1 and 2 are given as % W/W and are determined by HPLC analysis.
Tablets had been exposed to 115 kLux for 24 h (Dr Hönle SOL 2)
Certain tablets were selected as being representative of the exposed tablets, others were taken as the palest or yellowest samples obtained during the test.

7.4 Conclusions

The case studies presented in this chapter illustrate a consistent strategy for evaluating light stability of a compound during formulation development. This includes studies to evaluate mechanistically the effects of formulation and packaging on the stability of the final product.

The challenge to the product under likely in-use conditions is examined as an important aspect of understanding the photostability of the drug substance and its formulations.

Acknowledgements

The authors wish to acknowledge S. A. Hancock, A. J. Goodall, M. E. Morris and J. Oduro-Yeboah for their help in valuable discussion and preparing this presentation.

They also wish to acknowledge T. Franklin, M. Lau, B. Hall, S. Ritchie, R. Denton and the late M. Churchill, for much of the practical work contained in these studies.

References

ALLWOOD, M. C. and PLANE, J. H., 1986, The wavelength dependant degradation of Vitamin A exposed to ultraviolet radiation, *Int. J. Pharm.*, **31**, 1–7.

ANDERSON, N. H., JOHNSTON, D., MCLELLAND, M. A. and MUNDEN, P., 1991a, Photostability testing of drugs, *Manuf. Chemist*, **62**, 25.

1991b, Photostability testing of drug substances and drug products in UK pharmaceutical laboratories, *J. Pharm. Biomed. Anal.*, **9**, 443–9.

ASKER, A. F. and HABIB, M. J. 1989, Influence of certain additives on the photostatibility of colchicine solutions, *Drug Dev. and Ind. Pharm.*, **15** (5), 845–9.

ASKER, A. F. and HARRIS, C. W., 1988, Influence of certain additives on the photostability of physostigmine sulphate solutions, *Drug Dev. and Ind. Pharm.*, **14**, (5) 733–46.

AUGUSTINE, M. A., BERNSTEIN, D. F., NARURKAR, A. N. and SHEEN, P.-C., 1986, Studies on the light stability of flordipine tablets in amber blister packaging material, *Drug Dev. and Ind. Pharm.*, **12** (8 + 9), 1241–7.

BÉCHARD, S. R., KWONG, E. and QURAISHI, O., 1992, Film coating: effect of titanium dioxide concentration and film thickness on the photostability of nifidipine, *Int. J. Pharm.*, **87**, 133–9.

BHADRESA, B. and SUGDEN, J. K., 1981, Light Transmittance through Amber Glass Medicine Bottles, *Pharm. Acta Helv.*, **56** 4–5, 122.

COLARDYN, F., DE MUYNCK, C., REMON, J. P. and VANDENBOSSCHE, G. M. R., 1993, Light stability of molsidomine in infusion fluids, *J. Pharm. Pharmacol.*, **45**, 486–8.

JANSSON, I., LUNDREN, P. and NYQVIST, H., 1980, Studies on the physical properties of tablets and tablet excipients. II. Testing of light stability of tablets, *Acta Pharm. Suec.*, **17**, 148–56.

MOORE, D. E., 1987, Principles and practice of drug photodegradation studies, *J. Pharmaceutical and Biomedical Analysis*, **5**, 441–53.

NEMA, S., WASHKUIIN, R. J. and BEUSSINK, D. R., 1995, Photostability Testing: An Overview, *Pharm. Technol.*, **March**, 170.

NYQVIST, H., 1984, Light stability testing of tablets in the Xenotest and the Fadeometer, *Acta Pharm. Suec.*, **21**, 245–52.

THOMA, K. and KLIMEK, R., 1991, Photostabilization of Drugs in Dosage Forms without Protection from Packaging Materials, *Int. J. Pharm.*, **67**, 169–75.

TØNNESEN, H. H., 1991, Photochemical degradation of components in drug formulations, *Pharmazie*, **H4**, 263–5.

TØNNESEN, H. H. and MOORE, D. E., 1993, Photochemical Degradation of Components in Drug Formulations, *Pharm. Technol. Europe*, **5**, 27–33.

8

Light-activated Drugs and Drug Formulations in Drug Targeting

J. KARLSEN

8.1 Introduction

One of the main goals of chemical therapeutics and pharmaceutics is to deliver a pharmaceutical agent efficiently and specifically to the site of disease. In some cases this can be achieved by administering the drug directly. However, controlled-release drug delivery vehicles can lead to increased efficacy, reduction in toxicity and prolongation of therapeutic effect. This delivery to the site of action is called drug targeting (Chien, 1993).

The development of new, strong-acting drugs and drugs with a limited stability in biological fluids has been the driving force behind the concept of drug targeting. It is the powerful tools of recombinant DNA and monoclonal antibody technologies that have contributed to our knowledge of molecular recognition and protein structure-function relationships that has given us the possibilities of getting a drug molecule to a predestined site. The localization of the drug in the body and consequently better diagnostic or therapeutic effect has led to the use of compounds which otherwise could show too much of unwanted side-effects (Lee, 1990).

In traditional therapy where the organism is given a very large number of drug molecules and only a fraction of them eventually reach the active site, the better targeting of the drug molecule has become necessary (Fig. 8.1). For safety considerations the side-effects must be minimized too and a great number of targeting systems have been exploited with this aim. The use of light offers the experimental pharmacologist a physical method to trigger an intended reaction at a given site. This is done in photodynamic therapy (PDT) now used on a routine basis in hospitals. Although the catalytic action of light on certain chemicals in biological systems has been known for more than a century, exploitation of this mechanism for the purpose of drug action studies has developed rather slowly.

Scientists working in very different fields such as optics, natural products, perfumery, analytical chemistry, drug production, synthetic dyes and reaction kinetics have had scattered projects concerning the activity of light on compounds of their interest. This has resulted in a number of papers where many interesting

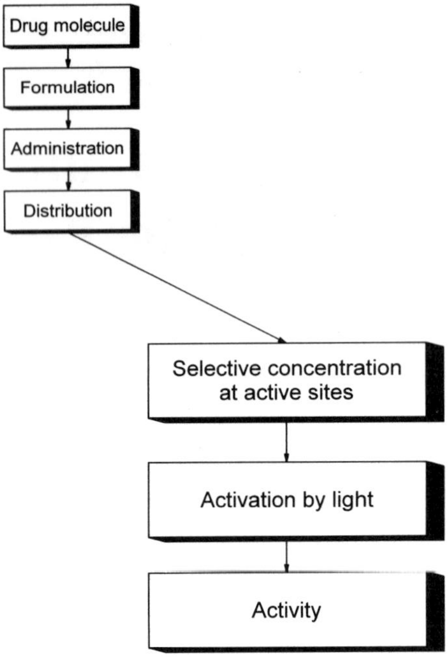

Figure 8.1 Pathway of drug molecule from compound to activity

results are reported but where the standardization of methods are often lacking. For scientists working in the medical field, standardization of procedures is a key issue and this may be the reason for the slow development of this very interesting field. To illustrate the use of light activation applied to drug action studies it will be necessary to define the field of interest, the intended effect and the problems encountered.

8.2 Drug targeting

To achieve better effect of drugs administered to the human body we need to bring the drug molecule to the site of action and bring about drug action on the spot where drug action is needed. This action can be a direct effect of the drug molecule on an active site, indirect effect of the drug molecule via a sensitizer which triggers a biological response or by an external parameter such as temperature, light or pH to initiate the biological response intended (Tyle, 1990).

The combination of light-induced drug reactions and real therapeutic value is only just beginning to take off as scientists working in the fields of photochemistry and photophysics are putting their knowledge to practical medical use. For some years, however, photodynamic therapy has been applied in therapeutics although this kind of application of light-induced biological effects can only be regarded as one of many aspects of light-induced drug targeting.

Before going into details of using light as an activator in drug targeting, it is necessary to point out the benefit when we are adding another complicating factor

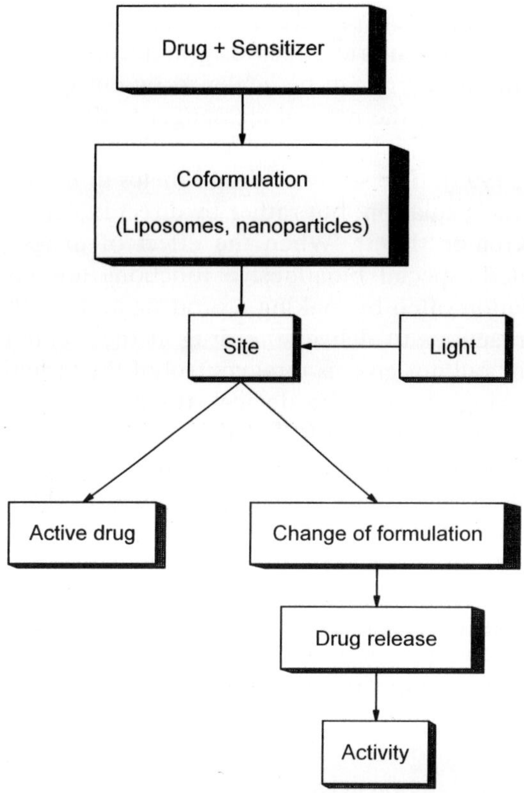

Figure 8.2 Possible formulation to achieve drug targeting

to the drug targeting system. When a drug is given to an organism, the molecules will distribute themselves according to solubility, polarity, dissolution, enzyme actions and other factors. The amount of drug molecules reaching the target will be very small compared with the number of molecules given. By formulating the drug molecule into tablets with controlled release patterns of the drug, aggregates with different solubility and encapsulate the drug into nanoparticles with a surface which can be changed to mention a few formulations, a much better targeting of the drug can be achieved (Fig. 8.2). However, we will still need an activation of the drug molecule at the site of action to achieve a real targeting effect (Gregoriades, 1995).

The most obvious aim of formulation scientists, biochemists, pharmacologists and others working with drug action projects is to try to achieve a release of the drug molecule at the site of action. In this way they can study the kinetics of drug action in basic research for future applications as well as use this knowledge to initiate drug targeting in real patients. When the drug is administered to a patient, added to a cell culture or an organ in a physiological bath, it will be distributed all over the cells in question before a tiny amount of the drug molecules actually will exert their therapeutic activity. This will naturally lead to unwanted side-effects of the drug and unwanted reactions in the *in vitro* experiments and obscure the kinetics of drug action. We are, therefore, always looking for methods where the drug release can be monitored exactly in amount, time and site. We are often looking for a direct stoichiometric effect/drug molecule-active site.

In some cases, site targeting can be achieved by encapsulating the drug in an envelope which is carried by the natural messenger to the organ where the activity is wanted. The envelope may be liposomes or other nanoparticles (Lesermann, *et al.*, 1993; Petrak, 1993). The study of drug kinetcs in *in-vitro* studies, however, cannot be done by encapsulation, but rather by direct injection methods (Sato and Sunamoto, 1993; Kreuter, 1995). When the effect of drugs directly on cellular membranes is wanted, special bioadhesive functions are attached to the drug molecule or formulation often by making a prodrug and we then make use of the normal enzymatic reactions to deliver the drug at the wanted site. The methods briefly described will still not give us total control of the targeting effect. However clever we are in making the 'homing device', the activity will only be a result of a random hit of the drug entity with the intended site (Meijer, 1994).

If we can release the drug at the site by an initiating 'bullet' to the drug delivery system a number of advantageous effects are achieved. Such a 'bullet' can be a certain amount of irradiation. The advantage by using light sources is that wavelength, intensity, time and site can be determined prior to the experiment. All the parameters mentioned above can be set with great accuracy and the kinetics of a drug reaction can be measured. The drug molecules which are released by the light radiation will be targeted to the site of action. What we are trying to do is to mimic nature by copying the activity of chloroplasts in plants when light quanta are used to initiate photosynthesis or the reactions which take place in the retina of the human eye, to mention a few examples (Fig. 8.3).

Liposomes or other nanoparticles can be used to transport a certain drug to a defined area in the body. A wealth of literature exists describing how targeted liposomes or homing liposomes can be made (Lasic, 1993). Drugs of different polarity and different concentration can be encapsulated by phospholipids and transported to the active site. However, with all the problems concerning a precalculated encapsulation being solved, we face the problem when we want to release the content at a predestined site. In many cases, the drug delivered to the active site is still inactive if it is contained within the liposome membrane – the drug remains entrapped and is not bio-available. The resulting slow release of the drug may be advantageous in some cases such as slow-growing tumours, but in many cases high local concentrations of drug are required immediately or in a short burst. Typical examples of the last type are bacterial infections where a minimal inhibitory concentration of an antibiotic is necessary.

There are many ways of changing the leakage of drugs from liposomes, and studies of this permeability effect become an important subject of research. Examples of the permeability as such from different phospholidic bilayers are shown in Fig. 8.4.

We normally measure leakage rate under different conditions rather than calculating a more general permeability coefficient. After all we are interested in leakage under different physiological conditions for the design of targeted drug formulations. Secondary treatments of administered liposomes to allow release of encapsulated drugs are necessary. One way of controlling the leakage is to let the temperature of the liposomes pass through the point of phase transition. The phase transition temperature is the temperature where the phospholipid bilayer of the liposomes passes from a gel state to a liquid crystalline state. At this point the

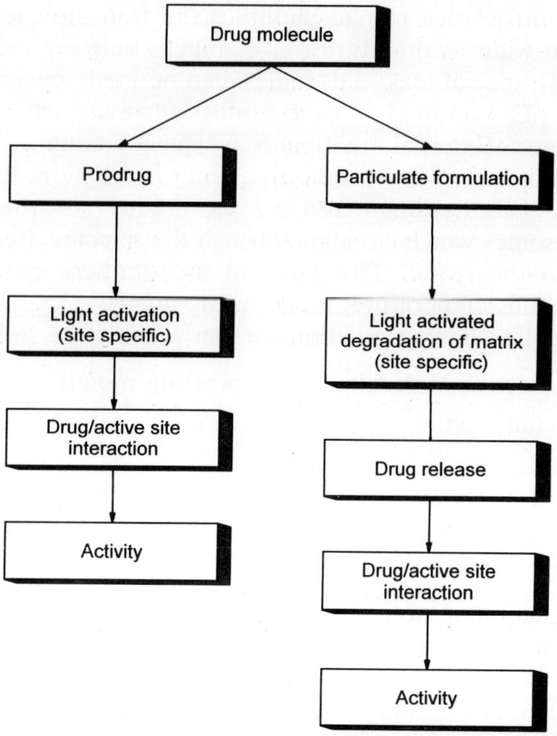

Figure 8.3 Schematic achievement of drug action by light-induced reaction

COMPOSITION	% RELEASED CARBOXYFLUORESCEIN				
	1 min	1 h	5 h	29 h	48 h
EPC	51.8	100			
HPC	6.6	2.6	4.6		
DSPC	3.9	2.6	7		
EPC/Chol	2.6	3.2	3.9		
DLPC/Chol	25.4	84.5			
DOPC/Chol	2.7	2.2	13		
DMPC/Chol	0.5	1.1	2.5		
SM/Chol	0.3	0.5	0.7	1.4	
DPPC/Chol	0	0	0.3	0.8	
DSPC/Chol	0	0	0	1	1.6
Abbreviations: HPC–hydrogenated egg phosphatidylcholine					

Figure 8.4 Leakage of carboxyfluorescein from liposomes with different lipid composition

liposome membrane becomes permeable to the molecules entrapped inside the liposome. By careful choice of phospholipids the transition temperature can be chosen between a wide temperature range, that is between −20 and +60 °C.

In other words, 'special leaky liposomes' can be made by triggering this effect by temperature, pH-, antibody-, or receptor-sensitivity as well as by making fusogenic liposomes. External treatments which are equally important include heating, ultrasound, microwaves, radio frequency irradiation, lasers and magnetic or electric fields. This treatment can act on already adsorbed or immobilized liposomes or liposomes which circulate through the selected area and already leak their content into the tissue. The practical use of these parameters, with the exception of pH and temperature, is still in its infancy.

By using controlled light irradiation we can achieve the following.

1 Release of the active molecule from a prodrug moiety.
2 Release of the active molecule from a targeted formulation like nanocapsules.
3 Transformation of inactive molecules into active ones.
4 Initiating a biological effect by light induction of bonding between xenobiotics and endogenous compounds.

By using light induction of biological effects we may improve the therapeutic index of certain drugs, achieve a much better targeting effect of a compound and achieve totally new ways of treatment. In particular, pharmacokinetic studies at cell level can be carried out very elegantly when the drug action can be induced and dosed by the light quantum given. It will not always be necessary to encapsulate the drug molecule in question. Another way of obtaining triggerable drugs is to make a prodrug which is inactive in circulation but can be activated by splitting it into a biologically active moiety and an inactive part. Figures 8.2 and 8.3 show schematically the different ways one can use light quanta to achieve a drug targeting effect.

8.3 Light sources

The light source is an important cause of concern as the wavelength of the light applied should be harmless to the living cell and initiate a specific reaction. This prerequisite usually limits the wavelength used to the visible part of the electromagnetic spectrum, although in some instances UV-light may also be permitted.

The use of laser sources greatly improves the experimental set-up since laser light of very different wavelength is now available. Laser-light sources covering the normal UV-VIS region are now available. In addition, computer techniques will also give us the possibility to administer the light in precise amounts of energy and time. Focusing the laser beam will also allow us to initiate a photoreaction in a single cell if so needed.

When using light as an activator in drug targeting in therapeutics we need to be able to relate the effect measured to the wavelength and the amount of light given in the reaction. The use of laser sources for the light inducement of reactions will allow close control of this parameter. It must also be mentioned that light-stimulated biological reactions are not only suited for direct treatment of patients but are even more an elegant way to follow biological effect in cells and cellgroups by three-dimensional focusing of the light beam. This will allow us to carry out investigation of biological effects under the microscope.

8.4 Photomodification of biological/artificial membranes. Sensitizers as drugs

The photomodification of cell membranes which is critical for membrane permeability and for cell destruction is governed by the properties of membrane-associated sensitizers. The heterogenous structure of biological membranes can be an important factor in photosensitization reactions. Sensitizers or protective agents may associate preferentially with the hydrophobic membrane core, accumulate at the aqueous interface or bind to membrane proteins. This localization effect combined with irradiation will indirectly give a targeting effect of the sensitizer or stabilizer given as a drug molecule. Membrane photomodification will be an issue of investigation for everyone studying the photomodification of cells, tissue and organisms. The active compound in membrane modifications is considered to be singlet oxygen and the study of the formation of singlet oxygen by different sensitizer molecules is therefore important. The cellular membranes present an exceptional environment for photosensitizers which generates singlet oxygen. Membrane differs from its environment in polarity, water content and dielectric constant. They are also heterogenous structures which present a variety of domains with which sensitizers can associate and from which they can act. The concept of the structure of the cell membrane is based upon the fluid mosaic model of Singer and Nicholson (1972). According to this model the phospholipids are arranged in a fluid bilayer. The hydrophobic, hydrocarbon tails of the fatty acids are oriented towards the centre of the bilayer exposing their polar head groups to the aqueous environment at either surface. This arrangement is stabilized by the hydrophobic forces between the phospholipids and does not include any covalent bonding. Most of the phospholipids are free to move within the plane of the bilayer, but with difficulty from one surface of the bilayer to the other. The phospholipid molecules are interspersed by proteins which partly span the bilayer or are partly located on the surface of the phospholipid membrane. The portion of the intrinsic proteins in contact with the lipophilic interior of the membrane is rich in hydrophobic amino acids and the portion on the outside of the membrane contains a high proportion of hydrophilic amino acids. Both the proteins and the phospholipids can have carbohydrates attached to them. The proteins are stabilized in their position in the membrane by hydrophobic interaction and by the surface proteins. The result of the membrane structure described is that the interior of the membrane has the characteristics of the interior of a lipid bilayer. The dielectric constant is very low in this region. Lipophilic solutes and thus drugs and sensitizer of low polarity can be expected to partition readily into this domain. Water is present in greatly reduced concentration, but the water permeability of most membranes is still quite high.

The other significant feature of the membrane is the interfacial region. This region extends roughly from the glycerol backbone of the phospholipids to the end of the attached carbohydrates. This region is of intermediate polarity between the lipophilic interior of the membrane and the aqueous surroundings. Due to the polar nature of the charged groups at the membrane surface, the surface of most biological membranes has a net negative charge. To study the biological membrane it is necessary to apply models. Model systems can be found in liposomes. The liposomes are formed by phospholipids and to a certain extent mimic the real cell membrane. To study the effect of sensitizers on membranes, liposomes are used in the introductory tests prior to the *in vivo* testing. The sensitizers damage biological target molecules by photosensitized oxidation. This causes enzyme deactivation (due to destruction of specific amino acids, in particular methionine, histidine and tryptophan), nucleic acid oxidation (primarily of guanine) and membrane damage (by oxidation of unsaturated fatty acids and cholesterol).

The two mechanisms of photosensitized oxidation are named Type I and Type II reactions. In the Type I process, substrate or solvent reacts with the sensitizer excited state (either singlet or triplet state) to give radicals or radical ions, respectively, by hydrogen atom or electron transfer. Reacting these radicals with oxygen gives oxygenated products. In the Type II process, the excited sensitizer reacts with oxygen to form singlet molecular oxygen which then reacts with the substrate to form the products (Fig. 8.5).

The reactions caused by the sensitizers can be used as such in therapy and through guided irradiation give a targeted drug. In case the reactions cause destruction of the drug formulation or release the drug molecule from a prodrug entity at a definite site or near predefined cells, the use of sensitizers is of interest.

To establish the mechanism of photosensitized oxidations in complex systems is a very difficult task. Kinetic tests and the use of inhibitors are more ambiguous than in homogeneous solutions. The reagents are compartmentalized, bound or localized and it is rarely possible to know the local concentrations of various reacting species, sensitizers, quenchers and traps. Many scientists have discussed specific quenchers for various reactive species including singlet oxygen, superoxide ion, hydroxyl radicals and peroxide radicals. Interconversions and interactions between reactive species complicate the reaction process even further. In both Type I and Type II reactions the initial products are often peroxides which can break down into free radical reactions. These secondary reactions are known to cause much of the damage observed in membranes under specific conditions. Methods for assessing the relative importance of various processes are needed.

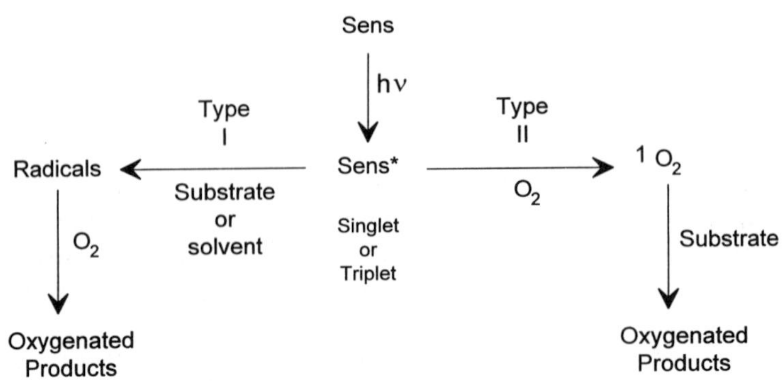

Figure 8.5 The two sensitizing reactions involving oxygen

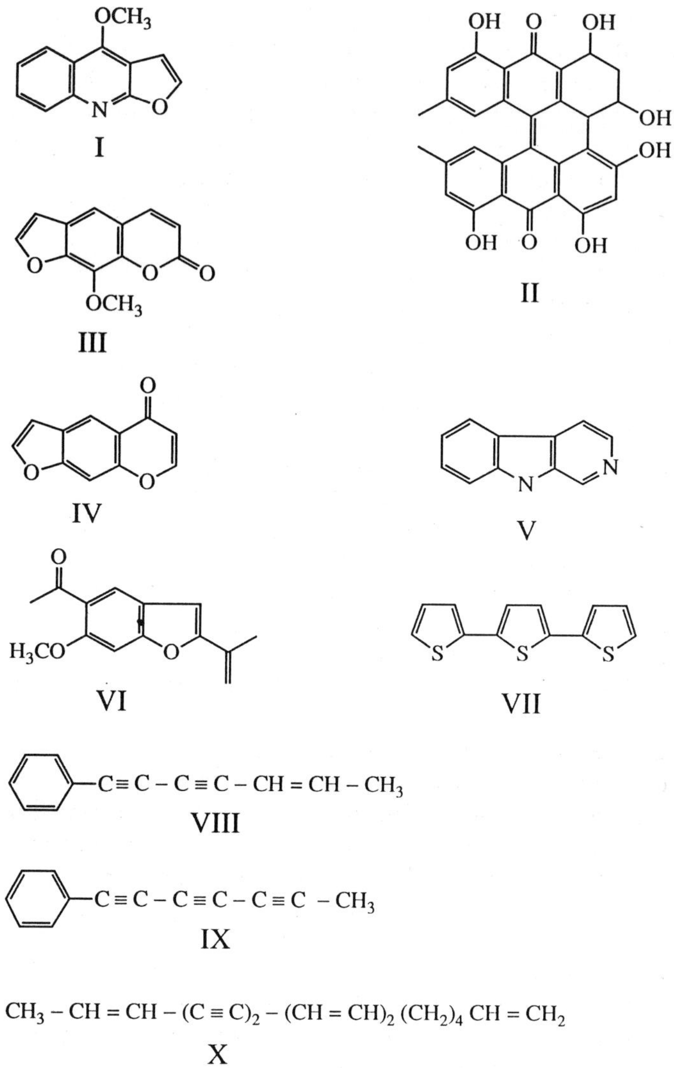

Figure 8.6 Some commonly occurring sensitizers

The detection of an intermediate is a necessary but not sufficient condition to conclude that this particular molecule has a decisive role in the reaction. Merely detecting the intermediate without being able to estimate what fraction of the overall reaction it causes is of little interest. This requires sophisticated instrumental analysis not always available. In heterogenous systems (physiological systems), this quantification is rarely realized. However, as long as one gets the biological reaction intended in the end, correct interpretation of the reaction kinetics can be left for the future. Some important sensitizers are shown in Fig. 8.6. The use of these compounds as direct drugs or more sophisticated as formulating agents of new drugs will be an interesting field in the years to come.

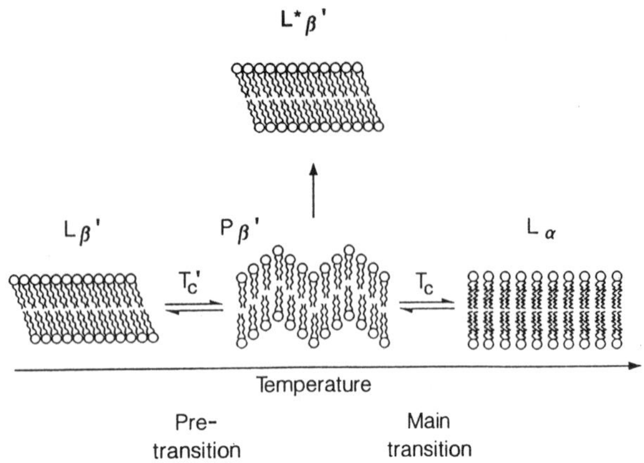

Figure 8.7 Change of the lipid bilayer when the transition temperature is passed

8.5 Triggered release of drugs from liposomes

Drug targeting by liposomal preparations is being investigated by a large number of scientists. Using various means like size, specific surface modifications and antibody attachments, the liposomes can be made to give a specific distribution in the body. If the target cannot be reached by this means one has the same problem as with normal drug treatment – the body is bombarded with a large number of molecules and, hopefully, a tiny fraction will reach the active site and work. Encapsulation in liposomes can, however, drastically reduce side-effects and improve pharmacokinetics of delivery to the target site and thereby improve the therapeutic index. A serious limitation to the therapeutic use of liposomes has been the difficulty of directing them to specific target sites. Liposomes administered intravenously are rapidly taken up by the reticuo-endothelial system and destroyed. High liposomal concentrations are thereby achieved in organs such as liver, spleen and bone marrow. Liposomes with encapsulated cytotoxic drugs can be effective in treating tumours that infiltrate these organs, but are much less useful in treating targeted tumours or other cell structures in other anatomical locations. A way of overcoming this problem might be to initiate drug 'leakage' at other sites in the blood circulation. In this case, if the drug is encapsulated before the distribution in the body and activated at the site of action we would have another example of drug targeting of obvious interest in patient treatment. If the encapsulation consists of phospholipids we have to trigger a release of the drug encapsulated by changing the liposome membrane at a given site. We know that the liposome membrane can be made sensitive to temperature, pH and light.

Several methods for initiating leakage have been invented. One approach has been to heat a liposomal saturated target (Fig. 8.7) site above a critical temperature range, the heating being done with radio frequency heating of target tissue (Fig. 8.8). Another way of organizing specific leakage is to make pH-sensitive liposomes which releases their content into low pH regions. Such areas of low pH are sometimes found in tumours.

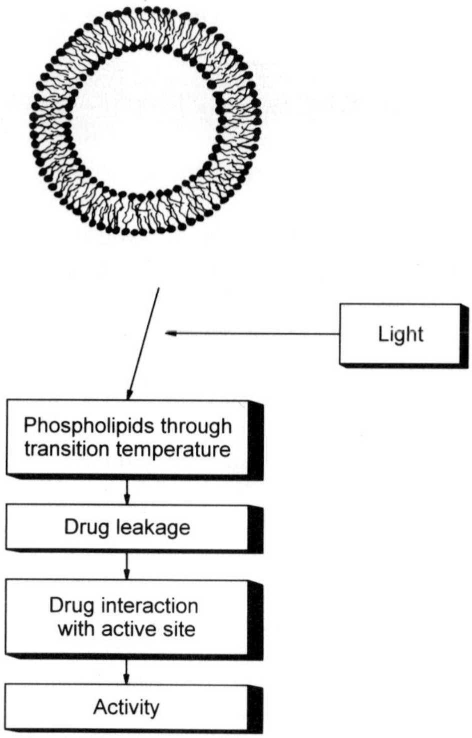

Figure 8.8 Schematic change of liposome to achieve drug action locally

Use of light-sensitive liposomes is reported where the fatty acid chain in the phospholipid undergo a trans to cis isomerization upon the irradiation with an appropriate wavelength of light to allow the fluid content of the liposome to diffuse through the membrane into the surrounding fluid. In this case UV light was used. A patent from the UK described an analogy with the use of sensitizers by incorporating a sensitizer into the liposome membrane. The sensitizer would alter the membrane and release the content when light was absorbed. Several other reports have been made describing the release of particular content using UV or visible light. However, none of these systems has used biologically compatible lipids.

The use of sensitizers in photodynamic therapy does still have problems. Very high systemic doses of the sensitizer must often be given to achieve therapeutic levels at the irradiated tumour site and therefore many sites in the body are non-selectively infiltrated by the sensitizer. The low solubility of the sensitizer reduces their practical usefulness because intravascular administration can cause thrombosis.

A novel system using biocompatible lipids in a liposome formulation has been described for light-triggered release of content which may give an interesting lead to further development (Thompson and Andersen, 1992). The phospholipids of the liposomal membrane contain a vinyl ether functionality which, when cleaved, causes local disruption of the liposomal membrane. The liposome also contains a substance with lysigenic properties that can cleave the vinyl ether as a response to an external triggering event such as illumination with light and change in pH.

When the liposomes are used in this case, the lysogenic substance (sensitizer) can be encapsulated either in the membrane or in the aqueous volume within the liposome. The sensitizer that cleaves the vinyl functionality preferably produces reactive oxygen or acid in response to illumination of a wavelength which is absorbed by the substance. The drug, cosmetic substance or the diagnostic substance can be carried in the liposome membrane or in the aqueous volume in the liposome. When the local disruption of the membrane is carried out, the drug can escape from the liposome and be delivered directly to the target. Liposomes can be administered to the patient as a topical preparation or as an intravenous preparation. The target site is then illuminated with light in the red or near infrared portion of the spectrum. Light with a wavelength above 640 nm or preferably above 800 nm will penetrate deeply into the tissue.

The vinyl functionality is broken by direct acid hydrolysis or acids generated by a sensitizer following illumination. The reaction mechanism of the cleavage of the vinyl ether functionality is described in a monograph by Snyder (1972).

We need sensitizers that generate reactive oxygen species by illumination at a wavelength above 6–700 nm or sensitizers that reduce the pH in response to photoillumination.

A number of sensitizers that react to high wavelength light are given in Fig. 8.6. Compounds that reduce pH as a response to illumination are 4-formyl-6-methoxy-3-nitrophenoxyacetic acid, triarylsulphonic acids and dibenzenesulphonyldiazomethane.

The reactivity of the vinyl structure in a phospholipid depends upon the distance from the glycerol backbone. The vinyl ether functionality should lie within four to six carbon atoms from the glycerol backbone. We also know that additional double bonds conjugated with the vinyl group will facilitate cleaving of the chain. A short alkyl chain will also break easier. However, more stable liposomes with lower leakage of the encapsulated drug are obtained with longer alkyl chains. The competing considerations of liposomal stability and rapid-triggered release may be balanced by making use of phospholipids with one short chain (12–14 carbons) and one long chain (18–24 carbons).

Drugs are loaded into the liposome by co-solubilizing the drug with the phospholipids at the formation of the liposomes. A hydrophobic drug is co-solubilized with lipid in an organic solvent. The drug will primarily be in the liposome membrane. Water-soluble drugs can be sequestered in the aqueous volume of the liposome by several cycles of freeze-thawing. Charged amphipatic drugs can be loaded into the liposome by using a transmembrane pH gradient (Bally *et al.*, 1985).

The choice of sensitizers is guided by the molar absorptivity at the wavelength chosen for irradiation. However, for drug delivery one may first want to consider the hydrophilicity or hydrophobicity of the drug. A sensitizer should then be chosen that has the opposite solubility in water or lipid. The available light source will, in addition, guide the choice of sensitizer. When the triggering is to be effected by radical oxygen species, a sensitizer must be preferred that has a high quantum yield ratio in transferring energy to oxygen and exciting it. When triggering is to occur by pH reduction, a sensitizer may be preferred that has a high quantum yield for production of hydrogen ions. Model experiments with photosensitized release

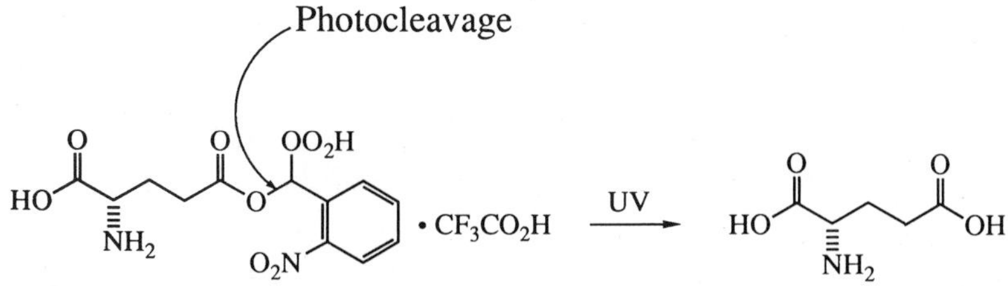

Figure 8.9 Model for photocleavage of prodrug

of glucose and of calcein from disrupted liposomes have been described. It seems that this way of targeting drugs shows a lot of promise.

8.6 Release of drugs from a prodrug compound (caged compounds)

The term caged compounds is given to synthetic molecules whose biological activity is controlled by light, usually by photolytic conversion from an inactive to an active form (Fig. 8.9). The term caged is based upon an early concept that small biologically active molecules can be trapped inside a large matrix that could be opened upon illumination. In almost all caged compounds so far, simple covalent bonding makes some feature important for biological recognition. Photochemical cleavage of that single bond releases the active compound. Again, the usefulness of light triggered release is due to the fact that light can be easily controlled in timing, location and amplitude. Abrupt or localized changes in concentrations of drugs can be generated with controlled amplitudes. This is again what we want to achieve in drug targeting. The use of caged compounds represents one of the best methods to study fast kinetics or spatial heterogeneity of biochemical responses in such systems. Drug targeting of microscale proportions can also use the same methods. The applications of caged compounds in cell biology and biochemistry have been the topic of many reviews during the last decade (Lester and Nerbonne, 1982; Gurney and Lester, 1987; Kaplan and Somlyo, 1989; Walker *et al.*, 1989; Homsher and Millar, 1990; Somlyo and Somlyo, 1990). These applications can often be seen as drug targeting on microscale possibly leading to practical therapy in the future.

The caged compounds are commonly designed by modifying the desired bioactive molecule with a suitable photoremovable protecting group or caging group. To have useful caged compounds for biological experiments there are several criteria these compounds must satisfy. These criteria are exactly like the ones used in prodrug studies for normal therapy.

1 The caged compound must be inert to the biological system used.
2 It should release the bioactive molecule in high yield at sufficient speed by photolysis at wavelengths of light that will not destroy or disturb the biological system.
3 Any photoproducts formed other than the intended biomolecule should not interact or interfere with the biological system.

The caged compounds (in this context) were introduced by Engels and Schläger

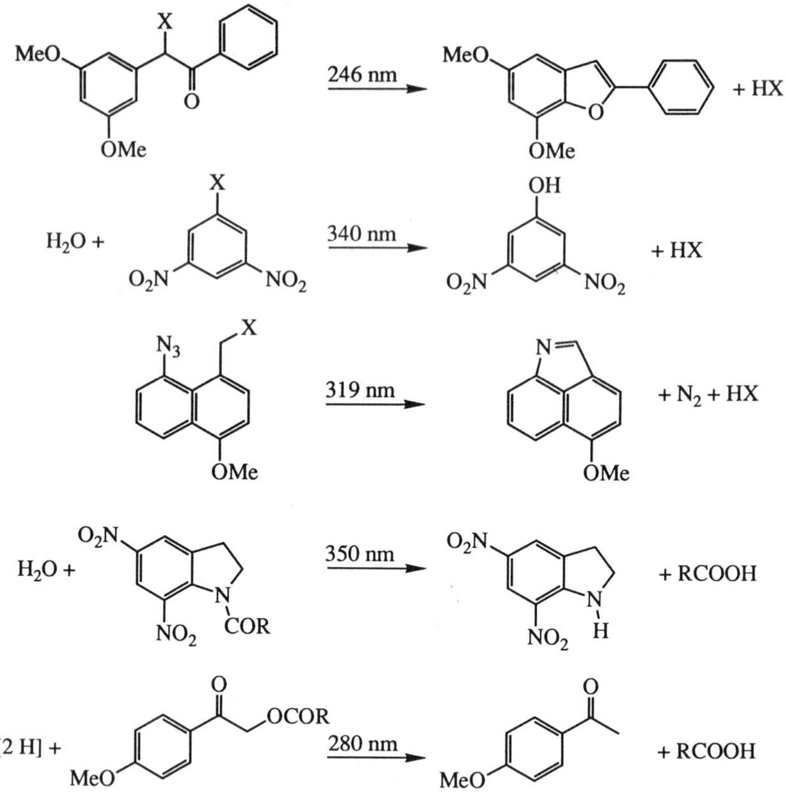

Figure 8.10 I. Functional groups used for caged compounds

(1977) and Kaplan *et al.* (1978). There exists today a large number of caging groups some of whom are shown in Figs 8.10 and 8.11.

The standard comment from biologists is why all the photosensitive groups used in caged compounds require ultraviolet wavelength for photolysis rather than visible wavelengths which would be more convenient optically and better able to penetrate biological tissue. Do we need to restrict the use of caged compounds to UV light? Photobiological research has shown that vision, ion-pumping by bacteriorhodopsin, photosynthesis and phytochrome signalling use visible or far IR photons to accomplish important photobiochemistry. These reactions are impressive and good models since their chromophores are recycled and the reactions of interest do not involve high-energy bonds like nitrogen–nitrogen or nitrogen–oxygen, bonds which we find in nearly all caged compounds. The problems stem from the aims of the scientists. Nature has strongly selected visible or near IR wavelengths since these are most abundant in sunshine while the organic chemist usually had the opposite idea. Molecules sensitive to visible light tend to have more complicated chromophores and are experimentally more difficult to work with than UV-sensing molecules. However, new compounds sensitive to visible light keep on coming to the market and excitation of UV chromophores by two photon-pulsed IR offers new inventives in this field. A very interesting and

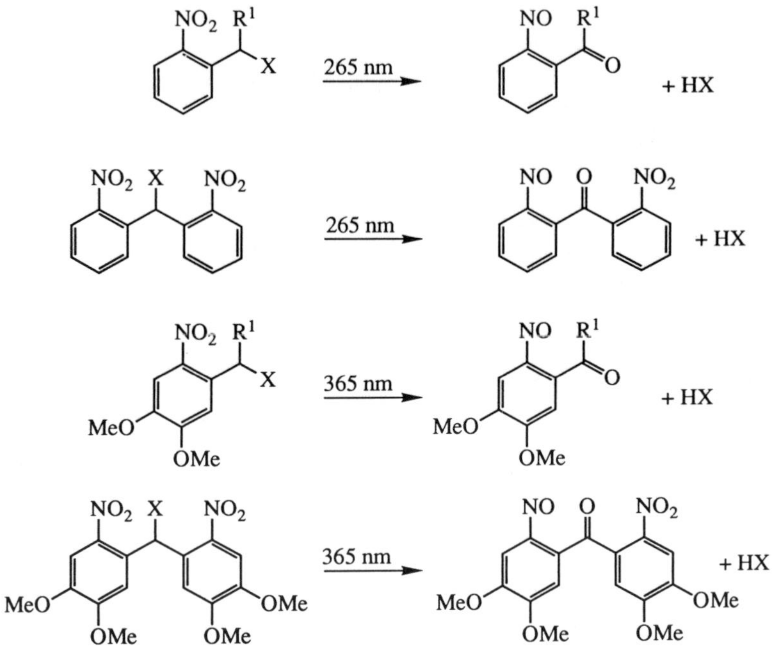

Figure 8.11 II. Functional groups used for caged compounds

stimulating review has been given by Adams and Tsien (1993) pointing out the possibilities of the caged compounds and the outlook for the future.

Most of the applications of caged compounds have emphasized the short temporal change in concentration of drug molecules without spatial control of the drug release. However, the new focusing techniques will allow spatial control of the relcasc too. This should come as a very welcome possibility in experimental medicine. It would then be possible to focus the targeting to very small items like agglomerates of cells or tiny parts of tissue and even down single cells. The biological activity can be followed by fluorescence, and some reports on the spational activation of fluorophores have been published by O'Neill *et al.* (1990) and Parker and Yao (1991). The movement of molecules in a biological system is normally measured by the technique called 'fluorescence recovery after photobleaching', FRAP (Jacobson *et al.*, 1989). For example, fluorescent-labelled proteins or phospholipids are incorporated into biological membranes or cells, allowed to equilibrate, photobleached in a defined region of interest and the movement of unbleached label into this area is then measured. Photoactivation of fluorescence (PAF) obtained by the photolysis of caged compounds attached to biologically active molecules and correspondingly measurement of its movements has been proposed as another interesting technique of a complimentary method (Krafft *et al.*, 1986; Denk *et al.*, 1990). The use of localized illumination of the normal caged compounds cannot be found in many papers probably due to the difficulty of adapting UV illumination through the microscopic lens. A theoretical advantage will be that not only can localized illumination reveal the spatial spread of or response heterogeneity of a cell or tissue, but also has

Dopamine
Epinephrine
Glutamic acid
Haloperidol
Isoproterenol
Norepinephrine
Phenylephrine
Propranolol
Serotonin

Figure 8.12 Drugs available on the market as caged compounds

the advantage that exhaustion of the caged compound is much slower since caged molecules will diffuse into the illuminated area. Although the photon illumination will not be restricted to just one plane (the light will illuminate the object as two cones of light with a focus), a technique such as confocal microscopy will restrict the read-out to that plane. In order to confine the irradiation as much as possible to the focal plane, the technique of two-photon absorption, where a quantum effect of two infrared photons is simultaneously absorbed by a chromophore to simulate a single UV photon of half the wavelength, can be applied (Piston *et al.*, 1992). This method sets high requirements to the laser to be used and restrictions as to the cost and delicacy of the pulsed laser. In view of future development a starting reference for the reader is made to Göppert-Mayer (1931). A possible alternative keeping the traditional wavelengths and the light sources may be to put more than one caging group on the bioactive molecule in such a way that all of them have to photolyse to generate the biological activity. In agreement with what has been proposed by others, the use of two-photon IR excitation seems to offer the most exciting future for microtargeting experiments. A table of selected caged compounds for biological activity studies is shown in Fig. 8.12.

8.7 Photodegradation of matrices containing drugs/temporal drug delivery

Some investigators have been concerned with the photoinduced change of matrices containing drugs to achieve temporal drug delivery. An interesting study of the degradation of hyaluronic acid as a matrix of lipid microspheres is reported by Yui *et al.* (1993). The authors claim that heterogeneous structured gel systems can be used to treat inflammation, for instance, to the eye as the gel will release the lipid spheres in response to surface-controlled degradation of the hyaluronic acid gels and be quantitatively inflammation responsive and biodegradable. Hyaluronic acid which is cross-linked with polyglycerol polyglycidylether is used as matrix in the study. The authors have used methylene blue as a sensitizer to check the release of the lipid spheres from the matrix. They found that the release of the lipid microparticles corresponded well to the degree of degradation of the matrix.

Visible light-induced drug delivery may be advantageous for the photochemical control of pulsatile drug release in many aspects. The photolysis of cross-linked hyaluronic acid gel might be used to achieve in ocular formulations pulsatile release of hypothalamic regulatory hormones to control the endocrine network.

References

ADAMS, S. R. and TSIEN, R. Y., 1993, Controlling Cell Chemistry with Caged Compounds, *Ann. Rev. Physiol.*, **55**, 755–84.

BALLY, N., HOPE, M. J., VAN ECHTELD, C. J. A. and CULLIS, P. R., 1985, Uptake of Safranine and other lipophilic cations into model membrane systems via response to a membrane potential, *Biochem. Biophys. Acta*, **812**, 66–76.

CHIEN, Y. W., 1993, *Novel Drug Delivery Systems* (2nd edn), in *Drugs and the Pharmaceutical Sciences* series, J. Swarbrick (ed.). New York: Marcel Dekker Inc.

DENK, W., STRICKLER, J. H. and WEBB, W. W., 1990, Two-photon laser scanning fluorescence microscopy, *Science*, **248**, 73–6.

ENGELS, J. and SCHLÄGER, E.-J., 1977, Synthesis, structure and reactivity of adenosine cyclic 3,5-phosphate benzyl triesters, *J. Med. Chem.*, **20**, 907–11.

GREGORIADES, S., 1995, Engineering liposomes for drug delivery *Trends Biotechnol.*, **13**, 527–37.

GURNEY, A. M. and LESTER, H. A., 1987, Light-flash physiology with synthetic photosensitive compounds, *Physiol. Rev.*, **67**, 583–617.

GÖPPERT-MAYER, M., 1931, Über elementarakte mit zwei quantensprüngen, *Ann. Phys.*, **9**, 273–94.

HARDING, C. V., COLLINS, D. S., SLOT, J. W., GEUZE, H. J. and UNANOUE, E. R., 1993, Liposome-encapsulated antigens are processed, recycled and presented to T-cells, *Cell*, **64**, 393.

HOMSHER, E. and MILLAR, N. C., 1990, Caged compounds and striated muscle contraction, *Ann. Rev. Physiol.*, **52**, 875–96.

JACOBSON, K., ELSON, E., KOPPEL, D. and WEBB, W., 1989, International workshop on the application of fluorescence photobleaching techniques to problems in cell biology, *Fed. Proc.*, **42**, 72–9.

KAPLAN, J. H., FORBUSH, B. and HOFFMANN, J. F., 1978, Rapid photolytic release of adenosine-5-triphosphate from a protected analogue, *Biochemistry*, **17**, 1929–35.

KAPLAN, J. H. and SOMYLYO, H. P., 1989, Flash photolysis of caged compounds, *Trends Neurosci.*, **12**, 54–9.

KRAFFT, G. A., CUMMINGS, R. T., DIZIO, J. P., FUROKAWA, R. H. and BRVENIK, L. J., 1986, Fluorescence photoactivation and Dissipation (FPD), in *Nucleocytoplasmic Transport*, Peters, R. and Trendelenburg, M. (eds), pp. 35–52. Berlin: Springer Verlag.

KREUTER, J., 1995, Nanoparticulate systems in drug delivery, *J. Drug Targeting*, **3**, 171–4.

LASIC, D. D., 1993, *Liposomes: From Physics to Applications*. Amsterdam: Elsevier.

LEE, V. H. L., 1990, *Peptide and Protein Drug Delivery*, Vol. 4, in *Advances in the Parenteral Sciences*, J. R. Robinson (ed.). New York: Marcel Dekker Inc.

LESERMANN, L., SUZUKI, H. and MACHY, P., 1993, Comments on the application of liposome technology, in *Liposome Technology*, Vol. III, G. Gregoriades (ed.), pp. 275–98. Boca Raton: CRC Press.

LESTER, H. A. and NERBONNE, J. M., 1982, Physiological and pharmacological manipulations with light flashes, *Ann. Rev. Biophys. Bioeng.*, **11**, 151–75.

MEIJER, D. K. F., 1994, Drug targeting with glycoproteins and other peptide carriers, in

Targeting of Drugs 4, Gregoriades, G., McCormack B. and Poste, G. (eds), pp. 1–30. NATO ASI Series, Vol. 273.

O'Neill, S. C., Mill, J. G. and Eisner, D. A., 1990, Local activation of Contraction in isolated rat myocytes, *Am. J. Physiol.*, C 1165–8.

Parker, I. and Yao, Y., 1991, Regenerative release of calcium from functionally discrete subcellular stores by inositol triphosphate, *Proc. R. Soc. London*, Ser. B **246**, 269–74.

Petrak, K., 1993, Design and properties of particulate carriers for intravascular administration, in *Pharmaceutical Particulate Carriers*, Rolland, A. (ed.), pp. 275–98. Drugs and Pharmaceutical Sciences Textbook Series #61. New York: M. Dekker.

Piston, D. W., Wu, E.-S. and Webb, W. W., 1992, Three-dimensional diffusion measurements in cells by two-photon excitation fluorescence photobleaching, *Biophys. J.*, **61**, A34.

Sato, T. and Sunamoto, J., 1993, Site Specific Liposomes coated with Polysaccharides, in *Liposome Technology Vol. III*, Gregoriades, G. (ed.), pp. 179–98. Boca Raton: CRC Press.

Singer, S. J. and Nicholson, G. L., 1972, The fluid mosaic model of the structure of cell membranes, *Science*, **175**, 720–31.

Snyder, F., 1972, *Ether Lipids: Chemistry and Biology*. New York: Academic Press.

Somlyo, A. P. and Somlyo, A. V., 1990, Flash photolysis studies of excitation-contraction coupling, regulation and contraction in smooth muscles, *Ann. Rev. Physiol.*, **52**, 857–74.

Thompson, D. H. and Anderson, V. C., 1992, International Patent PCT/US92/07674.

Tyle, P. and Ram, B. P., 1990, Targeted Therapeutic Systems, Vol. 3, in the series *Targeted Diagnosis and Therapy*, Rodwell, J. D. (ed.). New York: Marcel Dekker Inc.

Walker, J. W, Reid, G. P. and Trentham, D. R., 1989, Synthesis and properties of caged nucleotides, *Methods Enzymol.*, **172**, 288–301.

Yui, N., Okano, T. and Sakurai, Y., 1993, Photo-responsive degradation of heterogenous hydrogels comprising crosslinked hyaluronic acid and lipid microspheres, *J. Contr. Rel.*, **26**, 141–5.

9

Benefits and Adverse Effects from the Combination of Drugs and Light

J. MOAN

9.1 Introduction

The fact that some substances sensitize human skin to light has been known at least since the time of the Pharaohs. The old Egyptians used plant extracts from Ammi Majus (which contains psoralens) and sunlight to treat skin disorders such as leukoderma (vitiligo). In one of India's sacred books, *Atharva Veda* (1400 BC or earlier), photochemotherapy of leukoderma with extracts of Psoralea corylifolia, which contains furocoumarins, is carefully described.

In our century, a new branch of medicine has developed: that of photomedicine. Photomedicine is the science of how drugs, light and biomolecules interact, mainly in human skin. The skin is the largest organ of the human body, covering about 2 m^2 and accounting for about 15 per cent of the total body weight. It is an organ of protcction against external exposures of toxic compounds and intruders and is among the most important sites of the immunological processes in the body.

Every minute a significant fraction of the blood flows through the skin and can thus be exposed to small doses of substances as well as light that can penetrate down to the dermis. At the borderline between dermis and epidermis the basal cells as well as most of the melanocytes reside. Ultraviolet radiation can interact with these cells in such a way that they are transformed to cancer cells and give rise to melanoma- or non-melanoma skin cancer. In the UV-B range (280–320 nm) this interaction is mainly a direct one, i.e. absorption of radiation directly in DNA. However, in the UV-A (320–400 nm) and visible spectral range photosensitized reactions are believed to play the major role in photocarcinogenesis. The photosensitizers involved have not been identified; they may be endogenously formed or derived from food ingredients, topically applied substances such as cosmetic products, or medicines.

The photosensitized reactions in skin can be beneficial, such as in photochemotherapy of cancer (PCT) or psoriasis, or unwanted such as in the cases of photosensitivity induced by chlorpromazine, phenothiazine, tetracycline and a large number of other drugs. These reactions can be classified as photoallergic or phototoxic. Some photosensitized reactions are dependent of oxygen while others are not. Of the oxygen-dependent reactions, those which occur via the formation of singlet oxygen are termed photodynamic reactions or photosensitized reactions of Type II. All other reactions, oxygen-dependent as well as oxygen-independent ones, are termed Type I reactions. These reactions mainly occur via the formation of free radicals.

9.2 Penetration of light and UV radiation into human tissue

Proteins and nucleic acids are the main absorbers in human tissues in the UV-B range (290–320 nm). In the UV-A and visible range melanin and haemoglobin play the main roles as absorbers. All human tissues are inhomogeneous and contain high concentrations of light-scattering particles and boundaries. The concentration and type of absorbers as well as scatterers vary widely from tissue to tissue, and, therefore, also the optical penetration depth, δ, varies within the range 0.2–4 mm.

About 5 per cent of the incident influence is reflected from the stratum corneum and, due to backscattering, the space irradiance increases down to about 0.1 mm. Further down, the space irradiance below a skin surface exposed to a wide beam of parallel light decays approximately exponentially downwards and δ is defined as the distance over which the space irradiance decreases by a factor $e = 2.718$. The thickness of the stratum corneum is typically 10–150 μm and that of the dermis 1–4 mm.

The role of melanin as absorber in the skin is demonstrated by the fact that about 50 per cent of the incident UV-A may reach the basal layer of white skin while only 5–15 per cent may reach the same layer in black skin (Parrish *et al.*, 1978).

The penetration depth of visible light from a Xenon lamp in normal, white skin was measured as a function of wavelength by means of a conventional fluorescence spectrometer equipped with a fibre-optic system (Fig. 9.1). This figure demonstrates that the absorption of haemoglobin plays a significant role in the wavelength regions around 400 nm and 580 nm. Generally, the penetration increases with increasing wavelength up to about 700 nm. Thus, dyes absorbing in the red part of the spectrum can be excited more efficiently in tissue than dyes absorbing at shorter wavelengths. This is taken advantage of when porphyrins, chlorins, phthalocyanines and naphthalocyanines are used as sensitizers in PCT. These sensitizers absorb at wavelengths larger than 600 nm, i.e. outside the main bands of absorption of haemoglobin.

9.3 Type I and Type II reactions

In Type I reactions electrons or H-atoms are transferred between sensitizer molecules and substrate or solvent molecules. Oxygen may participate in subsequent reactions. The probability that a Type I reaction will occur increases with decreasing oxygen concentration and with increasing concentration of substrate. The use of psoralens plus UV-A radiation, PUVA, in the treatment of psoriasis and vitiligo is the best known application of a Type I reaction (Musajo *et al.*, 1974; Harber *et al.*, 1982). Psoralens intercalate in DNA even without light

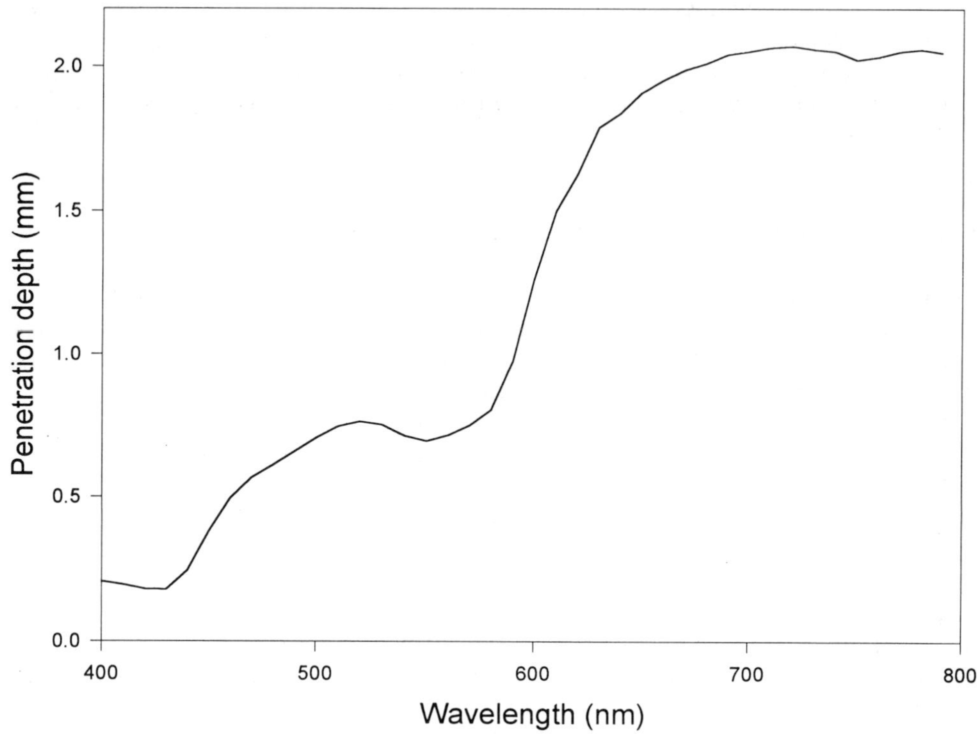

Figure 9.1 The penetration depth of light into living human tissue as a function of the wavelength. The space irradiance decreases approximately exponentially with the depth. The penetration depth is defined as the distance over which the space irradiance decreases by a factor of e = 2.7

exposure. Light exposure, within the absorption spectrum of the psoralens (300–400 nm) leads to a covalent binding of the psoralen to pyrimidine bases in DNA. Monofunctional psoralens, such as angelicines, bind to only one of the DNA strands, while bifunctional ones (linear psoralens) bind to both strands and, therefore, crosslink one strand to the other. This will slow down the rate of cell division, which is abnormally high in psoriatic skin. Psoralens may also act on membranes in a Type II process, and it is not known which of the two processes plays the main role in the treatment. PUVA can also increase the number and size of melanocytes in skin. This is the basis for its use in the treatment of vitiligo (Pathak *et al.*, 1974). It is debatable whether PUVA pigmentation protects DNA in the basal cells. Unpublished experiments carried out in our laboratory show that pigmentation generated by PUVA protects skin against DNA strand breaks caused by subsequent UV exposure (J. Kinley, personal communication). However, it seems that PUVA treatment may be associated with a significant carcinogenic risk (Lindelöf *et al.*, 1991).

Photophoresis is a variant of PUVA treatment (Edelson *et al.*, 1987). Leukocytes from patients with cutaneous T-cell lymphoma or autoimmune disorders (rheumatoid arthritis and systemic sclerosis) are exposed to UV-A radiation in the presence of a psoralen and given back to the patients. It is believed that PUVA changes some surface markers of T-cells and thus triggers the immune system. Thus, this treatment can be regarded as a beneficial application of a photoallergic reaction. Experiments are now going on with BPD (a benzoporphyrin derivative called benzoporphyrin derivative monoacid ring A or Verteporfin) in the treatment of immunologically related diseases (Levy *et al.*, 1994). Since BPD absorbs light in the wavelength region 650–700 nm, where the penetration depth into tissue is large (Fig. 9.1), transcutaneous light application may be efficient.

In Type II reactions (photodynamic reactions), singlet oxygen is formed by energy transfer from sensitizers in the lowest excited triplet state to oxygen molecules in the ground state. Thus, no Type II reactions can occur in the absence of oxygen. For porphyrins in cells energy transfer to oxygen is equally efficient as decay of the excited porphyrin molecules at an oxygen concentration of about 15 μM or 1 per cent partial pressure (Moan and Sommer, 1985). This oxygen dependency is quite similar to the oxygen dependency of gamma- and X-ray irradiation. Many sensitizers have a very high quantum yield of singlet oxygen, typically 0.2–0.6. The lifetime of singlet oxygen is about 10 times longer in D_2O. This fact is often used to check whether singlet oxygen is involved in a photosensitizer reaction. However, if singlet oxygen is generated in cell membranes, from lipophilic sensitizers, only a small negligible D_2O effect is expected, since singlet oxygen usually reacts while diffusing in the membrane and before it reaches the aqueous phase. Using two dyes in different concentrations, one generating singlet oxygen during light exposure and the other being degraded by singlet oxygen, we estimated the lifetime of singlet oxygen in cells to be about 10–40 ns, corresponding to a diffusion radius $\delta = (6Dt)^{1/2} = 0.01$–$0.02\ \mu$m (Moan and Berg, 1991). D is here the diffusion coefficient of oxygen and $t = 10$–40 ns. This radius is very small compared with a typical cell diameter ~10 μm. Even the size of cell organelles (~ 0.5–1 μm) is at least an order of magnitude larger than δ. We therefore conclude that photodynamic damage to cells is always induced close to cellular locations where the concentration of sensitizer is high. This is in agreement with the observations: Most of the lipophilic sensitizers used in photochemotherapy of cancer are not taken up by cell nuclei, and sensitize photodamage only to a small fraction of the DNA that is localized close to the nuclear membrane (Kvam *et al.*, 1992). In fact, using a method based on this observation, we were able to estimate the average length of DNA between each attachment point to the nuclear membrane in interphase cells to about 180 kilobases (Kvam *et al.*, 1992). One hundred and eighty kilobases correspond to 60 μm of the 30 nm fibre.

To prove that a photosensitizer reaction is a Type II reaction is not easy. Several scavengers, such as β-carotene, react efficiently with singlet oxygen and can be used to indicate its involvement in a reaction. A hydroperoxide formed when singlet oxygen reacts with cholesterol can be used as an indicator (Suwa *et al.*, 1978). However, no test is really conclusive since singlet oxygen may be formed in the membranes of cells (where there is no water available for the D_2O effect to be checked) or in the aqueous phase of the cytoplasm (where there is no cholesterol and where the lipophilic β-carotene does not localize). Furthermore, all compartments of cells contain large concentrations of molecules that react with singlet oxygen, shorten its lifetime and reduce its steady state concentration to below levels where its diagnostic 1240 nm luminescence can be detected.

Singlet oxygen reacts efficiently with and degrades a number of biomolecules, the most important ones being guanine, tryptophan, tyrosine, histidine, cysteine, cystine and cholesterol. It causes proteins crosslinks and DNA strand breaks.

9.4 Phototoxic-, photoallergic- and drug-induced photosensitivity reactions

In 1900 Oscar Raab reported in *Zeitschrift für Biologie* that paramecia were killed by exposure to acridine and light. Neither acridine alone nor light alone had any harmful effect on the organisms.

A few years later Tappeine and Jodlbauer (1904) demonstrated that oxygen was required for the biological response to acridine and light and termed the process 'photodynamic action'. These are the first reports of *phototoxic* reactions. Later it has been shown that some phototoxic reactions, notably those sensitized by psoralens, do not require oxygen.

Phototoxic reactions in humans occur exclusively on skin exposed to light. Their morphology and clinical symptoms may vary. In some cases a burning and painful sensation is felt during light exposure, while in other cases reactions such as erythema, oedema and vesiculation occur later. On the cellular level the phototoxic reactions may involve DNA, proteins, lipids, lysosomes, mitochondria and the plasma membrane.

Phototoxic reactions can be caused by a number of drugs: aminobenzoic acid derivatives, anthiaquinone dyes, chlorothiazides, chlorpromazine, anthracine, acridine, pyrene, nalidixic acid, phenothiazine, protriptyline, psoralens, sulfanilamide and tetracycline being among the most common ones (Harber *et al.*, 1982).

The phototoxic effects of chlorpromazine may be partly due to its metabolites (see references in Harber *et al.*, 1982). Both photodynamic and non-photodynamic processes have been demonstrated for this drug. Stable photoproducts of the drug may account for some of its effects. The action spectrum for chlorpromazine phototoxicity is redshifted (max.at 330 nm) compared with its absorption spectrum (max.at 305 nm). This may be due to the shape of the penetration spectrum of light into tissue and/or to metabolism producing compounds that absorb at longer wavelengths. Also, the action spectrum for tetracycline phototoxicity seems to be redshifted to a position with a peak value at about 400 nm, and demethylchlorotetracycline is more potent than the parent compound tetracycline.

On the other hand, the action spectrum for phototoxicity caused in human skin by sulphanilamide correlates well with the absorption spectrum of the drug (see references in Harber *et al.*, 1982). A photooxidation product, p-hydroxylaminobenzene, seems to have toxic effects. Sulphanilamide does not induce only phototoxic, but also *photoallergic* reactions. Photoallergic reactions require immunologic responses. The general scheme for such a reaction is that antigens that activate lymphocytes are produced by photosensitized reactions of proteins. The sensitizer in the skin is photoconverted to a product that reacts with proteins to form antigens. In contrast to phototoxic reactions, photoallergic reactions are systemic and can result in symptoms both in irradiated and unirradiated tissue. Most photoallergic reactions can be considered to involve delayed-type hypersensitivity immunologic mechanisms. Photoallergic reactions can be caused by low sensitizer concentrations, while larger sensitizer concentrations are usually needed to elicit phototoxicity. The most widely known sensitizers of photoallergy include: aminobenzoic acids, bithionol, chlorpromazine, chlorpropamide, fentichlor, salicylanilides, 6-methylcoumarin, promethazine, sulfonilamide and thiazides (Harber *et al.*, 1982).

Among the *drug-induced photosensitivity* diseases different variants of the porphyrias are best known. *Porphyria cutanea tarda* can be induced or aggravated by chlorophenols, oestrogens, ethanol, hexachlorobenzene and 2-benzyl-4, 6-dichlorophenol.

The 5-aminolevulinic acid (ALA) synthetase is the rate-limiting enzyme in heme biosynthesis. A number of chemicals, including hypnotics, sedatives, analgesics, antirheumatics, antibiotics, hypoglycemic agents and sex steroids, are known to induce ALA synthetase (Harber *et al.*, 1982). This enzyme seems to be controlled by an operator gene that can be regulated by an aporepressor. Certain chemicals can inhibit this repressor and thus result in an uncontrolled ALA synthetase activity and elevated porphyrin production. In heme biosynthesis both hydrophilic and lipophilic porphyrinogens are involved. When overproduced, the porphyrinogens are easily oxidized to photoactive porphyrins. As a general rule the hydrophilic porphyrins sensitize photodamage to lysosomes, while the lipophilic ones sensitize photodamage to mitochondrial membranes and other membrane systems (Sandberg, 1981; Sandberg and Romslo, 1982). As discussed below, application of ALA can be used to induce overproduction of protoporphyrin IX (Pp IX) in tumours; a phenomenon that is being applied in photochemotherapy.

In addition to porphyrias, lupus erythematosus and pellagra can be induced by photosensitization.

9.5 Treatment of adverse photosensitized reactions in humans

The main treatment of the diseases mentioned above is, whenever possible, to stop the exposure to photosensitizing substances or substances that initiate endogenous production of photosensitizers. In cases when this is impractical or impossible, quenchers of singlet oxygen, such as betacarotene, can help if photodynamic processes are involved. Other treatments of porphyrias, lupus erythematosus and pellagra are described in textbooks of dermatology.

9.6 Therapies based on the use of photosensitizing drugs

The use of psoralens in the treatment of psoriasis and vitiligo has been discussed above. Psoriasis is a very common disease and photochemotherapy is a frequently used therapy. Due to the carcinogenic effect of this treatment (Lindelöf *et al.*, 1991), photosensitizers other than DNA-intercalating psoralens are sought. It seems that ALA induction of Pp IX may offer a good alternative.

Photochemotherapy (PCT) of cancer is a rapidly developing field. Since the introduction of porphyrin derivatives in clinical PCT in the late 1970s, a number of sensitizers have been proposed for PCT: phthalocyanines, chlorins, benzoporphyrins, porphyrins and Pp IX induced by ALA. Some of these sensitizers are used in clinical trials. The most widely used one, a porphyrin mixture called Photofrin, is approved for PCT of cancer in Japan, Canada and The Netherlands. When injected in tumour-bearing animals (or humans), these sensitizers are taken up and/or retained in tumours with some selectivity with respect to normal tissues such as skin, muscles and brain (Moan and Berg, 1992). The reasons for the tumour selectivity are only partly understood. One explanation may be that these drugs, notably the lipophilic ones, tend to bind to lipoproteins in serum. Since tumours have a high concentration of receptors for low-density lipoproteins, drugs that bind to these lipoproteins may also be led to the tumours. Many of the photosensitizers mentioned aggregate in aqueous solutions. Since tumours usually contain many macrophages that may take up aggregates by endocytosis, aggregation may play a role in tumour selectivity. The low lymphatic drainage together with the leaky vascular system of tumours are other factors to consider. Finally, since tumours have a low pH, drugs that change from being hydrophilic to lipophilic upon a lowering of the pH from 7.5 to, say, 6.5 are expected to be tumour selective. By injecting glucose in tumour-bearing animals we have shown that the low tumour pH plays an important role for the selective uptake of Photofrin (Peng *et al.*, 1991). Glucose injection selectively lowers the tumour pH as well as the tumour uptake of Photofrin.

Recently, ALA–PCT has become widely used in the treatment of skin tumours, notably basal cell carcinomas (BCCs). Application of excess ALA leads to an accumulation of Pp IX with some tumour selectivity (Kennedy *et al.*, 1990; Pottier and Kennedy, 1993). The last step in heme synthesis, the action of the enzyme ferrochelatase (addition of iron to Pp IX to form heme), is not efficient enough – under such conditions. In contrast to heme, Pp IX is a very efficient photosensitizer. The tumour selectivity may have several reasons such as low ferrochelatase activity and a low concentration of ALA in BCCs. The tumour selectivity may partly be due to the fact that the skin overlaying BCCs often is more permeable than normal skin. We have shown that ALA–PCT can be made more efficient in a number of ways: by gently scraping the skin overlaying the tumours to remove the stratum corneum (T. Warloe, personal communication), by adding penetration enhancers such as dimethylsulfoxide, by applying iron chelators such as EDTA or drugs that inactivate the ferrochelatase and by applying ALA esters instead of ALA itself (work in progress). The ALA esters are less polar than ALA but are converted to ALA by esterases present in the tissues. Ultrasound can also enhance the ALA penetration. Thus, even nodular BCCs with depths of more than 2 mm can be treated. ALA–PCT gives very good cosmetic results.

So far, attempts to treat cutaneous malignant melanomas by PCT have not been successful, supposedly because the light absorption of melanin reduces the penetration depth of light into the melanomas.

9.7 The action mechanisms of PCT on the cellular level

Depending on the conditions, PCT destroys tumours either by a direct effect on the tumour cells or by an effect on the cells of the vascular system (Moan and Berg, 1992). The endothelial cells lining the vascular walls seem to take up large concentrations of photosensitizers, both after intravenous and intraperitoneal administration.

Cells are exposed to light in the presence of photosensitizers can be destroyed through a number of mechanisms. Large light exposures of cells incubated with lipophilic sensitizers lead to an immediate cell inactivation mainly caused by a destruction of membranes. Both mitochondrial membranes and the plasma membrane are destroyed. This damage is seen immediately after PCT: blebs are formed on the plasma membrane and the cells swell and disintegrate. Both lipids and proteins in the membranes are destroyed. The singlet oxygen photoproduct of cholesterol is often found. Under anoxic conditions, such as in the central parts of poorly vascularized tumours, or after prolonged light exposure times, free radical mechanisms (Type I reactions) may play a role. An indication of this is that photoprotoporphyrin is formed in small yields in ALA–PCT. This photoreaction is expected to be more important the lower the oxygen concentration is (Harel *et al.*, 1976).

At low-light exposures, a major cause of cell inactivation is destruction of free tubulin (Berg *et al.*, 1992). After such a destruction less microtubules are formed in the cells and the cells are unable to perform mitosis. This is seen as an accumulation of cells in the metaphase after PCT (Berg *et al.*, 1992). Furthermore, it seems that PCT may trigger apoptosis (Zaidi *et al.*, 1993). Cells surviving low exposures may possess damaged proteins on their surfaces. *In vivo*, such damaged proteins may be recognized by the immune system. This may explain the efficiency of PCT (notably that of photopheresis) as well as the low metastatic potential of tumours after PCT (Canti *et al.*, 1983; Edelson *et al.*, 1987; Sindelar *et al.*, 1991).

9.8 Photodegradation of dyes during PCT

Most photosensitizers are degraded during light exposure (Fig. 9.2). The photoabilities of dyes are widely different as demonstrated by this figure. Under anoxic conditions the quantum yield of degradation is usually smaller than in the presence of oxygen, and other products are formed. Form porphyrins, for instance, chlorins are formed under reducing conditions (Harel *et al.*, 1976). An example of this is the formation of photoprotoporphyrin during light exposure of Pp IX in deaerated solutions. Since oxygen is consumed in many photosensitized reactions, the reaction pathways may change during light exposure under conditions where the oxygen supply is limited such as in a tumour. Since chlorins absorb light at longer wavelengths than porphyrins, and since they are good photosensitizers themselves, one should pay attention to the above-mentioned photoconversion phenomenon, notably, when the optimal wavelength band for therapy is sought. For Pp IX one should use light of wavelengths between 600 and 700 nm in order to excite both the porphyrin and the chlorin. One should also consider that photoprotoporphyrin is more water soluble than Pp IX and therefore has different binding and retention properties in cells and tissues. We have shown that photoprotoporphyrin, formed in cell membranes during PCT, leaves the cell membranes and diffuses out into the surrounding fluid (unpublished data). Many sensitizers are photodegraded much faster in the presence of proteins than in pure buffer solutions. Type I reactions are more likely to occur in the presence of

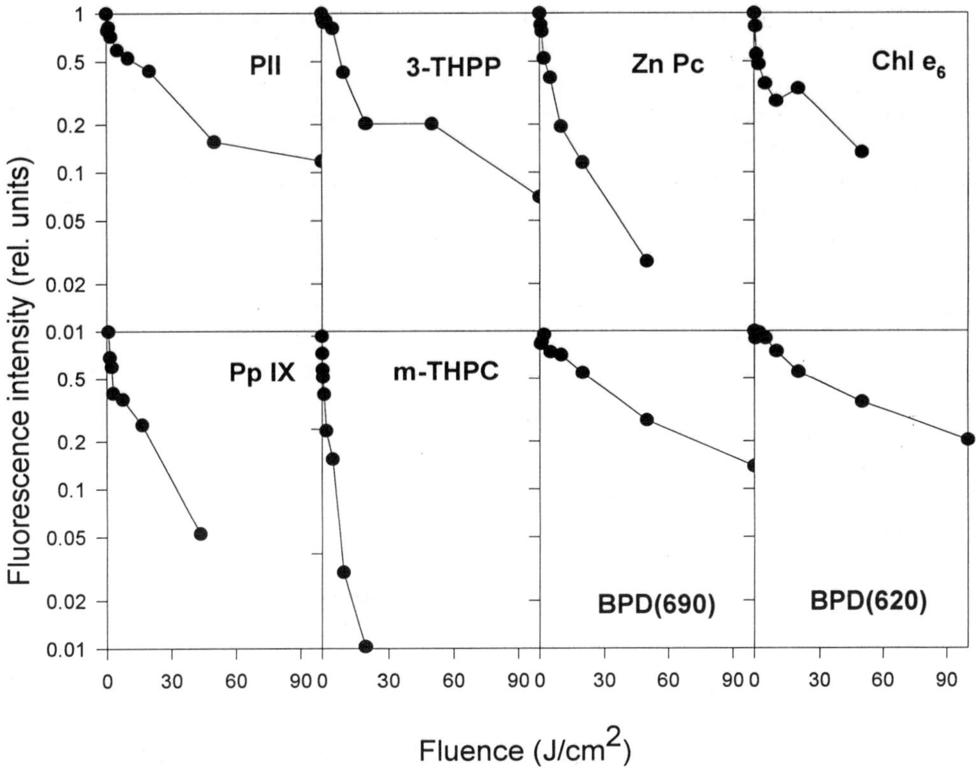

Figure 9.2 The photodegradation of a number of dyes in mouse skin (balb/c nu/nu nude mice), measured fluorimetrically during laser exposure *in vivo* at a fluence rate of 100 mW/cm^2 and at the wavelengths given below. The mice were injected i.p. with the dye 24 hours before laser exposure. PII: Photofrin (10 mg/kg, Quadra Logic, $\lambda = 630$ nm), 3-THPP: 3-tetrahydroxyphenyl porphine (Porphyrin Products, $\lambda = 645$ nm), Zn Pc: Zink-phthalocyanine 10 mg/kg, (Ciba Geigy, $\lambda = 660$ nm), Pp IX: protoporphyrin IX (generated *in vivo* from 5-aminolevulinic acid, $\lambda = 635$ nm), m-THPC: meso-tetra-hydroxy phenyl chlorine (1 mg/kg, Scotia, $\lambda = 670$ nm), BPD: Benzoporphyrin derivative, monoacid ring A (Quadra Logic, $\lambda = 690$ nm, or 620 nm)

proteins, and protein photooxidation products, formed in reactions between singlet oxygen and proteins, may react with and degrade the photosensitizers. Notably, for sulphur-containing amino acids and proteins the latter reaction pathway is known to be important (Krieg and Whitten, 1984a and b). In tissue, of course, this is of utmost importance. Light exposure may result in a degradation of the binding sites of sensitizers on proteins (Moan *et al.*, 1988). Thus, the sensitizer is liberated and may move to other binding sites. It has been shown that during light exposure Pp IX molecules in red cells of patients with erythropoietic protoporphyria move from their binding sites on globin to the cell membrane and further to surrounding cells in contact with the red cells (Brun *et al.*, 1990). This 'photomovement' has been proposed to explain some of the skin pathogenesis seen in these patients. Similarly, light exposure of cells containing hydrophilic sensitizers in their lysosomes leads to a permeabilization of the lysosomes, followed by a transfer of the sensitizers from the lysosomes to the cell nucleus (Moan *et al.*, 1989; Moan *et al.*, 1994). This transfer can take place in cells after sub-lethal PCT. When sensitizers leak out of the lysosomes, they become diluted. This dilution leads to an increase in the fluorescence quantum yield as well as an increase in the quantum yield for cell inactivation (Berg *et al.*, 1991; Berg and Moan, 1994). The increase can be demonstrated *in vivo* as shown in Fig. 9.3. Under other *in vivo* conditions one may observe a transient reduction of sensitizer fluorescence of tumours during PCT, an indication of a transient transfer of the fluorescing molecules to binding sited in the tissue with a lower quantum yield of fluorescence (Moan and Anholt, 1990). Changes in the quantum yield for photoinactivation of cells has been observed also for extralysosomally localized sensitizers and are probably due to a relocalization of the dye during PCT (Moan, 1987). Exposure of a tumour to a small light dose at a time when a significant amount of a sensitizer is in the circulation sometimes leads to an increase in the tumour uptake of the sensitizer (Ma *et al.*, 1992).

9.9 Advantages and disadvantages of using photolabile dyes in PCT

Photosensitizers are widely different with respect to photolability (Figs 9.2 and 9.3). The disadvantage of using a photolabile dye in PCT is that one needs to use large fluences of light for tumour destruction, especially if low drug doses are applied. However, the advantages of using photolabile dyes may be larger than the disadvantages, since the lability can be taken advantage of to obtain a larger therapeutic ratio. With most cancer treatments normal tissue damage limits the therapeutic dose one can apply. For PCT, skin photosensitivity, which may last for weeks after drug application, is the main adverse effect. If one uses a low sensitizer dose, the sensitizer in the skin and in the muscles surrounding the tumour may be photodegraded to non-phototoxic levels before an unacceptable photodamage has been induced in these tissues. The tumour, however, may be inactivated, since it contains 2–10 times more of the photosensitizer than the normal tissues. Such a situation is analysed in Fig. 9.4. As a crude approximation one may assume that the dye is photodegraded according to first-order exponential kinetics (Fig. 9.2, Fig. 9.4 upper panel). This is not far from the truth since the rate of photodegradation (*in vivo* and *in vitro*) is almost independent on the sensitizer concentration (Mang *et al.*, 1987). The rate constant applied in the figure ($1.0\,\mathrm{min}^{-1}$) is typical for photodegradation of porphyrins at fluence rates used in PCT (100–150 $\mathrm{mW/cm}^2$). Two typical survival curves for cells under conditions where no photodegradation occurs, are shown; one for a three times higher concentration than the other. These curves are constructed under the assumption that their shapes are the same as measured curves for high sensitizer concentrations, when photodegradation plays no role since only small fluences are needed for cell inactivation. Such real survival curves are well described according to simple target theory: $Y = \ln\,[1 - (1 - e^{-At})^B]$ (Fig. 9.1). $Y = \ln S$, where S is the surviving fraction. B is a constant and in the present case the best curve fit is obtained with $B = 54$. A is another constant, proportional to the concentration. Yc and Y3c correspond to $A = 4$ and $A = 12$, respectively, in the curves shown in Fig. 9.4. To obtain the survival curves Y′ under conditions of photodegradation

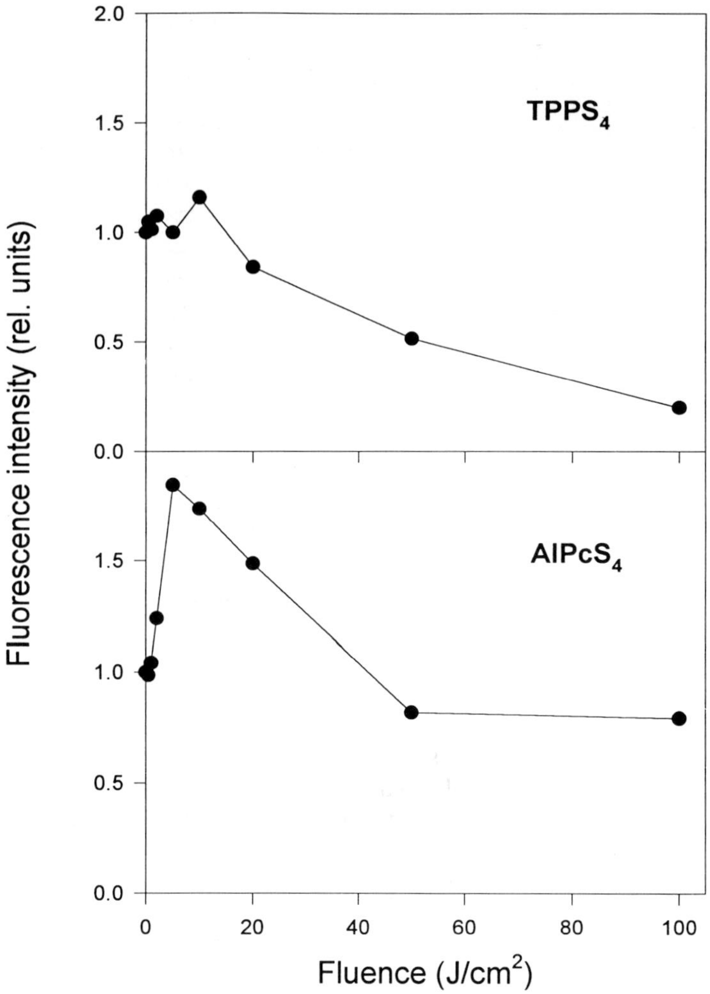

Figure 9.3 The fluorescence of $TPPS_4$ (tetraphenyl porphine tetrasulfonate, (Porphyrin Products), $\lambda = 645$ nm) and $AlPcS_4$ (aluminium phthalocyanine tetrasulfonate, (Porphyrin Products), λ 670 nm) in mouse skin during laser exposure, 100 mW/cm^2, at the wavelengths given above. The dyes were injected i.p. in the mice, 10 mg dye/kg bodyweight, 24 hours before laser exposure

one assumes that the photochemical dose is proportional to the number of emitted fluorescence quanta:

$$D = \int I_o e^{-kt}\,dt = I_o/k\,(1 - e^{-kt}) \quad (9.2)$$

Thus, $Y'(t)$ is obtained simply by replacing t in equation (9.1) with the expression $1/k(1 - e^{-kt})$. The assymptotic value for $Y'(t)$ when $t \to \infty$ is easily calculated: $Y'c\ (\infty) = 0.45$ and $Y'_{3c}\ (\infty) = -8 \cdot Y'_c = -0.45$ corresponds to a survival of $\exp\ (-0.45) = 0.6$ and $Y'_{3c} = -8$ corresponds to a survival level of 3.4×10^{-4}. Normal tissue can supposedly regenerate if only 40 per cent of the cells have been destroyed, while the tumour will probably disappear when only 3.4×10^{-4} of its cells remain alive, since one may assume that the tumour

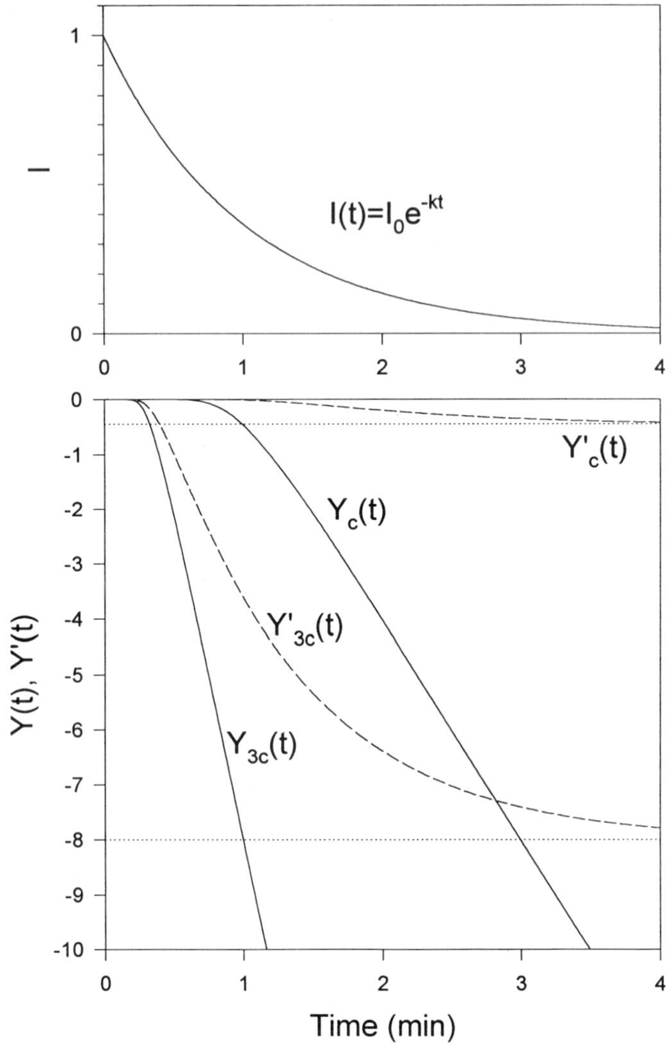

Figure 9.4 Upper panel: Typical *in vivo* photodegradation of a dye, for instance Photofrin, during PCT. To a first approximation the degradation follows exponential kinetics. Lower panel: Typical survival curves for cells in the presence of the same dye in two concentrations, one three times larger than the other. (Y(t) = ln surviving fraction under conditions when no photodegradation occurs; Y′(t) = ln surviving fraction under conditions when photodegradation occurs according to the fluorescence decay curve in the upper panel

vasculature is heavily damaged and cannot supply the tumour with nutrients at such a low survival level. In principle one can use as large light exposure as one needs to kill a tumour at any depth without damaging normal tissue too much. In reality, however, the exposures are limited by the fact that hyperthermia will start to inactivate normal tissue at fluence rates larger than about 150 mW/cm^2. With this limitation the exposure times will be very large for tumours of more than a few mm. With a penetration depth of 1 mm and an exposure time of 1 min to inactivate the cells at the top of the tumour, one would need an exposure time of 2.5 hours to inactivate the cells 5 mm further down.

Acknowledgements

The author is indebted to Vladimir Iani and Mette Jebsen for help with preparation of the figures and typing of the manuscript. The work was supported by the Association for International Cancer Research.

References

Berg, K., Madslien, K., Bommer, J. C., Oftebro, R., Winkelman, J. W. and Moan, J., 1991, Light-induced relocalization of sulfonated meso-tetraphenylphorphines in NHIK 30205 cells and effects of dose fractionation, *Photochem. Photobiol.*, **53**, 203–10.

Berg, K. and Moan, J., 1994, Lysosomes as photochemical targets, *Int. J. Cancer*, **59**, 814–22.

Berg, K., Steen, H. B., Winkelman, J. W. and Moan, J., 1992, Synergistic effects of photoactivated tetra (4-sulfonatophenyl) porphine and nocodazole on microtubuli assembly accumulation of cells in mitosis and cell survival, *J. Photochem. Photobiol.*, **13**, 59–70.

Brun, H., Western, A., Malik, Z. and Sanberg, S., 1990, Erythropoietic protoporphyria: Photodynamic transfer of protoporphyrin from intact erythrocytes to other cells, *Photochem. Photobiol.*, **51**, 573–7.

Canti, G., Ricci, L., Cantone, V., Franco, P., Marelli, O., Andreoni, A., Cubeddu, R. and Nicolin, A., 1983, Hematoporphyrin derivative photoradiation therapy in murine solid tumors, *Cancer Lett.*, **21**, 233–7.

Edelson, R., Berger, C., Gasparro, F., Jegasothy, B., Heald, P., Wintroub, B., Vonderheid, E., Knobler, R., Wolff, K., Plewig, G., McKiernan, G., Christiansen, I., Oster, M., Honigsmann, H., Wilford, H., Koroschka, E., Rehle, T., Perez, M., Stingl, G. and Laroche, L., 1987, Treatment of cutaneous T-cell lymphoma by extracorporeal photochemotherapy, *N. Engl. J. Med.*, **316**, 297–303.

Harber, L. C., Kochevar, I. E. and Shalita, A. R., 1982, Mechanisms of photosensitization to drugs in humans, in *The Science of Photomedicine*, Regan, J. D. and Parrish, J. A. (eds), p. 323–47. New York and London: Plenum Press.

Harel, Y., Manassen, J. and Levanon, H., 1976, Photoreduction of porphyrins to chlorins by tertiary amines in the visible spectral range. Optical and EPR Studies, *Photochem. Photobiol.*, **23**, 337–41.

Kennedy, J. C., Pottier, R. H. and Pross, D. C., 1990, Photodynamic therapy with endogeneous protoporphyrin IX: Basic principles and present clinical experience, *J. Photochem. Photobiol., B:Biol.*, **6**, 143–8.

Krieg, M. and Whitten, D. G., 1984a, Self-sensitized oxidation of porporphyrin IX and related free-base porphyrins in natural and model membrane systems. Evidence for novel photooxidation pathways involving amino acids, *J. Am. Chem. Soc.*, **106**, 2477–9.

1984b, Self-sensitized photo-oxidation of protoporphyrin IX and related porphyrins in erythrocyte ghosts and microemulsions: A novel photo-oxidation pathway involving singlet oxygen, *J. Photochem.*, **25**, 235–52.

KVAM, E., STOKKE, T., MOAN, J. and STEEN, H. B., 1992, Plateau distributions of DNA fragment lengths produced by extended light exposure of extranuclear photosensitizers in human cells, *Nucleic Acid Res.*, **20**, 6687–93.

LEVY, J. G., CHOWDHARY, R., RATKAY, L., WATERFIELD, D., OBOCHI, M., LEONG, S., HUNT, D. and CHAN, A., 1994, Immune modulation using transdermal photodynamic therapy, in *Photodynamic Therapy of Cancer II*, Brault, D., Jori, G., Moan, J. and Ehrenberg, B. (eds), SPIE Proceedings Vol. 2325. Bellingham, WA. 98227-0010 USA: The International Society for Optical Engineering, 155–65.

LINDELÖF, B., SIGURDGEIRSSON, B., TEGNER, E., LARKÖ, O., JOHANNESSON, A., BERNE, B., CHRISTENSEN, O. B., ANDERSSON, T., TÖRNGREN, M., MOLIN, L., NYLANDER-LUNDQUIST, E. and EMTESTAM, L., 1991, PUVA and cancer: a larger scale epidemiological study, *The Lancet*, **338**, 91–3.

MA, L. W. , MOAN, J. and PENG, Q., 1992, Effects of light exposure on the uptake of Photofrin II in tumors and normal tissues, *Int. J. Cancer*, **52**, 120–3.

MANG, T. S., DOUGHERTY, T. J., POTTER, W. R., BOYLE, D. G., SOMMER, S. and MOAN, J., 1987, Photobleaching of porphyrins used in photodynamic therapy and implications for therapy, *Photochem. Photobiol.*, **45**, 501–6.

MOAN, J., 1987, A change in the quantum yield of photoinactivation of cells observed during photodynamic treatment, *Lasers in Med. Sci.*, **3**, 93–7.

MOAN, J. and ANHOLT, H., 1990, Phthalocyanine fluorescence in tumors during PDT, *Photochem. Photobiol.*, **51**, 379–81.

MOAN, J. and BERG, K., 1991, The photodegradation of porphyrins in cells can be used to estimate the lifetime of singlet oxygen, *Photochem. Photobiol.*, **53**, 549–53.

1992, Photochemotherapy of cancer: Experimental research, *Photochem. Photobiol.*, **55**, 931–48.

MOAN, J., BERG, K., ANHOLT, H. and MADSLIEN, K., 1994, Sulfonated aluminium phthalocyanines as sensitizers for photochemotherapy. Effects of small light doses on localization, dye fluorescence and photosensitivity in V79 cells, *Int. J. Cancer*, **58**, 865–70.

MOAN, J., BERG, K., KVAM, E., WESTERN, A., MALIK, Z., RÜCK, A. and SCHNECKENBURGER, H., 1989, Intracellular localization of photosensitizers, in *Photosensitizingcompounds; their Chemistry, Biology and Clinical Use*, pp. 95–111. Harnett, S. (ed.), Wiley, Chichester (Ciba Foundation Symposium 146).

MOAN, J., RIMINGTON, C. and MALIK, Z., 1988, Photoinduced degradation and modification of Photofrin II in cells *in vitro*, *Photochem. Photobiol.*, **47**, 363–7.

MOAN, J. and SOMMER S., 1985, Oxygen dependence of the photosensitizing effect hematoporphyrin derivative in NHIK 3025 Cells, *Cancer Res.*, **45**, 1608–10.

MUSAJO, L., RODIGHIERO, G., CAPORALE, G., DALL'ACQUA, F., MARCIANI, S., BORDIN, F., BACCICHETTI, F. and BEVILAQUA, R., 1974, Photoreactions between skin-photosensitizing furocoumarins and nucleic acids, in *Sunlight and Man*, Fitzpatrick, T. B., Pathak, M. A., Harber, L. C., Seiji, M. and Kukita, A. (eds), pp. 369–87. Tokyo, Japan: University of Tokyo Press.

PARRISH, J. A., ANDERSON, R. K., URBACH, F. and PITTS, D., 1978, UV-A. *Biological Effects of Ultraviolet Radiation with Emphasis on Human Responses to Longwave Ultra Violet.* Chapter 4. New York and London: Plenum Press.

PATHAK, M. A., KRÄMER, D. M. and FITZPATRICK, T. B., 1974, Photobiology and photochemistry of furocoumarins (psoralens) in *Sunlight and Man*, Fitzpatrick, I. B., Pathak, M. A., Harber, L. C., Seiji, M. and Kubita, A. (eds), pp. 335–68. Tokyo, Japan: University of Tokyo Press.

PENG, Q., MOAN, J. and CHENG, L.-S., 1991, The effect of glucose administration on the uptake of Photofrin II in human tumor xenograft, *Cancer Lett.*, **58**, 29–35.

POTTIER, R. H. and KENNEDY, J. C., 1993, A new approach to photodynamic tumour therapy: Forcing the tumour to do most of the work, *Biologija*, **3**, 47–9.

RAAB, O., 1900, Über die Wirkung fluorescierenden Stoffe auf Infusorien, *Z. Biol.*, **39**, 524.

SANDBERG, S., 1981, Protoporphyrin-induced photodamage to mitochondria and lysosomes from rat liver, *Clin. Chimica Acta*, **111**, 55–60.

SANDBERG, S. and ROMSLO, I., 1982, Phototoxicity of uroporphyrin as related to its sub-cellular localization in rat livers after feeding with hexachlorobenzene, *Photobiochem. Photobiophys.*, **4**, 95–106.

SINDELAR, W. F., DEHANY, T. F., TOCHNER, Z., THOMAS, G. F., DACHOWSKI, L. J., SMITH, P. D., FRIANF, W. F., COLE, J. M. and GLATSTEIN, E., 1991, Techniques of photodynamic therapy for disseminated intraperitoneal malignant neoplasm, *Arch. Surg.*, **126**, 318–24.

SUWA, K., KIMURA, T. and SCHAAP, A. P., 1978, Reaction of singlet oxygen with chloresterol in liposomal membranes. Effect of membrane fluidity on the photo-oxidation of cholesterol, *Photochem. Photobiol.*, **28**, 469–73.

TAPPEINER, H. V. and JODLBAUER, A., 1904, Die sensibilisierende Wirkung fluorescierender Substanzer, *Dtsch. Arch. Klin. Med.*, **80**, 524.

ZAIDI, S. I. A., OLEINICK, N. L., ZAIM, T. and MUKHTAR, H., 1993, Apoptosis during photodynamic therapy-induced ablation of RIF-I tumors in C3H mice: Electron microscopic, histopathologic and biochemical evidence, *Photochem. Photobiol.*, **58**, 771–6.

10

Screening of New Drugs for Ocular Phototoxicity

J. E. ROBERTS

10.1 Introduction

The eye is constantly subjected to and interacts with ambient radiation. Over time this can lead to age-related changes and damage in the lens and retina of the eye. (Andley, 1987; Dillon, 1991; Zigman, 1993). This danger is enhanced with increased exposure to intense light because of high altitudes (Hu *et al.*, 1989), outdoor employment (Rosenthal *et al.*, 1988), the use of sun beds or during phototherapy for seasonal depression (Roberts *et al.*, 1992a; Terman *et al.*, 1990). Furthermore, certain dyes and drugs have the potential to induce damage to the lens and retina in the presence of ambient light (Roberts, 1988).

10.2 Light exposure and the eye

10.2.1 Eye risk from the sun

Ambient radiation from the sun can contain varying amounts of ultraviolet (UV)-C (100–280 nm), UV-B (280–320 nm), UV-A (320–400 nm) and visible light (400–760 nm). Most UV-C and some UV-B is filtered by the ozone layer. Ultraviolet light contains shorter wavelengths of light than visible. The shorter the wavelength, the greater the energy and the greater the potential for biological damage.

10.2.2 Optical properties of the eye

Each wavelength of light will affect different areas of the eye. This is due to the unique filtering characteristics of the primate/human eye as shown in Fig. 10.1. The human cornea cuts off all light below 295 nm. This means that the shortest, most energetic wavelengths of light (all UV-C and some UV-B) are filtered from the lens. The adult human lens absorbs the remaining UV-B and all UV-A.

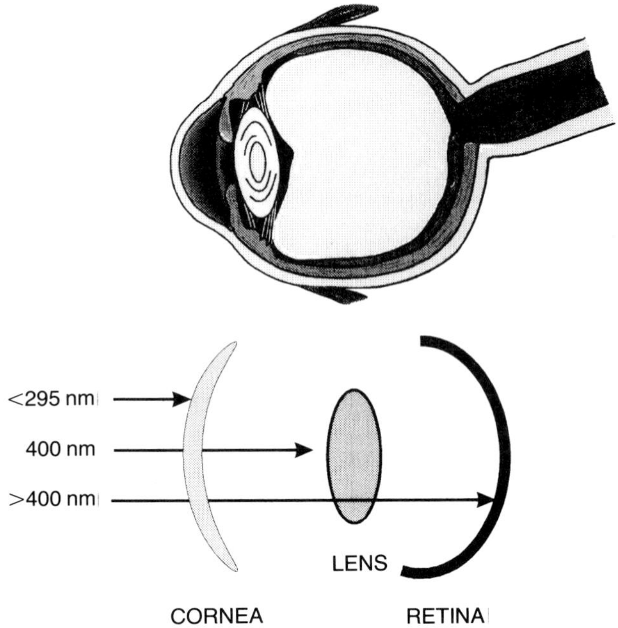

Figure 10.1 The eye as an optical system. The cornea cuts out all light below ca 295 nm, while the lens absorbs most light below 400 nm

Therefore, only visible light reaches the retina. This differs with age. The young human lens transmits a small window of UV-B light to the retina (Barker and Brainard, 1993) and the elderly lens filters out much of the short blue visible light. Aphakia (removal of the lens) and certain forms of blindness may also change the wavelength characteristics of light reaching the retina.

10.2.3 ***Light damage in the eye***

Short ultraviolet light exposure to the cornea leads to an inflammation reaction. This is very painful and similar to a sunburn. However, these corneal wounds will heal. On the other hand, damage to the lens and retina is painless, accumulative and permanent.

Over time, the natural protection against ocular damage induced by ultraviolet radiation is lost. The lens produces enzymes (superoxide dismutase (SOD); catalase) and antioxidants (glutathione) which protect the eye against UV-B light. But these protectants diminish with age. As a result everyone who lives long enough (60–80 years) will get a cataract.

There is also an age-related retinal disease known as macular degeneration. This disease is exacerbated by light and results in damage that can lead to permanent blindness. Again, this age-related disorder is, in part, a result of our protective enzyme and antioxidant systems not working efficiently as we age. Macular degeneration is particularly enhanced in the elderly by ultraviolet light, which may reach the retina after a cataract is removed (aphakia), and short (blue) visible light

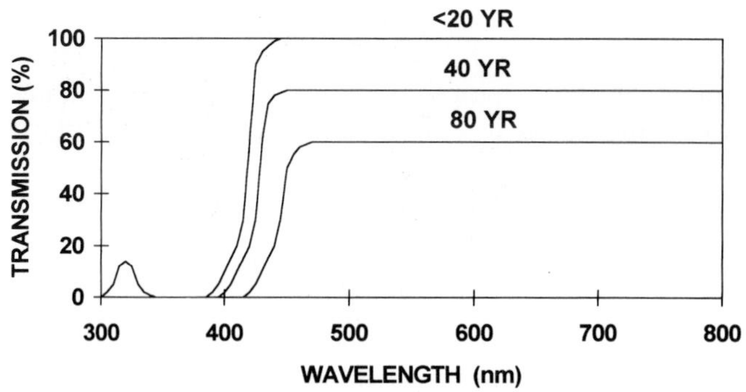

Figure 10.2 The transmission characteristics of the human lens with age showing an increase as a result of an age-related yellowing. The young human lens has a window of light at 320 nm that is transmitted to the retina. The adult human lens transmits only visible light (above 400 nm) to the retina. The aged human lens filters out most of the blue light (400–450 nm) from the retina

Table 10.1 Ocular sites at risk with increased UV-B (depletion of the ozone layer)

Tissue	Pathology		Effect
Cornea	Photokerititis		+/−
Lens	Cataract	Direct	+
		Drugs	++
Retina	Macular	Old	−
	Degeneration	Young	+
		Aphakic	+

which is normally filtered from the elderly lens (Fig. 10.2), but now may be present after a cataract operation.

The ocular injury described above will be specifically increased if there is a decrease in the ozone layer. Ozone filters out most short ultraviolet wavelengths of light. Without the earth's natural protection the human eye, as well as the skin and immune system, will be subjected to severe damaging effects (WHO, 1995). The expected ocular effects are summarized in Table 10.1.

10.2.4 Light damage from artificial sources

In addition to sunlight, artificial light must also be responsible for the increased light damage to the eye even in a hospital setting. Lights used as illumination for the light microscope to remove cataracts may leave the patient with light damage to the retina (Khwarg, 1987). Premature infants in intensive care can suffer permanent damage to their retinas leading to blindness if their eyes are not protected from fluorescent lighting overhead which is not filtered to remove ultraviolet light.

The most recent source of potential light damage to the eye comes from light sources used to treat Seasonal Affective Disorder (SAD). The lamps that are being sold are presumably releasing 'only' intense visible light (above 400 nm), when actually they are not. The lamps are not sufficiently filtered to remove UV-A light released from most of these lamps and this can lead to damage to the eye. Furthermore, even the intense visible (blue 400–450 nm) light may lead to retinopathy and maculopathy (Terman *et al.*, 1990).

Some lamps commonly used in the home, school or office such as quartz-halogen, mercury-metal halide and even 'visible' fluorescent sources have been found to emit significant amounts of harmful UV radiation. Using the ACGIH (American Conference of Governmental Industrial Hygienists) TLV (Threshold Limit Values) maximum exposure times of as short as 10 minutes for some sources and as long as 8 h for others were measured (Sayre and Roberts, 1995).

In summary, the exposure of the eye to direct sunlight or lamps containing UV light is very damaging. It leads to inflammation of the cornea (photokeratitis), clouding of the lens (cataracts) and visual loss due to retinal damage (macular degeneration).

10.3 Drug-induced ocular phototoxicity

Most of the damage to the eye caused by direct irradiation from the sun or artificial sources is from ultraviolet radiation. However, in the presence of a light-activated (photosensitized) diagnostic dye or drug, patients are in danger of enhanced ocular injury to both ultraviolet and visible light.

There are certain factors that allow for the prediction of the potential ocular phototoxicity of a substance.

10.3.1 *Molecular mechanisms of ocular phototoxicity*

The molecular mechanism (Fig. 10.3) by which ocular (and dermal) phototoxicity occurs is through a photosensitized oxidation reaction (Straight and Spikes, 1985). In this reaction, certain wavelengths of light are absorbed by an endogenous sensitizer (i.e. 3-hydroxykynurenine glucoside) (Dillon, 1991) or exogenous sensitizer (diagnostic dyes or drugs) (Fraunfelder, 1982; Lerman, 1986; Dayhaw-Baker, 1987; Roberts, 1988) which is promoted from the ground state to an excited single state. This short-lived excited singlet goes to a longer-lived triplet state through intersystem crossing (ISC) (Turro, 1965). From the triplet state of the sensitizer, there are two major mechanisms which eventually lead to the modification of the substrate (ocular tissues) molecule.

1. *Type I mechanism.* A photoredox reaction of the triplet sensitizer involving electron transfer from the substrate (ocular tissue) to the sensitizer (dye or drug) with the production of free radicals. These can either react rapidly with oxygen to oxidize the substrate or form covalent cross-links between sensitizer and substrate (i.e. 8-methoxypsoralen and DNA) (Gasparro *et al.*, 1985).

2. *Type II mechanism.* The sensitizer triplet reacts directly with oxygen either transferring its energy to ground state (triplet) oxygen to generate singlet oxygen and ground state (non-activated) sensitizer. Singlet oxygen is a very active intermediate and it oxidizes the ocular tissues. A less likely reaction is the

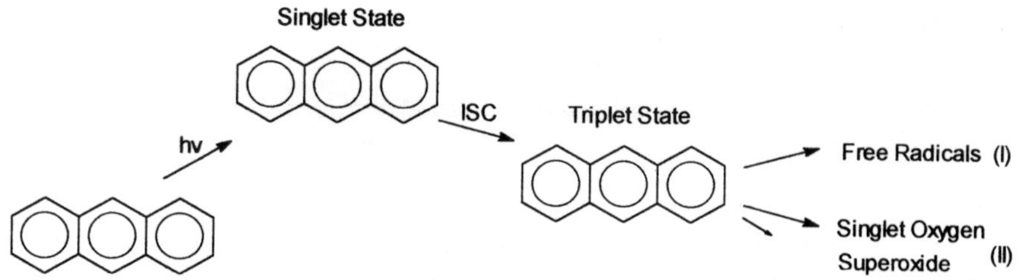

Figure 10.3 The molecular mechanism involved in the phototoxic damage induced in the eye is through a photosensitized oxidation reaction depicted here. This involves the absorption of light by the sensitizer which excites the compound to the singlet state and then, through intersystem crossing (ISC), goes to the triplet state. It is the excited triplet state of the drug/dye that then proceeds to form either free radicals (Type I) or singlet oxygen or superoxide (Type II) which cause the eventual biological damage

transference of an electron to oxygen forming a less reactive intermediate, superoxide.

In the complex biological system of the eye, these photooxidation reactions can occur by either Type I or Type II mechanism or both concurrently. These mechanisms are described in Fig. 10.3.

10.3.2 ***Factors involved in ocular phototoxicity***

The extent to which a particular dye or drug is capable of producing phototoxic side-effects in the eye is dependent on several parameters including:

1 the chemical structure;
2 the absorbance spectra of the drug;
3 the age/and or disease state of the patient;
4 binding of the drug to ocular tissue;
5 oxygen levels of ocular sites;
6 the ability to cross blood–ocular barriers;
7 the site of damage;
8 the effectiveness of the natural defence systems in the eye.

1. *Chemical structure*. The chemical structure of a drug gives the first clear indication of potential phototoxicity. The most potent photosensitizers usually have structures which are heterocyclic, tricyclic or porphyrin-related ring systems (Figs 10.4 and 10.5). This is because compounds with these structural characteristics have long-lived triplet states, low oxidation potential (favouring Type I reactions) and/or their triplet state energy is such that transfer of energy to ground state oxygen is possible (favouring a Type II reaction) (Straight and Spikes, 1985).

2. *Absorbance spectra*. In order for a chemical compound (diagnostic dye, drug, endogenous sensitizer) to induce a phototoxic response in any biological tissue,

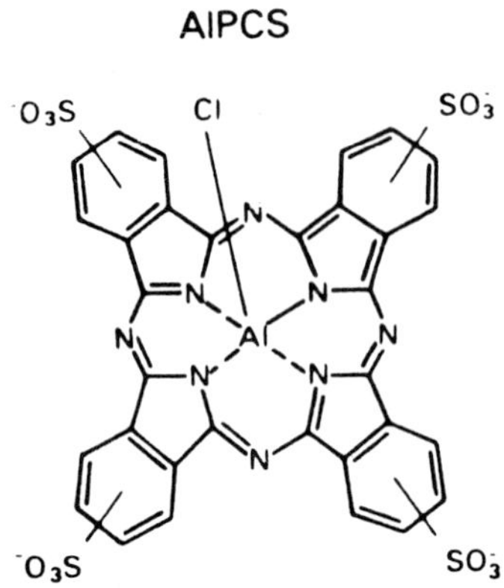

Figure 10.4 The structure of Aluminium Phthalocyanine. A typical porphyrin-type ring system

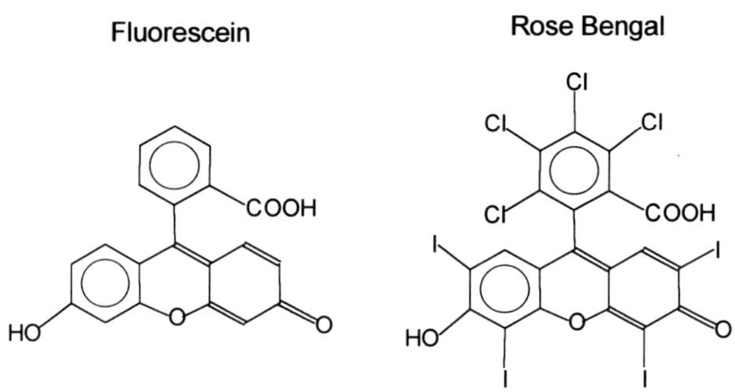

Figure 10.5 The structures of fluorescein and Rose Bengal. They have hetercyclic ring systems

it must first absorb light. This absorption is limited by the filtering characteristics of the biological tissues involved. To damage the human aqueous or the lens, a drug need only have a UV spectrum that consists of absorptions longer than 295 nm (Bachem, 1956). This includes drugs such as chlorpromazine (max, 310), tetracycline (max, 365) and the porphyrins (392; (500–650 nm)). An example of this is seen in Fig. 10.6 (Roberts *et al.*, 1992a).

In the older human, only drugs/dyes which have absorptions above 400 nm could produce phototoxic damage to the retina. However, in aphakic (patients with their lenses removed after a cataract operation) and very young humans (Fig. 10.2) all drugs/dyes absorbing above 295 nm are potential photosensitizers in both the retina as well as the lens.

3. *Age/disease state of the patient.* As described above, the transmission characteristics of the human lens vary with age (Dillon, 1991; Dillon and Atherton, 1990; Dillon *et al.*, 1990) and certain disease states (i.e. porphyria; Roberts *et al.*, 1991a and b). Therefore, a drug with an absorbance in the near UV or visible range is a potential photosensitizer of the lens or retina or both. The presence

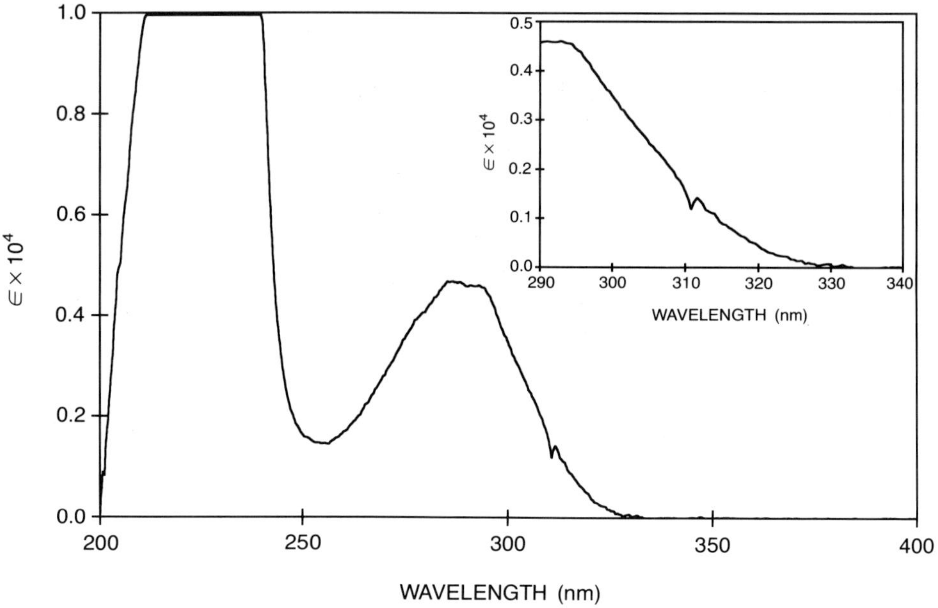

Figure 10.6 Absorbance Spectra of iprindol, a prescribed psychoactive drug. Since this drug has an end point absorbance of above 330 nm, transmitted light can activate the drug in the lens, therefore it is a potential photosensitizing drug in the eye

of an efficient endogenous sensitizer (i.e. porphyrins) may act synergistically with other exogenous photosensitizing drugs or dyes.

4. *Binding studies*. Binding a drug to ocular tissues (DNA–cornea, lens; proteins–lens; melanin–retina; Roberts, 1988; Steiner and Buhring, 1990; Sarna, 1992) would increase its retention time in the eye. Furthermore, binding of a photosensitizing substance to macromolecules increases the lifetime of its triplet state. It is the triplet state of the dye or drug that leads to further oxidative and free radical reactions (Fig. 10.3). Therefore, drug/dyes that bind to ocular tissues are very likely to induce phototoxic damage (Roberts *et al.*, 1991a) in that organ.

5. *Oxygen tension*. The cornea is highly oxygenated. The retina is supplied by the blood so it has varying but high oxygen content in different portions of retinal tissues. The aqueous and the lens have low oxygen content but it is sufficient for photooxidation to occur (Roberts *et al.*, 1992b; Kwan *et al.*, 1971).

6. *Blood–ocular barriers*. These barriers should be amphiphilic or lipophilic in order to cross blood–retinal and/or blood–lenticular barriers (Roberts, 1991a, Lerman, 1986). The site of damage will be determined by the hydrophobicity or hydrophilicity of the photosensitizing dye or drug.

7. *Potential sites of phototoxic damage in the eye*. There are numerous substrates for phototoxic damage in the eye. The site of damage is determined by the penetration of the drug and the transmission of the appropriate wavelengths of light to that site.

Cornea. Corneal epithelial and endothial cells may be easily damaged leading to keratitis (Pitts *et al*., 1976; Hull *et al*., 1983) However, these cells have a very efficient repair mechanism and the damage is rarely permanent.

Lens. The epithelial cells of the lens have direct contact with the aqueous. Their function is to control transport to the lens. They are most vulnerable to phototoxic damage. Damage to these cells would readily compromise the viability of the lens (Roberts *et al*., 1994a). The lens fibre membrane can be photochemically damaged through damage to the lipids and/or the main intrinsic membrane protein (Roberts *et al*., 1985). This will result in a change in the refractive index causing an opacification. Phototoxic reactions can lead to a modification of certain amino acids (histadine, tryptophan, cysteine) (Roberts, 1984; Roberts and Dillon, 1987; McDermott *et al*., 1991) and/or a covalent attachment of sensitizer to cytosol lens proteins. In either case, this changes the physical properties of the protein leading to aggregation and finally opacification (cataractogenesis). The covalently bound chromophore may now act as an endogenous sensitizer producing prolonged sensitivity to light. Since there is little turnover of lens proteins this damage is cumulative.

Retina. Phototoxic damage can occur in retinal pigment epithelial tissues, the choroid and the rod outer segments which contain the photoreceptors. If the damage is not extensive, there are repair mechanisms to allow for recovery of retinal tissues. However, extensive phototoxic damage to the retina can lead to permanent blindness (Dayhaw-Barker and Barker, 1986; Ham *et al*., 1982).

8. *Natural defence systems in the eye*. The eye is under constant oxidative and photo-oxidative stress; however, it uses several systems to protect itself (Roberts, 1988; Handelman and Dratz, 1986). The extent to which an exogenous photosensitizing drug or dye will cause phototoxic side-effects in the eye depends upon the efficiency of endogenous scavengers and quenchers to prevent irradiation damage by interrupting transient excited state intermediates which cause the ocular damage.

There are a number of endogenous photoprotective systems to control these photochemical processes. Quenchers which can negate specific reactive intermediates are important as a major defence mechanism against light insult to the eye. Glutathione, due to the low energy of the SH (thiol) bond (65 kcal) is an efficient free-radical scavenger and singlet oxygen quencher. Ascorbic acid quenches free radical and superoxide reactions. The alpha-tocopherol quenches both singlet oxygen and free radicals. There are also various antioxidant enzymes present in the eye.

Cornea. Although the cornea contains quenchers such as glutathione, ascorbic acid and alpha-tocopherol, its primary defence mechanism against radiative damage is through a very efficient repair system of the corneal epithelial and endothelial layer.

Aqueous. The tissues of the cornea and lens are fed by the aqueous and it contains numerous antioxidant enzyme systems (catalase, superoxide dismutase, SOD) as well as high concentrations of quenchers such as glutathione, ascorbate and alpha-tocopherol.

Lens. The adult primate lens contains a yellow substance, 3-hydroxy-kynurenine glucoside which diffuses direct photolytic energy (Dillon, 1991). Glutathione is present throughout the lens, particularly in the epithelial cells, but decreases with age (Sethna *et al*., 1983); the cytosol contains ascorbic acid while alpha-tocopherol is present in the lens membrane. There are also antioxidant enzyme systems such as catalase, glutathione peroxidase and reductase and superoxide dismutase in the lens (Sethna *et al.*, 1983).

Retina. Pigments such as melanin and lutein may play an important role in the photoprotection of the retina (Handelman and Dratz, 1986). Melanin is thought to protect the retinal pigment epithelial cells (Sarna, 1992) while lutein located in the fovea or macular region serves as an effective screening filter of the photoreceptors from blue light damage. Further protection of the retina is provided by glutathione, ascorbic acid, alpha-tocopherol, beta-carotene (Kirschfeld, 1982) and the antioxidant enzymes mentioned above.

10.3.3 *Summary*

The extent to which a particular photosensitizer will affect the human cornea, lens and/or retina *in vivo* depends upon: (1) its residence time in the eye, which is determined by its structure; (2) the photoefficiency of the particular sensitizer which is due to its ability to absorb the appropriate wavelengths of light that have been transmitted to the lenticular or retinal substrates. These transmission characteristics change with age. (3) The mechanism by which it causes that damage including possible binding of the dye to ocular constituents. Binding can not only alter the photochemical mechanism but can increase the retention time of the sensitizer in the eye; (4) and finally, the presence of endogenous quenchers or free radical scavengers and enzyme detoxification systems that could stop or retard these photochemical reactions.

10.4 Techniques to predict ocular phototoxicity

10.4.1 *The short screen*

Based on the theoretical considerations stated above, it is relatively easy to predict if a drug will not cause ocular damage through a photoinduced event. The short screen given in Table 10.2 will dramatically reduce the number of potential substances needed to be considered for ocular phototoxicity.

Examine the chemical structure. Most drugs must have a tricyclic, heterocyclic

Table 10.2 Short screen for potential phototoxicity in the lens/retina

1. Chemical structure
   (a) heterocylic, tricylic, porphyrin
   (b) amphiphilic/lipophilic
2. Absorbance spectra
   (a) above 295 nm (lens); 400 nm (retina)
   (b) binds to DNA, lens protein, melanin
3. Skin phototoxicity

or porphyrin ring (Figs 10.4 and 10.5) to fulfil the energy requirements to produce a stable, long-lived triplet. The addition of a halide group can enhance intersystem crossing from the singlet to the triplet (Turro, 1978). As seen in Fig. 10.5, Fluorescein and Rose Bengal have similar structures except for the attached iodine groups. Fluorescein is highly fluorescent (singlet state) but is not very efficient at reaching the triplet state (poor quantum yield for the triplet). The singlet state of Rose Bengal would easily go through an intersystem crossing to the triplet (good quantum yield for the triplet). By simple inspection of the structures of these two diagnostic dyes we would conclude that Rose Bengal has a much greater potential to produce phototoxic damage to the eye than Fluorescein.

Test the solubility properties. Dyes or drugs to be examined should be tested for their partitioning in protic and aprotic solvents. Their hydrophobicity will indicate potential for crossing blood–ocular barriers and probably site of damage (membranes). More hydrophilic substances are less likely to cross blood–ocular barriers. Fluorescein and tetrasulphonato-phenylporphyrin (TPPS) are examples of dyes or drugs that can pass all ocular barriers.

Absorbance spectra. A comparison of the transmission characteristics of the eye (Fig. 10.1) with the absorbance spectra of the drug may be used as a quick screen for phototoxicity. A photon of light must be absorbed for a photosensitized event to occur. If the drug has an absorbance of below 295, transmitted light will not activate the drug in the lens or retina, below 400 nm the drug will not be activated in the retina. However, any absorbance, even end-point absorbance as seen with iprindol in Fig. 10.6 at 330 nm, can make a dye or drug a potential photosensitizer in the eye.

Binding experiments. Additional information can be obtained by measuring the absorption spectra of the drug in the presence and absence of lens proteins, DNA and/or melanin (Roberts and Dillon, 1987; Roberts *et al.*, 1990b; Steiner and

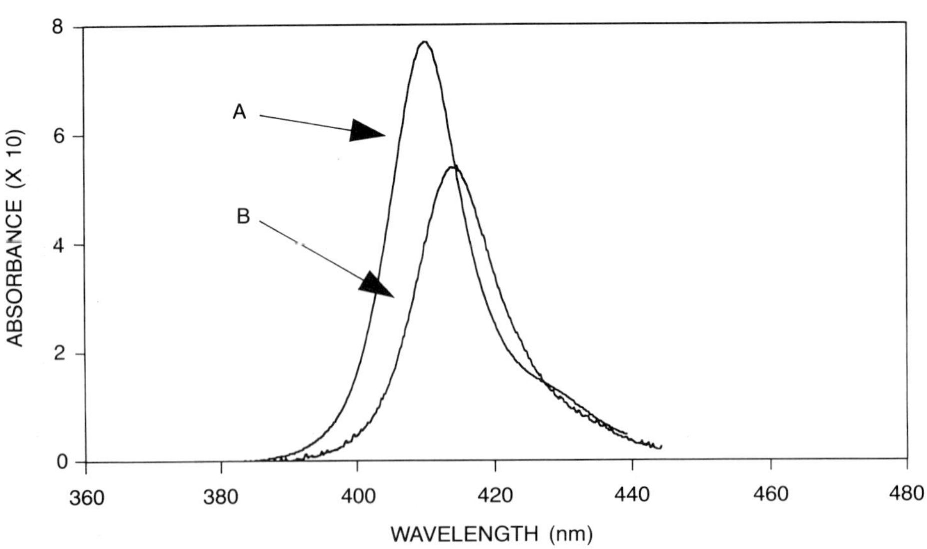

Figure 10.7 The Soret band of a porphyrin (A) and with the addition of lens protein (B). The decrease in the absorbance with the shift to the red seen here is indicative of binding of a drug to the macromolecule

Buhring, 1990). A red shift in the absorbance spectra of the drug in the presence of any of these biomolecules is an indication of binding of the sensitizer to the biomolecule. An example of this red shift is shown in Fig. 10.7. It gives the Soret band for a porphyrin (TPPS) alone and in the presence of cytosol lens proteins. The red shift of the peak is indicative of binding of the sensitizer to lens protein.

Skin phototoxicity. Finally, any reports of skin phototoxicity for a particular drug should provide a clear warning of potential ocular phototoxicity. Skin phototoxicity is more readily apparent than ocular phototoxicity although it is induced by compounds with similar chemical features (Oppenländer, 1988).

10.4.2 *Detailed techniques for screening for ocular phototoxicity*

The targets of photooxidative reactions may be proteins, lipids, DNA, RNA, and/or cell membranes (Straight and Spikes, 1985). *In vitro* tests can be designed to determine the specific site(s) of damage to the various ocular compartments (i.e. lens and retinal epithelial cells and photoreceptor cells) and the products of those reactions. Photophysical studies can be used to determine short-lived excited state intermediates.

Photochemical reactions in tissues

Excited states→	Intermediates→	Radicals→	Damage
Singlet	.. Singlet oxygen	.. Polymers	
Triplet	.. Superoxide	.. Lipid peroxides	
	.. OH·, ROO·	.. Cross-links	

Table 10.3 presents a summary of additional biochemical and photophysical techniques that can be performed to more accurately predict the potential for and extent of *in vivo* phototoxicity.

10.4.2.1 In vitro *techniques*

Cell culture/whole tissues. The first reported assay for phototoxicity in human cells (Roberts, 1981) measured changes in macromolecular synthesis in the presence and absence of a light-activated drug. More recent studies have assessed damage to corneal, lenticular and retinal cells by measuring pump function and enzyme activities both *in vitro* and *in situ* (Handelman and Dratz, 1986; Lou and Zigler, 1986; Andley, 1987; Dayhaw-Barker, 1987; Dorey *et al.*, 1990; Rao and Zigler, 1992; Andley, 1994a and b; Organisciak and Winkler, 1994; Roberts *et al*:., 1994a; Wang *et al.*, 1995).

Gel electrophores, amino acid analysis. Gel electrophoresis has been used to monitor polymerization of ocular proteins (Zigler *et al.*, 1982; Roberts, 1984, 1992a; Kristensen *et al.*, 1995). Photopolymerization is one of the most apparent changes in ocular protein induced by photosensitizing dyes and drugs. Quantitative changes can be measured by scanning the gel and determining relative reaction rates as seen in Fig. 10.8. Specific amino-acid modifications can be determined using amino-acid analysis (Roberts, 1984; Roberts *et al.*, 1985, 1986; Roberts and

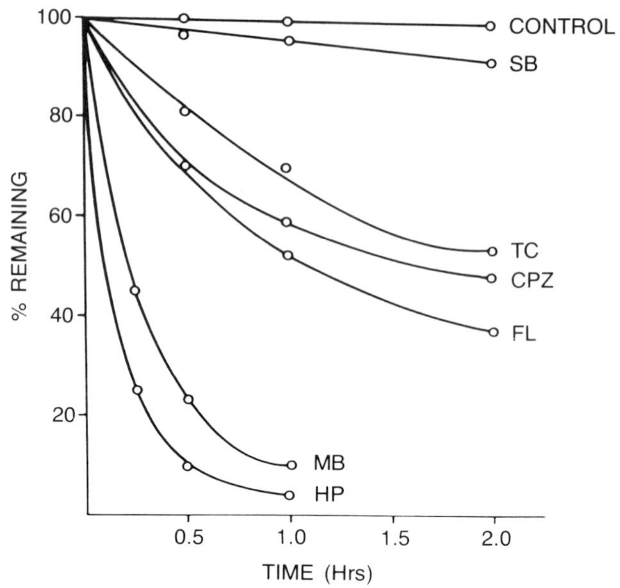

Figure 10.8 Cytosol proteins from the cortex of calf lenses (2 mg/ml in 0.1 M phosphate buffer) in the presence of 0.1 mmole of drug or diagnostic dye: sorbinil (SB); tetracycline (TC); chlorpromazine (CPZ); fluorescein (FL); methylene blue (MB) and hematoporphyin (HP), were irradiated using a Hanovia 450 W medium-pressure mercury lamp which was placed in an immersion well and cooled with 5% copper sulfate solution. The transmission characteristics of the solution mimic that of the cornea of the eye. These photolyses were analysed using SDS-PAGE gel electrophoresis. The first order rates of these reactions were calculated from the loss of the material in the 20 k dalton region. These are relative rates that have not been normalized

Dillon, 1987; Roberts *et al.*, 1992a and b). A recent paper by Zhu and Crouch (1992) illustrates the variety of classical protein analysis techniques (gel electrophoresis, amino-acid analysis, sequencing, isoelectric point determination, western blot, ELISA) that can be used to investigate phototoxic damage induced by dyes and drugs.

Mass spectrometry. Recent innovations in the field of mass spectrometry (liquid secondary ion mass spectrometry (LSIMS) and electrospray ionization (ESI)) have allowed for the identification of specific amino-acid modifications within large proteins through molecular weight mapping. A typical scan is given in Fig. 10.9. These techniques have been applied to determine the specific sites of photooxidative damage in corneal and lenticular proteins (Finch *et al.*, 1993; Schey *et al.*, 1995; Beischel *et al.*, 1995). These studies can serve as a model for defining damage from any potential phototoxic agent in the eye.

Thin layer chromatography. This technique is particularly effective at separating triacyclycerol, free fatty acid and phospholipids from lens (Fleschner, 1995) and retinal (Organisciak *et al.*, 1992) membranes. TLC/GC/mass spec may be used to

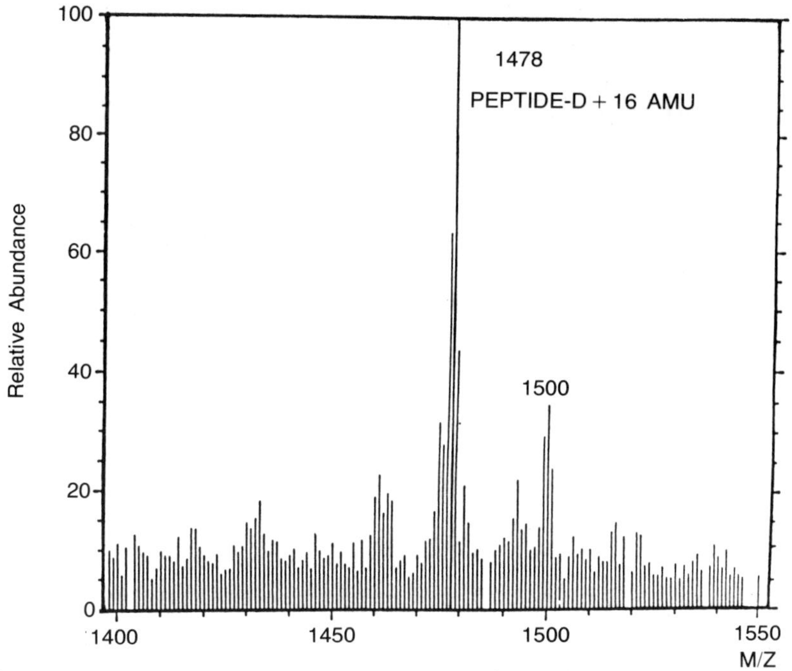

Figure 10.9 The mass spectrum of a tryptic digest peptide photoproduct of calf lens protein that was photooxidized with in the presence of a porphyrin. The addition of 16 AMU to peptide D is indicative of photooxidation of an amino acid residue with singlet oxygen. In this case a methionine was converted to methionine sulfoxide

measure lenticular or retinal lipid modifications (Handelman and Dratz, 1986). Specific lipids may be modified in the presence of photosensitizing agents and separated on TLC plates. The plates can then be scanned for quantitative analysis of these specific changes.

High-pressure liquid chromatography (HPLC). HPLC is particularly effective at separating and identifying lipid peroxides from the retina (Akasaka *et al.*, 1993; Organisciak *et al.*, 1992). It has also been used to identify adducts formed between DNA nucleotides and phototoxic agents (Oroskar *et al.*, 1994). HPLC has also been used to assess the rates of photooxidation of lens proteins in the presence of a sensitizer. Using this technique, which is demonstrated in Fig. 10.10, it is possible to determine the induced amino-acid modification within the protein, their location and to detect possible binding of sensitizering drugs to specific lens crystallins (McDermott *et al.*, 1991).

Normalization for photons absorbed. Whatever the target tissue or extent of damage, the toxic effects of these dyes and drugs are the result of photochemical reactions. As such, their rate of efficiency is dependent on the number of photons absorbed by the sensitizer in the biological tissue. Therefore, in order to get an accurate comparison of the photosensitizing potency of various dyes and drugs with

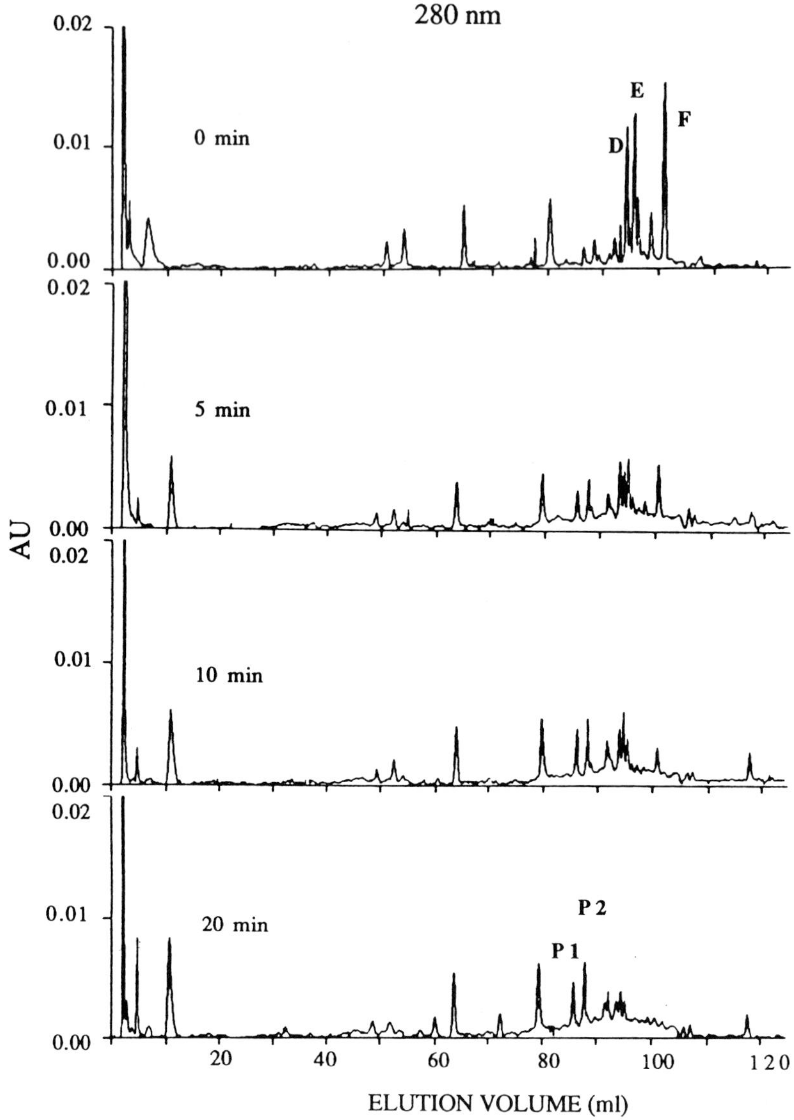

Figure 10.10 The HPLC elution pattern of a tryptic digest peptide of calf lens protein that was photooxidized with porphyrin for 0, 5, 10 and 20 mins. The eluant was monitored at 280 which primarily detects tyrosine and tryptophan residues. There is a clear loss of peptides D, E and F which contain tryptophan and the formation of at least two photoproduct peptides P1 and P2

different structures and absorptive characteristics it is essential to normalize for the number of photons absorbed by each drug in a particular system (Roberts and Dillon, 1986; Andley *et al.*, 1994c; Kristensen *et al.*, 1995).

This can be done with a simple computer-generated mathematical formula (Roberts and Dillon, 1986) which takes into account the absorbance spectra of the drug, the output of the lamp source used in the experiments and the optical

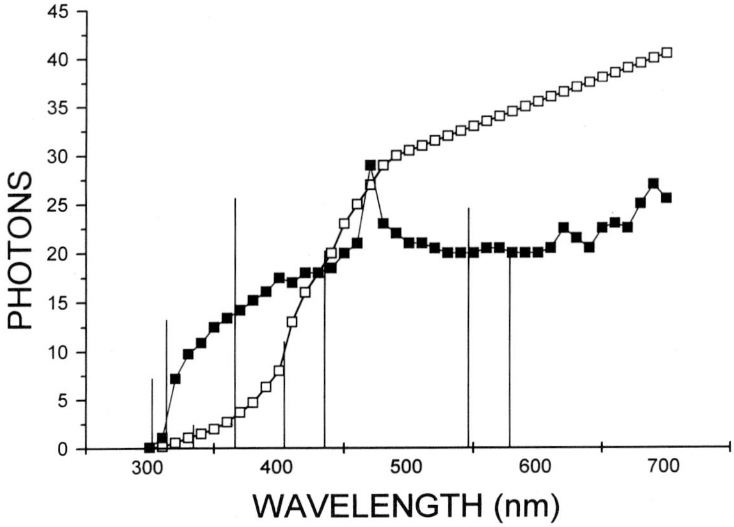

Figure 10.11 This gives the outputs of typical lamps used in these studies. A xenon lamp (■) almost approximates the output of the sun (□), whereas a medium-pressure mercury lamp consists totally of spikes (straight vertical lines)

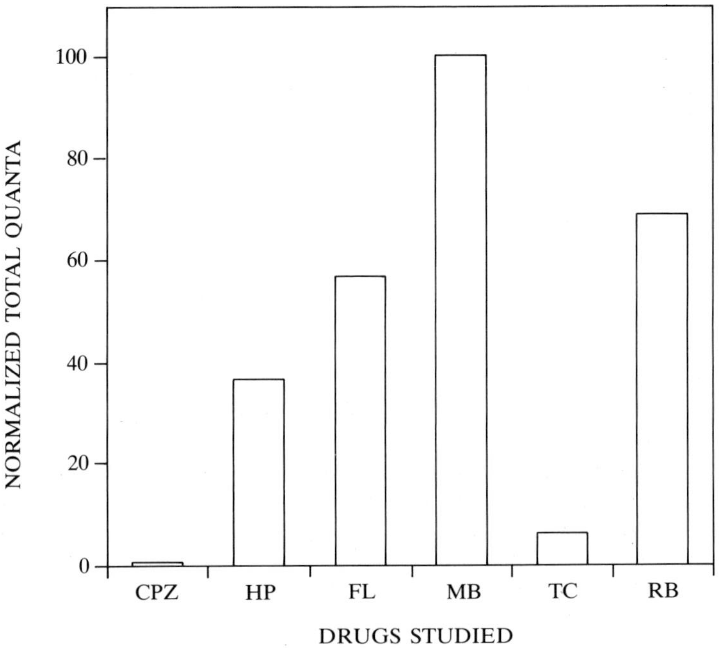

Figure 10.12 The number of photons (quanta) absorbed by drugs studied (chlorpromazine, CPZ; hematoporphyrin, HP; fluorescein, FL; tetracycline, TC and Rose Bengal, RB) using a xenon lamp. In a typical calculation for the number of photons absorbed by a drug, the absorption spectra of that compound, is multiplied, every 10 nm by i) the output of the xenon lamp reaching the solution and the relative number of photons at those wavelengths. The relative number of photons absorbed by a drug is then the area under the product curve

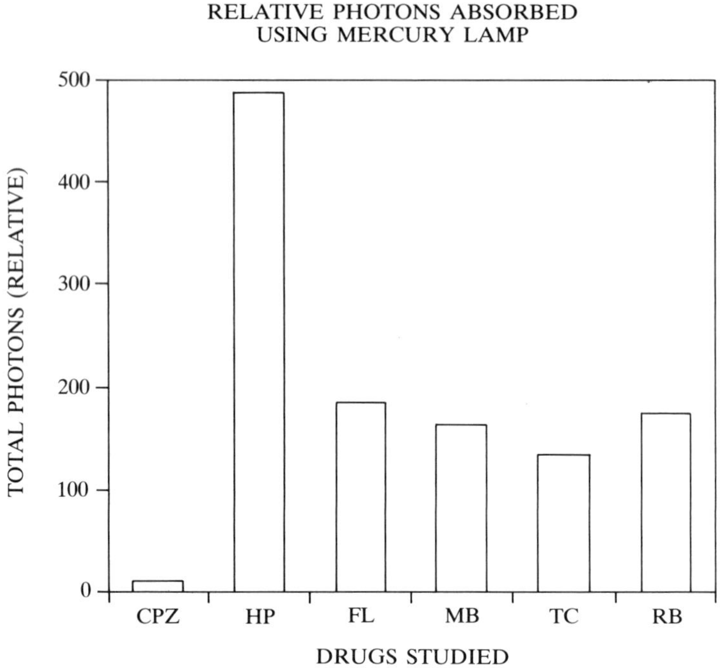

Figure 10.13 The number of photons (quanta) absorbed by drugs studied (chlorpromazine, CPZ; hematoporphyrin, HP; fluorescein, FL; tetracycline, TC and Rose Bengal, RB) using a mercury lamp. In a typical calculation for the number of photons absorbed by a drug, the absorption spectra of that compound, is multiplied, every 10 nm by (i) the output of the mercury lamp reaching the solution and the relative number of photons at those wavelengths. The relative number of photons absorbed by a drug is then the area under the product curve

properties of the eye. The total relative number of photons absorbed by a drug under particular experimental conditions is the area under the product curve. The rates of each photooxidative event are then adjusted accordingly for each sensitizer. This in turn can be corrected for the actual transmission characteristics of the cornea and/or lens and the output of the sun to predict *in vivo* effects. These calculations are essential since without them the results obtained from *in vitro* assays of photooxidative destruction of ocular tissue may not be biologically relevant. A specific example of this normalization is given in Figs 10.11 to 10.14.

10.5 In *vitro* techniques: summary

In vitro techniques determine the potential damage done to an ocular substrate, which gives information about the photoefficiency of a drug should it be taken up into the various compartments of the eye. Additional information about the site of potential damage can be predicted based on which ocular substrate (DNA, protein lipid) is effected.

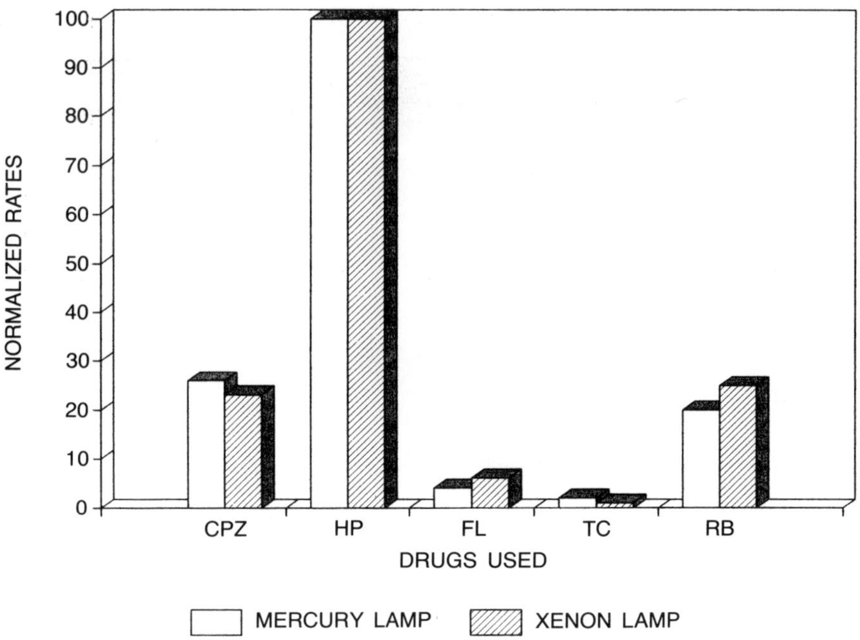

Figure 10.14 Each of the rates for the sensitizers listed above were adjusted for photons absorbed for either lamp. The normalized rates are similar for both studies

10.5.1 ***Biophysical techniques***

10.5.1.1 Fluorescence

After a dye or drug has absorbed light and is excited to the singlet state it can decay to the ground state and is accompanied by the emission of light. This is known as fluorescence.

Uptake Fluorescence spectroscopy off the surface of intact lenses (Roberts *et al.*, 1991b) can be used to determine uptake of sensitizers into lenses. An example is found in Fig. 10.15, which gives the fluorescence of TPPS in an intact calf lens. It may also be used to monitor interactions of sensitizers with retinal pigments (Docchio *et al.*, 1991). Since most photosensitizers are fluorescent, reflective fluorescence provides an accurate means of measuring uptake of a sensitizer into ocular tissue, that is simpler, less expensive and arduous than using radiolabelled materials. This technique may also be used non-invasively, *in vivo*, using a slit lamp to detect uptake of sensitizers into the human eye (Docchio, 1989; Kampfer *et al.*, 1989).

Binding Additional information can be obtained by measuring the fluorescence excitation spectra of the drug in the presence and absence of lens proteins, DNA and/or melanin (Roberts and Dillon, 1987; Roberts *et al.*, 1990b; Steiner and Buhring, 1990). A decrease in intensity and a slight shift to the red in the excitation

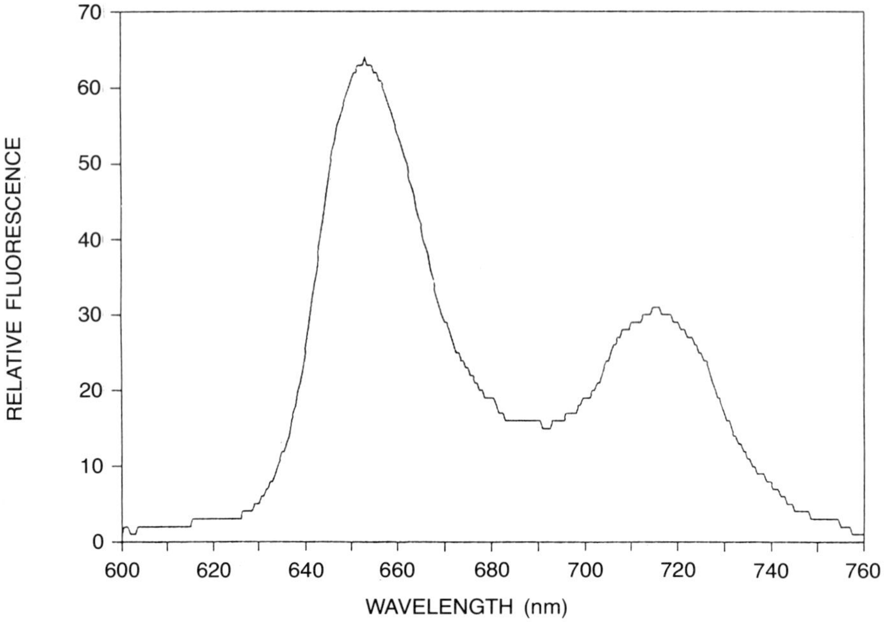

Figure 10.15 The fluorescence of TPPS off the surface of an intact lens, with an excitation at 355 nm

spectra of the drug in the presence of any of these biomolecules are an indication of binding of the sensitizer to the biomolecule.

Quantum yields. In predicting the phototoxicity of a dye or drug it is important to determine what amount of photons leads to a benign (fluorescence, singlet state) event and what amount of photons lead to a potentially destructive event (phosphorescence, triplet state). The efficiency of a photoinduced process may be expressed as its quantum yield (Q).

$$Q = \frac{\text{the number of photons used to produce an event}}{\text{the number of photons originally absorbed}}$$

It gives a measure of how likely a photochemical event is to occur. The quantum yield is often expressed as a percentage. For instance, the quantum yield for fluorescence of fluorescein is 0.92 and that of Rose Bengal is 0.08 (Fig. 10.5). This means that most of the light absorbed by fluorescein is given off in the form of fluorescence energy. This makes it a relative safe diagnostic dye for the eye. On the other hand, whereas the triplet quantum yield for fluorescein is 0.03 and for Rose Bengal is 0.60. This indicates that very little of the singlet of fluorescein energy is transformed into a triplet while most of the energy of Rose Bengal will be available for intersystem crossing (Fig. 10.3) and reach the triplet state from whence it can produce ocular damage. Therefore, fluorescein is appropriate but Rose Bengal would be an inappropriate diagnostic dye for the eye.

10.5.1.2 *Laser flash photolysis*

This method uses a pulse of monochromatic light to promote a specific dye or drug to an excited state (Rodgers, 1985).

Triplet detection. Time-resolved techniques (either absorption spectroscopy or diffuse reflectance) allows for the detection of the triplet state of the excited chromophore even in intact tissues. This technique has been used to determine the absence (Dillon and Atherton, 1990a) of triplet formation by the endogenous 3-hydroxykynurenine as well as the presence of triplets from sensitizing drugs in intact lenses (Roberts *et al.*, 1991a).

Lifetimes/binding. All ocular damage from photooxidation reactions occurs through the triplet state of the drug (Fig. 10.3). The longer the lifetime the greater the potential for damage. The lifetimes of the triplet from sensitizing drugs were found to be greater when bound to macromolecules or in an intact organ, than when free in solution as depicted in Fig. 10.13 (Roberts *et al.*, 1991b). The presence of a triplet and its increase in lifetime when bound to intact ocular tissue is predictive of a drug causing photooxidative damage to the eye *in vivo* (Roberts *et al.*, 1991a).

Quantum yields. The quantum yield of the triplet can be determined by the comparison method (Bensasson and Land, 1978; Nord *et al.*, 1994) which consists of comparing the triplet spectra of an unknown dye or drug with that of a known compound (QYK) under conditions in which both are excited by the same wavelength and lamp intensity. Then the quantum yield for the dye or drug (QYD) will be equal to:

$$QYD = QYK \times Ad/Ak \times ODd/ODk$$

where QYK is the quantum yield of the known compound, Ad or Ak is the integrated area of the triplet spectra and ODd or ODk is the absorbance at the exciting wavelength of the drug (d) and known substance (k) respectively.

If the triplet quantum yield is above 0.5 (i.e. over 50 per cent of the exciting photons have been used to form the triplet) it means that an excited dye or drug can easily reach the triplet state and therefore has the potential to do ocular phototoxic damage *in vivo* (Roberts *et al.*, 1991a).

10.5.1.3 *Luminescence*

Singlet oxygen. The presence and lifetime of singlet oxygen can be determined using time-resolved infrared luminescence measurements at 1270 nm (Rodgers and Snowden, 1982). Using this technique, it can be determined if and how efficiently a dye or drug can produce singlet oxygen (Roberts *et al.*, 1991a). Since singlet oxygen is the most powerful oxidant in a photooxidation reaction, a dye or drug that is an efficient producer of singlet oxygen could be predicted to induce phototoxic damage if present in the eye.

Oxygen tension. Oxygen tension varies drastically throughout the eye which contains one of the most (i.e. rod outer segments) and the least (i.e. lens)

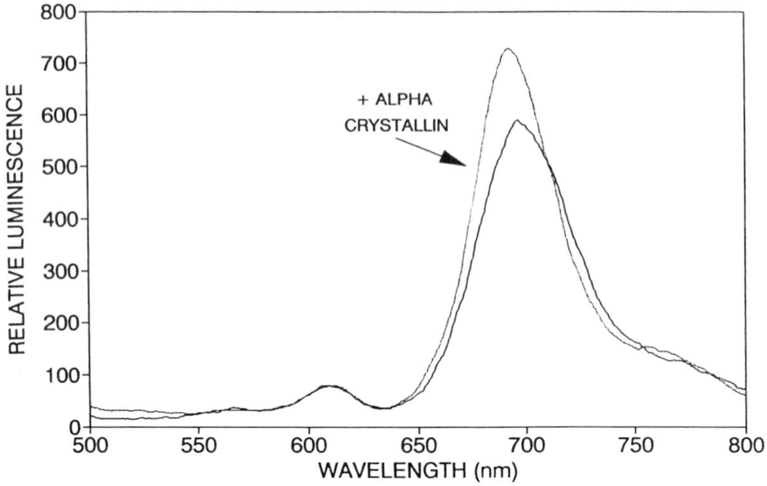

Figure 10.16 Gives the luminescence spectra from the palladium derivative of tetrasulphonatophenylporphyrin PdTSPP, in the presence and absence of calf lens alpha crystallin. This data is used to determine oxygen concentration changes with binding to a protein

metabolically energetic tissues in the body. The palladium derivative of TPPS has been used to determine the changes in oxygen concentration with binding of a sensitizer to a macromolecule (Fig. 10.16). Using palladium (II) uroporphyrin (PdUp), which has both an intense fluorescence and phosphorescence at room temperature and does not bind to ocular tissues, the oxygen content of intact lenses has been measured (Roberts *et al.*, 1992b). With a modified ocular fluorophotometer, the oxygen content of different chambers of the human eye in the presence and absence of potential photosensitizing dyes or drugs may be measured in a non-invasive fashion. The modification of oxygen in various chambers of the eye will add to the accuracy of prediction of ocular phototoxicity for a new dye or drug.

10.5.1.4 Pulse radiolysis

Pulse radiolysis consists of the delivery of a very short intense pulse of ionizing radiation to a sample, the resultant changes in light absorption of the sample being followed by a very fast spectrophotometer (Land, 1985). The technique may be used to detect the formation short-lived radical species of a dye or drug. In addition, the interaction of a dye and drug with excited oxygen intermediates (hydroxyl radical, superoxide, peroxy radicals) (Land *et al.*, 1983) which are cleanly generated in this system, allow for an understanding of a possible mechanism of *in vivo* photooxidative ocular damage. For example, the endogenous lenticular chromophore, 3-hydroxykynurenine was found to be very efficient at interaction with hydroxyl radicals using pulse radiolysis techniques (Atherton *et al.*, 1993).

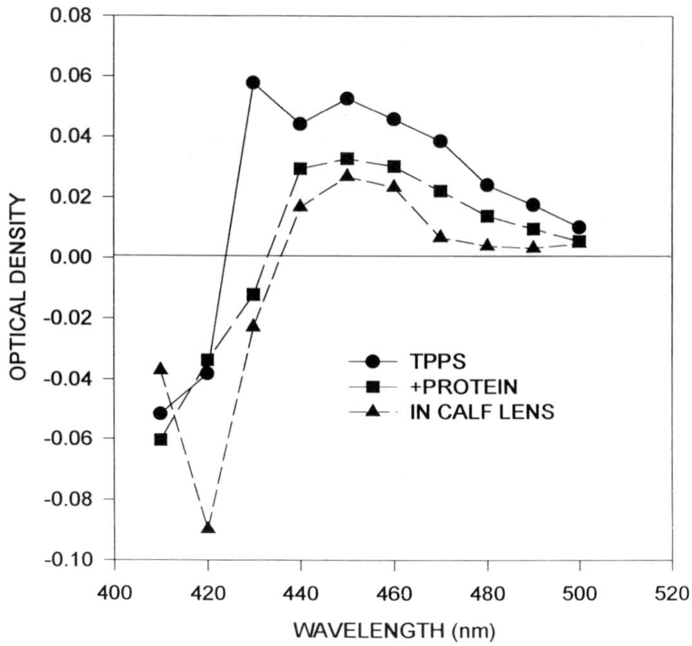

Figure 10.17 The transient (triplet) spectra from a solution of TPPS, with calf protein and within an intact lens. There is a lengthening of the lifetime with binding and within the intact organ

10.5.1.5 *Electron spin resonance*

Electron spin resonance (ESR or EPR) spectroscopy detects and characterizes species containing an odd number of electrons, namely free radicals and paramagnetic metal ions. The photooxidation reactions responsible for the phototoxic responses in the eye involve free radicals which are formed via electron transfer (electron exchange) between the sensitizing drug in an excited state and a substrate from the ocular tissues. Although these radicals are very short lived, they can be observed with ESR *in situ*, during their photogeneration. For instance, illumination of Rose Bengal (RB), an ophthalmic diagnostic dye, in the presence of an electron donor such as NADH, affords a radical anion of the dye $RB^{\cdot -}$, that can be directly measured using ESR (Sarna *et al.*, 1991).

Radicals which are too reactive, and therefore which do not accumulate in detectable quantities, can frequently be detected by ESR using spin trapping techniques. In this approach, an agent called a spin trap reacts with the short-lived radical $R^{\cdot}$ to give a spin adduct $R\text{-}T^{\cdot}$, which has a much longer lifetime. The original radical, $R^{\cdot}$ is identified by the characteristic ESR spectrum of the $R\text{-}T^{\cdot}$. Carbon, nitrogen, sulfur as well as all of the important oxygen-centred radicals (hydroxyl, superoxide, alkoxyl and peroxyl) can be identified using ESR either directly or in combination with the spin trapping technique.

Using these techniques, the photosensitized generation of superoxide in protic (Reszka *et al.*, 1992) and aprotic media (Reszka *et al.*, 1993) was monitored. These are model systems for the hydrophilic (aqueous) and hydrophobic (ROS membranes) portions of the eye. ESR has recently been used to define the photochemical mechanisms involved in the light activation of endogenous pigments in the lens (Reszka *et al.*, 1995a) and the retina (Reszka *et al.*, 1986). With these systems defined, ESR can be used to predict potential phototoxic events induced by exogenous photosensitizing dyes and drugs, and natural pigments.

10.5.2 *Biophysical techniques: summary*

The molecular mechanism involved in the phototoxic damage induced in the eye is through photosensitized oxidation reactions. This involves the absorption of light by the sensitizing compound (endogenous pigment, dye or drug) which promotes the compound to an excited singlet state (short lived) and then, through intersystem crossing, goes to the triplet state (long lived). The excited triplet state of the drug/dye that then proceeds either via a Type I (free radical) or Type II (singlet oxygen) mechanism causing the eventual biological damage (Straight and Spikes, 1985). Therefore information about the efficiency and excited state intermediates for a phototoxic reaction in the eye obtained by using photophysical techniques (fluorescence, flash photolysis, pulse radiolysis, esr) can be predictive of phototoxicity *in vivo*. We have confirmed that photophysical studies collate well with *in vivo* data (Roberts *et al.*, 1991a). For instance, tetrasulphonatophenylporphyrin (TPPS), which binds to lens proteins, shows a long-lived triplet in the intact calf and human lens and produces singlet oxygen efficiently causes photooxidative damage *in vivo* in pigmented mouse eyes. Whereas uroporphyrin (URO), which produces an efficient triplet but does not bind to ocular tissues, does not cause photooxidative damage *in vivo*.

10.6 *In vivo* testing

These short or more detailed (Tables 10.2 and 10.3) screens will not totally eliminate the need for accurate *in vivo* experiments. The function of these studies is to limit the need for *in vivo* testing for ocular phototoxicity of large numbers of drugs. Those drugs found in screening to be highly likely to produce phototoxic side-effects in the eye, should be tested further in animal studies to determine the exact site and extent of damage to be expected in humans.

Prolonged use of a phototoxic drug is most probably of greater long-term risk to the eye than short-term dosage because of accumulative damage. Since there is an active repair system in the cornea, there should be little or no long-term side-effects of phototoxicity. However, since there is no turnover in the lens constituents, any modification in that tissue will tend to stay and accumulate with age. Thus cataractogenesis may not develop until much later than the initial insult. In addition phototoxic damage to the lens may not only cause direct damage to cell viability but may undermine its defence system (i.e. synthesis of glutathione), so that gross morphological effects may appear much later than the original insult. Although there is some repair in the retinal tissue gross damage can lead to blindness. The environmental lighting, particularly the constant presence of intensive ambient light, must also be taken into account when assessing potential *in vivo* ocular phototoxicity of a drug.

Table 10.3 Techniques used in studying photosensitizers in the eye

Technique	Use
1. *In vitro* studies	
(a) Cell culture	DNA, RNA, protein,
Enzyme assays	antioxidant enzymes
Histology	endothelial, epithelial photoreceptor cell damage
(b) Gel electrophoresis amino acid analysis	protein changes
(c) Mass spectrometry	lipid changes peptide maps
(d) TLC	lipid oxidation
(e) HPLC	lipid peroxides DNA adducts protein modification
(f) Normalization for photons absorbed	
2. Biophysical techniques	
(a) Fluorescence	(i) uptake, (ii) binding (iii) quantum yields
(b) Laser flash photolysis	(i) triplet detection (ii) binding/lifetime (iii) quantum yields
(c) Luminescence	(i) singlet oxygen (ii) oxygen tension
(d) Pulse radiolysis	radical and oxyradical intermediates
(e) ESR	radical and oxyradical intermediates

10.7 Protection

Even if a drug has the potential to produce phototoxic side-effects in the eye, no damage will be done if the specific wavelengths of optical radiation absorbed by the drug are blocked from transmittance to the eye. This can be easily done with wrap-around eyeglasses (Narayanan *et al.*, 1995) which incorporate specific filters. Furthermore non-toxic quenchers and scavengers could be given in conjunction with the phototoxic drug to negate its ocular side-effects while allowing for the primary effect of the drug. (Roberts, 1981; Roberts *et al.*, 1991a and b; Roberts and Mathews-Roth, 1993; Roberts *et al.*, 1994b and c).

10.8 Conclusion

With simple, inexpensive *in vitro* testing, compounds can be checked at their developmental stage for potential ocular phototoxicity. It may be that a portion of the molecule can be modified to reduce phototoxicity while leaving the primary drug effect intact. This may reduce the necessity of future, more costly, drug recalls.

Acknowledgements

This work was supported by funds from the Hugoton Foundation and a NATO Collaborative Research Grant D.880144. All graphics were designed by Dr James Dillon, Columbia University, New York.

References

AKASAKA, K., OHRUI, H. and MEGURO, H., 1993, Normal-phase high-performance liquid chromatography with a fluorimetric postcolumn detection system for lipid hydroperoxides, *J. Chromatog.*, **628**, 31–5.

ANDLEY, U. P., 1987, Photodamage to the eye, *Photochem. Photobiol.*, **46**, 1057–60.

ANDLEY, U. P., RHIM, J. S., CHYLACK, L. T. Jr and FLEMING, T. P., 1994a, *Invest. Ophthal. Vis. Sci.*, **35**, 3094–102.

ANDLEY, U. P., HEBERT, J. S., MARRISON, A. R., REDDAN, J. R. and PENTLAND, A. P., 1994b, Modulation of lens epithelial cell proliferation by enhanced prostaglandin synthesis after UVB exposure, *Invest. Ophthalmol. Vis. Sci.*, **35**, 374–81.

ANDLEY, U. P., LEWIS, R. M., REDDAN, J. R. and KOCHEVAR, I. E., 1994c, Action spectrum for cytotoxicity in the UVA and UVB wavelength region in culture lens epithelial cells, *Invest. Ophthalmol. Vis. Sci.*, **35**, 367–73.

ATHERTON, S. J., DILLON, J. and GAILLARD, E. R., 1993, A pulse radiolysis study of the reactions of 3-hydroxykynurenine and kynurenine with oxidizing and reducing radicals, *Biochimica Biophysica Acta*, **1158**, 75–82.

BACHEM, A., 1956, Ophthalmic action spectra, *Am. J. Ophthal.*, **41**, 969–75.

BARKER, F. M and BRAINARD, G., 1993, Transmission of the human lens, *Photochem. Photobiol. Abs*, Chicago, Ill.

BARKER, F. M., BRAINARD, G. C., DAYHAW-BARKER, P. and HANIFIN, J. P., 1992, Age-related changes in human lenticular transmission, *SLTBR Abstracts*, **4**, 11.

BEISCHEL, C. J., SCHEY, K. L., CROUCH, R. K., DILLON, J. and KNAPP, D. R., 1995, Mass spectral analysis of bovine lens MP 26, *Invest. Ophthal. Vis. Sci.*, **36**, 262.

BENEDEK, G. B., 1971, Theory of transparency of the eye, *Appl. Optics*, **10**, 459–73.

BENSASSON, R. and LAND, E. J., 1978, Physical properties of excited states: A general method for measuring triplet-triplet extinction coefficients, singlet-triplet intersystem crossing efficiencies and related parameters, *Photochem. Photobiol. Reviews*, **3**, 163–9.

DAYHAW-BARKER, P., 1987, Ocular photosensitization, *Photochem. Photobiol.*, **46**, 1051–6.

DAYHAW-BARKER, P. and BARKER, F. M., 1986, Photoeffects on the eye, in Jackson, E. M. (ed.), *Photobiology of the Skin and the Eye*, pp. 117–47. New York: Marcel Dekker,Inc.

DILLON, J., 1991, Photophysics and photobiology of the eye, *J. Photochem. Photobiol. B. Biol.*, **10**, 23–40.

DILLON, J. and ATHERTON, S. J., 1990, Time resolved spectroscopic studies on the intact human lens, *Photochem. Photobiol.*, **51**, 465–8.

DILLON, J., WANG, R. H. and ATHERTON, S., 1990, Photochemical and photophysical studies on human lens constituents, *Photochem. Photobiol.*, **52**, 849–54.

DOCCHIO, F., 1989, Ocular fluorometry: principles, fluorophores, instrumentation and clinical applications, *Lasers Surg. Med.*, **9**, 515–32.

DOCCHIO, F., BOULTON, M., CUBEDDU, R., RAMPONI, R. and DAYHAW-BARKER, P., 1991, Age-related changes in the fluorescence of melanin and lipofuscin granules of the retinal pigment epithelium: A time-resolved fluorescence spectroscopy study, *Photochem. Photobiol.*, **54**, 247–53.

FINCH, J. W., CROUCH, R. K., KNAPP, D. R. and SCHEY, K. L., 1993, Mass spectrometric identification of modifications to human serum albumin treated with hydrogen peroxide, *Archives of Biochem. Biophys.*, **305**, 595–9.

FLESCHNER, C. R., 1995, Fatty acid composition of triacylglycerols, free fatty acid and phospholipids from bovine lens membrane fractions, *Invest. Ophthal. Vis. Sci.*, **36**, 261.

FRAUNFELDER, F. T., 1982, *Drug-Induced Ocular Side Effects and Drug Interactions*, 2nd edn. Philadelphia: Lea & Febiger.

GASPARRO, R. P., BAGEL, J. and EDELSON, R. L., 1985, HPLC Analysis of 8-MOP photoadducts in calf thymus DNA, poly (dAdT.dAdT), poly (dA.dT) and poly (dT), *Photochem. Photobiol.*, **42**, 98–101.

HAM, W. T., MUELLER, H. A., RUFFOLO, J. J., GUERRY, D. and GUERRY, R. K., 1982, Action spectrum for retinal injury from near ultraviolet radiation in the aphakic monkey, *Amer. J. Ophthal.*, **93**, 299–305.

HANDELMAN, G. J. and DRATZ, E. A., 1986, The role of antioxidants in the retina and retinal pigment epithelium and the nature of prooxidant induced damage, *Adv. in Free Radical Biology and Medicine*, **2**, 1–89.

HU, T.-S., ZHEN, Q., SPERDUTO, R. D., ZHAO, J.-L., MILTON, R. C., NAKAJIMA, A. and the TIBET EYE STUDY GROUP, 1989, Age-related cataract in the Tibet Eye Study, *Arch. Ophthalmol.*, **107**, 666.

HULL, D. S., CSUKAS, S. and GREEN, K., 1983, Trifluoperazine: corneal endothelial phototoxicity, *Photochem. Photobiol.*, **38**, 425–8.

KAMPFER, T., WEGENER, A., DRAGOMIRESCU, V. and HOCKWIN, O., 1989, Improved biometry of the anterior eye segment, *Ophthalmic Res.*, **21**, 239–48.

KHWARG, S. G., LINSTONE, F. A. and DANIELS, S. A., 1987, Incidence, risk factors and morphology in operating microsope light retinopathy, *Am. J. Ophthalmol.*, **103**, 255–63.

KIRSCHFELD, K., 1982, Carotenoid pigments: their possible role in protecting against photooxidation in eyes and photoreceptor cells, *Proc. R. Soc. Lond*, **216**, 71–85.

KRISTENSEN, S., WANG, R. H., TØNNESEN, H. H., DILLON, J. and ROBERTS, J. E., 1995, Photoreactivity of biologically active compounds. VII. Photosensitized polymerization of lens proteins by antimalarial drugs *in vitro*, *Photochem. Photobiol.*, **61**, 124–30.

KWAN, M., NIINIKOSKE, J. and HUNT, T. K., 1971, Oxygen tension in the aqueous and the lens, *Invest. Ophthal.*, **11**, 108–11.

LAND, E. J., 1985, Pulse radiolysis, in Bensasson, R. V., Jori, G., Land, E. J. and Truscott, T. J., (eds), *Primary Photoprocesses in Biology and Medicine*. pp. 35–44. North Holland: Elsevier.

LAND, E. J., MUKHERJEE, T., SWALLOW, A. J. and BRUCE, J. M., 1983, One-electron reduction of adriamycin: properties of the semiquinone, *Arch. Biochem. Biophys.*, **225**, 116–21.

LERMAN, S., 1986, Photosensitizing drugs and their possible role in enhancing ocular toxicity, *Ophthalmol.*, **93**, 304–18.

LOU, M. F. and ZIGLER Jr, J. S., 1986, The effect of photooxidation on protein mixed sulfides in lens, *Abs: Seventh International Congress of Eye Research*, Nagoya, Japan.

MCDERMOTT, M., CHIESA, R., ROBERTS, J. E. and DILLON, 1991, Photooxidation of specific residues in a-crystallin polypeptides, *Biochemistry*, **30**, 8653–60.

NARAYANAN, P., MERRIAM, J., VAZQUEZ, M. and DILLON, J., 1995, The spatial distribution of light on the human lens, *Invest. Ophthalmol. Vis. Sci.*, (in press).

NORD, K, KARLSEN, J. and TØNNESEN, H. H., 1994, Photochemical stability of biologically active compounds. IX. Characterization of the spectroscopic properties of the 4-aminoquinolines, chloroquine and hydroxychloroquine and of selected metabolites by absorption, fluorescence and phosphorescence measurements, *Photochem. Photobiol.*, **60**, 427–31.

OPPENLÄNDER, T., 1988, A comprehensive photochemical and photophysical assay exploring the photoreactivity of drugs, *Chimia*, **42**, 331–42.

ORGANISCIAK, D. T., DARROW, M. A., JIANG, Y.-L., MARAK, G. E. and BLANKS, J. C., 1992, Protection by dimethylthiourea against retinal light damage in rats, *Invest. Ophthalmol. Vis. Sci.*, **33**, 1187–92.

ORGANISCIAK, D. T. and WINKLER, B. S., 1994, Retinal light damage: practical and theoretical considerations, *Progress Retinal Research*, **13**, 1–29.

OROSKAR, A., OLACK, G., PEAK, M. J. and GASPARRO, F. P., 1994, 4′-Aminomethyl-4,5″,8-trimethylpsoralen photochemistry: the effect of concentration and UVA fluence on photoadduct formation in poly(dA-dT) and calf thymus DNA, *Photochem. Photobiol.*, **60**, 567–73.

PITTS, D. G., CULLEN, A. P. and PARR, W. H., 1976, Ocular ultraviolet effects in the rabbit eye, *DHEW (NIOSH)*, Publication No. 77:130.

RAO, D. M. and ZIGLER Jr, J. S., 1992, Levels of reduced pyridine nucleotides and lens photodamage, *Photochem. Photobiol.*, **56**, 523–8.

REME, C. E., BRASCHLE, U. F., ROBERTS, J. and DILLON, J., 1991, Light damage in the rat retina: effects of a radioprotective agent (WR 77913) on acute rod outer segment disk disruptions, *Photochem. Photobiol.*, **54**, 137–42.

RESZKA, K., LOWN, J. W. and CHIGNELL, C. F., 1992, Photosensitization by anticancer agents. 10. ortho-semiquinone and superoxide radicals produced during anthapyrazole-sensitized oxidation of catechols, *Photochem. Photobiol.*, **55**, 359–66.

RESZKA, K., BILSKI, P., SIK, R. H. and CHIGNELL, C. F., 1993, Photosensitized generation of superoxide radical in aprotic solvents: an EPR and spin trapping study, *Free Rad. Res. Comms.*, **19**, S33–S44.

RESZKA, K., BILSKI, P., CHIGNELL, C. F. and DILLON, J., 1995a, Photosensitization by the human lens component kynurenine: An EPR and spin trapping investigation, *Free Rad. in Biol. Med.* (in press).

RESZKA, K., ELDRED, G., WANG, R. H., CHIGNELL, C. F. and DILLON, J., 1995b, Photochemistry of human lipofuscin as studied by EPR, *Photochem. Photobiol.* (in press).

ROBERTS, J. E., 1981, The effects of photooxidation by proflavin in HeLa cells I. The molecular mechanisms, *Photochem. Photobiol.*, **33**, 55–60.

1984, The photodynamic effect of chlorpromazine, promazine and hematoporphyrin on lens protein, *Invest. Ophthalmol. Vis. Sci.*, **25**, 746–50.

1988, Ocular phototoxicity, in Moreno, G., Pottier, R. H. and Truscott, T. G. (eds), *Photosensitization: Molecular, Cellular and Medical Aspects*. pp. 325–30. Berlin: Springer-Verlag, pp. 325–30.

ROBERTS, J. E., ATHERTON, S. J. and DILLON, J., 1990, Photophysical studies on the binding of tetrasulphonatophenylporphyrin (TPPS) to lens proteins, *Photochem. Photobiol.*, **52**, 845–8.

1991b, Detection of porphyrin excited states in the intact bovine lens, *Photochem. Photobiol.*, **54**, 855–7.

ROBERTS, J. E. and DILLON, J., 1986, Phototoxicity in the lens, in Cronly-Dillon, J., Rosen, E. S. and Marshall, J. (eds), *Hazards of Light, Eye and Skin*, Vol. 1, pp. 237–45. Oxford: Pergamon Press.

1987, *In vitro* studies on the photosensitized oxidation of lens proteins by porphyrins, *Photochem. Photobiol.*, **46**, 683–8.

1990, Screening for potential *in vivo* phototoxicity in the lens/retina, *Lens and Eye Toxicity Research*, **7**(3&4), 655–66.

1994c, Improvement of photodynamic therapy by selective protection of normal tissues, in Jung, E. G. and Holick, M. F. (eds), *Biological Effects of Light 1993*, pp. 393–9. Berlin, New York: Gruyter & Co.

Roberts, J. E., Harriman, A., Atherton, S. J. and Dillon, J., 1992b, A non-invasive method to detect oxygen tensions and other environmental factors in the lens, *Int. Soc. Ocular Phototox.*, Sedona, Arizona, 29.

Roberts, J. E., Kinley, J., Young, A., Jenkins, G., Atherton, S. J. and Dillon, J., 1991a, *In vivo* and photophysical studies on photooxidative damage to lens proteins and their protection by radioprotectors, *Photochem. Photobiol.*, **53**, 33–8.

Roberts, J. E. and Mathews-Roth, M., 1993, Cysteine ameliorates photosensitivity in erythropoietic protoporphyria, *Archives Dermatology*, **129**, 1350–1.

Roberts, J. E., Reme, C., Terman, M. and Dillon, J., 1992a, Exposure to bright light and the concurrent use of photosensitizing drugs, *New Eng. J. Med.*, **Vol. 326**, 1500–1.

Roberts, J. E., Roy, D. and Dillon, J., 1985, The photosensitized oxidation of the calf lens main intrinsic protein (MP26) with hematoporphyrin, *Curr. Eye Res.*, **4**, 181–5.

Roberts, J. E., Schieb, S., Garner, W. H. and Lou, M., 1994a, Development of a new photo-oxidative induced cataract model using TPPS in an intact lens, *Invest. Ophthal. Vis. Sci.*, **35**, 2137.

Roberts, J. E., Selman, S., Morgan, A. and Dillon, J., 1994b, Selective protection against the side-effects of photodynamic therapy by radioprotectors, in Jung, E. G. and Holick, M. F. (eds), *Biological Effects of Light, 1993*, pp. 400–6. Berlin, New York: Gruyter & Co.

Rodgers, M. A. J., 1985, Instrumentation for the generation and detection of transient species, in Bensasson, R. V., Jori, G., Land, E. J. and Truscott, T. J. (eds), *Primary Photoprocesses in Biology and Medicine*, pp. 1–24. North Holland, Elsevier.

Rodgers, M. A. J. and Snowden, P. T., 1982, Lifetime of O_2 (singlet delta) in liquid water as determined by time resolved infra-red luminescence measurements, *J. Am. Chem. Soc.*, **104**, 5541–3.

Rosenthal, F. S., Phoon, C., Bakalian, A. E. and Taylor, H. R., 1988, The ocular dose of ultraviolet radiation to outdoor workers, *Invest. Ophthal. Vis. Sci.*, **29**, 649–56.

Sarna, R., Zajac, J., Bowman, M. K. and Truscott, T. G., 1991, Photoinduced electron transfer reactions of rose bengal and selected electron donors, *J. Photochem. Photobiol. A: Chem.*, **60**, 295–310.

Sarna, T., 1992, Ocular melanins, *J. Photochem. Photobiol.*, **12**, 215–58.

Sayre, R. and Roberts, J. E., 1995, Risk of ocular damage from common light sources, *Photochem. Photobiol. Abs. Washington, D.C.*, **61**, 25.

Schey, K. L., Finley, E., Crouch, R. K., Wang, R. H. and Dillon, J., 1995, Site and extent of oxidative damage to lens alpha crystallins, *Invest. Ophthal. Vis. Sci.*, **36**, 409.

Sethna, S. S., Holleschau, A. M. and Rathbun, W. B., 1983, Activity of glutathione synthesis enzymes in human lens related to age, *Curr. Eye Res.*, **2**, 735–42.

Steiner, K. and Buhring, K. U., 1990, The melanin binding of bisoprolol and its toxicological relevance, *Lens and Eye Toxicity Research*, **7**(3 and 4), 319–33.

Straight, R. and Spikes, J. D., 1985, Photosensitized oxidation of biomolecules, in Frimer, A. A. (ed.), *Singlet O, Vol. IV. Polymers and Biopolymers*, pp. 91–143. Boca Raton, Fl. CRC Press.

TERMAN, M., REME, C. E., RAFFERTY, B., GALLIN, T. F. and TERMAN, J. S., 1990, Bright light therapy for winter depression: potential ocular effects and theoretical implications, *Photochem. Photobiol.*, **51**, 781.

TURRO, N., 1978, *Modern Molecular Photochemistry*. Menlo Park, California: Benjamin/Cummings Publishing Co.

WANG, G.-M., RAGHAVACHARI, N. and LOU, M. F., 1995, Effect of oxidation on lens thiol transferase, *Invest. Ophthalmol. Vis. Sci.*, **36**, 885.

WORLD HEALTH ORGANIZATION REPORT, 1995, *Environmental Health Criteria–Ultraviolet Radiation*. Geneva, Switzerland.

ZHU, L. and CROUCH, R. K., 1992, Albumin in the cornea is oxidized by hydrogen peroxide, *Cornea*, **11**, 567–72.

ZIGLER, J. S. Jr, JERNIGAN, H. M. Jr, PERLMUTTER, N. S. and KINOSHITA, J. H., 1982, Photodynamic cross-linking of polypeptides in intact rat lens, *Exp. Eye Res.*, **35**, 239–49.

ZIGMAN, S., 1981, Photochemical mechanisms in cataract formation, in Duncan, G. (ed.), *Mechanisms of Cataract Formation in the Human Lens*, pp. 117–49. New York: Academic Press.

1993, Ocular light damage, *Photochem. Photobiol.*, **57**, 1060–8.

11

The Contribution of Organic Photochemistry to Investigations of Phototoxicity

T. OPPENLÄNDER

11.1 Introduction

The contribution of organic photochemistry to investigations of phototoxicity combines the interdisciplinary aspects of the interaction of light with drugs in biological systems with respect to photophysics, toxicology, photomedicine and photobiology. Its spectrum covers the underlying photophysical principles and photochemical mechanisms (Turro, 1978; Klessinger and Michl, 1989; Albini, 1991; Gilbert and Baggott, 1991; Kopecky, 1992) of drug degradation *in vitro* and *in vivo* and it includes the utilization of the experimental and analytical techniques of organic photochemistry (Moore, 1987; Braun *et al.*, 1986). Furthermore, attention must be paid to the molecular targets of a biological system, to the cellular and medical aspects of phototoxicity (Moreno *et al.*, 1988; Kochevar, 1993) and to the ultimate structure–phototoxicity relationships (Oppenländer, 1988, 1990).

The research that is concerned with drug-induced photosensitization phenomena in humans should take into account meteorological, climatic and environmental factors. For example, the stratospheric ozone depletion (Kerr and McElroy, 1993) leads to an increase of the UV-B solar radiation level and hence to an increase of skin cancer incidence in humans (Black, 1992). On the other hand, the public's obsession with sun bathing or exposure to artificial light sources like solaria, television, home therapy units or daylight lamps is responsible for the dramatic increase of photosensitization reactions to the human skin or to the eye during the past decade.

The interests of the pharmaceutical industries in the photochemistry of drugs and other pharmaceuticals should be guided first of all by *drug safety* and the minimization of light-induced skin photosensitizations which contribute about 3–10 per cent to secondary effects of drugs (Zürcher and Krebs, 1980). Secondly, the design of novel photosensitizers for photochemotherapy and the development of novel sunscreens for skin protection are challenging activities for pharmaceutical industries. Additionally, photochemical drug degradation studies may lead to the synthesis of sophisticated novel compounds using light as a highly selective reagent. Last but not least, the contamination by toxic photo-degradation products during the drug-development process should be avoided. Therefore, standardized protocols for phototoxicity testing of pharmaceuticals have to be established.

11.2 Structure-function relationships in phototoxicology

11.2.1 Drug-induced photosensitization

The requirements of a drug to cause photosensitization to the skin or to the eye are shown in Fig. 11.1.

Above all, the most trivial requirement is the absorption of light by the drug in the UV-B/UV-A or visible part of the solar spectrum to generate an electronically excited species in its singlet or triplet state which subsequently may cause biological damage. The resulting adverse effects are usually classified as either *phototoxicity* or *photoallergy*. On the other hand, photochemotherapy utilizes the controlled application of photosensitizers to treat skin diseases like psoriasis (PUVA therapy) or several malignant tumours by photodynamic therapy (PDT). Important factors which influence the photosensitizing potential of pharmaceuticals are the optical properties of the skin, the distribution of the drug in the skin, its bioavailability and its biological half-life. Furthermore, the photoreaction or energy transfer must be sufficiently fast to compete with radiationless decay and excited-state quenching mechanisms to cause biological damage without existing biological repair mechanisms. Additionally, metabolization of the drug has to be considered.

Essential information about the optical properties of the human skin (Ippen, 1969) is presented in Fig. 11.2.

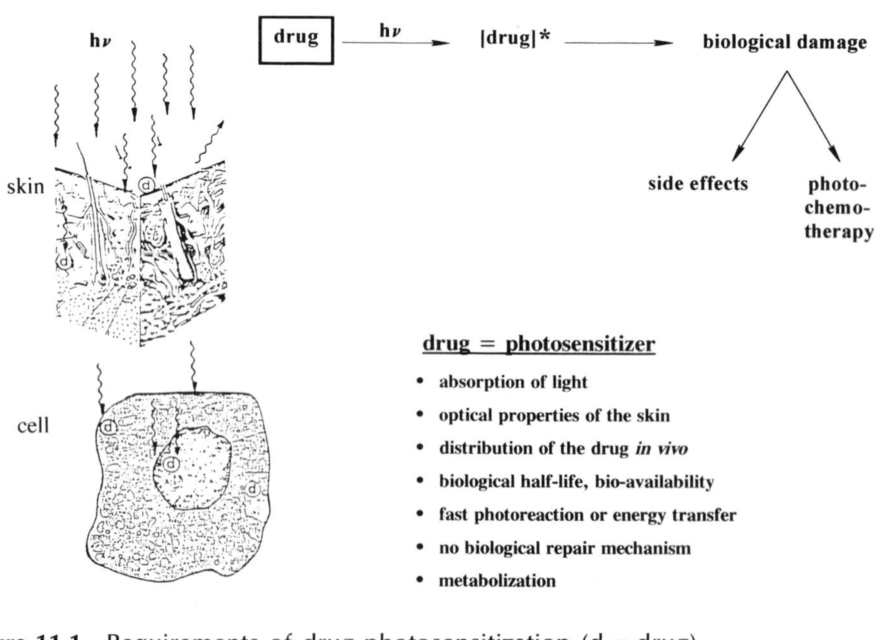

Figure 11.1 Requirements of drug photosensitization (d = drug)

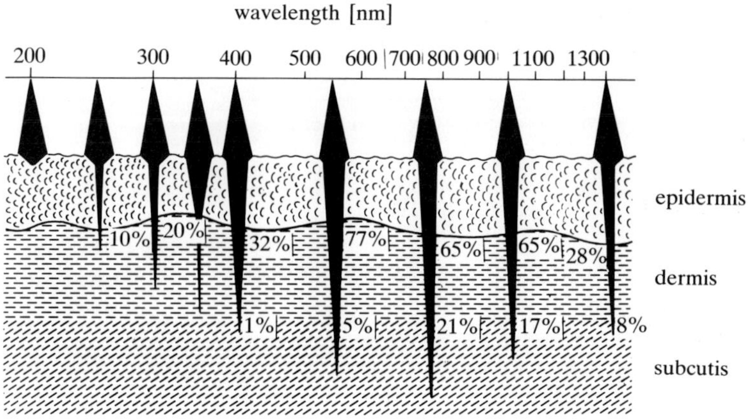

Figure 11.2 Light skin penetration as a function of wavelength

Short-wavelength UV radiation is almost completely absorbed by the epidermis. However, the depth of light penetration to the skin increases with increasing wavelength. Therefore, visible light and infrared radiation reach the dermis and subcutis, respectively. UV-B and UV-A radiation reach the *stratum papillare* and the *stratum reticulare* in the dermis with its capillary blood vessels which is crucial for drug photosensitization.

11.2.2 ***Definition of terms in photochemistry, photobiology and photomedicine***

To avoid misinterpretations in this interdisciplinary field of research it is important to unequivocally define the term 'photosensitization' which is used in organic photochemistry (Braslavsky and Houk, 1988) and in photobiology or photomedicine (Fig. 11.3).

A photosensitizer in organic photochemistry is strictly defined as a ground state molecule (= sensitizer) which is transferred to its electronically excited singlet state (^{1}S) by the absorption of light according to the first law of photochemistry. After intersystem crossing to the excited triplet state (^{1}T) of the sensitizer subsequent energy transfer via exciplexes to a lower–lying ground state of a substrate molecule generates an electronically excited state of this molecule according to the spin conservation rules (Turro, 1978). Simultaneously, the ground state of the sensitizer is restored and therefore the photosensitizer is not consumed in the reaction. The primary and secondary photophysical processes are illustrated by the Jablonski diagram (Pfoertner, 1991).

In contrast to this photophysical definition, a photosensitizer in photobiology or photomedicine is a ground state molecule that is incorporated into a biological system prior to light absorption (Fig. 11.3). The absorption characteristics of this molecule may be changed by the interaction with cell components or by filter effects that are exerted by the biochemical environment. After light absorption photochemical and/or photophysical alterations may lead to biological damages. However, in photomedicine two broad categories of drug-induced photosensitiza-

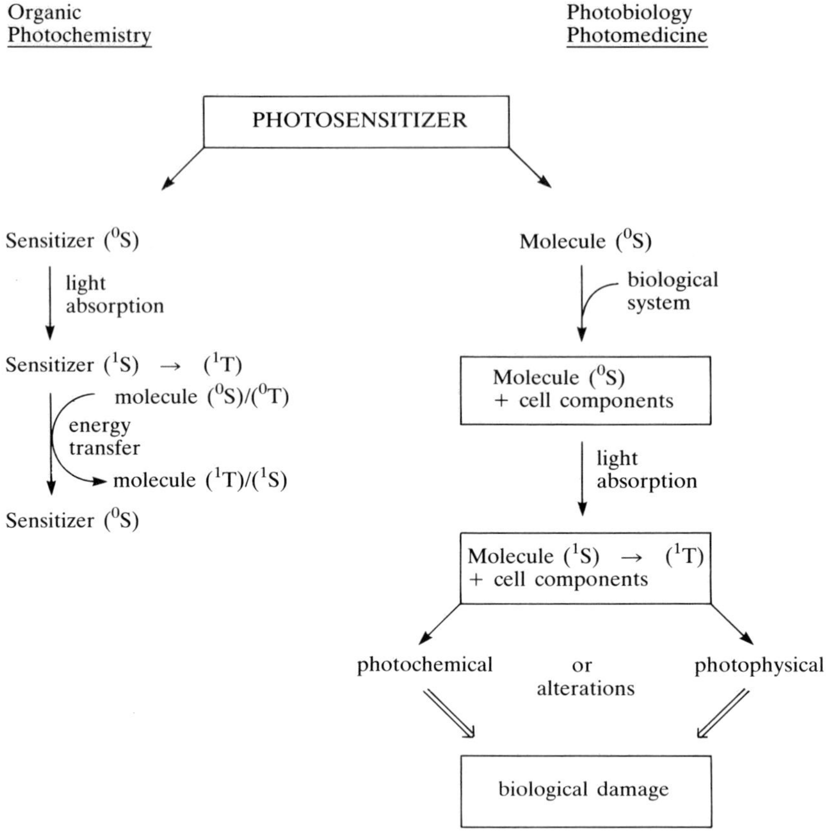

Figure 11.3 Definition of terms in organic photochemistry and photobiology/-medicine

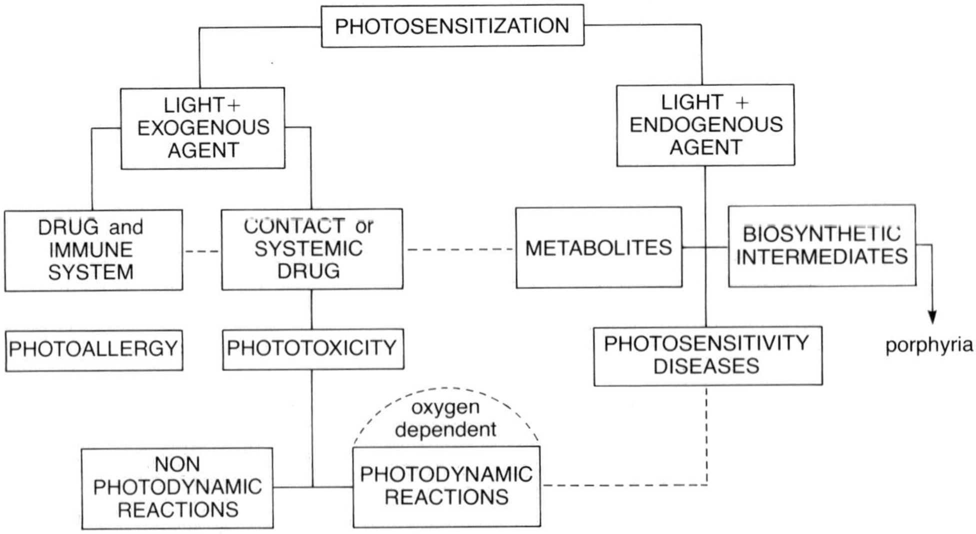

Figure 11.4 Photomedicinal classification of drug-induced photosensitization reactions

tion reactions in humans are considered (Fig. 11.4): the interaction of light with an exogenous or an endogenous agent.

Endogenous photosensitizers may be accumulated *in vivo* due to drug-induced disorders in metabolic or biosynthetic pathways. This may cause severe *photosensitivity* diseases like porphyrias. On the other hand, the interaction of light with exogenous agents such as contact photosensitizers or systemically administered drugs may cause either *photoallergy* which involves the immune system or non-immunologic *phototoxic* responses. Phototoxic reactions are subdivided into non-photodynamic and oxygen-dependent photodynamic processes.

11.2.3 Structure-activity relationships

The drugs that cause cutaneous or ocular phototoxicity belong to many different structural families of organic compounds (see Appendix: Molecular Structures of Phototoxic Pharmaceuticals). Representative phototoxic pharmaceuticals include tetracycline and fluoroquinolone antibiotics, sulfonamide diuretics, polyhalogenated salicylanilides which are used as disinfectants and many other chemicals like cosmetics, detergents, dyes, food additives and naturally occurring compounds. The action spectrum of phototoxic compounds usually covers the UV-A or visible light range. However, several drugs become phototoxic with UV-B radiation. As can be deduced from the compilation of molecular structures of

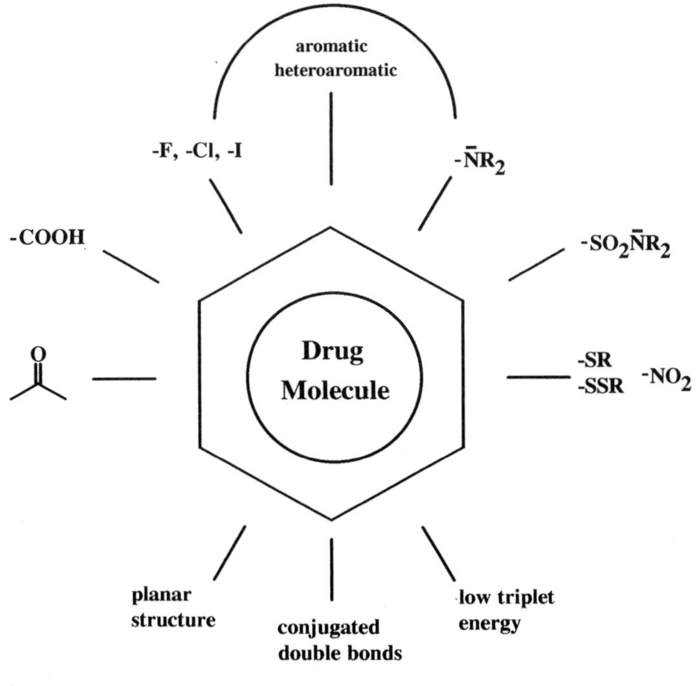

Figure 11.5 Chromophores and functional group interactions that are responsible for phototoxicity

phototoxic pharmaceuticals, a limited number of chromophores and functional group interactions (Fig. 11.5) seems to be responsible for phototoxicity.

The essential functionalities of aromatic or heteroaromatic phototoxic drugs are identical with the most important pharmacophoric groups like the carbonyl group, the halogen atoms, the amino group and the sulfonamide substituent. Of minor importance are thiols, sulfides, disulfides and nitro compounds. Most detrimental is the inter- or intramolecular combination of halogen substituents with the amino group because of photoinduced electron transfer mediated dehalogenation reactions (Pac and Ishitani, 1988; Fox, 1990).

Further structural properties of phototoxic compounds include either planar moieties or conjugated double bonds which are responsible for low triplet energies. Although it is well known that the physiological action of enantiomers and diastereoisomers is well differentiated with regard to the beneficial pharmacological action and to toxic side-effects (Stinson, 1992; Crossley, 1992), there are no data available about enantiospecific or diastereospecific phototoxicity. To demonstrate the diversity of functional group interactions and of the underlying photochemical and photophysical principles of phototoxicity, several examples from the current literature are selected and consecutively discussed in the following section.

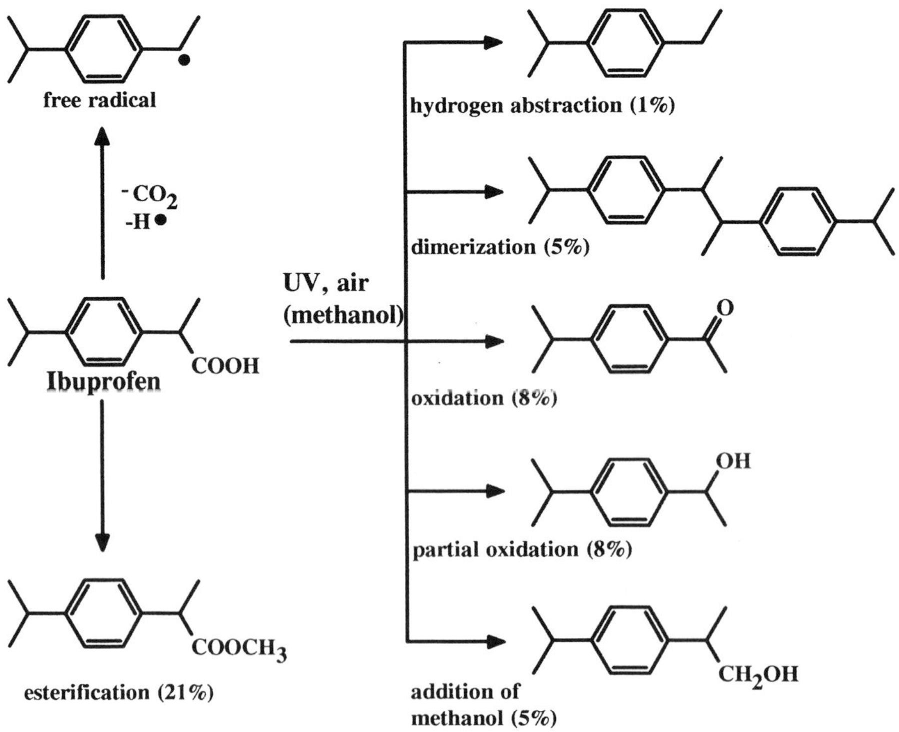

Figure 11.6 Organic photochemistry of Ibuprofen

11.2.4 *A survey of photochemical and photophysical mechanisms of phototoxicity*

The group of non-steroidal anti-inflammatory drugs (NSAID) (Moore and Chappuis, 1988) like benoxaprofen, naproxen, ketoprofen and ibuprofen is structurally characterized by a secondary carboxyl group in benzylic position of an aromatic ring system (Fig. 11.6). Typically for this class of compounds, the photochemical behaviour of ibuprofen (Castell *et al.*, 1987) can be reduced to the initial decarboxylation with formation of a benzylic radical intermediate. The analysis of the photoproducts under aerobic conditions in methanol as solvent reveals the spectrum of pure radical chemistry with respect to hydrogen abstraction, dimerization, oxidation, partial oxidation and the addition of methanol. The competitive esterification of the carboxylic acid is observed in 21 per cent chemical yield. Thus, ibuprofen is a *photolabile drug*.

The combination of organic photochemistry and photophysical effects is observed with the antibacterial drug nalidixic acid (Moore *et al.*, 1984; Dayhaw-Parker and Truscott, 1988; Fernandez and Cardenas, 1990) which consequently is classified as a photolabile drug and a photosensitizer (Fig. 11.7).

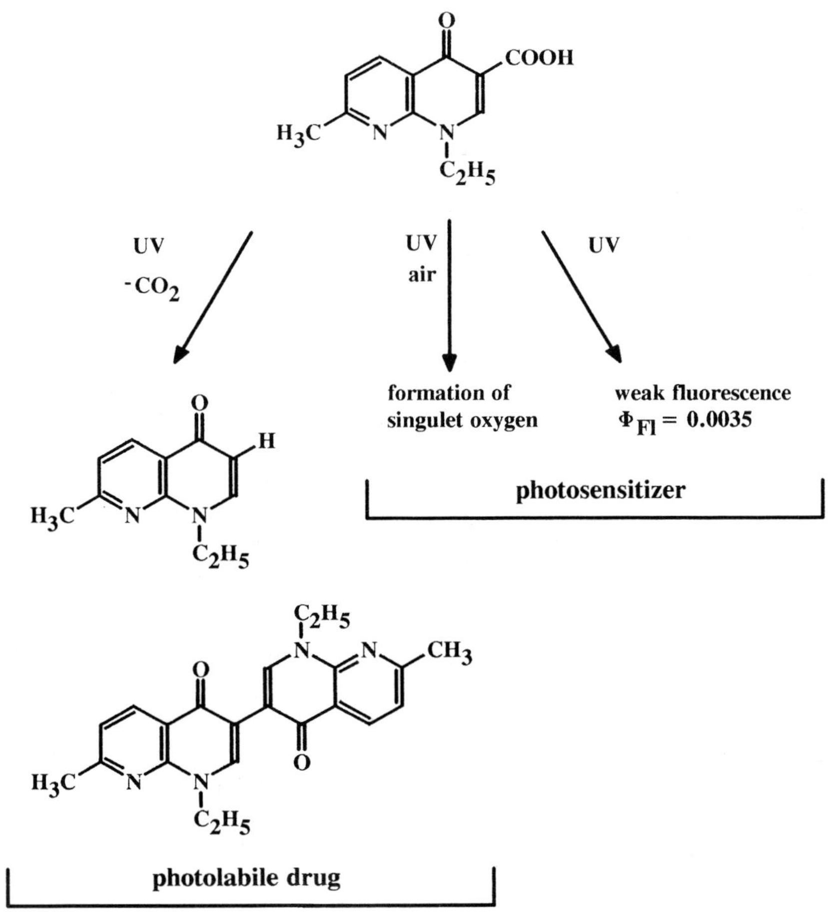

Figure 11.7 Organic photochemistry and photophysics of nalidixic acid

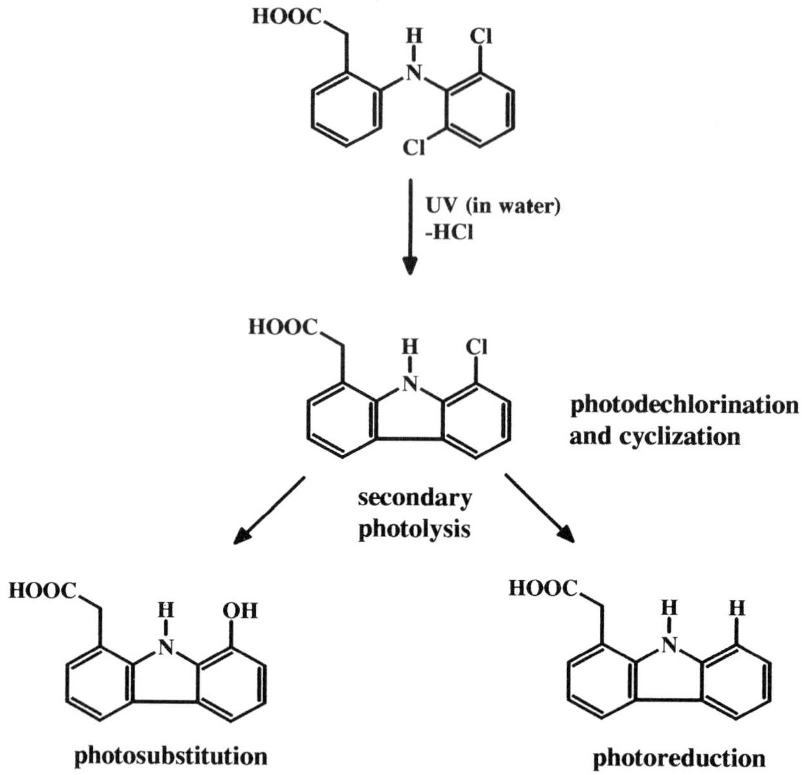

Figure 11.8 Organic photochemistry of diclofenac

Three reaction channels of the electronically excited nalidixic acid account for its photoreactivity: decarboxylation to the parent compound accompanied by dimerization; formation of singlet oxygen under aerobic conditions, and the weak fluorescence decay. Nalidixic acid is structurally related to fluoroquinolone antibiotics which are photolabile and cause photosensitization reactions in humans (Hidalgo *et al.*, 1993).

In contrast, the organic photochemistry of the anti-inflammatory drug diclofenac (Moore *et al.*, 1990) which again contains a carboxyl functionality in combination with aromatic chlorine substituents (Fig. 11.8) is characterized by the alternative pathway of dechlorination and cyclization to a carbazole derivative. Secondary photolysis of the primary photoproduct in water leads to the formation of the phenolic substitution product and to photoreduction. The carboxyl group is retained in the photoproducts. Thus, the free radical nature of the dehalogenation processes is postulated to be the primary act in diclofenac-initiated phototoxicity. Additionally, the carbazole formation causes a significant bathochromic shift in the UV spectrum. Another prominent example of photodechlorination is documented by chlorpromazine (Kochevar, 1987; Schoonderwoerd *et al.*, 1989) which causes phototoxic effects in humans (Fig. 11.9).

The organic photochemistry of this compound is related to the formation of a free-radical intermediate which is generated via photodechlorination. Photo-

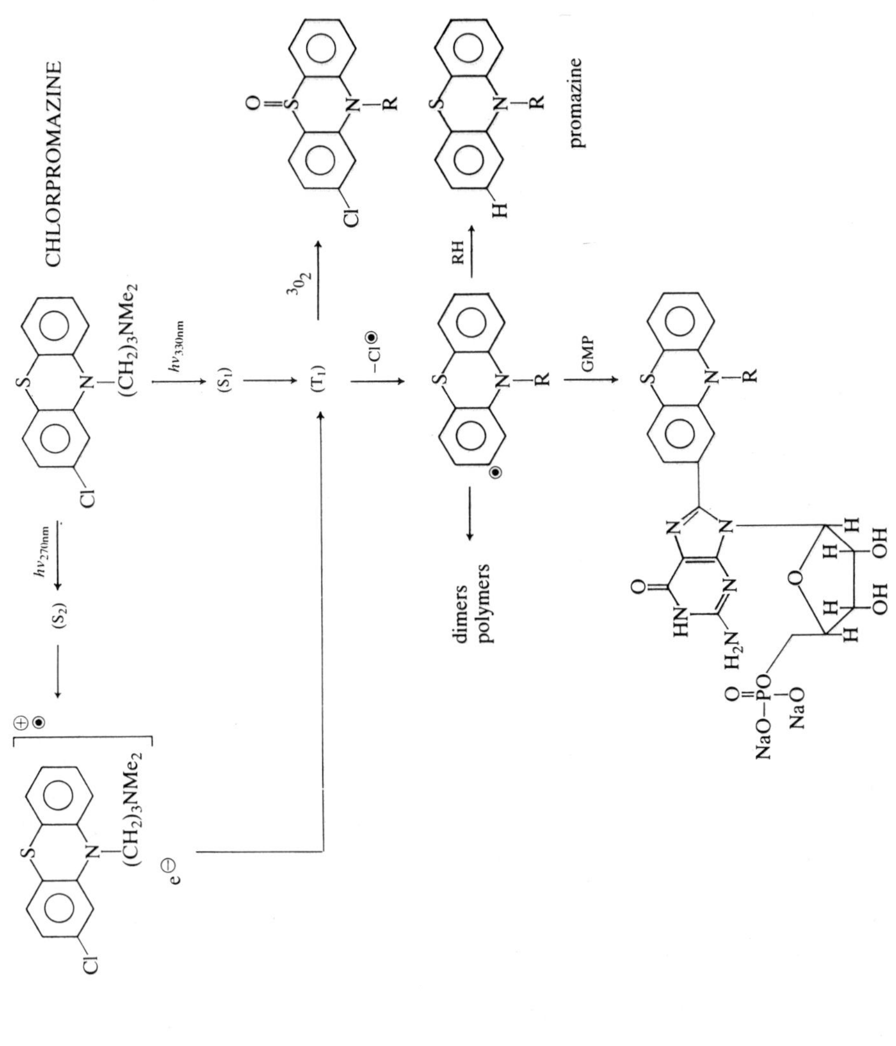

Figure 11.9 Organic photochemistry of chlorpromazine

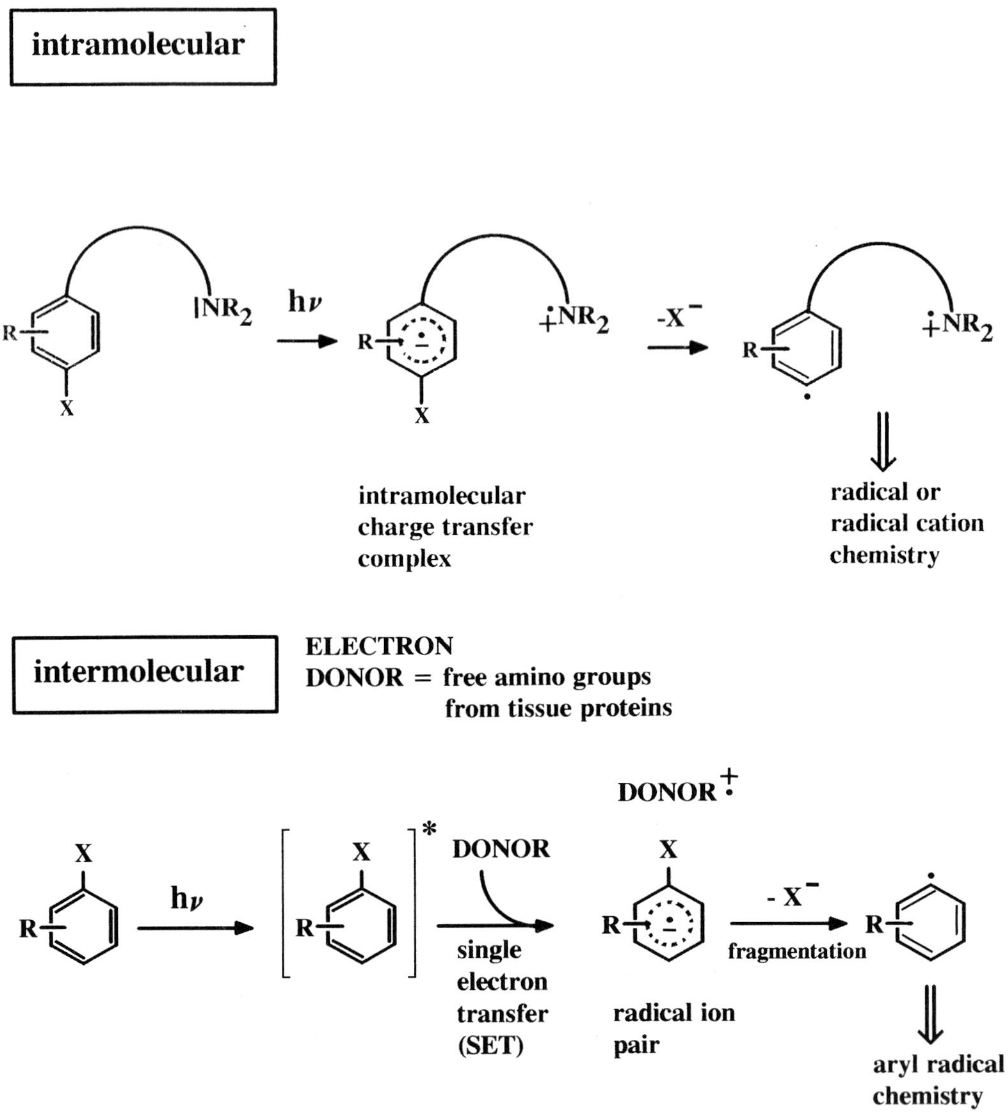

Figure 11.10 Dehalogenation of aryl halides by photoinduced electron transfer as primary event in phototoxicity

ionization from the second excited singlet state does only occur at short-wavelength irradiation ($\lambda < 270$ nm). Therefore, the involvement of the radical cation of chlorpromazine in photosensitization is unlikely. The promazyl radical is the precursor of dimers and polymers and the parent promazine. This radical adds to biomolecules like guanosine-monophosphate (GMP) and consequently represents the active species in chlorpromazine-induced photosensitization. Modern mechanistic concepts of the dehalogenation of aryl halides by photoinduced electron transfer (Pac and Ishitani, 1988; Fox, 1990) via intramolecular or intermolecular processes account for a better understanding of the primary events in phototoxicity (Fig. 11.10).

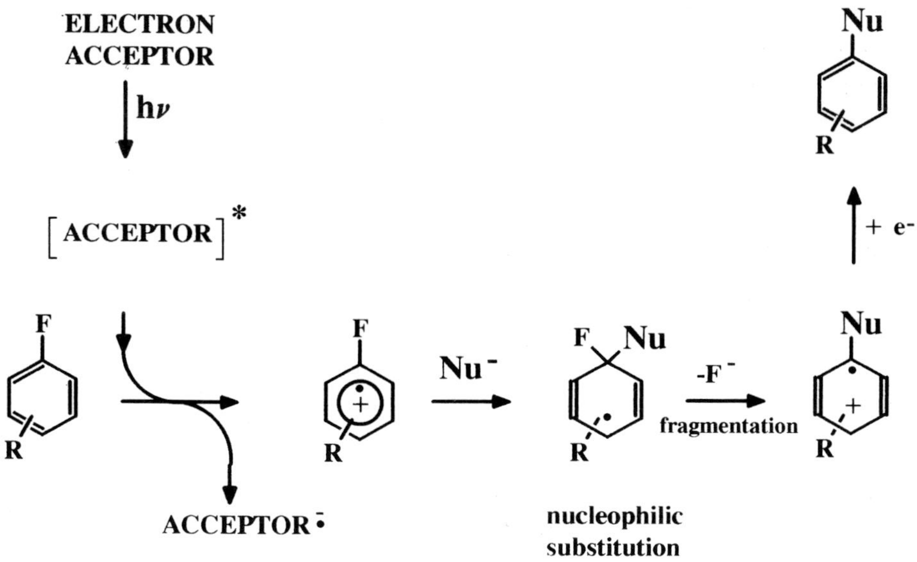

Figure 11.11 Photoinduced electron transfer catalysed defluorination of fluoroaryl compounds

The electron-donating amino group transfers an electron to the aromatic acceptor part of the molecule. Subsequently, the intramolecular charge transfer complex eliminates a halide anion with formation of an arene radical and an amine radical cation moiety. This may lead to radical or amine radical cation (Pandey, 1992) chemistry.

On the other hand, it is well established that the photodehalogenation of aryl halides is accelerated by the presence of amines. The intermolecular process is considered to involve single electron transfer and halide-ion expulsion from the radical ion pair which leads to aryl radical chemistry. In human photosensitization free amino groups from tissue proteins may serve as electron donors.

However, photoinduced electron transfer catalysed defluorination of electron-rich fluoroaryl compounds (Fig. 11.11) may be described by formation of an aryl radical cation in the presence of suitable electronically excited acceptors (Julliard and Chanon, 1983). Subsequently, nucleophilic substitution is followed by one-electron reduction and yields the defluorinated derivative. A series of different reactive intermediates may interfere with biological systems.

Another important photolabile chromophore is the sulfonamide functionality (Fig. 11.12). Sulfonamides are antibacterial chemotherapeutics and are rapidly photodegraded by alpha, alpha′ or β cleavage as was demonstrated by irradiation of sulfadoxine (Oppenländer, 1988). A series of possible radical intermediates is responsible for the formation of a complex reaction mixture. However, we were able to isolate the products of S–N bond cleavage (α), 4-amino-5,6-dimethoxypyrimidine in 29 per cent yield and traces of sulfanilic acid.

The most prominent example of planar structure-related phototoxicity is 8-methoxy-psoralen (8-MOP) which is applied in the photochemotherapy of psoriasis. This furocoumarine is a naturally occurring photosensitizer and is the active ingredient of the common cow parsnip (Fig. 11.13) (Frohne and Pfänder,

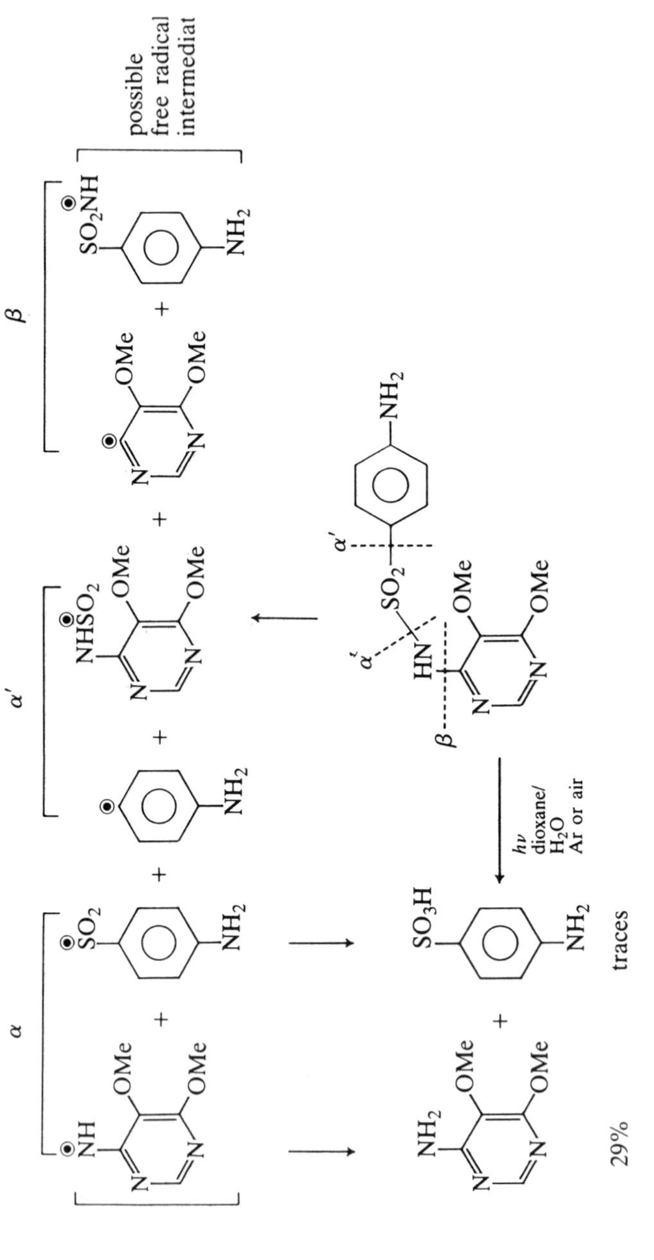

Figure 11.12 Photodegradation of sulfadoxine

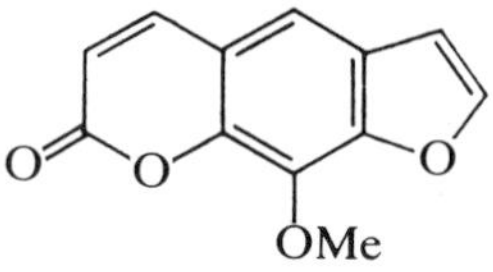

8-MOP

Figure 11.13 8-Methoxypsoralen as a natural constituent of Heracleum sphondylium (common cow parsnip); reproduced with permission of the Wissenschaftsverlag, Stuttgart (Frohne *et al.*, 1987)

1987). On contact with 8-MOP and after a short period of sunshine the phototoxic response manifests itself by induction of a sunburn and after three days an erythema with marked blisters is formed (Fig. 11.14).

The photoreactivity of 8-MOP (Fig. 11.15) is characterized by its planar structure which allows intercalation into DNA-double strands and the formation of a non-covalent dark complex. Irradiation of this complex leads to the formation of cyclobutane-type mono- or diadducts via [2 + 2]-cycloaddition to the DNA-pyrimidine bases.

Additionally, this non-photodynamic mechanism is accompanied by the photodynamic generation of singlet oxygen. Both mechanisms seem to be responsible for the PUVA-induced photosensitization reactions which are observed in human skin.

These examples conveniently verify the complexity of photochemical and photophysical aspects which are involved in photosensitization reactions. Thus, the interaction of light with drugs (Fig. 11.16) may produce a variety of short-lived reactive intermediates, radical anions, radical cations or cytotoxic oxygen species like singlet oxygen or the superoxide radical anion in addition to a series of substituted oxygen radicals. Furthermore, light-induced oxidative bioactivation may lead to an oxidized derivative of the drug ($drug_{ox}$) which may induce phototoxic or photoallergic effects. The photoinduced [2 + 2]-cycloaddition of appropriate planar substrates to pyrimidine nucleotides may lead to cyclobutane-type products which inhibit the DNA replication. On the other hand, photorearrangements and photofragmentations, as well as energy or electron-transfer

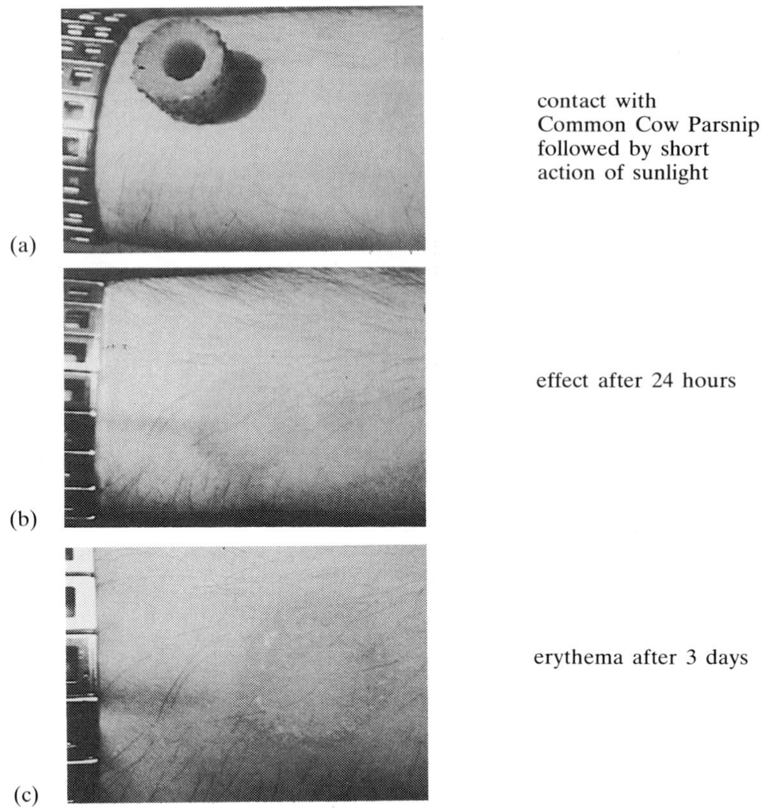

Figure 11.14 Phototoxic response to 8-MOP; reproduced with permission of the Wissenschaftsverlag, Stuttgart (Frohne *et al.*, 1987)

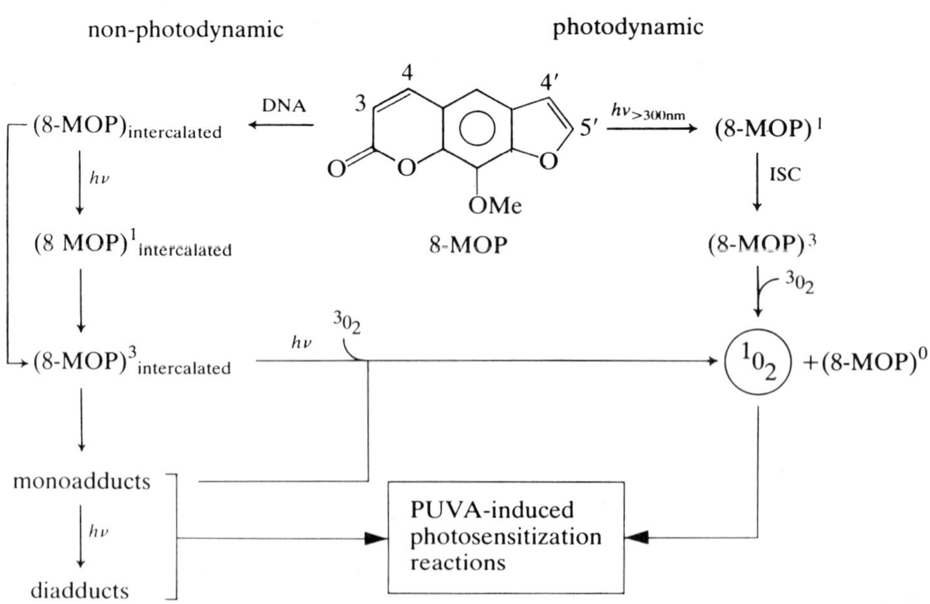

Figure 11.15 Photoreactivity of 8-methoxypsoralen

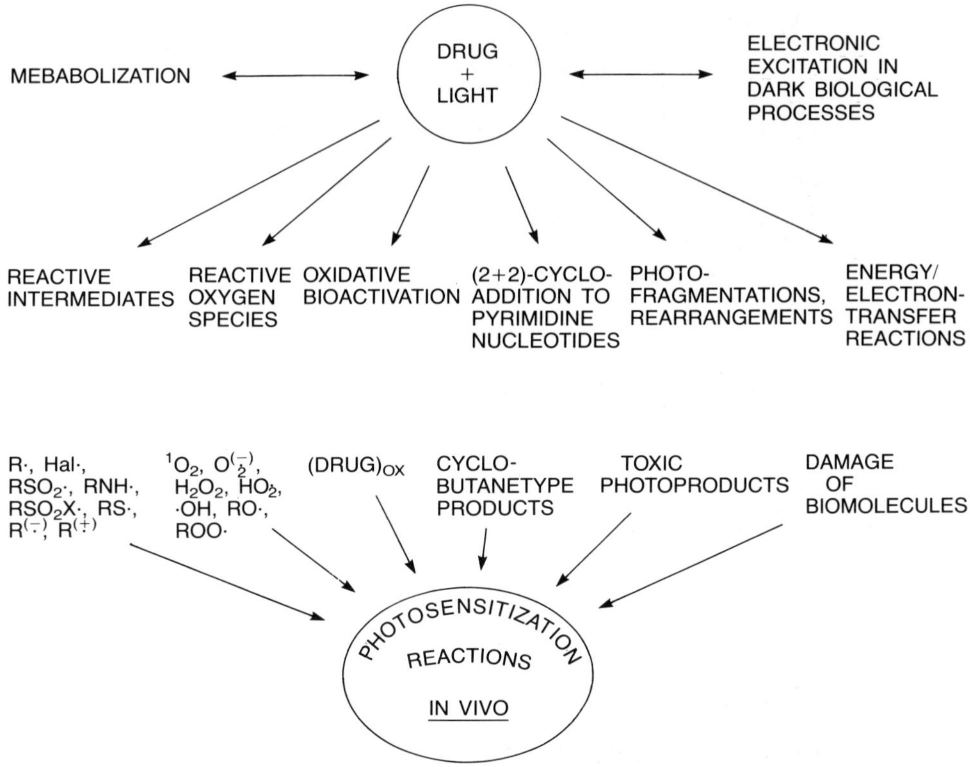

Figure 11.16 Photoreactions which are involved in photosensitization

induced reactions have to be considered to cause *in vivo* photosensitization reactions. This scheme of interactive photoreactions might even be complicated by metabolization of the drug to other photoactive compounds or by the biochemical generation of electronically excited species in so-called dark biological processes (Adam and Cilento, 1982). In conclusion, a wide variety of different photochemical reactions and/or photophysical effects have to be considered as primary events of phototoxicity *in vivo*.

11.3 *In vitro* degradation studies

11.3.1 A comprehensive photochemical and photophysical in vitro *assay*

To establish the crucial elements of drug phototoxicity the *in vitro* photostability *and* the *in vitro* photoreactivity of a drug must be considered (Fig. 11.17). Both kinds of drugs may exhibit *no in vivo* light-induced effect due to their pharmacokinetic and/or pharmacodynamic behaviour. However, a photostable drug may act as an efficient photosensitizer via a series of photophysical effects or a photolabile drug may cause *in vivo* phototoxicity by the formation of reactive

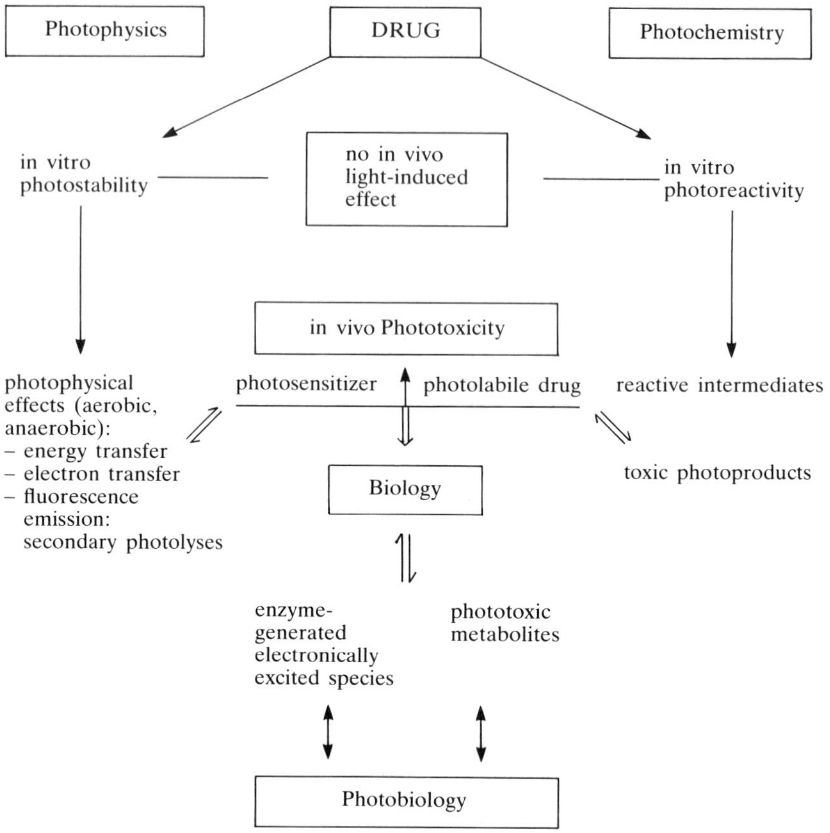

Figure 11.17 Crucial elements of drug phototoxicity

intermediates or toxic photoproducts. In any case, biological and photobiological investigations have to be performed to clarify the mechanism of phototoxicity. As a consequence, a comprehensive photochemical and photophysical *in vitro* assay was proposed (Oppenländer, 1988) for the standardization of the photoreactivity testing of drugs which includes correlations to *in vivo* toxicological studies and photobiological or photomedicinal investigations. The criteria of subjecting a drug or a potential drug candidate to this photoassay (Fig. 11.18) that includes photochemical and photophysical followed by photobiological, toxicological and photomedicinal studies are obvious. First, the light-absorbing potential of the drug in the UV-B, UV-A or visible range of the electromagnetic spectrum has to be established by UV/Vis absorption spectroscopy. This criterion by itself justifies the application of the *in vitro* photoassay because it bears the risk of phototoxicity or photoallergy in humans. Secondly, compounds which are known to form light absorbing metabolites via fast metabolization or long-term medication in combination with high-dose application that generates a high plasma concentration of the drug must be considered. Even environmental conditions like the medication in countries near the equator bear the risk of light-induced side-effects.

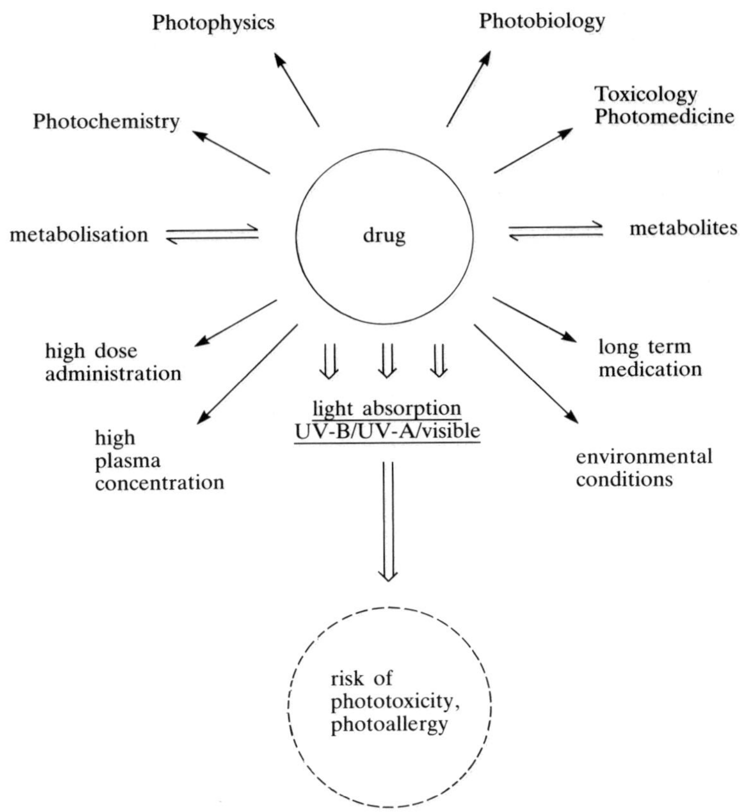

Figure 11.18 *In vitro* photoassay and the criteria for its application

11.3.2 Examples of drug degradation studies (Oppenländer, 1993, 1994a)

An attractive candidate for the application of the photoassay represents the yellow-coloured benzoquinolizin derivative (Fig. 11.19) that exhibits an absorption maximum at $\lambda = 435$ nm ($\varepsilon = 17\,700\,\mathrm{l\,mol^{-1}\,cm^{-1}}$) and additionally substantial absorption in the UV-A and UV-B range. This drug was developed as a potential cerebral circulation improver. Stimulated by photobiological experiments (Fig. 11.20) which established the photomutagenic activity of this compound in *salmonella typhimurium* (TA 102), we investigated its photochemistry and its photophysics. The emission characteristics of this compound are the following: fluorescence (at 293 K), $\lambda_{max} = 506$ nm and phosphorescence (at 77 K), $\lambda_{max} = 510$ nm. The energy of the first excited singlet state (^1S) was calculated from the ½ Stokes' shift and the triplet energy (^1T) was obtained from the phosphorescence onset at $\lambda = 490$ nm.

The singlet and triplet energies are relatively low with 256 and 244 kJ/mol, respectively. Irradiation of this compound in polar solvents led to undefined decomposition with formation of a complex reaction mixture. This photolysis

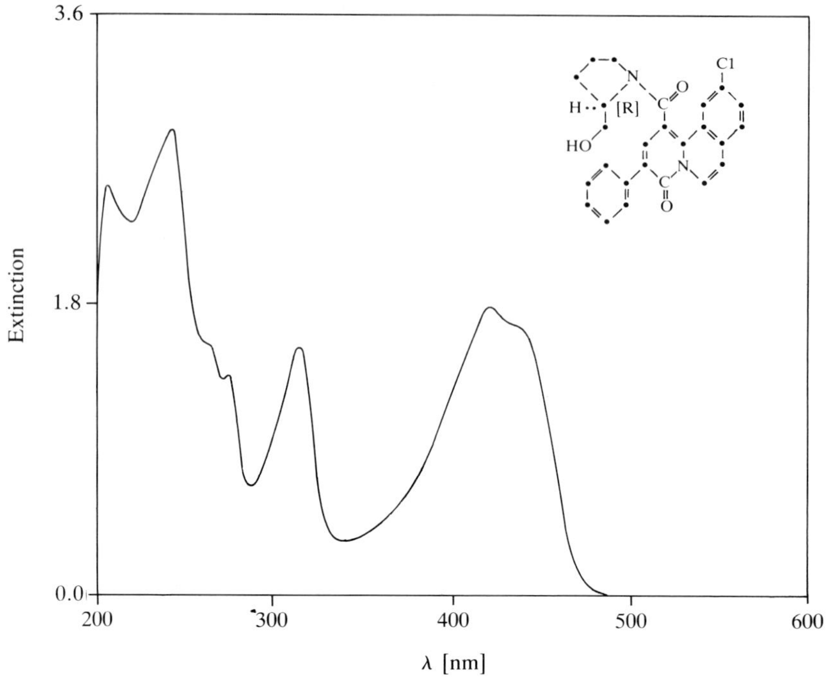

Figure 11.19 UV/Vis absorption spectrum of the yellow-coloured benzoquinolizin derivative (systematic name: R-1-[(10-chloro-4-oxo-3-phenyl-4H-benzo-[a]quinolizin-1-yl)carbonyl]-2-pyrrolidinemethanol)

mixture exhibited no mutagenic activity and hence, no mutagenic photoproducts were formed during light-induced decomposition. On the other hand, the efficient formation of singlet oxygen was detected during aerobic irradiation of the compound in ethanol with visible light, but *S. typhimurium* shows no singlet oxygen mutagenicity when $^1O_2$ is generated outside the bacteria (Piette, 1990). The photosensitized oxidation of the compound in the presence of singlet oxygen led to the formation of a novel chlorinated isoquinoline derivative. This compound was not photomutagenic in *S. typhimurium*.

Therefore, none of these mechanisms can be responsible for the photomutagenic activity of the benzoquinolizin and it may be concluded that light-induced dissociation of the carbon-chlorine bond accompanied by the formation of aromatic radical intermediates may account for the observed photomutagenicity. Further research involving DNA-binding studies is necessary to evaluate this complex light-induced behaviour.

From the organic photochemist's point of view the photo-oxygenation of the benzoquinolizin is most interesting (Fig. 11.21). In the first step the formation of an unstable endoperoxide via the Diels-Alder reaction with singlet oxygen as dienophile may be postulated (George and Bhat, 1979). Subsequently, the

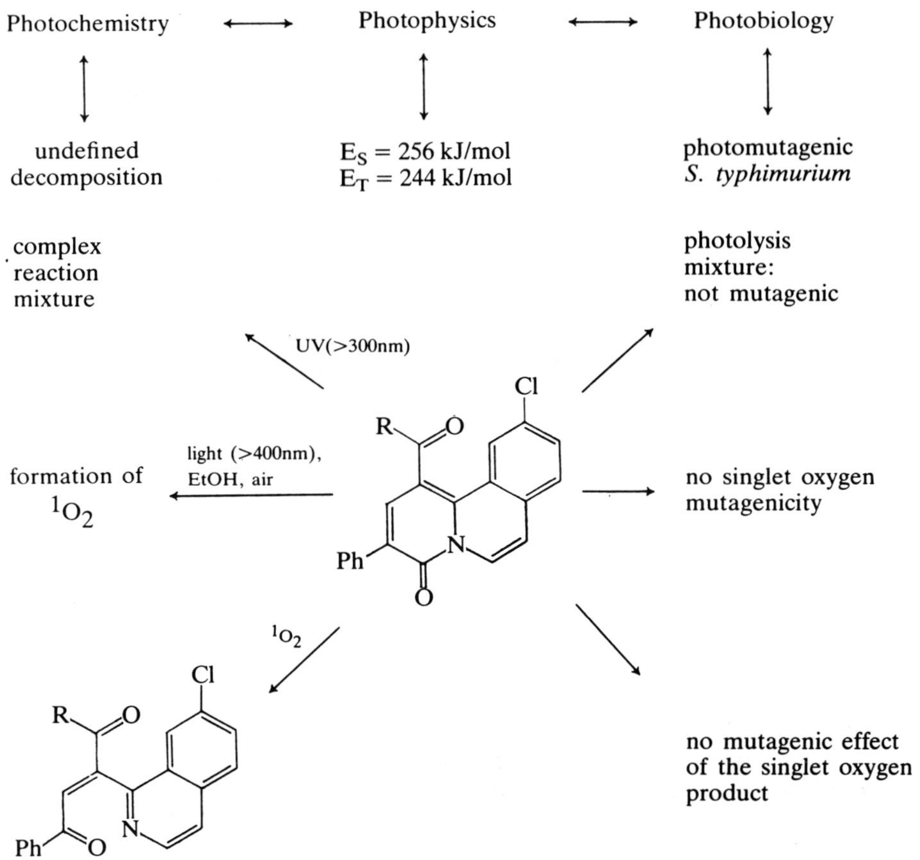

Figure 11.20 *In vitro* photoassay describing the light-induced reactivity of the benzoquinolizin derivative

formation of an unstable dioxetane via an intermediate zwitterionic species is presumably followed by thermal cleavage of the dioxetane and decarboxylation to yield the novel unsaturated diketone.

The photodynamic potential of the benzoquinolizin derivative was established by performing an intermolecular competition experiment which applies 2,3-dimethyl-2-butene (tetramethylethene, TME) as an efficient singlet oxygen trap (Fig. 11.22) (Oppenländer, 1988). The formation of the hydroperoxydimethylbutene during aerobic irradiation of the drug in the presence of TME, which is diagnostic for singlet oxygen involvement, was monitored quantitatively by gaschromatography and compared with the Rose Bengal sensitized photo-oxygenation of TME. Thus, the benzoquinolizin seems to be a medium efficient singlet oxygen sensitizer (Fig. 11.22) (Oppenländer, 1988–1991). Singlet oxygen formation and simultaneous reactivity against singlet oxygen is called 'self-sensitization' (Fig. 11.23).

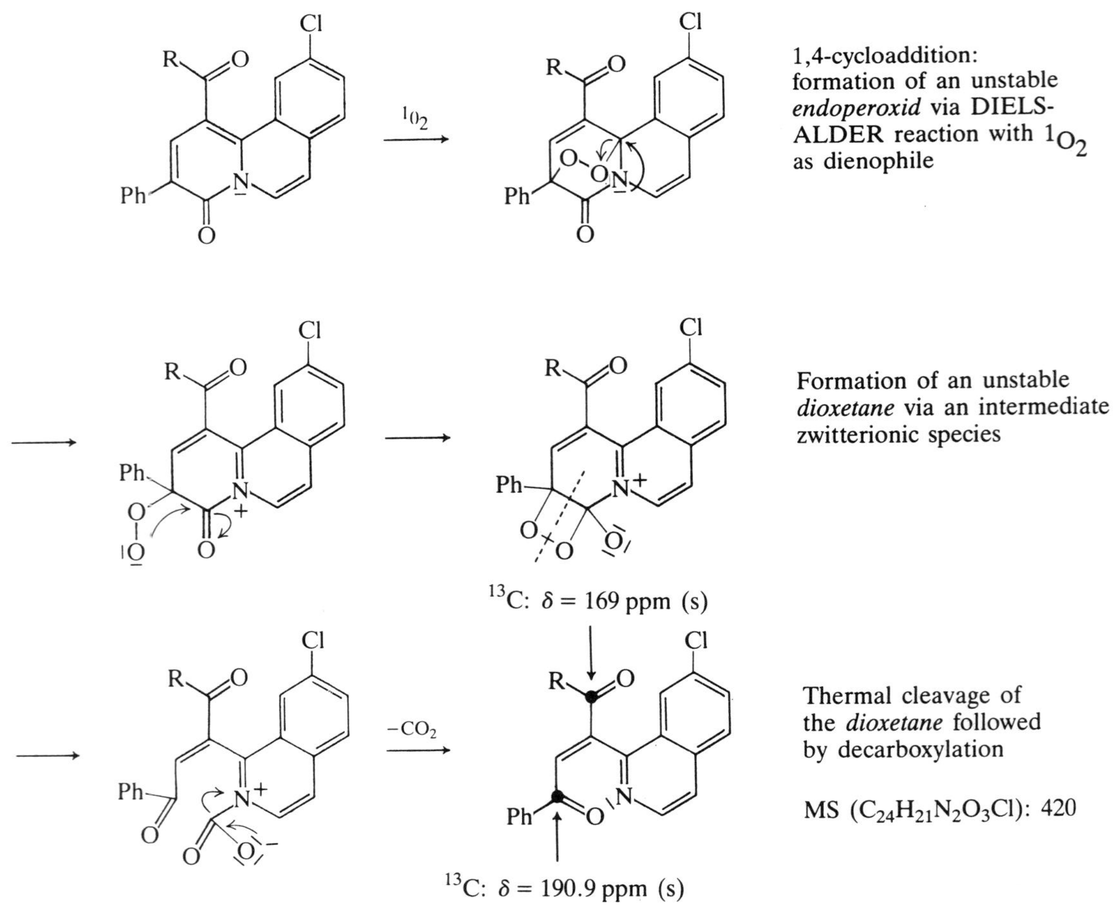

Figure 11.21 Reactivity of the benzoquinolizin against singlet oxygen: possible mechanism

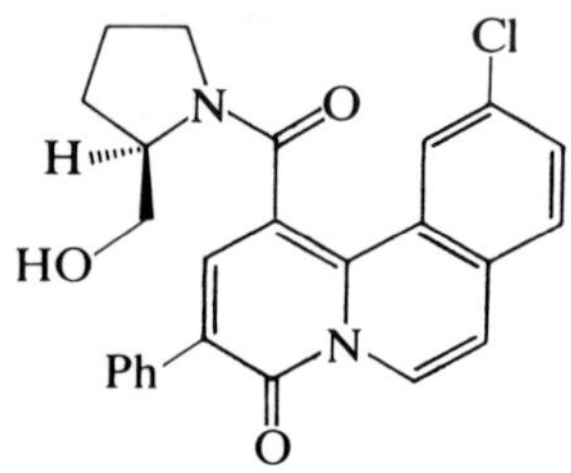

Ro 19-8022

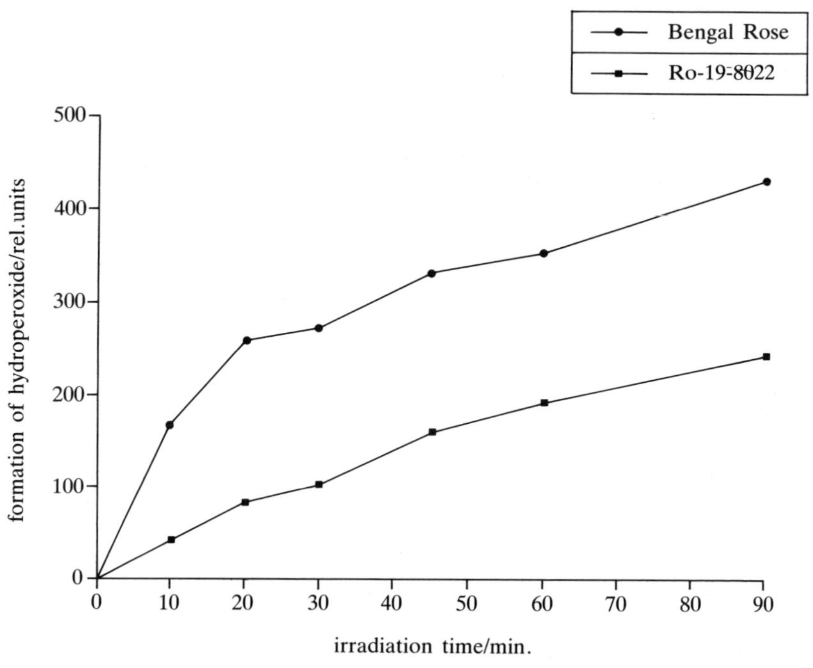

Figure 11.22 Qualitative method to estimate the photodynamic potential of a drug

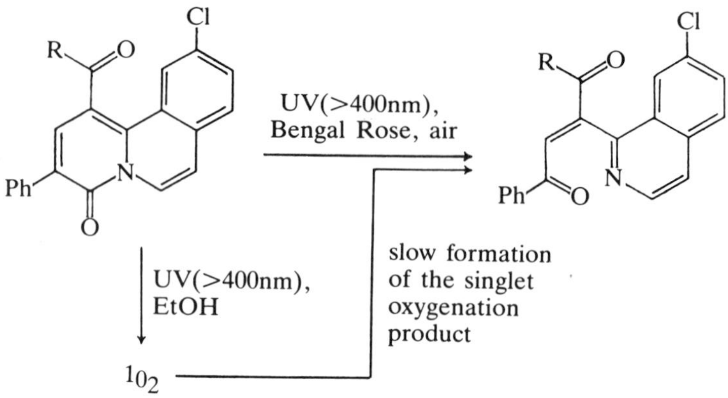

Figure 11.23 Singlet oxygen formation and selfsensitization

The self-sensitizing potential of the benzoquinolizin derivative could be demonstrated by its irradiation with visible light in ethanol under aerobic conditions. The slow formation of the singlet oxygenation product could be verified. A typical example of oxygen-dependent photoreactivity of a drug is demonstrated in Fig. 11.24 (Oppenländer, 1988–91).

Under anaerobic irradiation conditions no alteration of the isoquinoline derivative could be detected. However, irradiation of the free base under aerobic conditions with the Pyrex filtered output of a mercury high-pressure lamp led to fast decomposition with formation of a yellow-coloured solution. Similar results were obtained for the monomaleate salt. Three products could be isolated from the complex photolysis mixture by column chromatography. Direct photolysis and photo-oxygenation ($^1O_2$) of the free base led to the formation of an isoquinoline aldehyde in 10 per cent and 45 per cent chemical yield, respectively. Therefore, oxidative fragmentation of the piperazinyl part of the molecule seems to be the main reaction path. Similar oxidative photofragmentations of 2-aryl-4-quinoline-methanols have been reported (Epling *et al.*, 1984). On the other hand, the direct photolysis of the monomaleate salt led in 1 per cent isolated yield to the isoquinoline aldehyde. Its photo-oxygenation ($^1O_2$) in ethanol gave 6 per cent of the overoxidized product and 2 per cent of acetal formation.

The comparison of the UV/Vis spectra of the free base and its monomaleate formulation (Fig. 11.25) demonstrates the bathochromic effect of nitrogen protonation that leads to an extension of the UV absorption up to 400 nm. Both compounds absorb in the UV-A and UV-B range and it is recommended to handle solutions of these isoquinolines in brown glass bottles or ampoules to avoid photooxidation during the drug development process.

An example of wavelength-dependent photoreactivity is demonstrated by a tertiary amine (Fig. 11.26) (Oppenländer 1988–1991) which was developed as a potential thermogenic compound. Its UV spectrum consists of two distinct absorption maxima at $\lambda = 214$ nm ($\varepsilon = 24\,685\,\mathrm{l\,mol^{-1}cm^{-1}}$) and at $\lambda = 274$ nm ($\varepsilon = 2074\,\mathrm{l\,mol^{-1}cm^{-1}}$), the latter extending up to 295 nm.

Consequently, the irradiation of this compound under anaerobic or aerobic conditions in dioxane-water (8:2) with the Pyrex-filtered output of a mercury high-pressure lamp ($\lambda > 300$ nm: long wavelength UV-B, UV-A) showed no

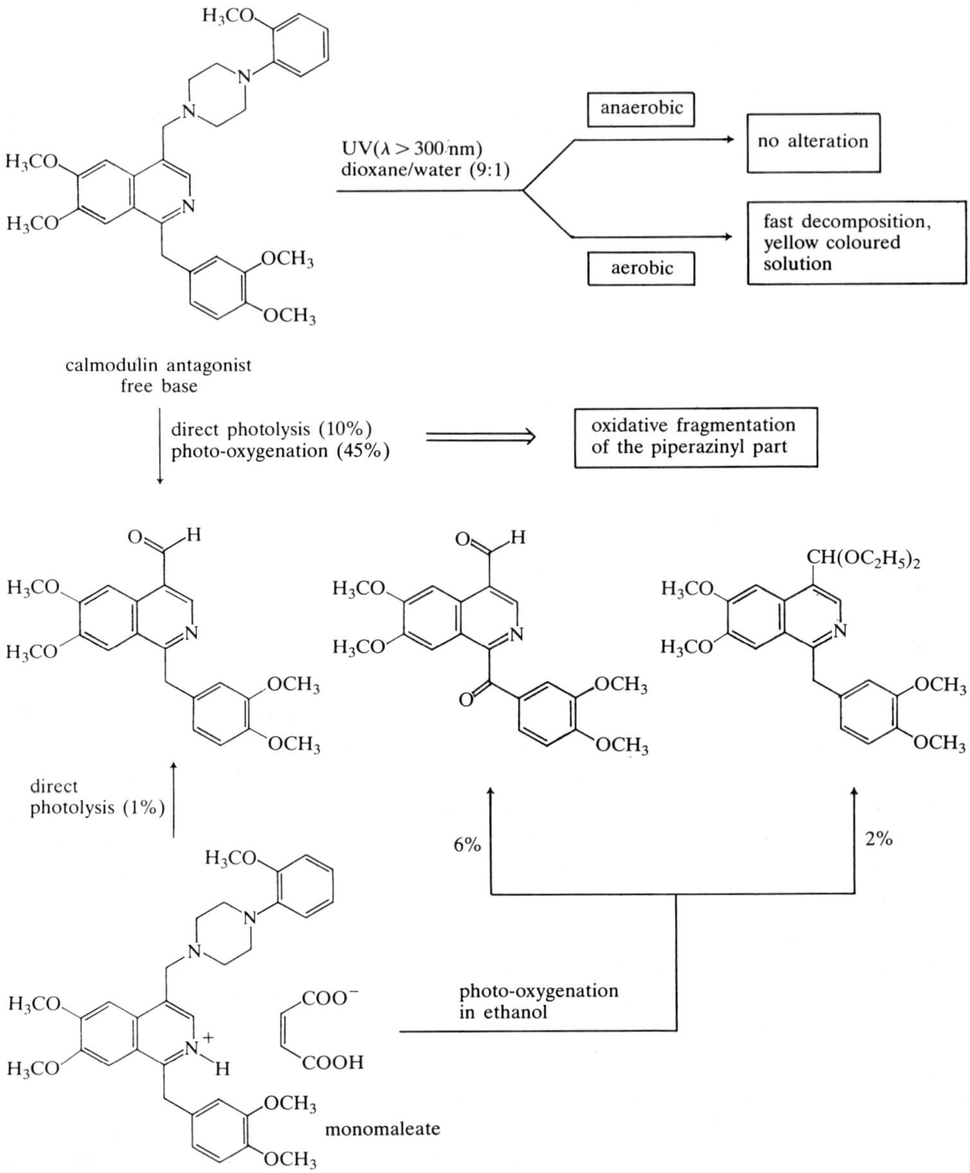

Figure 11.24 Typical example of oxygen-dependent photochemistry

alteration of the compound after four hours of photolysis (Fig. 11.27). On the other hand, the irradiation of the compound with the unfiltered output of the mercury high-pressure lamp (= UV-C, UV-B, UV-A and visible) led after 2.5 h to the complete decomposition of the substrate and the formation of a complex reaction mixture. From this mixture the formamide fragmentation product could be isolated in 40 per cent chemical yield. In contrast to this light-induced oxidative cleavage reaction the photo-oxygenation of this compound in ethanol with hematoporphyrin as singlet oxygen sensitizer led slowly to benzylic cleavage and to the formation

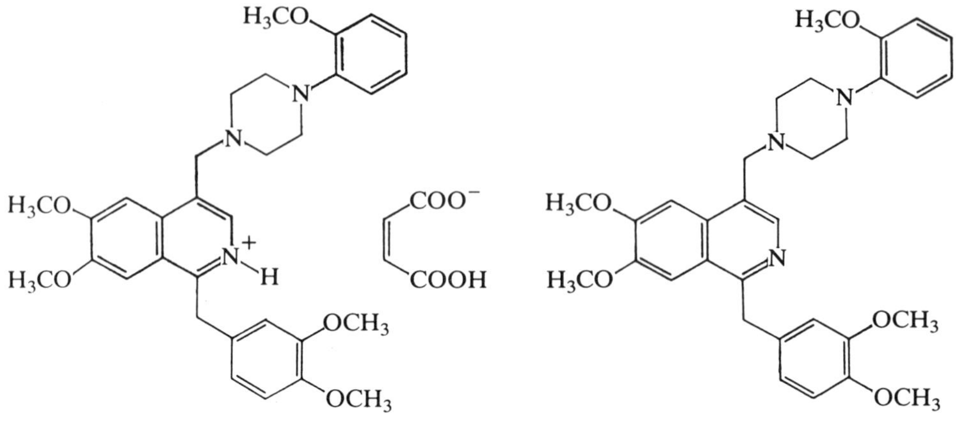

1

monomaleate

2

free base

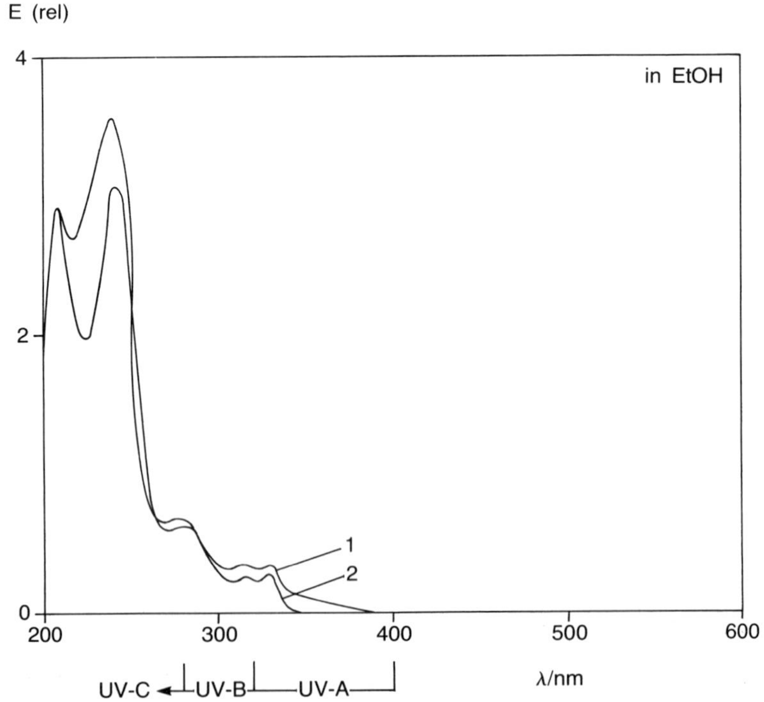

Figure 11.25 UV/Vis spectra of the isoquinoline free base and of its monomaleate

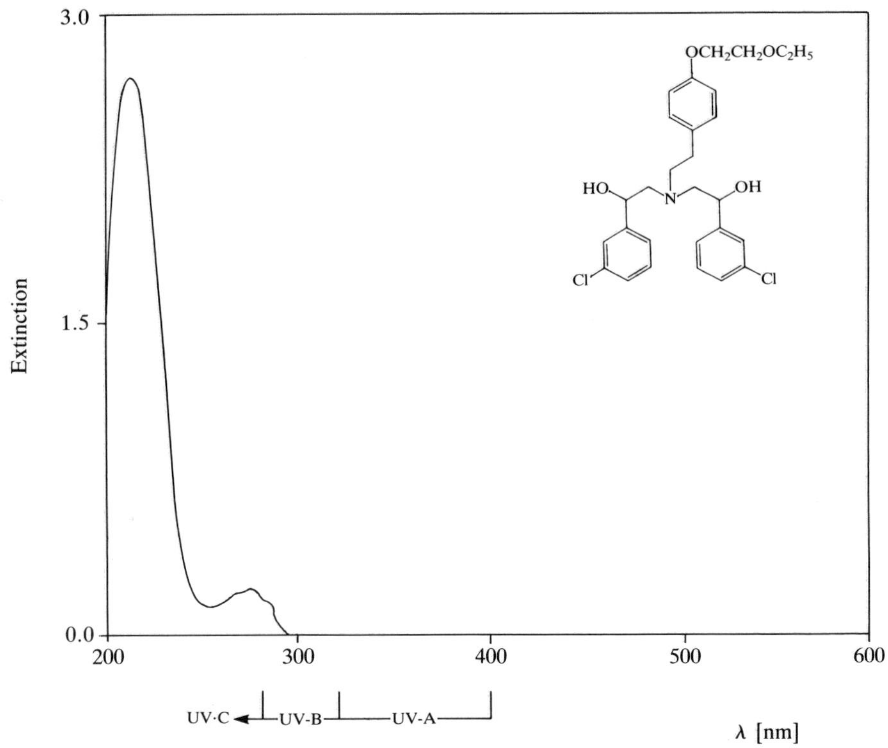

Figure 11.26 UV spectrum of a tertiary amine

of the para substituted benzaldehyde in 27 per cent isolated yield in addition to a series of unidentified decomposition products.

As a consequence, the parent drug should not lead to light-induced adverse effects during its application *in vivo*. Usually, the action spectra of photosensitizers in humans correlate well with their absorption spectra and most of the drug-induced photosensitization reactions are mediated or activated by UV-A radiation (Harber *et al.*, 1982; Pathak *et al.*, 1982).

In conclusion, the above examples of drug degradation studies impressively demonstrate the diversity of photochemical and photophysical effects which have to be considered in phototoxicity testing protocols.

11.4 UV light sources in photochemistry and photobiology

Commercially available UV lamps are of two broad categories: medium- or high-pressure mercury arc lamps and fluorescent tubes (Green *et al.*, 1992). The typical output spectrum of a medium-pressure mercury lamp covers the UV-C, UV-B and the UV-A range and it extends to the visible and infrared part of the electromagnetic spectrum. It consists of distinct lines which correspond to the electronic transitions of excited mercury atoms. The output spectrum of a UV fluorescent

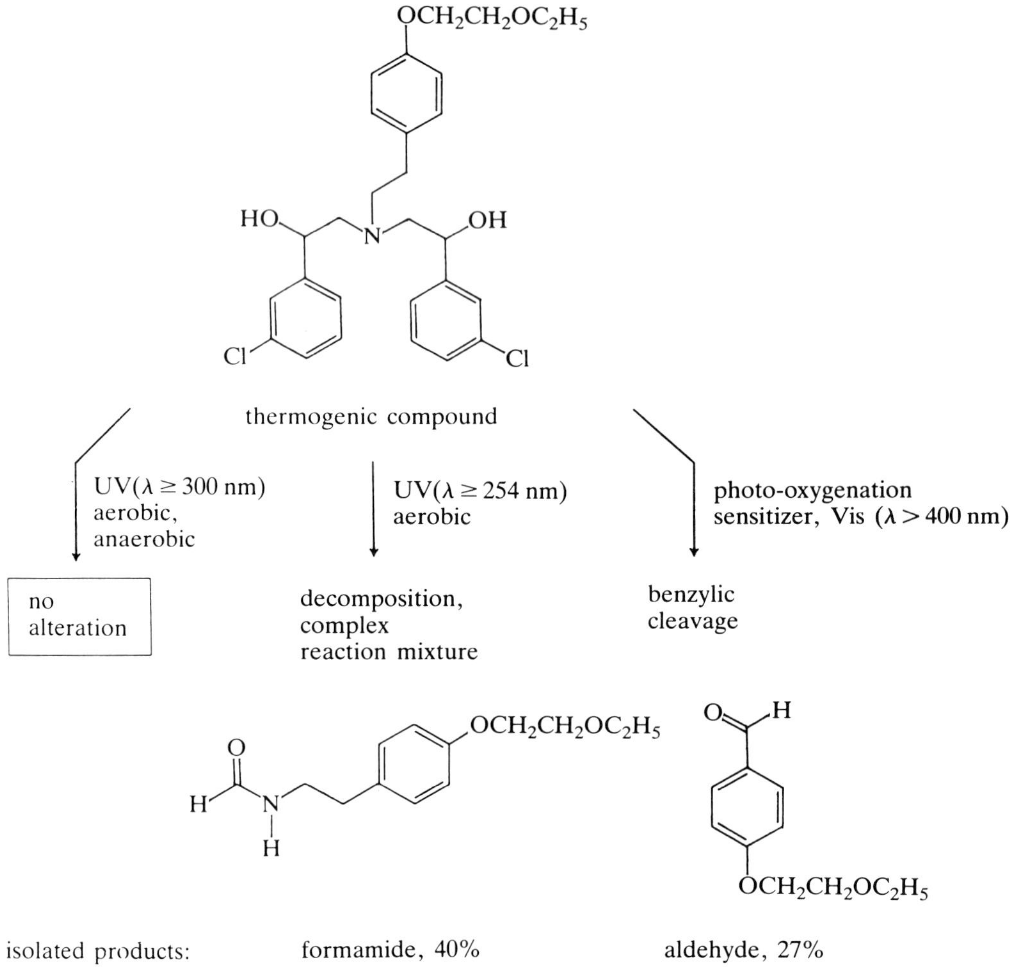

Figure 11.27 Example of wavelength-dependent photoreactivity

lamp exhibits a broad emission maximum at 300 nm. The obvious disadvantage of these conventional UV sources is their polychromatic radiation.

11.4.1 Novel incoherent excimer UV sources

However, novel excimer technologies which were developed recently (Kogelschatz, 1993) allow the selective generation of incoherent and almost monochromatic radiation. The schematic representation of a cylindrical excimer UV source demonstrates the principles of UV generation via silent electrical discharges (Fig. 11.28). The hermetically sealed discharge gap contains a gas or gas mixture which is capable of excimer formation known from laser technologies. The inner and outer electrode are connected to a high-frequency/high-voltage generator. The direct connection to a standardized flange facing on both ends of the quartz tube allows the coupling of different excimer UV sources to irradiation units (Oppenländer, 1994).

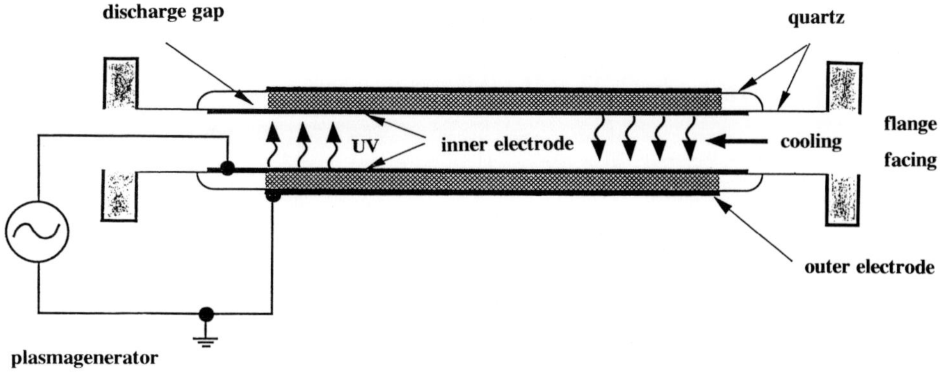

Figure 11.28 Schematic representation of a cylindrical excimer UV source with flange facings

11.4.2 ***Potential applications of the excimer UV sources***

This innovative technology is used for the generation of narrow-band UV radiation in the vacuum-UV, UV-C, UV-B and UV-A region of the electromagnetic spectrum depending on the gas or gas mixture which is applied (Fig. 11.29).

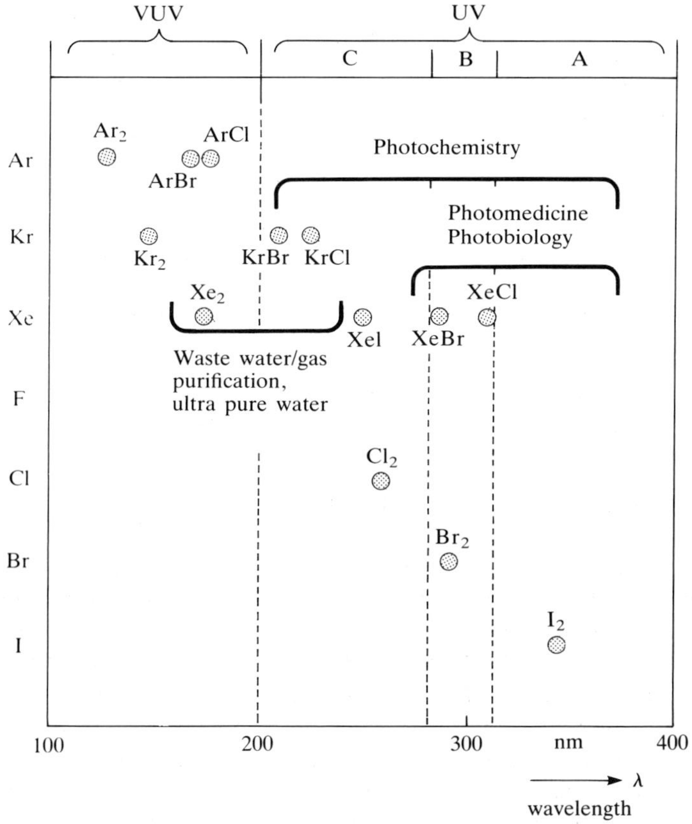

Figure 11.29 Strategic important excimers and their applications

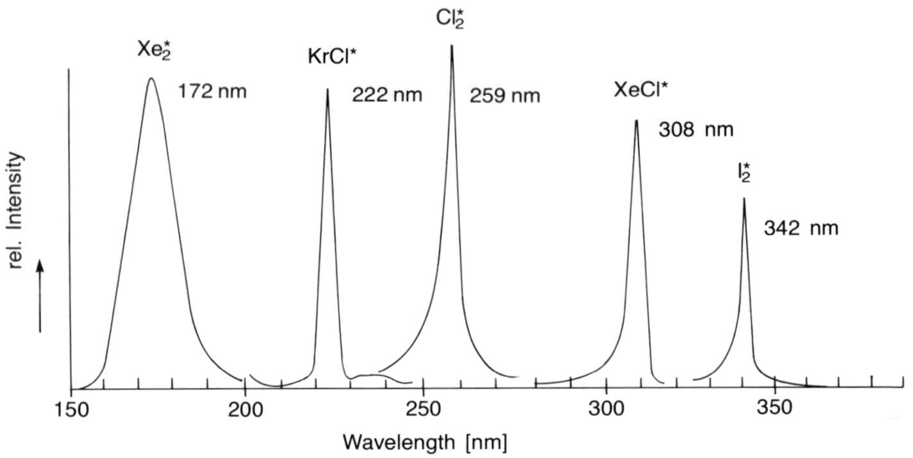

Figure 11.30 Emission spectra of strategic important incoherent excimer lamps

Potential applications include the waste water (Oppenländer and Baum, 1994) and gas purification and the production of ultra-pure water. Strategic important excimers for photochemistry, photobiology and for photomedicine involve the XeCl* excimer with an emission maximum at 308 nm and the halogen excimers. The emission characteristics of typical excimer UV sources (Fig. 11.30) usually exhibit no other emissions in the spectral range and have a half width of several nanometres (Kogelschatz, 1993).

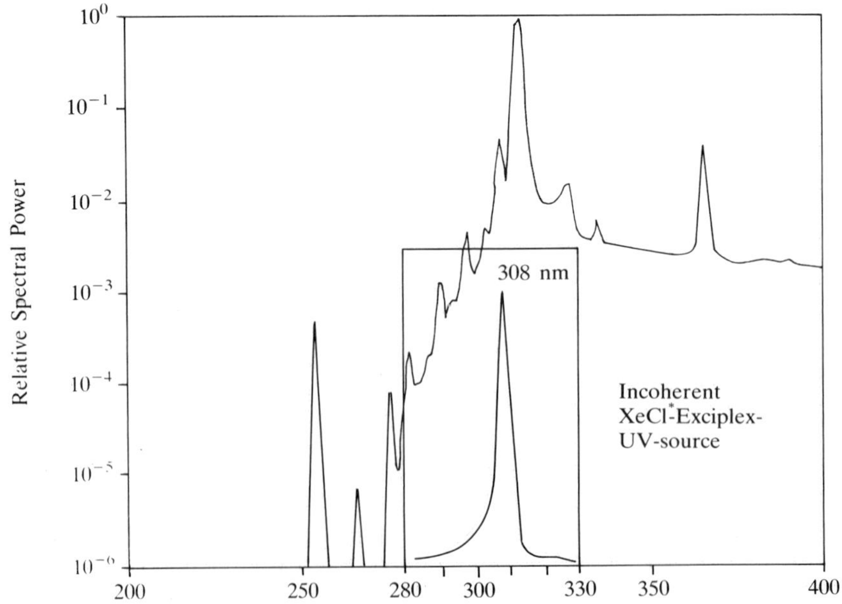

Figure 11.31 Emission characteristics of a narrow-band UV-B fluorescent lamp (Philips Tl01) and of the XeCl*-excimer UV source

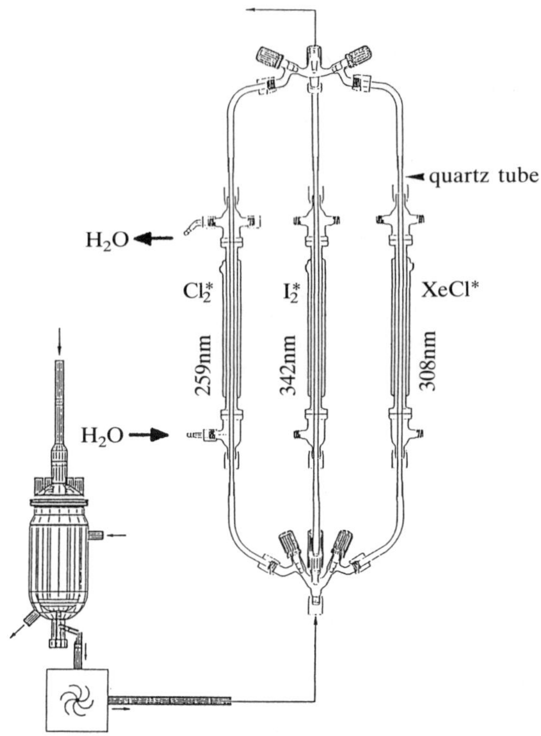

Figure 11.32 Excimer UV photoreactor for wavelength-selective irradiations

A comparison of the emission characteristics of a UV-B fluorescent lamp (Green *et al.*, 1992) and the XeCl* excimer UV source (Kogelschatz, 1993) (Fig. 11.31) clearly demonstrates the advantage of this novel technology over conventional UV sources for wavelength-selective photomedicinal and photobiological studies because almost no emissions below 300 nm are evident.

Therefore, we developed a novel excimer irradiation unit (Oppenländer, 1994) which consists of a thermostatic cooling unit for the UV source, an external pulse generator, the power supply and the XeCl* source ($\lambda = 308$ nm) covered by a PMMA security box for UV protection. The pulse generator guarantees UV dosage control via external triggering of the power supply down to the 10 ms scale. This pulse technique may overcome the undesired side-effects of phototherapy of psoriasis such as skin phototoxicity and the risk of skin cancer (Dall'Acqua and Jori, 1989). Furthermore, the flange technique (Oppenländer, 1994) allows the change between the UV-B (XeCl* = 308 nm) and the UV-A ($I_2^* = 342$ nm) lamps and the development of 'whole-body' irradiation units.

For photochemical drug degradation studies a novel excimer UV photoreactor for wavelength-selective irradiations was developed (Fig. 11.32). It consists of three different excimer UV sources with emissions at 259, 308 and 342 nm. Each individual irradiation unit is coupled via Teflon valves and is connected to a laboratory vessel. The substrate solution is continuously circulated by a Teflon pump through quartz tubes which are placed concentrically within the water cooling flow of the excimer UV sources. Potential applications include the photostability testing of pharmaceuticals, cosmetics, dyes, detergents and food additives in solution. Furthermore, this type of excimer-irradiation unit was applied to the efficient synthesis of previtamin D_3 by wavelength-selective double irradiation of 7-dehydrocholesterol (Oppenländer and Hennig, 1995).

Acknowledgements

The author thanks the company F. Hoffmann-La Roche (Switzerland). He is grateful to Dr U. Kogelschatz (ABB, Switzerland) for his generous support and for stimulating discussions about the novel excimer UV technology.

Appendix 11.A.1: Molecular structures of phototoxic pharmaceuticals

Afloqualone

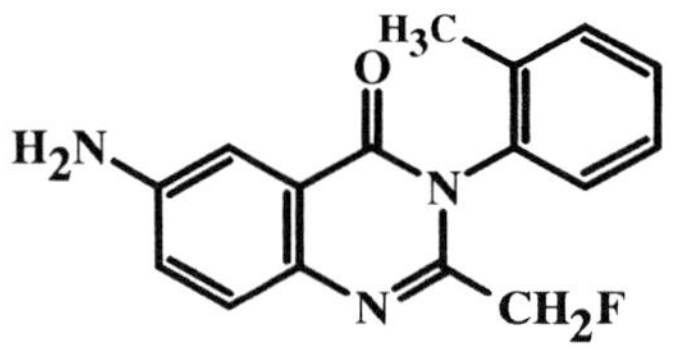

Amiodarone

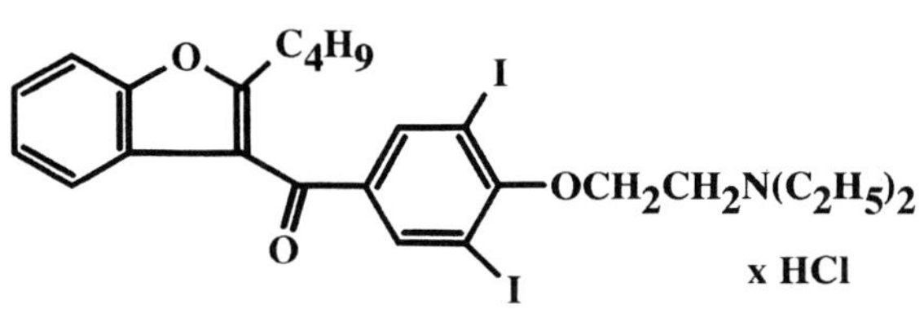

Angelicin

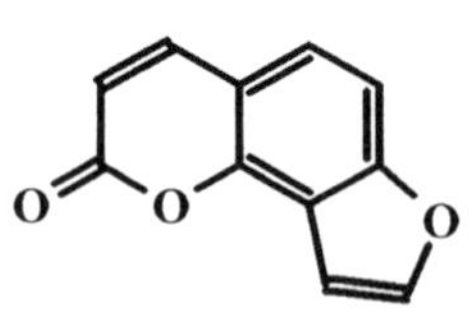

Anthracene

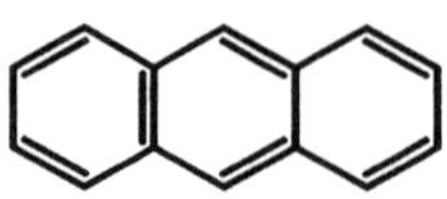

Anthraquinone

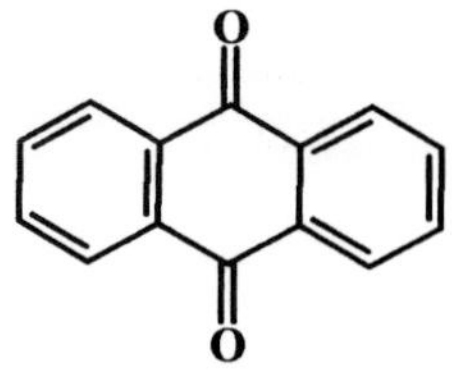

2-Arylquinolinemethanols

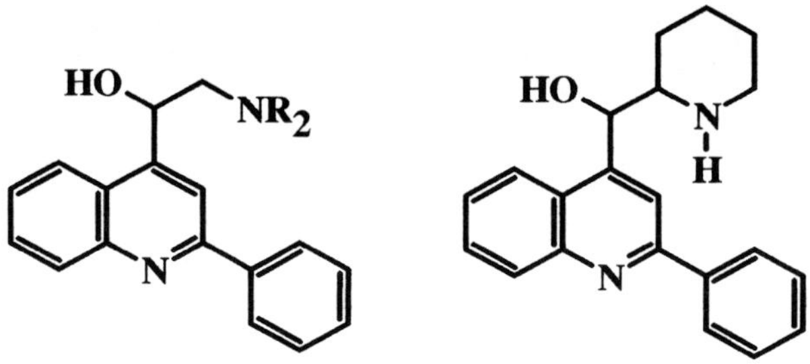

Azapropazone

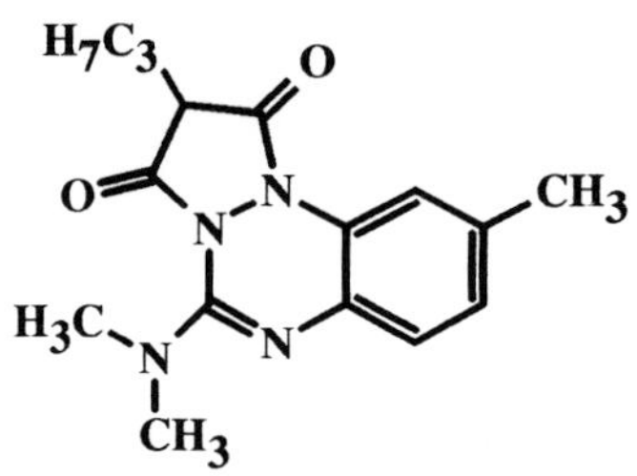

Azathioprine

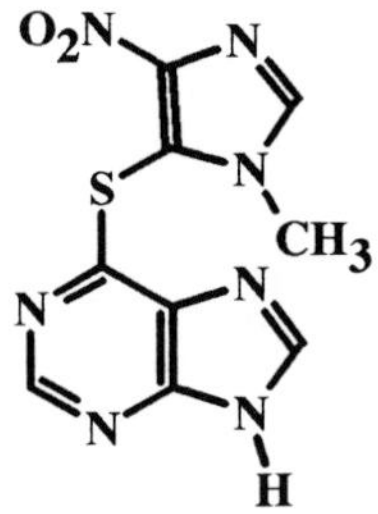

Azulene

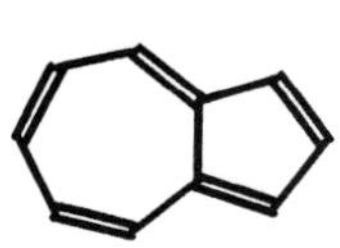

Benoxaprofen

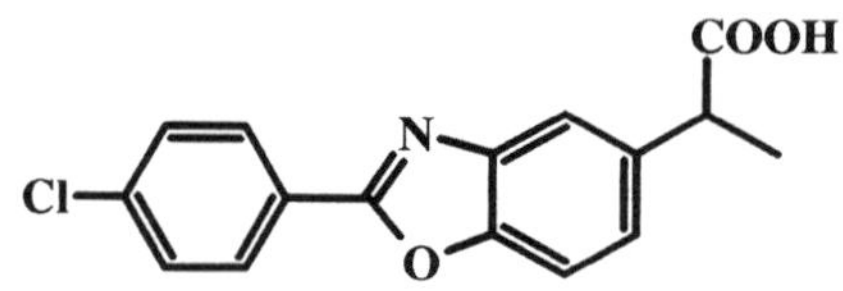

Benzoylperoxide

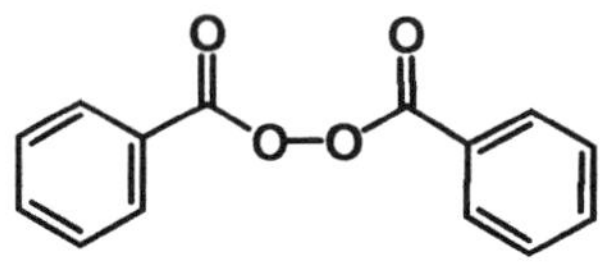

Benzpyren

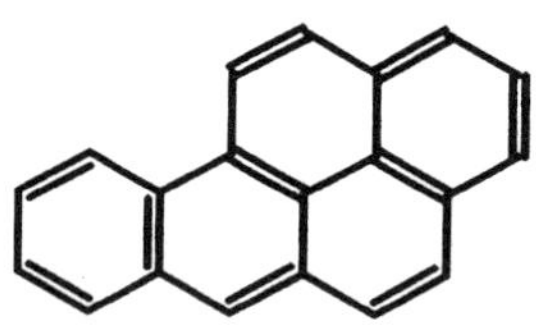

Carbamazepine

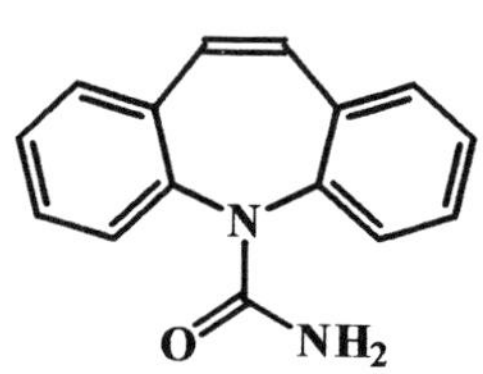

Carbutamide

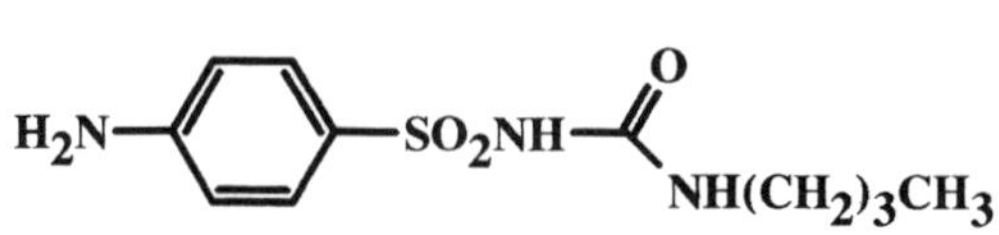

Carprofen

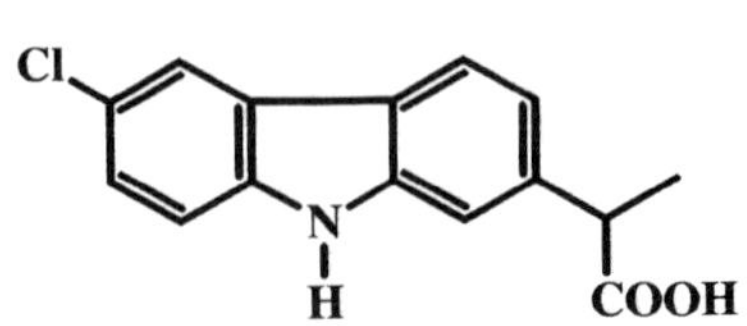

CD 271

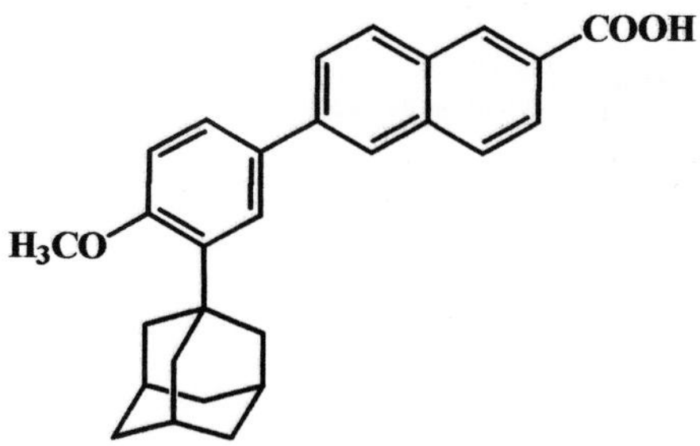

Chlordiazepoxide

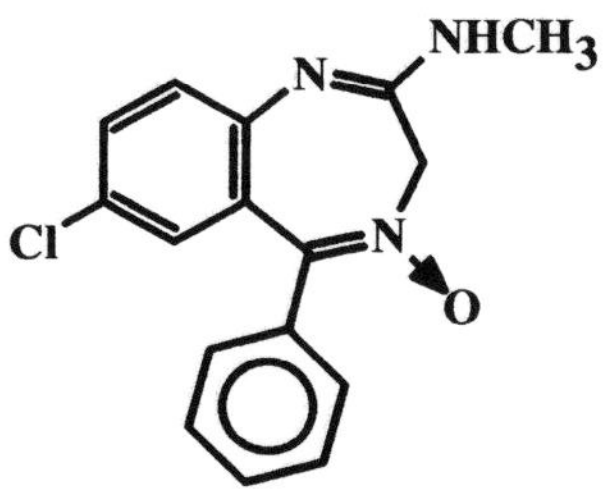

Chlorothiazide

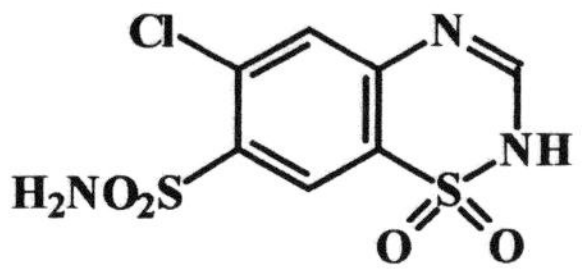

Cinoxacine

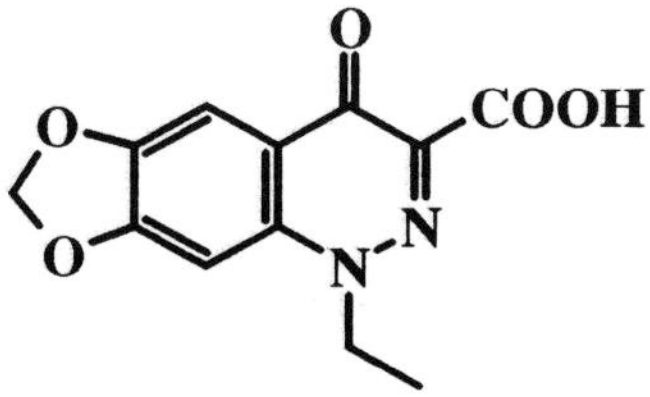

Clofazimine

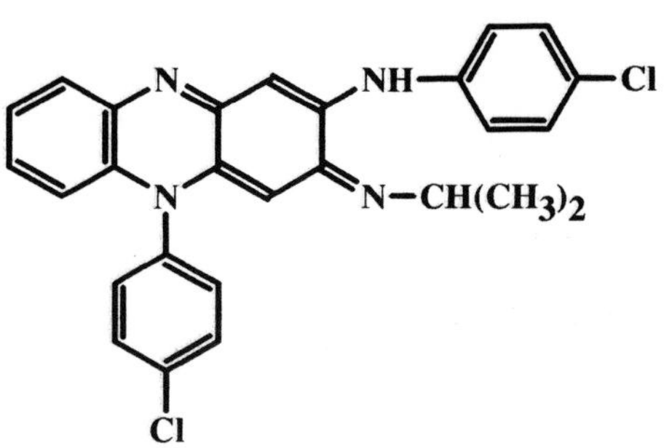

Clofibrate

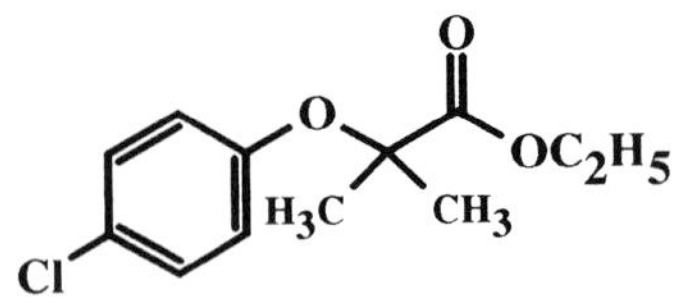

Dacarbazine

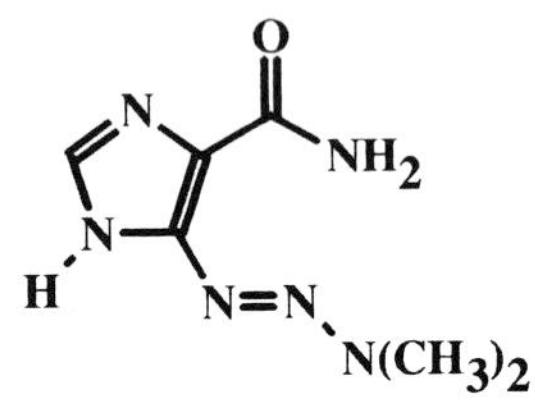

Diclofenac

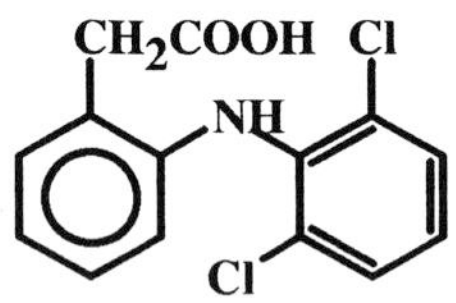

Diphenylhydantoin

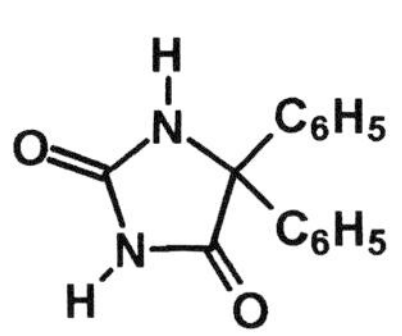

Dyes

Acridine (Aminoacridine)

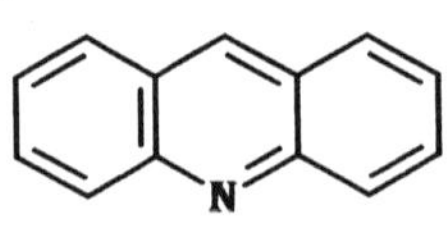

Acriflavine

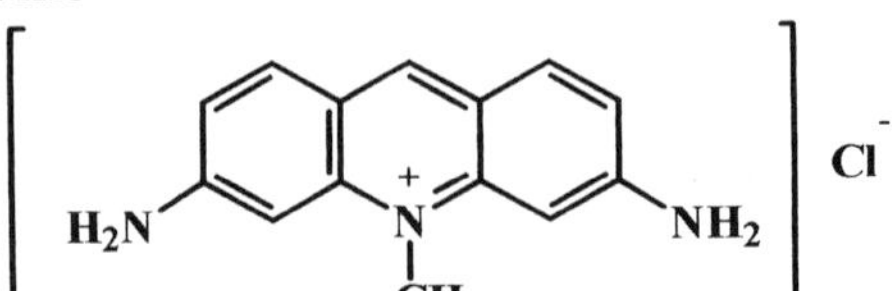

Benzanthrone

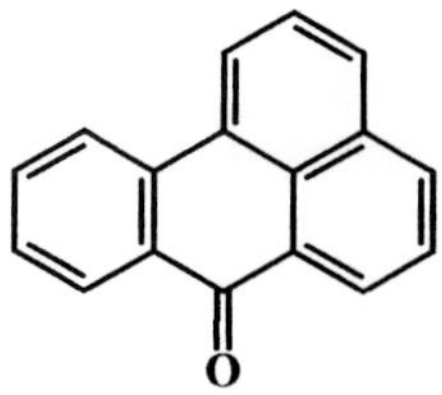

Cadmiumsulfide CdS

Eosin

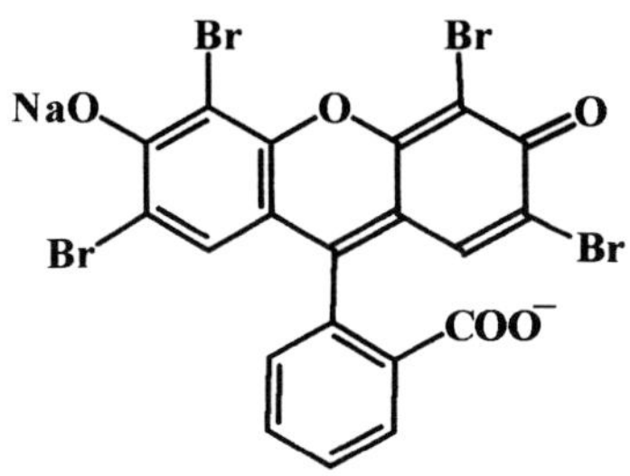

Erythrosin

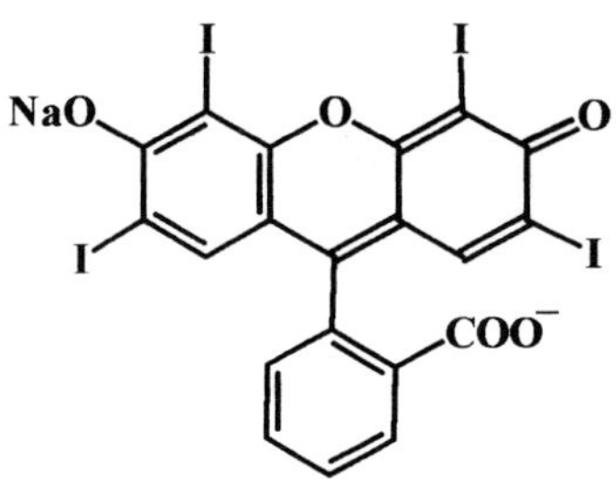

Fluorescein
(Bromofluorescein,
Dibromofluorescein)

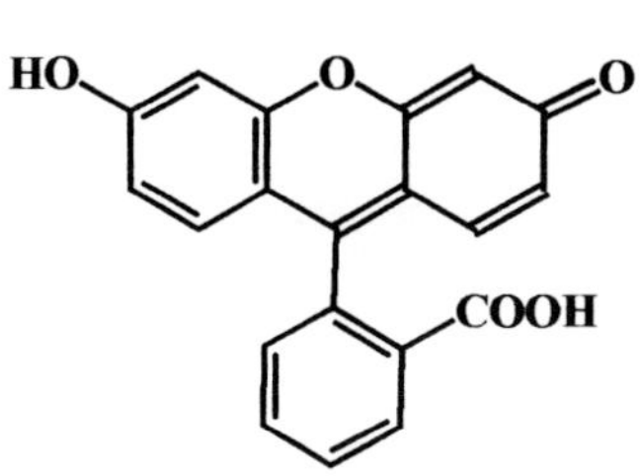

Methylene Blue

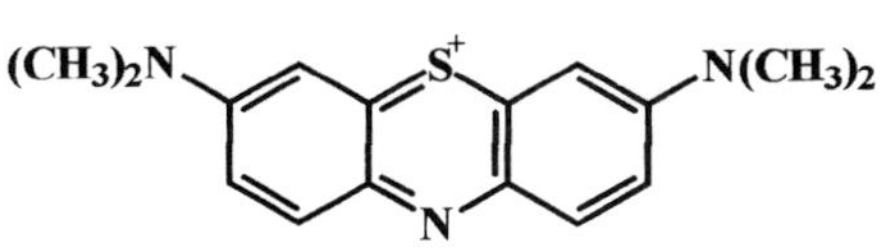

Rose Bengal

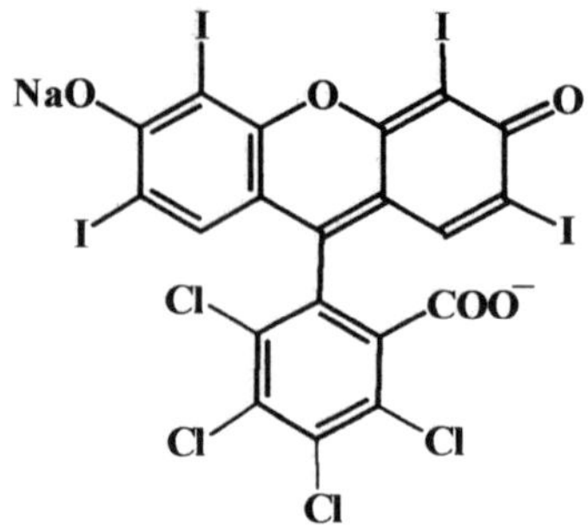

Paraphenylene diamine

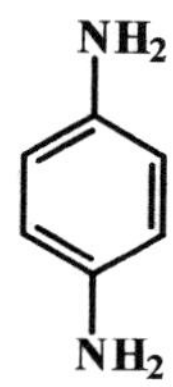

Toluidine Blue

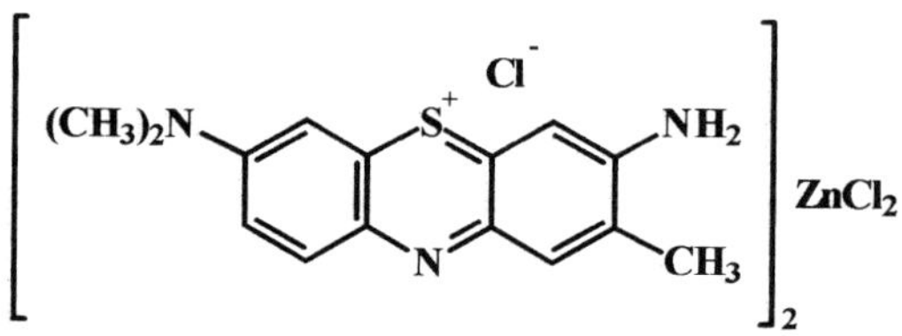

Trypan Blue

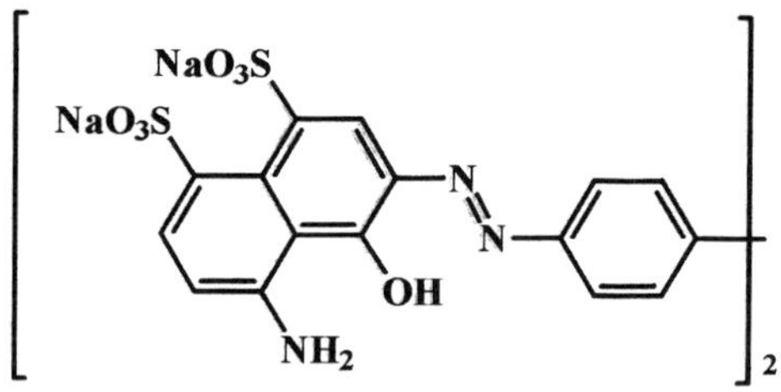

Etretinate

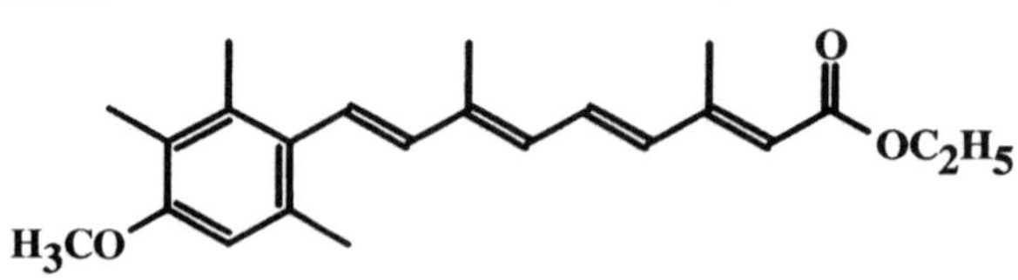

Fenofibrate

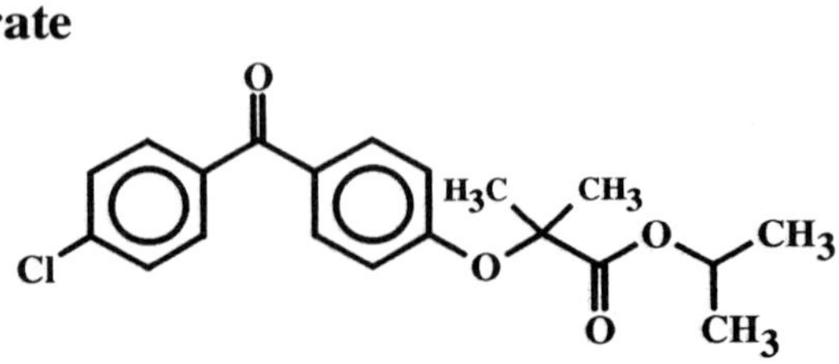

Fentichlor

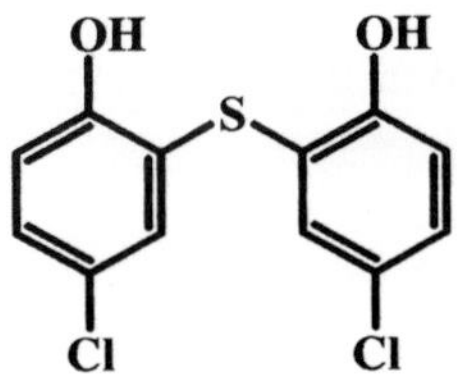

Fluoroquinolones

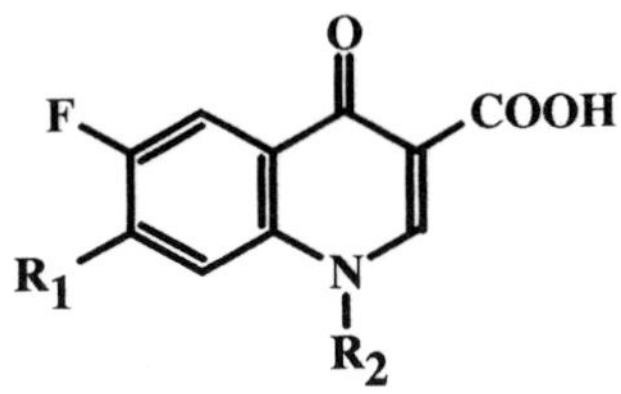

	R_1	R_2
Ciprofloxacine	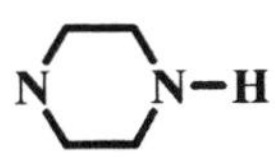	
Norfloxacine	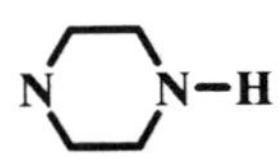	CH_2CH_3

Enoxacine

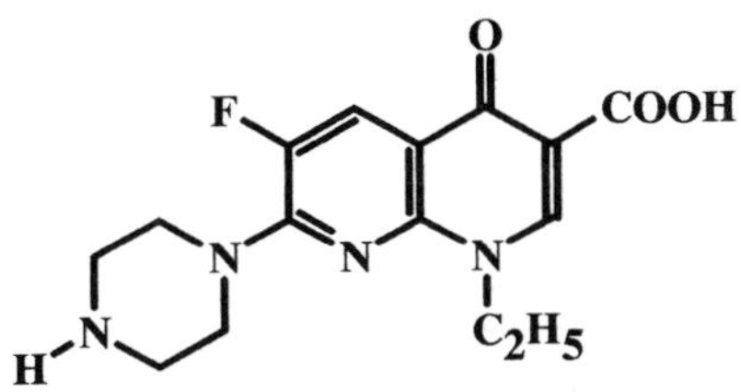

Fleroxacin

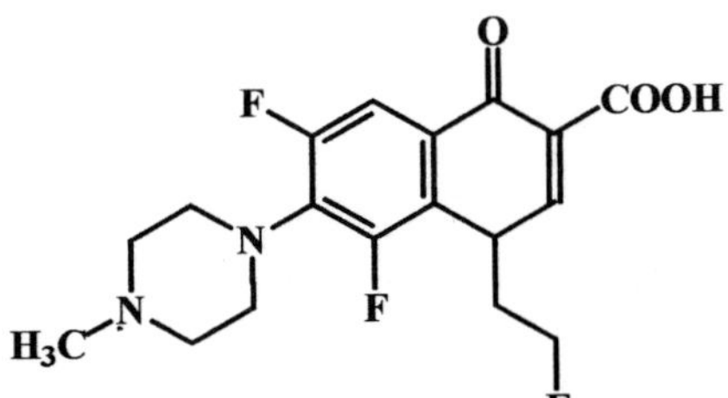

Lomefloxacin

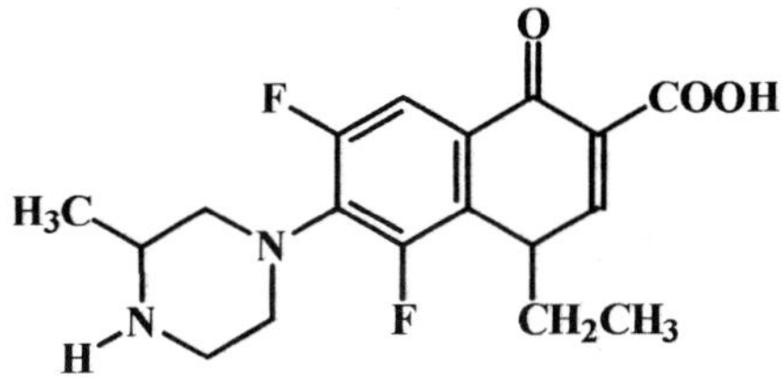

Ofloxacine

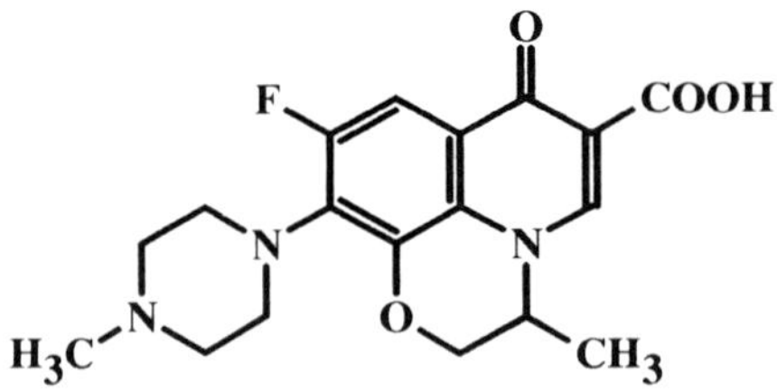

5-Fluorouracil

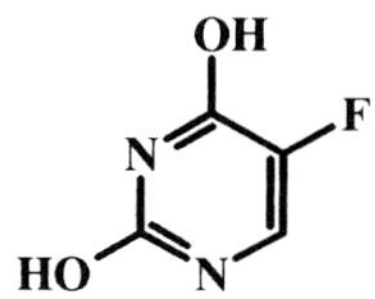

Frusemide

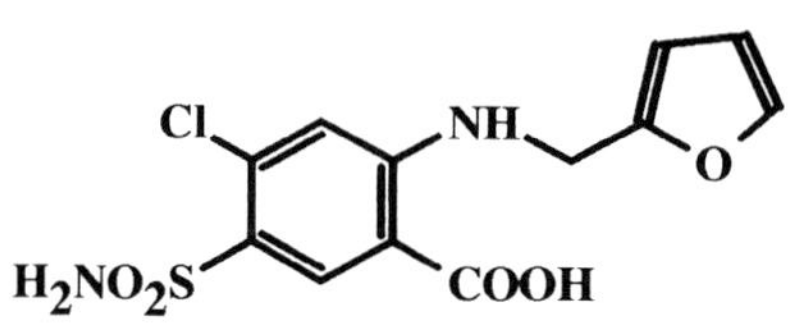

Griseofulvin

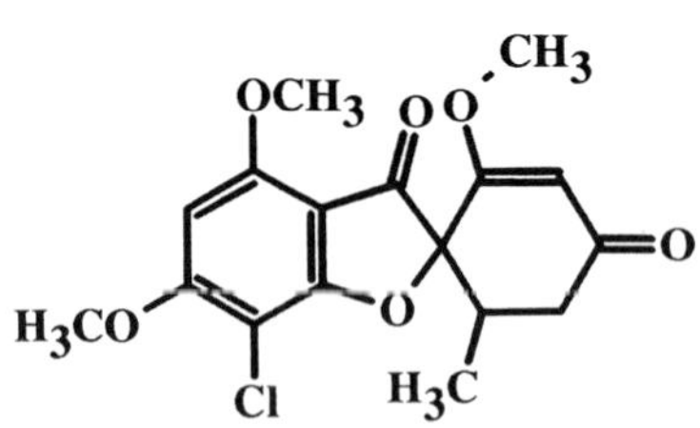

Hexachlorophene

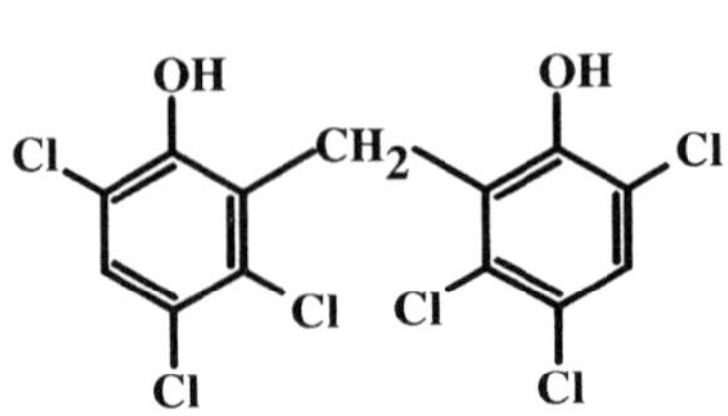

Hydrochlorothiazide

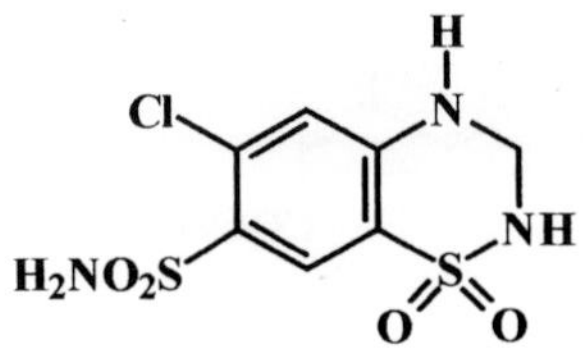

Hypericin

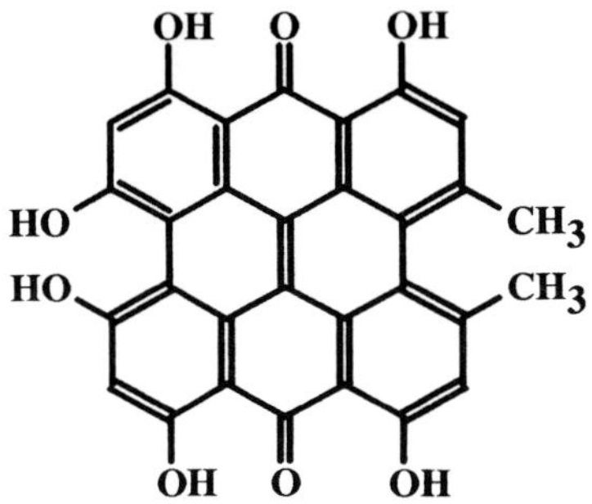

Ibuprofen

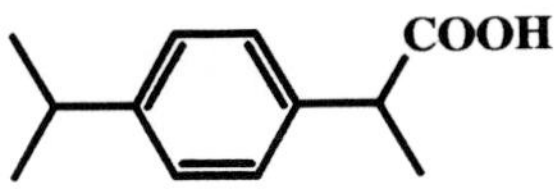

Indomethacine

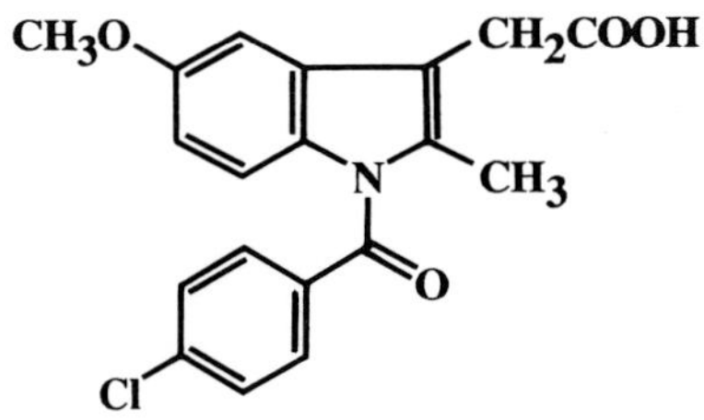

Isoniazide

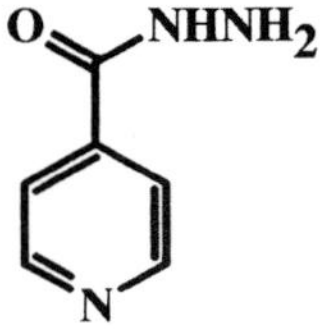

Isotretinoin

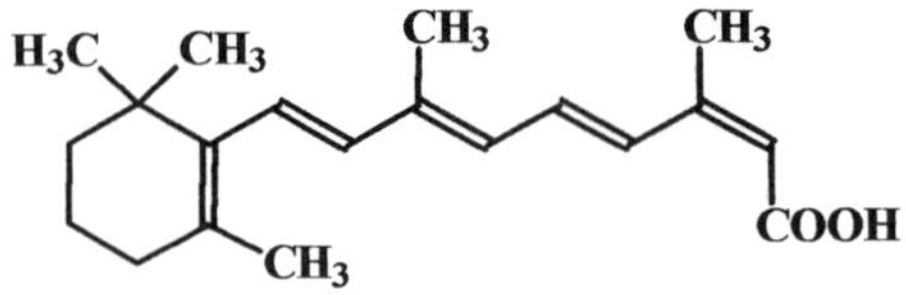

Ketoprofen

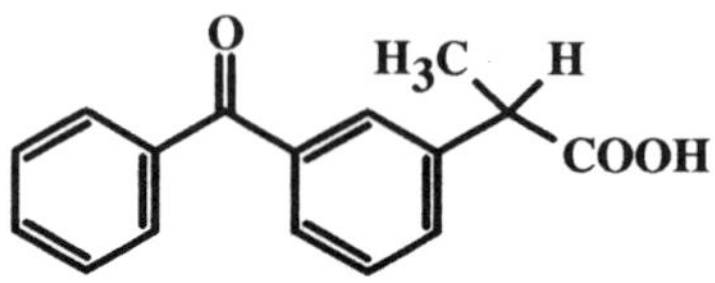

Methotrexate

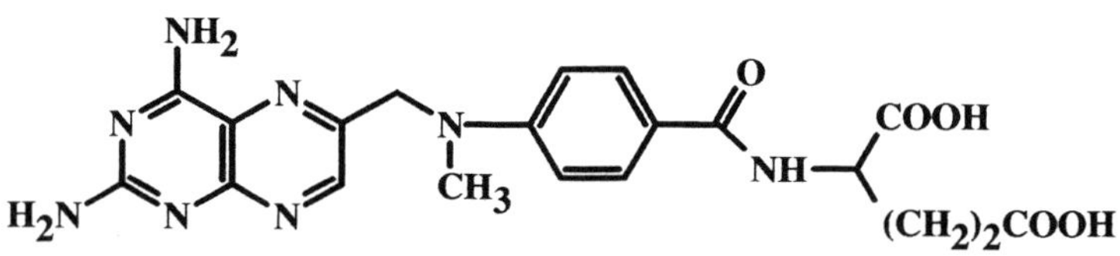

8-Methoxypsoralen

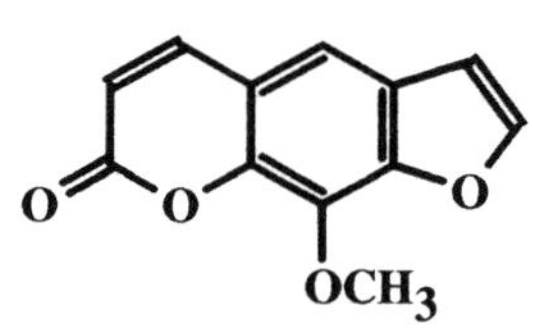

5-Methoxypsoralen

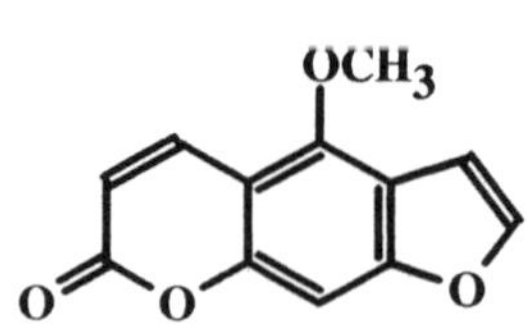

Methyldopa

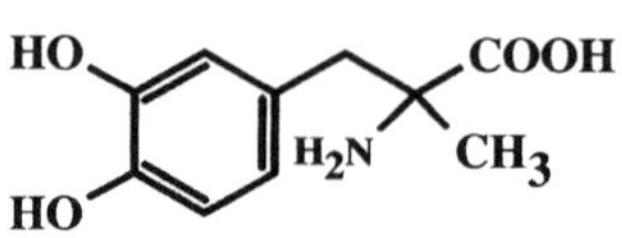

Metronidazole

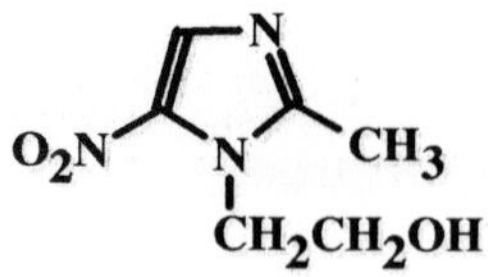

Nalidixic acid

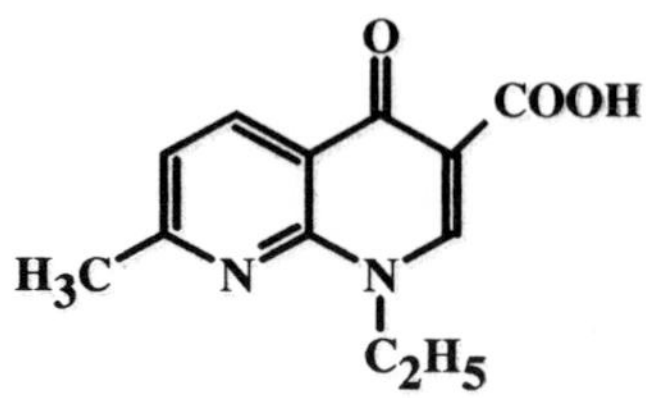

Naproxen

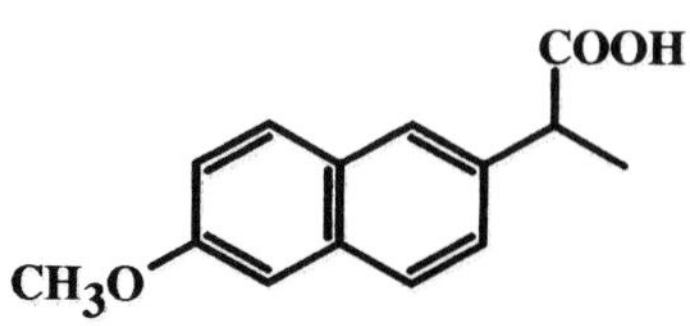

Nortriptyline

para-Aminobenzoic acid (PABA), esters

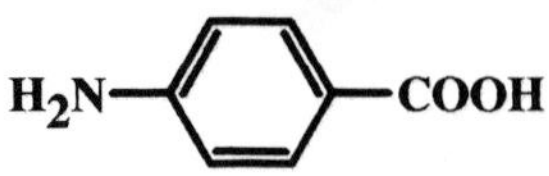

Penflutizide

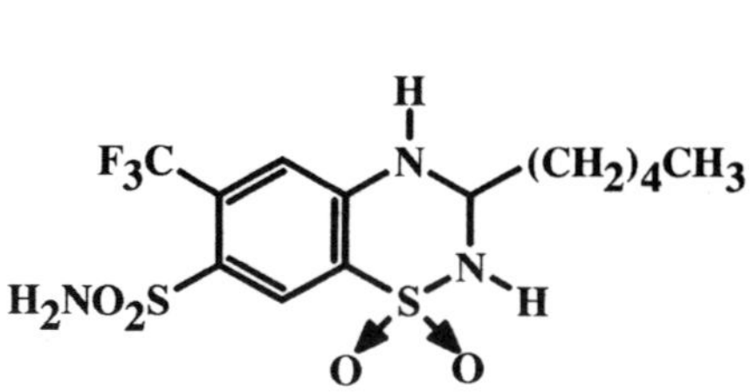

Phenanthrene

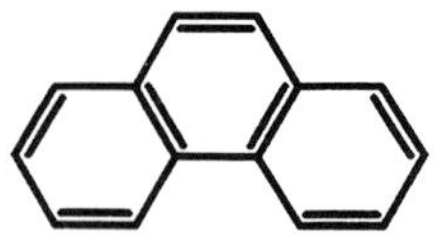

Phenothiazines

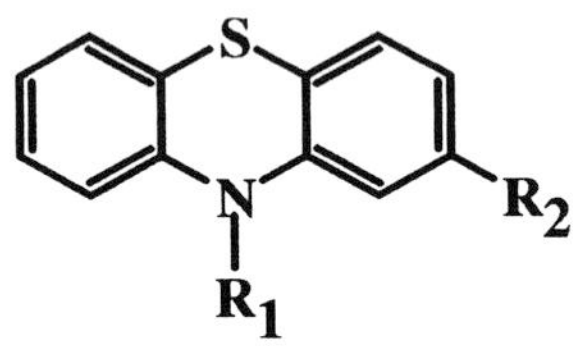

	R_1	R_2
Alimemazin	$CH_2CH(CH_3)CH_2N(CH_3)_2$	H
Chlorpromazin	$CH_2CH_2CH_2N(CH_3)_2$	Cl
Levomepromazin	$CH_2CH(CH_3)CH_2N(CH_3)_2$	OCH_3
Perazin	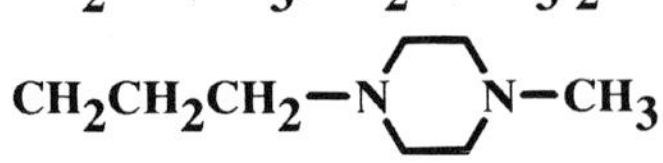 $CH_2CH_2CH_2$–N N–CH_3	H
Promazin	$CH_2CH_2CH_2N(CH_3)_2$	H
Promethazin	$CH_2CH(CH_3)N(CH_3)_2$	H
Thiopropazate	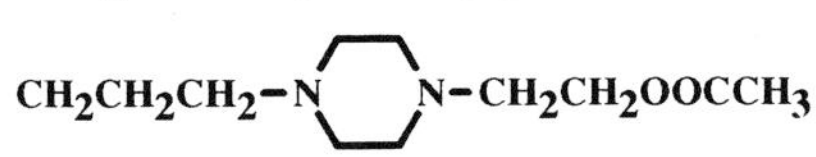 $CH_2CH_2CH_2$–N N–$CH_2CH_2OOCCH_3$	Cl
Thioridazin	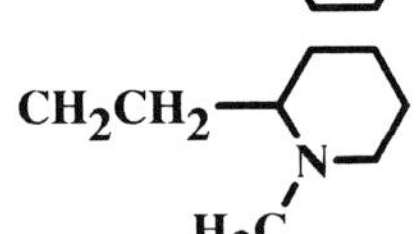 CH_2CH_2– N H_3C	SCH_3
Trifluorpromazin	$CH_2CH_2CH_2N(CH_3)_2$	CF_3

Piroxicam

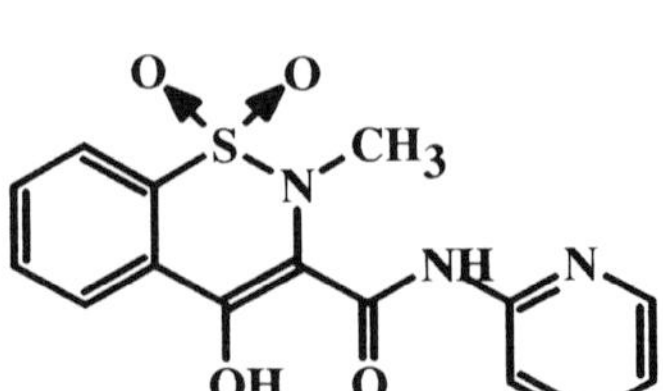

Porphyrins

Hematoporphyrin

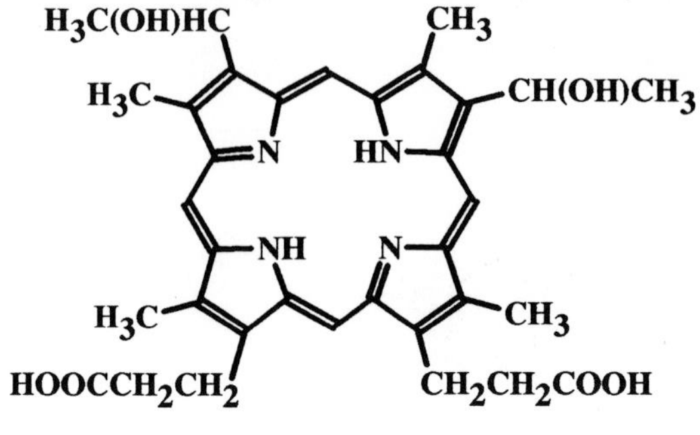

Propranolol

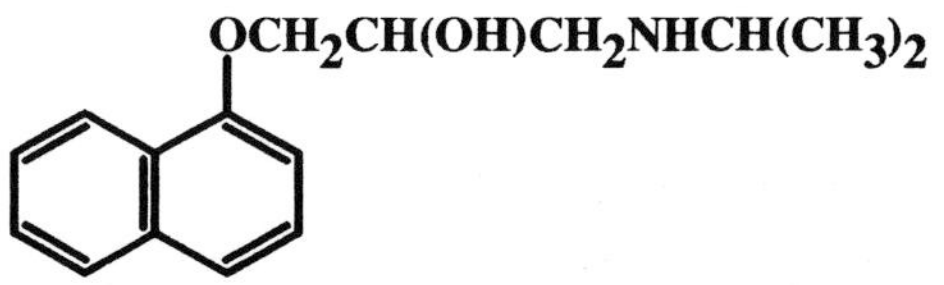

Protriptyline

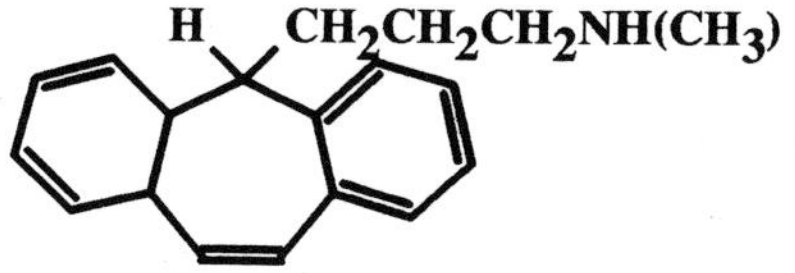

Psoralen

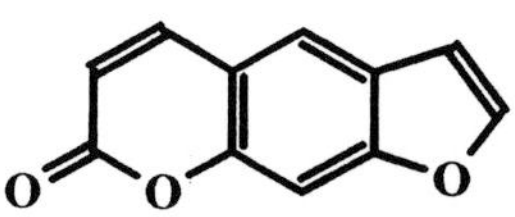

Quinidine

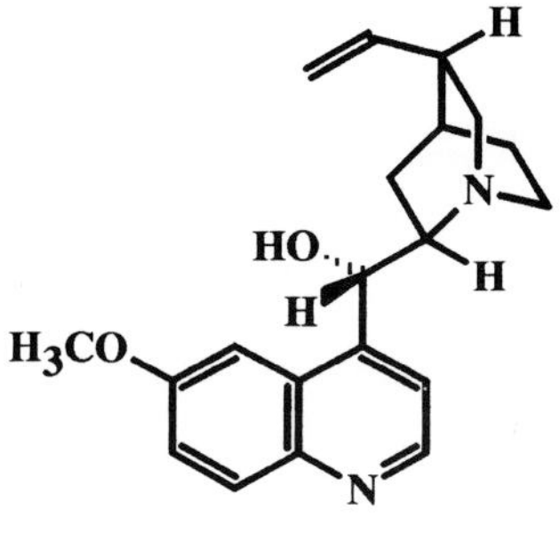

Retinoic acid

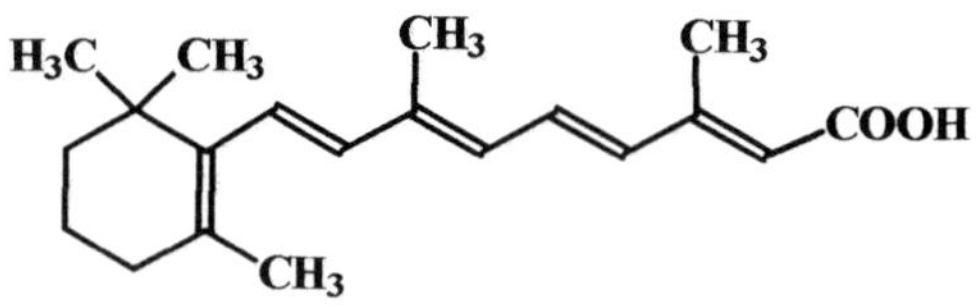

Sanguinarine chloride

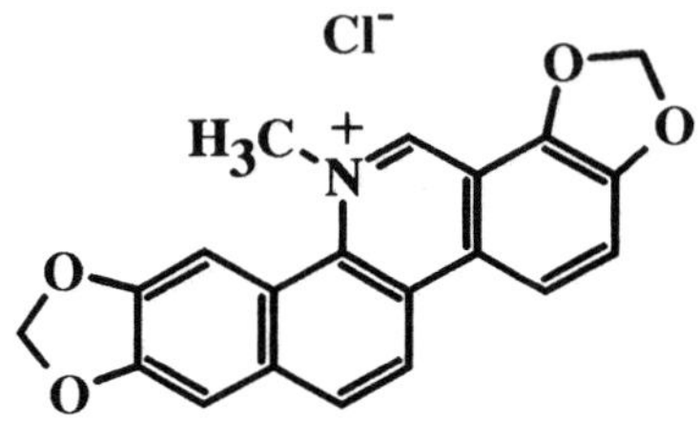

Sulfacetamide

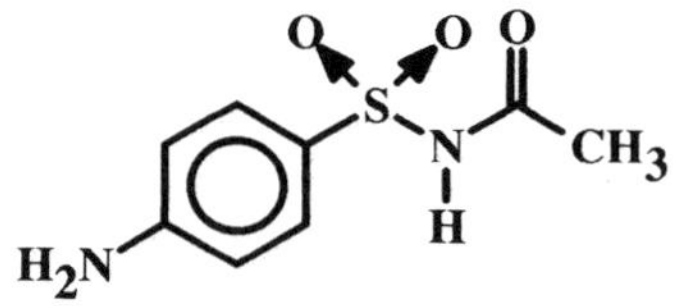

Sulfonamides

Sulfanilamide

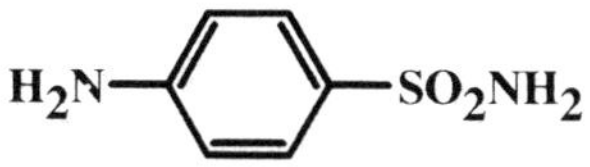

Sulfadiazine

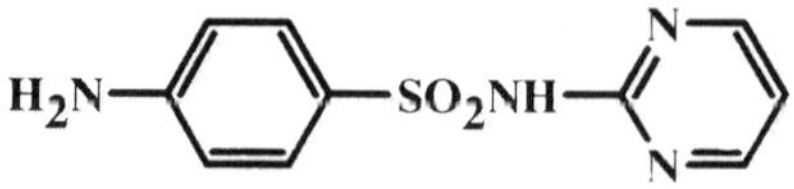

Sulfadimethoxine

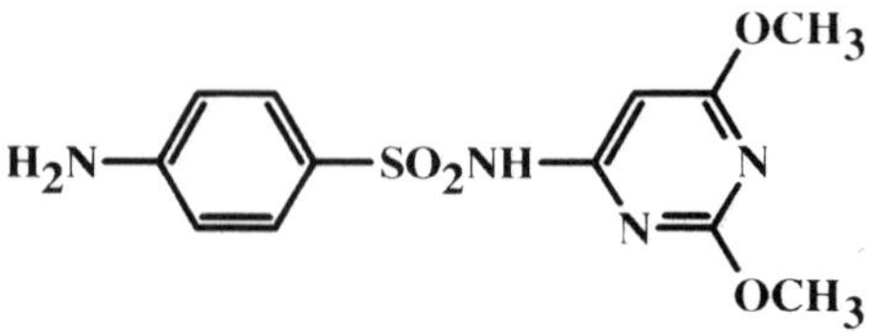

Sulfaguanidine

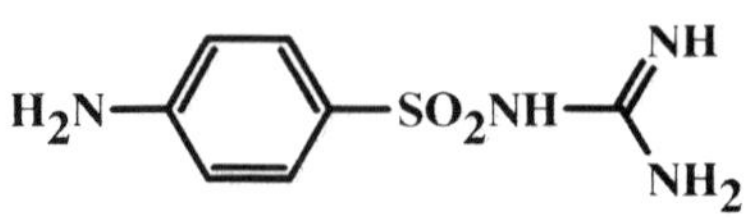

Sulfamethazine

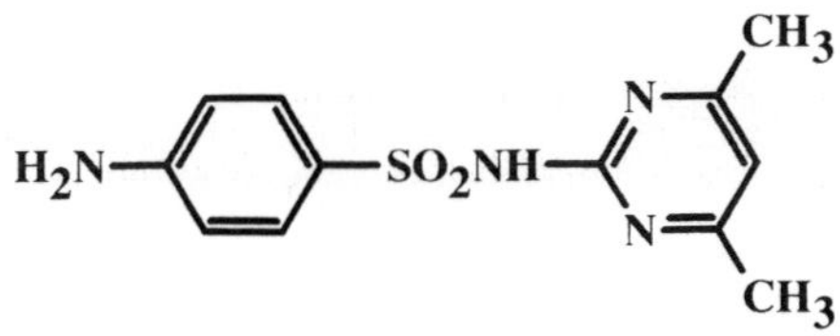

Sulfapyridine

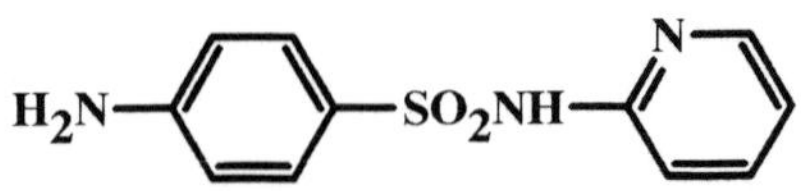

Sulfathiazole

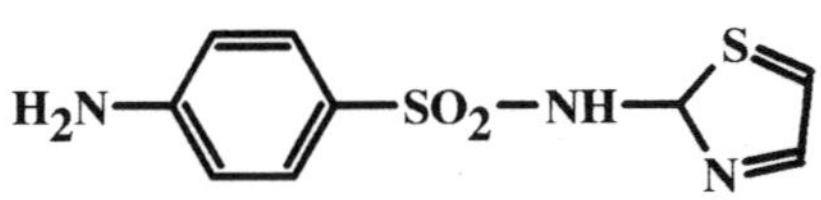

Sulfisoxazole

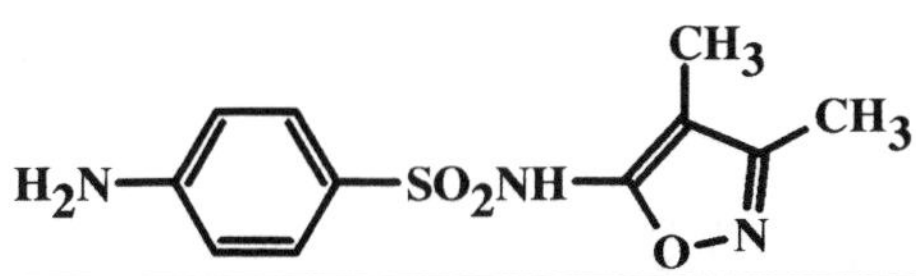

Thiambutosine

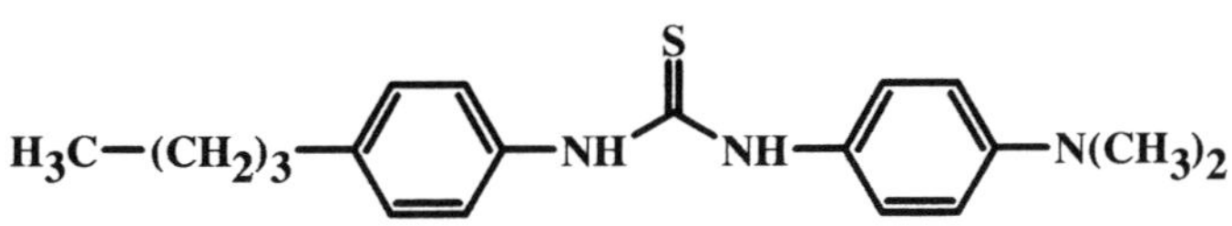

Tetrachlorosalicylanilide

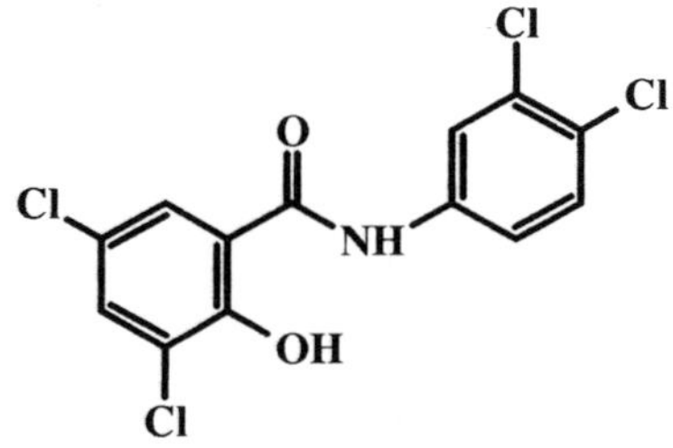

Tetracyclines

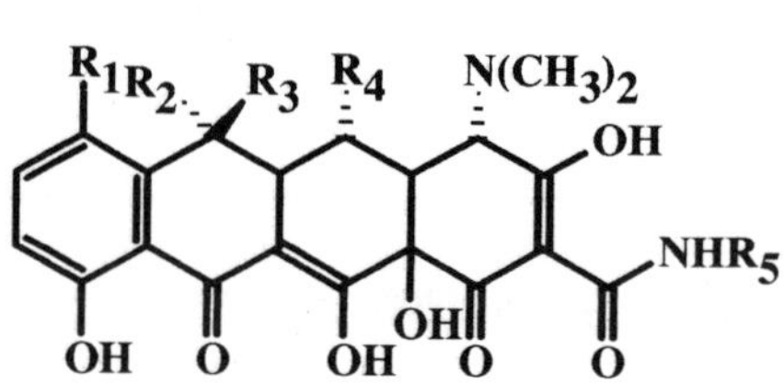

	R_1	R_2	R_3	R_4	R_5
Chlortetracyclin	-Cl	-CH3	-OH	-H	-H
Demethylchlor-tetracyclin	-Cl	-H	-OH	-H	-H
Doxycyclin	-H	-CH3	-H	-OH	-H
Methacyclin	-H	[$=CH_2$]		-OH	-H
Minocyclin	-N(CH3)2	-H	-H	-H	-H
Oxytetracyclin	-H	-CH3	-OH	-OH	-H
Tetracyclin	-H	-CH3	-OH	-H	-H

Tetramethylthiuram-disulfide

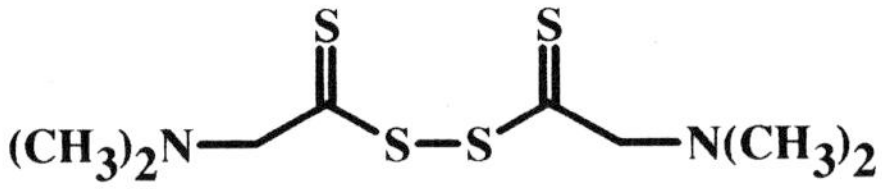

Tolbutamide

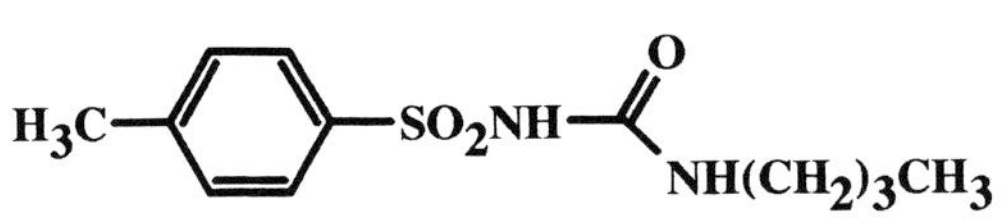

Trimethoprim

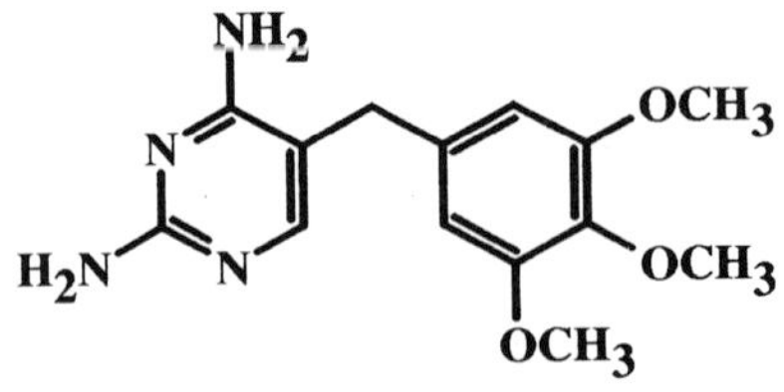

References

Adam, W., and Cilento, G. (eds), 1982, *Chemical and Biological Generation of Excited States*. New York: Academic Press.

Albini, A., 1991, Fotochimica di sostanze farmaceutiche, *Boll. Chim. Farmaceutico*, **130**, 393.

BLACK, H. S., 1992, Stratospheric ozone depletion and skin cancer incidence, *The Spectrum*, **5**, 13.

BRASLAVSKY, S. E. and HOUK, K. N., 1988, Glossary of terms used in photochemistry, *Pure Appl. Chem.*, **60**, 1055.

BRAUN, A. M., MAURETTE, M. T. and OLIVEROS, E., 1986, *Technologie Photochimique*. Lausanne: Presses Polytechniques Romandes.

CASTELL, J. V., GOMEZ-L., M. J., MIRANDA, M. A. and MORERA, I. M., 1987, Photolytic degradation of Ibuprofen. Toxicity of the isolated photoproducts on fibroplasts and erythrocytes, *Photochem. Photobiol.*, **46**, 991.

CROSSLEY, R., 1992, The Relevance of chirality to the study of biological activity, *Tetrahedron*, **48**, 8155.

DALL'ACQUA, F. and JORI, G., 1989, Photochemotherapy, in Foye, W. O. (ed.), *Principles of Medicinal Chemistry*, p. 803. Philadelphia: Lea & Febiger.

DAYHAW-BARKER, P. and TRUSCOTT, T. G., 1988, Direct detection of singlet oxygen sensitized by nalidixic acid: The effect of pH and melanin, *Photochem. Photobiol.*, **47**, 765.

EPLING, G. A., YOON, U. C. and AYENGAR, N. K. N., 1984, Photoreactivity of phototoxic antimalarial compounds, *Photochem. Photobiol.*, **39**, 469.

FERNANDEZ, E. and CARDENAS, A. M. J., 1990, The mechanism of photohaemolysis by photoproducts of nalidixic acid, *Photochem. Photobiol., B: Biology*, **4**, 329.

FOX, M. A., 1990, Photoinduced electron transfer, *Photochem. Photobiol.*, **52**, 617.

FROHNE, D. and PFÄNDER, H. J., 1987, *Giftpflanzen, Ein Handbuch für Apotheker, Ärzte, Toxikologen und Biologen, Wissenschaftliche Verlagsgesellschaft mbH* (3rd edn). Stuttgart; English edition: 1984, *A Colour Atlas of Poisonous Plants*, London: Wolfe.

GEORGE, M. V. and BHAT, V., 1979, Photooxygenations of nitrogen heterocycles, *Chem. Rev.*, **79**, 447.

GILBERT, A. and BAGGOTT, J., 1991, *Essentials of Molecular Photochemistry*. London: Blackwell Scientific Publications.

GREEN, C., DIFFEY, B. L. and HAWK, J. L. M., 1992, Ultraviolet radiation in the treatment of skin disease, *Phys. Med. Biol.*, **37**, 1.

HARBER, L. C., KOCHEVAR, I. E. and SHAHITA, A. R., 1982, in Regan, J. D. and Parrish, J. A. (eds), *The Science of Photomedicine*, p. 329. New York: Plenum Press.

HIDALGO, M. E., PESSOA, C., FERNANDEZ, E. and CARDENAS, A. M., 1993, Comparative determination of photodegradation kinetics of quinolones, *J. Photochem. Photobiol., A: Chem.*, **73**, 135.

IPPEN, H., 1969, in Urbach, F. (ed.), *The Biological Effects of Ultraviolet Radiation*, p. 681. Oxford: Pergamon.

JULLIARD, M. and CHANON, M., 1983, Photoelectron-transfer catalysis: Its connections with thermal and electrochemical analogues, *Chem. Rev.*, **83**, 425.

KERR, J. B. and MCELROY, C. T., 1993, Evidence for large upward trends of ultraviolet-B radiation linked to ozone depletion, *Science*, **262**, 1032.

KLESSINGER, M. and MICHL, J., 1989, *Lichtabsorption und Photochemie organischer Moleküle*. Weinheim: VCH Verlagsgesellschaft mbH.

KOCHEVAR, I. E., 1987, Mechanisms of drug photosensitization, *Photochem. Photobiol.*, **45**, 891.

1993, Basic principles in photomedicine and photochemistry, *Clin. Dermatol.*, **6**, 1.

KOGELSCHATZ, U., 1993, UV Production in Dielectric Barrier Discharges for Pollution Control, in Non-Thermal Plasma Techniques for Pollution Control, *NATO ASI Series*, **G34**, Part B, Springer-Verlag, Berlin, p. 339.

KOPECKY, J., 1992, *Organic Photochemistry: A Visual Approach*. New York: VCH Publishers, Inc.

MOORE, D. E., 1987, Principles and practice of drug photodegradation studies, *J. Pharm. Biomed. Anal.*, **5**, 441.

MOORE, D. E. and CHAPPUIS, P. P., 1988, A comparative study of the photochemistry of the non-steroidal anti-inflammatory drugs, Naproxen, Benoxaprofen and Indomethacin, *Photochem. Photobiol.*, **47**, 173.

MOORE, D. E., HEMMENS, V. J. and YIP, H., 1984, Photosensitization by drugs: Nalidixic acid and oxolinic acids, *Photochem. Photobiol.*, **39**, 57.

MOORE, D. E., ROBERTS-THOMSON, S., ZHEN, D. and DUKE, C. C., 1990, Photochemical studies on the anti-inflammatory drug diclofenac, *Photochem. Photobiol.*, **52**, 685.

MORENO, G., POTTIER, R. H. and TRUSCOTT, T. G. (eds), 1988, *Photosensitisation – Molecular, Cellular and Medical Aspects, NATO ASI Series H: Cell Biology, Vol. 15*, Berlin: Springer-Verlag.

OPPENLÄNDER, T., 1988, A comprehensive photochemical and photophysical assay exploring the photoreactivity of drugs, *Chimia*, **42**, 331.

1988–1991, unpublished results presented with permission of F. Hoffmann-La Roche, AG (Switzerland); the author thanks Drs U. Widmer, M. Völker, E. Gocke and K.-H. Pfoertner (F. Hoffmann-La Roche AG) for their kind support of this research. The photobiological studies and the photophysical measurements were performed by Dr E. Gocke and Dr M. Völker, respectively. The photoproducts were characterized by standard spectroscopical techniques and elemental analyses. The author thanks the analytical department of F. Hoffmann-La Roche AG for the measurements and the interpretation of ^{1}H-NMR, ^{13}C, IR and MS spectra and for CHN analyses.

1990, Photoreactivity of drugs: A challenge for pharmaceutical industry?, ESP Newsletter, *J. Photochem. Photobiol. B: Biology*, **5**, 528.

1993, Interaction of Light with Drugs, *Book of Abstracts*, Fifth Congress of the European Society for Photobiology, Sept. 19–26, Philipps-Universität, Marburg (Germany), p. 182.

1994a, Novel incoherent excimer UV irradiation units for the application in photochemistry, photobiology/-medicine and for waste water treatment, *EPA Newslett.*, **No. 50**, 2.

1994b, Structure-Function Relationships in Phototoxicology and *In Vitro* Drug Degradation Studies and Photoxicity, Workbook: Phototoxicology in the Regulatory Arena, April 13–15, The Hyatt Regency Crystal City Hotel, Arlington, Virginia, Chap. D, I.

OPPENLÄNDER, T. and BAUM, G., 1994, Ein Modularer Excimer-Durchflußreaktor zur Reinigung belasteter Abwässer durch Vakuum-UV/UV-Doppelbestrahlung ohne Oxidationsmittelzusatz, *Chem. Ing. Tech.*, **66**, 1523.

OPPENLÄNDER, T. HENNIG, T., 1995, Produktion von Prävitamin D_3 durch wellenlängenselektive Doppelbestrahlung von 7-Dehydrocholesterol mit inkohärenten Excimerstrahlern, *Chem. Ing. Tech.*, **67**, 594.

PAC, C. and ISHITANI, O., 1988, Electron-transfer organic and bio-organic photochemistry, *Photochem. Photobiol.*, **48**, 767.

PANDEY, G., 1992, Synthetic perspectives of photoinduced electron transfer generated Amine Radical Cations, *Synlett*, 546.

PATHAK, M. A., FITZPATRICK, T. B. and PARRISH, J. A., 1982, in Regan, J. D. and Parrish, J. A. (eds), *The Science of Photomedicine*, p. 446. New York: Plenum Press.

PFOERTNER, K.-H., 1991, *Ullmann's Encyclopedia of Industrial Chemistry, Photochemistry*, **Vol. A19**, p. 573. Weinheim: VCH Verlagsgesellschaft mbH.

PIETTE, J., 1990, Mutagenic and genotoxic properties of singlet oxygen, *J. Photochem. Photobiol., B: Biology*, **4**, 335.

SCHOONDERWOERD, S. A., BEIJERSBERGEN VAN HENEGOUWEN, G. M. J. and VAN

Belkum, S., 1989, *In vivo* Photodegradation of chlorpromazine, *Photochem. Photobiol.*, **50**, 659.

Stinson, S. C., 1992, Chiral drugs, *Chem. Eng. News*, **46**.

Turro, N. J., 1978, *Modern Molecular Photochemistry*. Menlo Park: The Benjamin/Cummings Publishing Co.

Zürcher, K. and Krebs, A., 1980, *Hautnebenwirkungen interner Arzneimittel*. Basel: S. Karger. Chap. C.

References to appendix

Arnason, J. T., Guerin, B., Kraml, M. M., Mehta, B., Redmond, R. W. and Scaiano, J. C., 1992, *Photochem. Photobiol.*, **55**, 35.

Bordin, F., Dall'Acqua, F. and Guiotto, A., 1991, *Pharmac. Ther.*, **52**, 331.

Braunwald, E., Isselbacher, K. J., Petersdorf, R. G., Wilson, J. D., Martin, J. B. and Fauci, A. S., 1987, *Harrison's Principles of Internal Medicine*, 11th edition. New York: McGraw-Hill Book Company.

Carpenter, S. and Kraus, G. A., 1991, *Photochem. Photobiol.*, **53**, 169.

Castell, J. V., Gomez-L., M. J., Miranda, M. A. and Morera, I. M., 1987, *Photochem. Photobiol.*, **46**, 991.

Epling, G. A. and Sibley, M. T., 1987, *Photochem. Photobiol.*, **46**, 39.

Forth, W., Henschler, D., Rummel, W. and Starke, K., 1992, *Allgemeine und spezielle Pharmakologie und Toxikologie*. Mannheim: Wissenschaftsverlag.

Johnson, B. E., 1988, in Moreno, G., Pottier, R. H. and Truscott, T. G. (eds), *Photosensitisation, Molecular, Cellular and Medical Aspects*, p. 253. NATO ASI Series H: Cell Biology, Vol. 15.

Kelly, G. E., Meikle, W. D. and Moore, D. E., 1989, *Photochem. Photobiol.*, **49**, 59.

Kochevar, I. E., 1987, *Photochem. Photobiol.*, **45**, 891.

Moore, D. E. and Champuis, P. P., 1988, *Photochem. Photobiol.*, **47**, 173.

Moore, D. E., Roberts-Thomson, S., Zhen, D. and Duke, C. C., 1990, *Photochem. Photobiol.*, **52**, 685.

Mutschler, E., 1981, *Arzneimittelwirkungen*. Stuttgart: Wissenschaftliche Verlagsgesellschaft mbH.

Sasaki, M., Matsuo, I. and Fujita, H., 1991, *Photochem. Photobiol.*, **53**, 385.

Schauder, S. and Ippen, H. 1988, in Fuchs, E., Schulz, K. H. (eds), *Manuale allergologicum*, Dustri-Verlag Dr K. Feistle, Deisenhofen, pp. 1–30.

Serrano, G., Fortea, J., Latasa, J., Millan, F., Janes, C., Bosca, F. and Miranda, M. A., 1992, *J. Am. Acad. Dermatol.*, **27**, 204.

Zürcher, K. and Krebs, A., 1980, *Hautnebenwirkungen interner Arzneimittel*. Basel, Switzerland: S. Karger.

12

In Vitro Screening of the Photoreactivity of Antimalarials: A Test Case

H. HJORTH TØNNESEN, S. KRISTENSEN and K. NORD

12.1 Introduction

It is essential to obtain information about the photoreactivity* of a drug molecule at the formulation development stage. A photochemical and photophysical *in vitro* assay should be designed to specify the photoreactivity of the compound in order to understand implications for lack of efficacy or possible photosensitizing activity *in vivo*. This represents an important field of interdisciplinary research involving photochemistry, photophysics, photomedicine, photobiology and pharmacokinetics. The photoreactivity of a molecule can change dramatically when the substance is introduced into a biological system. The selection of experimental conditions (including photon source) is therefore critical in order to obtain a correlation between *in vitro* reactivity and *in vivo* phototoxicity. This seems often to be neglected in the discussion of results obtained from *in vitro* experiments. In our laboratory we have applied an *in vitro* assay in the evaluation of photoreactivity of a number of antimalarial drugs. The aim of this work was to determine the underlying cause of ocular and cutaneous side-effects resulting from treatment with the most frequently used antimalarials. Changes in skin pigmentation and bleaching of the hair and corneal opacity, cataracts in the lens and visual disturbances, are frequently observed during medication with these drugs. However, irreversible retinal damage (retinopathy) and blindness are the most serious adverse effects, observed after long-term medication or after high accumulative doses of some of these compounds (Fraunfelder and Meyer, 1989; Moore and Hemmens, 1982; Tanenbaum and Tuffanelli, 1980). The antimalarial substances investigated are mainly structural analogues (Fig. 12.1). They fill almost every criteria needed to subject a drug to a complete photoassay (chapter 1), they are as follows:

- *Accumulation in skin, eye and hair.* Several of the antimalarial compounds possess a large distribution volume, an extremely long half-life and strong

*In this context the term 'photoreactivity' is used to describe how a compound responds to light exposure both *in vitro* and *in vivo*. This includes degradation reactions, other processes like radical formation, energy transfer and luminescence, and reactions with endogenous compounds.

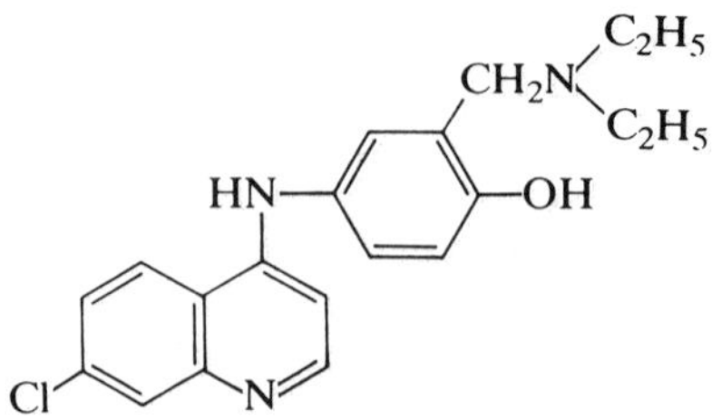

AQ

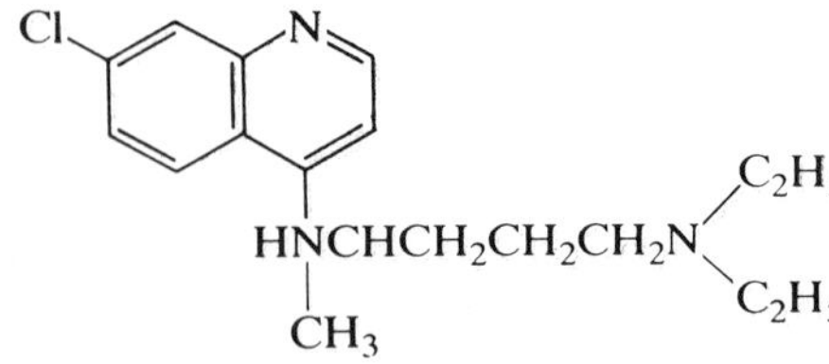

CQ

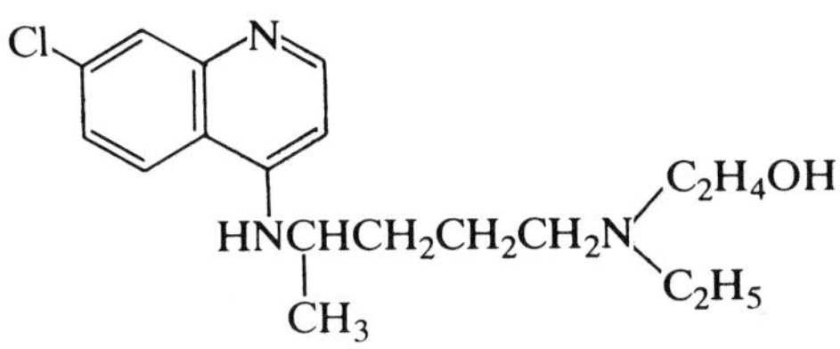

HQ

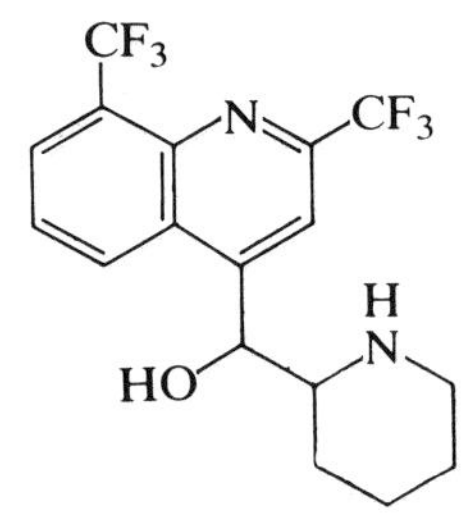

MQ

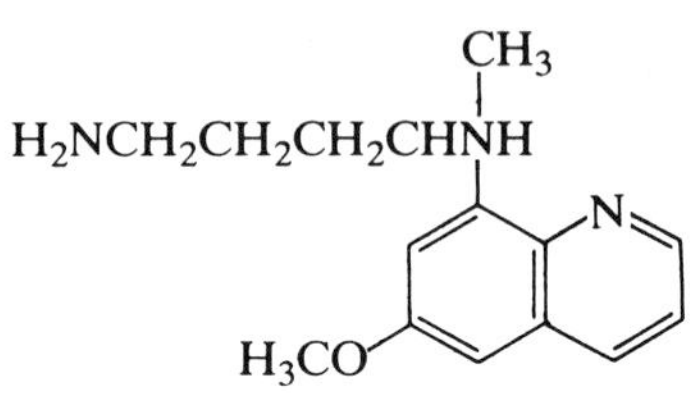

PQ

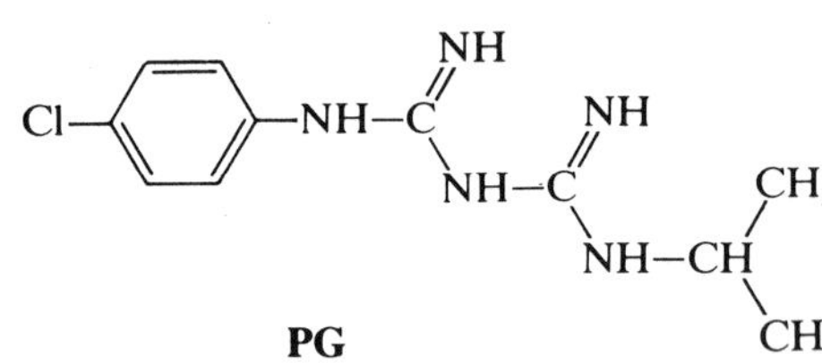

PG

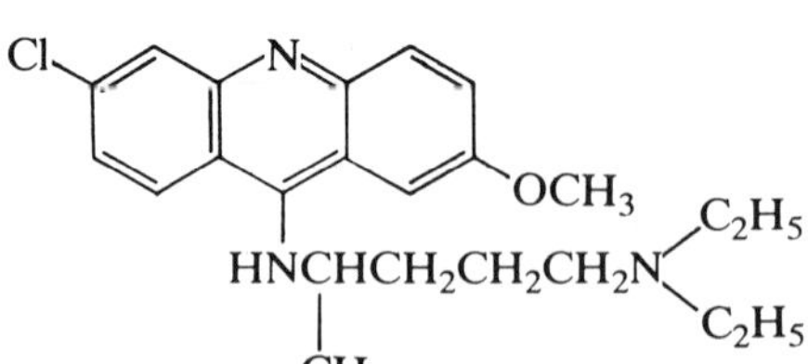

QC

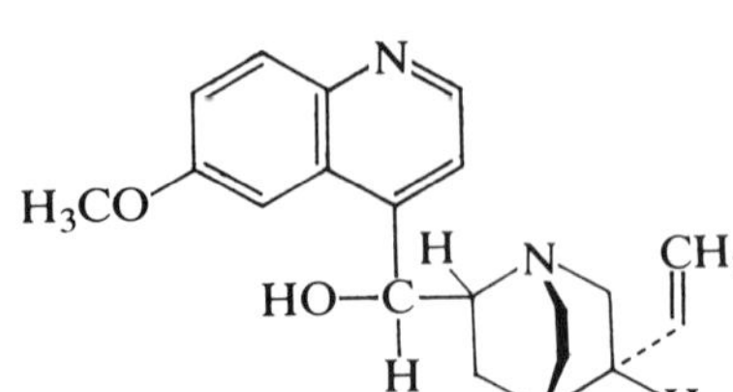

QU

Figure 12.1 Chemical structures of the antimalarials investigated. AQ = amodiaquine; CQ = chloroquine; HQ = hydroxychloroquine; MQ = mefloquine; PQ = primaquine; PG = proguanil; QC = quinacrine and QU = quinine interaction with melanin *in vitro*. These drugs are known to accumulate in melanin-rich areas of the body (eye, skin, hair) (Tanenbaum and Tuffanelli, 1980). Due to low turnover of melanin in the eye, a long-term retention (even for years) may be observed for drugs with high melanin affinity (Lindquist, 1986).

- *Administration at a high accumulative dosage or used in long-term medication.* The antimalarial drugs are often administered over a long period of time. Some of the drugs also possess anti-inflammatory effects although in much larger doses than those employed in prophylaxis and treatment of malaria (Webster, 1990). Accumulative doses of 200–300 g can in certain cases be reached (Ehrenfeld *et al.*, 1986).
- *Photolabile in vitro*. Several of the drugs have been demonstrated to decompose in solution during photoirradiation (Kristensen *et al.*, 1993; Laurie *et al.*, 1986, 1988; McHale *et al.*, 1989; Nord *et al.*, 1991; Taylor *et al.*, 1990; Tønnesen *et al*, 1988; Tønnesen and Grislingaas, 1990).
- *Forms photolabile degradation products or* in vivo *metabolites*. This is demonstrated for some of the drugs (Laurie *et al.*, 1988; Nord *et al.*, 1991; Tønnesen *et al.*, 1988).
- *Molecules contain essential functionalities* (aromatic moiety, amino group, halogen atoms etc.).
- *Environmental criteria*. Prophylactic use of antimalarial drugs is usually combined with a life in areas with high solar irradiance.

The antimalarial compounds are clearly qualified to undergo a complete evaluation according to the strategy demonstrated in Fig. 12.2. The *in vitro* photoassay used in this study is illustrated in Fig. 12.3. Such an assay should include evaluation of absorption and emission characteristics, quenching and sensitizing properties, photolysis experiments, determination of quantum yields, rate constants, action spectra and evaluation of radical formation.

12.1.1 Absorption characteristics

The photoreactivity of a pure drug substance will follow the basic law of photochemical absorption, i.e. a photochemical (or subsequent photobiological) reaction cannot occur unless non-ionizing radiation is absorbed by the compound. The determination of the absorption spectrum of a drug will immediately establish whether the substance will absorb the optical radiation penetrating the viable layers of the skin or various segments of the eye and is also an *in vitro* estimation of its potential to photosensitize. The absorption spectrum will, in most cases, not reveal the presence of other compounds that can influence the photoreactivity of the drug. Such compounds can be impurities from the synthesis or (photo)degradation products formed during the development process. These products will in most cases have absorption characteristics close to the main compound and be present only in small amounts. They are therefore difficult to detect unless a separation technique is applied. Although they are present in most cases only in trace amounts, they can have a catalytic effect on the degradation of the parent compound or cause phototoxic reactions *in vivo*.

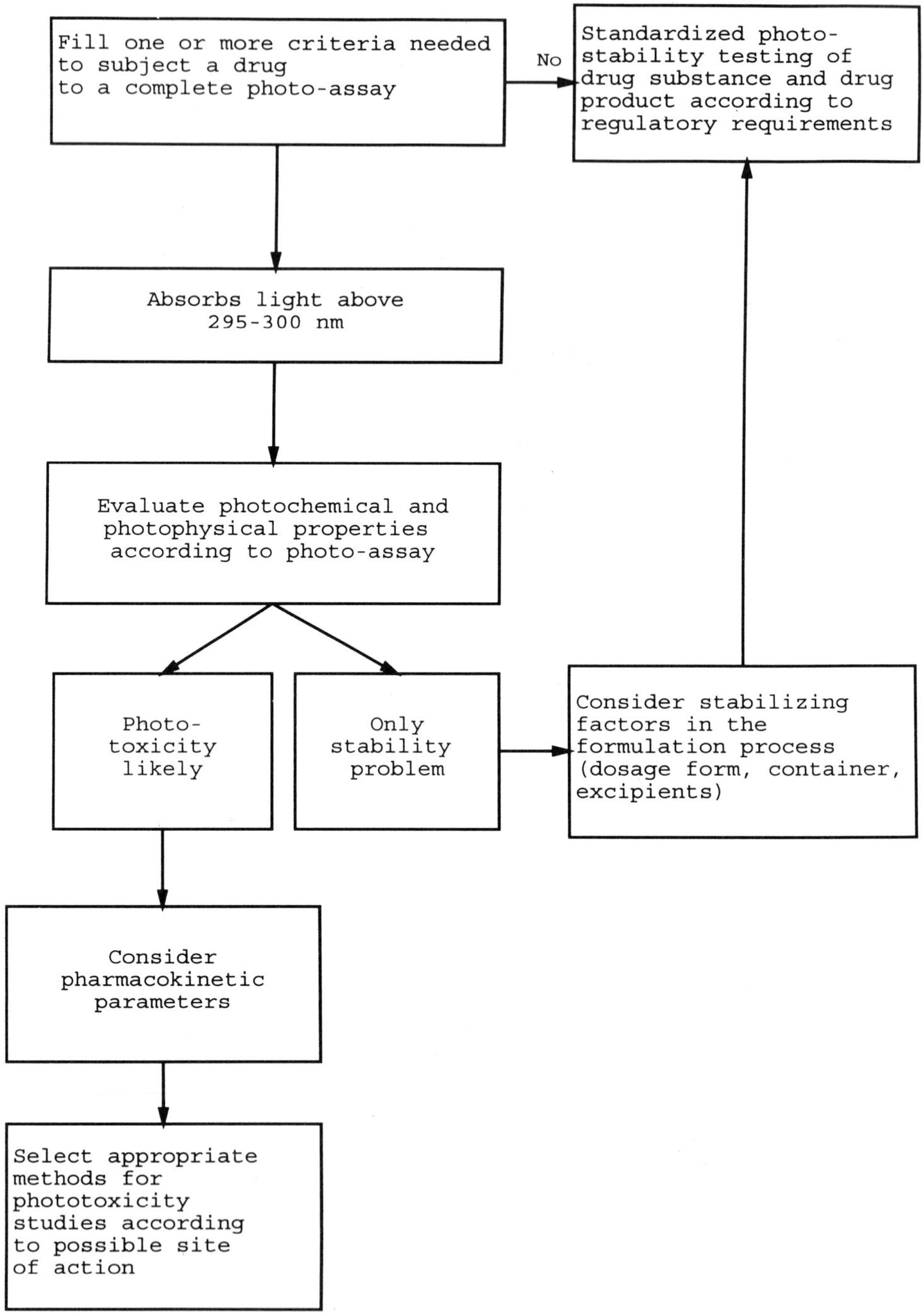

Figure 12.2 Strategy for the evaluation of photoreactivity of drugs as a part of the preformulation work

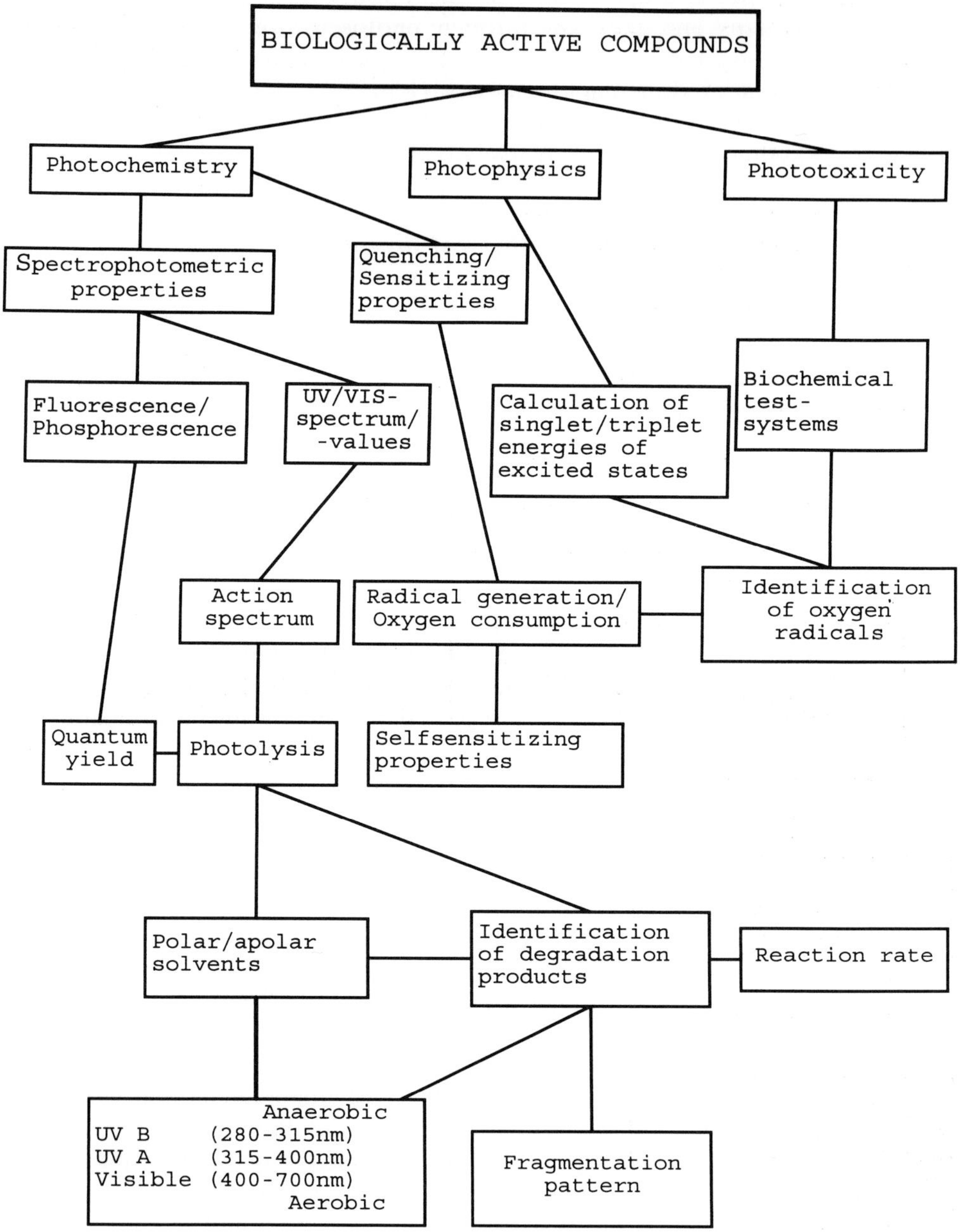

Figure 12.3 *In vitro* photoassay of biologically active compounds (e.g. drugs)

Light below approximately 295 nm is cut off by the cornea and does not reach the lens (Bachem, 1956). Light above 300 nm can reach the dermis with its capillary blood-vessels. Hence a drug has to absorb light above these wavelengths in order to damage the lens and the skin through photosensitized reactions. All the antimalarials investigated absorb light above 295–300 nm. According to the strategy for the evaluation of drug photoreactivity (Fig. 12.2) the photochemical and photophysical properties of the antimalarials should now be considered in order to find out whether phototoxicity is likely or not.

12.1.2 *Photochemical and photophysical properties*

12.1.2.1 Reaction medium and photon source

The reaction medium and the photon source must be selected with care in order to get a realistic description of the photoreactivity of a certain drug. In practice, the solubility of the substance may be a limiting factor in the choice of experimental conditions. Phosphate buffer, pH 7.4 is often used to 'mimic' biological conditions. The aqueous medium can contain micelles, liposomes, cyclodextrines or organic co-solvents dependent on the drug solubility. Photolysis in polar or apolar organic media can provide additional information about fragmentation pattern and reaction mechanisms. One should, however, keep in mind that organic solvents can participate in the reaction and lead to formation of 'non-physiological' degradation products. This is clearly demonstrated by the photolysis of chloroquine in isopropanol (Nord *et al.*, 1991). The photochemical decomposition of chloroquine was apparently more efficient in isopropanol than in water. Isopropanol was therefore chosen as reaction medium for the isolation of the degradation products. Isopropanol itself forms several photodecomposition products of low molecular weight (Pacakova *et al.*, 1985) under the actual conditions. Several condensation products formed from fragments of chloroquine and the reaction medium was identified.

Most photosensitized reactions in biological systems require the involvement of molecular oxygen. It is therefore natural to use oxygen-containing media in photoreactivity studies. Anaerobic conditions can, however, provide valuable information in mechanistic studies of light stability. There are also some exceptions in which oxygen-independent processes have an important place in the overall picture of photosensitization. It has been shown that chlorpromazine can photosensitize the haemolysis of erythrocytes under anaerobic conditions (Kochevar and Lamola, 1979) and that furocoumarins bind covalently to DNA upon irradiation in the absence of oxygen (Rodighiero and Dal'Acqua, 1976). Photoreactivity studies carried out under anaerobic conditions should therefore be considered as a part of the *in vitro* assay. There is, however, not always a correlation between results obtained in deaerated organic solvents and in aqueous medium at physiological pH. Dechlorination of chloroquine and hydroxychloroquine plays an important role in the photolysis of these compounds under anaerobic conditions (Moore and Hemmens, 1982) while the chlorine atom seems to be intact in all the photodegradation products identified under aerobic conditions (Nord *et al.*, 1991; Tønnesen *et al.*, 1988).

Many drug molecules are weak acids or weak bases. The pH of the reaction

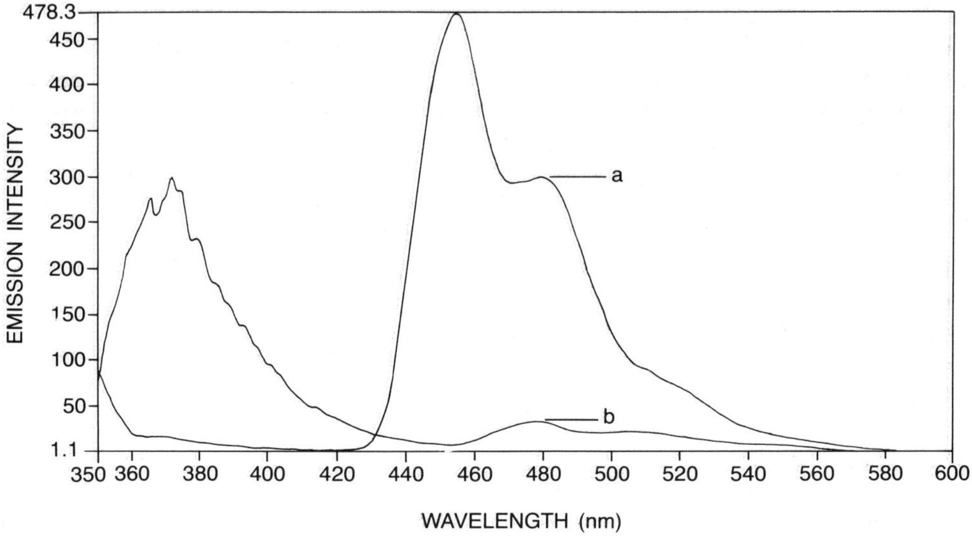

Figure 12.4 The observed luminescence spectra at 77 K of desethylhydroxychloroquine when changing the solvent from EPA to EPA + 0.5 per cent 1 M HCl at excitation wavelengths 328 nm and 346 nm respectively. Peaks between 350 nm and 430 nm correspond to the fluorescence emission and peaks between 430 nm and 600 nm correspond to the phosphorescence emission. The relative areas under the two uncorrected phosphorescence emission spectra are 1:10; (a) desethylhydroxychloroquine in EPA + 0.5 per cent 1 M HCl (i.e the heterocyclic N is ionized); (b) desethylhydroxychloroquine in EPA (i.e. the heterocyclic N is deionized)

medium may in that case strongly influence the results obtained. The photosensitized oxidation of mefloquine and other antimalarials shows a clear pH dependency (Tønnesen and Moore, 1991). Valuable information can be lost if experiments are carried out only in one medium, i.e. organic solvent that cannot differentiate between protonated and deprotonated forms of the molecules, or at non-physiological pH. This is clearly demonstrated for the antimalarials of 4-aminoquinoline structure (Nord *et al.*, 1994). These compounds are strongly fluorescent in the deprotonated form while phosphorescence dominates when the compounds are protonated leading to different photoreactivity in different media (Fig. 12.4). This would suggest that phototoxicity exhibited through triplet state energy transfer to form singlet oxygen may be greater if the drug exists in hydrophilic regions of the cell such as the cytosole.

It is of great importance to know the spectral distribution and intensity of the photon sources used in the evaluation of photoreactivity of drugs, both with respect to photostability testing according to regulatory requirements and in the evaluation of photoreactivity as a part of the preformulation work. The importance of using a standardized light source is frequently neglected in the discussion of drug photoreactivity. Kinetic data reported in the literature often have one thing in common: the lack of standardized experimental conditions, leading to discrepancy in the results. Expressions such as the sample was placed '0.5 m from a north-facing window so that constant daylight illumination was obtained' (Kirk, 1987) or '0.9 m from a continuously illuminated fluorescent bulb (30 W)' (Newton *et al.*, 1981) are typical. These expressions ignore the fact that the intensity of 'daylight' is dependent on factors like altitude, latitude, season and time of the day and the year and that the spectral distribution and intensity of the photon source varies with the type and age of the lamp. As a result the half-life reported in the literature for a given compound can differ with several hundred per cent (Bosanquet, 1986). The precautions to be taken in a hospital ward or in a hospital pharmacy in order to protect a certain drug formulation against light will consequently show great variations. In the evaluation of photoreactivity it will be of interest to use physiological important wavelengths of irradiation like UV-A, UV-B and visible light which also cover likely 'in use' conditions (Tønnesen and Karlsen, 1995). Light below 280 nm (UV-C) can be convenient for qualitative purposes due to a faster degradation rate.

Only light that is *absorbed* by a sample can cause photochemical reactions. The overlap between the absorption spectrum of the product and the emission spectrum of the lamp will determine the photoreactions that will occur in the sample. The emission intensity at the absorbed wavelengths will determine the rate at which the products will form. To ensure the formation of all possible degradation products including products formed in sensitized reactions, the sample must be irradiated at all absorbing wavelengths. This can be achieved by the use of a broad-spectrum light source. The qualitative and quantitative aspects of the process will be changed if the emission spectrum of the lamp only partly overlaps the absorption spectrum of the sample as may be the case if a combination of light sources is used.

12.1.2.2 Photolysis and identification of degradation products

Studies of drug photolysis will provide valuable information in the formulation process, i.e. in the selection of dosage form, containers and excipients. Pathways for the formation of radicals can be proposed from photodegradation studies. Free radicals generated can play an important role in phototoxic and photoallergic effects oberved after medication. Adverse effects may also occur as a consequence of photodecomposition of the drug leading to toxic or photoreactive degradation products. When, for instance, the photolysis occurs on or near the epidermal layer lipophilic products would be expected to partition into cell membranes and be cleared only slowly from the system. If the chromophore is retained, the photosensitivity may be observed on further radiation exposure. For many compounds it is found that phototoxic degradation products may be formed by metabolic or photodecomposition pathways. Isolation and identification of degradation products combined with photoreactivity studies of these compounds are therefore of importance. Photodegradation studies have been carried out for several of the antimalarial compounds (Kristensen *et al.*, 1993; Nord *et al.*, 1991; Tønnesen *et al.*, 1988; Tønnesen and Grislingaas, 1990). The main degradation products of chloroquine and hydroxychloroquine corresponds to their *in vivo* metabolites. These products are also demonstrated to be photolabile (Nord *et al.*, unpublished results). Some of the photodegradation products of hydroxychloroquine seem to be more potent as photosensitizers than the parent drug (Kristensen *et al.*, 1994a).

12.1.2.3 Degradation rate

Although great effort should be taken to stabilize the formulation itself in such a way that the shelf-life becomes independent on the storage conditions, it might still be necessary to take precautions to exclude or minimize the amount of light reaching the formulation. The precautions taken are often related to the photochemical half-life of the drug substance. As discussed above, a kinetic approach to the photochemical degradation of drugs is without any value unless the light source is specified. For the absolute determination of the extent of a photoreaction, the spectral distribution and intensity of the light source must be known. Determination of reaction quantum yield using monochromatic light and actinometry provides the *exact kinetic data* for a specific photochemical reaction and is usually obtained at the absorption maximum of the compound being tested. The *observed* decomposition rate of a drug substance or drug product can be obtained by the use of a light source simulating 'indoor light' or sunlight. The intensity of the light source must be related to the 'likely exposure' or to the actual intensity of sunlight for calculating the accelerating effect. This can be based on a defined value for the irradiance of sunlight at a specific location or as a global average (Tønnesen and Moore, 1993). The order of the reaction can only be determined if the reaction is followed up to more than 50 per cent degradation (Sande and Karlsen, 1993).

The temperature and humidity conditions in the test chamber should be kept at a level to ensure that the effects of changes in the physical state of the samples such as sublimation, evaporation or melting are minimized. Rate constants for photochemical reactions depend on the temperature because of secondary thermal reactions of the parent compound or primary products. Thermal stability of the material should be independently determined through accelerated stability testing. The ambient temperature and the temperature of the samples during irradiation are related to the photon source used for testing and the intensity and distance of the sample from the photon source. Infrared (IR) radiation has a significant effect on 'heating' of materials. The amount of IR radiation will affect both the ambient temperature and the temperature of the sample, the latter also being strongly influenced by the colour of the material. An increase in surface temperature is likely to be observed in dark samples with low thermal conductivity. The use of foil-wrapped dark control samples will normally be adequate. For products with low thermal stability, it may be necessary to determine the surface temperature of the light-exposed samples and place the dark control samples in a chamber under temperature and humidity conditions corresponding to the surface temperature and humidity of the irradiated drug. The surface temperature of a light sample (e.g. white tablet) and a dark sample can be compared with the temperature of a white standard thermometer and a black standard thermometer respectively at equal distance from the photon source. The importance of a dark control is often neglected in the discussion of photochemical stability data. In most cases the photolytic rate constants are much larger than the rate constants of the thermal reactions. For some compounds, e.g. primaquine, the influence of the dark reactions cannot be ignored (Kristensen *et al.*, in press).

The humidity in the test chamber can influence the photochemcial stability of certain solid samples. This is demonstrated for mefloquine. The photoinduced yellowing of uncoated mefloquine tablets is accelerated by an increase in humidity.

These tablets are mainly used in tropical countries and the real 'in use' conditions will include a high relative humidity. In such cases the influence of the humidity on the photostability must be taken into account.

12.1.2.4 Fluorescence and phosphorescence

Most photosensitized reactions originate from the triplet state of the molecules (Roberts, 1988), but there seems also to be an apparent relation between the fluorescence quantum yield *in vitro* and the photosensitizing ability *in vivo* (Moore and Hemmens, 1982). The production of singlet oxygen is found to occur by energy transfer both from the singlet and triplet excited states of the sensitizer, although the reaction from the triplet state is highly preferred since singlet–triplet interaction is of very low probability (Stevens *et al.*, 1981). The excited triplet state, because of its longer lifetime (milliseconds to several seconds), may diffuse a significant distance in fluid media and therefore has a much higher probability of interaction with other molecules. Those compounds with long-lived triplet states are therefore more likely to be photosensitizers. The lifetime of a molecule in the excited singlet or triplet states is related directly to the fluorescence and phosphorescence lifetimes respectively. Information about the possible photoreactivity of the drug substance and its degradation products or metabolites can therefore be obtained from quantum yield measurements and lifetime studies. Additionally, the singlet and triplet energies of the compound are of considerable interest in the study of the mechanism of the photoreaction especially with respect to energy transfer. The photosensitizing potential of the 4-aminiquinolines is emphasized by the results obtained from the fluorescence and phosphorescence measurements (Nord *et al.*, 1994). These compounds have long-lived triplet states. Triplet-state energy transfer to form singlet oxygen is likely to occur and they are good candidates for potential photosensitizing activity *in vivo*.

12.1.2.5 Quantum yield

The quantum yield provides information about the 'effectiveness' of a certain light-induced process. The quantum yield of loss of starting material or product formation will provide valuable information about structure/activity relationships. Fluorescence and phosphorescence quantum yields will indicate the fraction of molecules likely to be found in the excited singlet and triplet state. The quantum yield is a useful parameter to predict the importance of a certain reaction, for example an isolated degradation product can have a long phosphorescence lifetime and should therefore be considered as a possible sensitizer. If the quantum yield of formation of this product is very low, however, it is less likely to be formed in a biological active concentration and may therefore play a minor role in phototoxicity reactions. For the determination of quantum yield the change in drug concentration should not exceed 15 per cent during the irradiation in order that the photodecomposition remains linear with irradiation time (Moore, 1987). By comparing the reaction quantum yield for the photodecomposition of mefloquine with its fluorescence quantum yield at physiological pH it is demonstrated that deactivation of the excited state by formation of degradation products is strongly favoured compared with radiative deactivation (Tønnesen and Moore, 1991). This may be of importance since the main degradation product is shown to be photoreactive.

12.1.2.6 Action spectrum

The action spectrum of a drug substance or drug product consists of those wavelengths which elicit degradation *in vitro* or a certain clinical reaction *in vivo*. An action spectrum is obtained by measuring the radiation dose required to evoke the same degree of degradation or the same biological response at different wavelengths. The action spectrum for a pure compound would normally overlap the absorption spectrum of the compound. It should be noted that in formulations in which sensitized reactions are possible or where the absorption of radiation by inactive chromophores and radiation scattering is considerable, there is not always a direct relationship between the absorption spectrum of the formulation and the action spectrum for the photochemical degradation of the active compound. *In vivo* the action spectrum will not parallel the absorption spectrum *in vitro* unless the action mechanism and the quantum yield are the same at all wavelengths and the absorption spectrum *in vitro* and *in vivo* are identical. The action spectra of several of the antimalarial compounds are under investigation.

12.1.2.7 Photooxidation measurements and identification of oxygen radicals

It is widely accepted that the toxicity of many endogenous compounds arises from free-radical intermediates. Generally, reactions which lead to more permanent biological damage involve free-radical intermediates (Kensler and Taffe, 1986). Because of the oxygen content of blood and tissue, adverse reactions might be ascribed to photosensitized oxidation reactions. There are two mechanisms of photosensitized oxidation. The Type I mechanism is a free-radical process and the Type II mechanism involves excited molecular oxygen. For drugs that produce free radicals as well as singlet molecular oxygen, both mechanisms may be observed in photooxidation (Spikes, 1977). The photooxidation potential can be determined by means of an oxygen electrode and oxidable substrates with or without specific quenchers. Histidine and 2,5-dimethyl furane are substrates for singlet oxygen. The latter can be used independent of pH, while histidine is suitable at pH above 7.5. L-tryptophane is a substrate for superoxide. Suitable quenchers are DABCO and azide (singlet oxygen), mannitol (hydroxyl radical) and 2-mercapto ethylamine and glutathione (other radicals). The (lack of) specificity of the various substrates and quenchers should be considered. Other methods such as ESR (electron spin resonance) must be applied for the exact identification of radicals, but very useful information can be obtained from photooxidation studies by a combination of various substrates and quenchers. In order to compare the rates of photooxidation of various compounds, correction has to be made for the differences in absorption spectra between the compounds and different ionization states of the same compound. Thus the rates should be normalized to a constant amount of light absorbed, calculated from the area under the absorption curve (i.e. the overlap-integral between the absorption spectrum of the sample and the emission spectrum of the photon source). The photo-oxidizing potential of several antimalarial compounds has been investigated (Moore and Hemmens, 1982; Tønnesen and Moore, 1991).

Evaluation of self-sensitizing properties, i.e. the ability to induce formation of radicals and react with the same radicals in a secondary reaction, is of interest from a formulation viewpoint. The stability of primaquine in solution is strongly influenced by self-sensitized reactions involving oxygen radicals (Kristensen *et al.*, 1993 and in press). A stabilizing effect could be obtained by addition of appropriate radical quenchers to the samples.

12.2 Pharmacokinetic parameters

Based on the results obtained so far the antimalarial compounds are likely to cause phototoxic reactions *in vivo*. Pharmacokinetic aspects must now be taken into account, including information about *in vivo* metabolites and degradation products (if available). Phototoxicity is generally dose dependent, i.e. dependent on both the concentration of the drug sensitizer and the intensity of the incident radiation at the site of action. Prior to a phototoxic reaction the sensitizer must be distributed to tissues that are exposed to light and further absorb the light that penetrates these tissues. If no pharmacokinetic data are available the possibility of being retained in a lipophilic medium can to some extent be predicted from the pK_a value of the compound. For the antimalarial drugs, however, essential data are available in the literature. Many of the antimalarials investigated have a large apparent distribution volume (V_d) and a long elimination half-life ($t_{1/2}$) (Table 12.1). This reflects accumulation of the drugs in body tissues. Most of the adverse effects associated with the use of antimalarials are related to the eye and skin. The retina is richly supplied with blood vessels, but the blood–retinal barrier is normally very tight and therefore restricts the movements of substances from the capillaries. The lens has no blood supply and drugs cannot be distributed directly from the systemic circulation to this tissue. On certain occasions the permeability of the blood–retinal barrier can be altered. Drugs will, to a larger extent, then enter the retinal pigmented epithelium (RPE) and retina. Drugs accumulated in the RPE can further pass through the vitreous cavity and reach the lens.

Quinacrine, which is tricyclic and highly lipophilic, will easily penetrate cell membranes, but hardly diffuse through the hydrophilic vitreous cavity. Quinacrine

Table 12.1 Pharmacokinetics and absorption characteristics (phosphate buffer, pH = 7.4) of antimalarial drugs
V_d: apparent volume of distribution
$t_{1/2}$: elimination half-life

Compound	Absorption maximum (nm)	V_d (l/kg)	$t_{1/2}$
Amodiaquine	341	Widely	5 hours
Chloroquine	329, 343	200	25–60 days
Hydroxychloroquine	329, 343	Widely	40 days
Mefloquine	284	13–29	19.5 days
Primaquine	352	3–4	7 hours
Proguanil	253	33	16–20 hours
Quinacrine	423, 443	500	5 days
Quinine	331	2	9 hours

also forms a stable complex with melanin and will therefore be retained in the RPE. Quinacrine has to be considered a potent photosensitizer in the retina due to the absorption maximum of this drug in the visible region of the spectrum (Table 12.1). The more hydrophilic compounds chloroquine and hydroxychloroquine will more easily be transported to the lens. They are known to accumulate in the eye and induce toxic reactions. Primaquine is not likely to be distributed to the eye to any extent due to low distribution volume and fast elimination.

Several of the antimalarial compounds seem also likely to accumulate in the lipophilic or melanin-containing tissues of the skin. Adverse dermal effects observed after medication with antimalarials are probably a result of phototoxic reactions.

12.3 *In vitro* phototoxicity studies

12.3.1 *Examples of methods*

The possible site of action for the antimalarials as potential photosensitizers will be various parts of the eye and the skin. Photohaemolytical properties, photosensitized polymerization of lens proteins and interaction with melanin was selected as *in vitro* test methods for phototoxicity screening of these compounds. According to their photohaemolytical capability quinacrine, quinine and primaquine have to be considered as potential *in vivo* photosensitizers (Kristensen *et al.*, 1994a). It is widely accepted that photoinduced lipid peroxidation has detrimental effects on cell membranes and therefore plays an important role in skin phototoxicity. The observed photohaemolysis induced by these antimalarials may be an indication of extensive photoperoxidation of the membrane lipids. Photo-oxidation experiments with linoleic acid should be performed in addition to the photohaemolysis studies.

Photosensitized polymerization of lens proteins can be selected as a measure of eye phototoxicity. Proteins isolated from calf lens are suitable to simulate the conditions in the human eye. The reaction mechanisms can be evaluated by addition of various quenchers to the reaction medium during irradiation. Chloroquine, hydroxychloroquine, mefloquine and quinacrine-induced polymerization under the given experimental conditions. These compounds have a large apparent distribution volume and a long elimination half-life and must therefore be considered as potential photosensitizers in the eye (Kristensen *et al.*, 1995). Primaquine and quinine were also shown to induce polymerization of lens proteins *in vitro*, but are less likely to reach the eye *in vivo* due to fast elimination from the body.

Some of the antimalarial compounds accumulate in melanin-rich areas of the body (skin, eye and hair). The turnover of melanin in the body is very low, except for epidermal melanin. Compounds with high melanin affinity can be retained in melanin-containing tissues for years (Lindquist, 1986). Information about the interaction between drugs and melanin is therefore of great importance in the evaluation of drug phototoxicity. The total binding of the drug substance to melanin is the main concern. The surface characteristics of the melanin selected for the experiments must be taken into account as the chemical composition of melanin isolated from different tissues will vary. Binding to melanin was demonstrated for all the antimalarial compounds investigated (Kristensen *et al.*, 1994b). Chloroquine and hydroxychloroquine possess a large distribution volume and an extremely long half-life in addition to a strong interaction with melanin *in vitro*. Both compounds are therefore distinguished as particularly good candidates for melanin binding *in vivo*. It should, however, be kept in mind that the conditions *in vivo* are quite different from the simple *in vitro* system used in the assay. Interactions between the drug molecule and biological macromolecules like proteins can occur leading to a change in solubility and absorption characteristics of the drug.

12.3.2 Normalization of the results

One essential aspect in the study of photochemical and photobiological responses is the number of molecules available for light absorption (Megaw and Drake, 1986). Hence, comparison of biological effects induced by the various drugs foresees the drug concentration given in moles/litre. It is, however, a considerable oversimplification to assume that the sensitizer producing a given effect at the lowest molar concentration is the most effective compound. The biological effects should always be normalized to the number of photons absorbed by the sensitizer *in the actual medium* (Pooler and Valenzeno, 1982). This is often neglected in phototoxicity studies. The effect of normalization is clearly demonstrated in Figs 12.5–12.7. Figure 12.5 shows the photohaemolytic effect of the antimalarial drugs

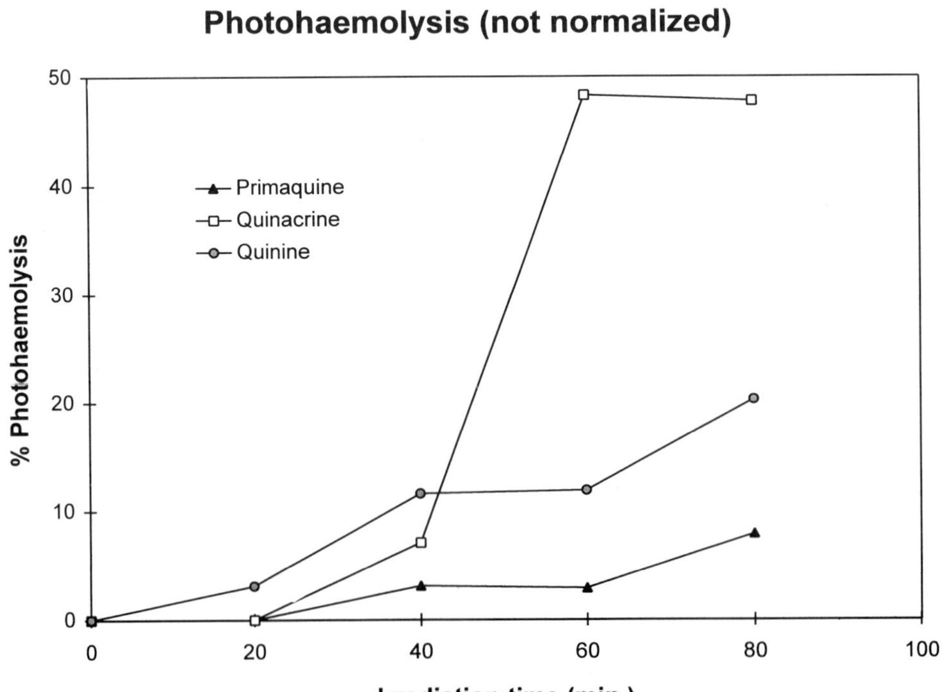

Figure 12.5 Photohaemolytic effect of primaquine, quinacrine and quinine at the same molar concentration (data not normalized)

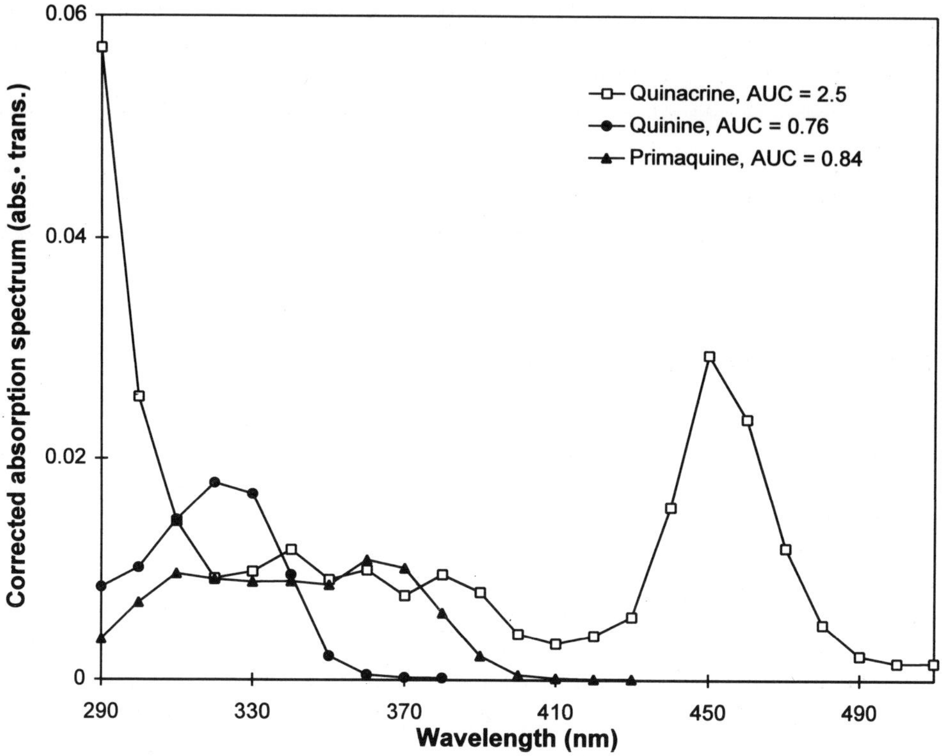

Figure 12.6 Corrected absorption spectra of primaquine, quinacrine and quinine

quinine, quinacrine and primaquine at the same molar concentration (without normalization). The medium contains red blood cells (RBC) and absorbs light in the UV-Vis region of the spectrum. A corrected absorption spectrum for each drug compound can be made from the amount of light *available* for the drug substances in the medium (Fig. 12.6) and the photohaemolytic effect can be estimated from the per cent photohaemolysis and the relative number of photons absorbed by the drug (Kristensen *et al.*, 1994a). When the drugs were evaluated without normalization of the results, quinacrine seemed to be the most potent compound (Fig. 12.5). However, compared on a normalized scale quinacrine turned out to be in the same range as primaquine while quinine was the most powerful photohaemolytic agent (Fig. 12.7). This emphasizes the importance of normalizing biological effects to the relative number of photons absorbed by the drug in order to evaluate their relative photosensitizing potencies.

12.3.3 ***Influence of degradation products***

The photochemical degradation of the drugs in the medium will run parallel to the phototoxicity reaction studied. The extent of degradation in the actual test medium cannot always be estimated from the degradation rate in a pure solution due to filter effect from compounds present in the medium (e.g. RBC). The concentration of the parent compound should be maintained at a level of >85 per

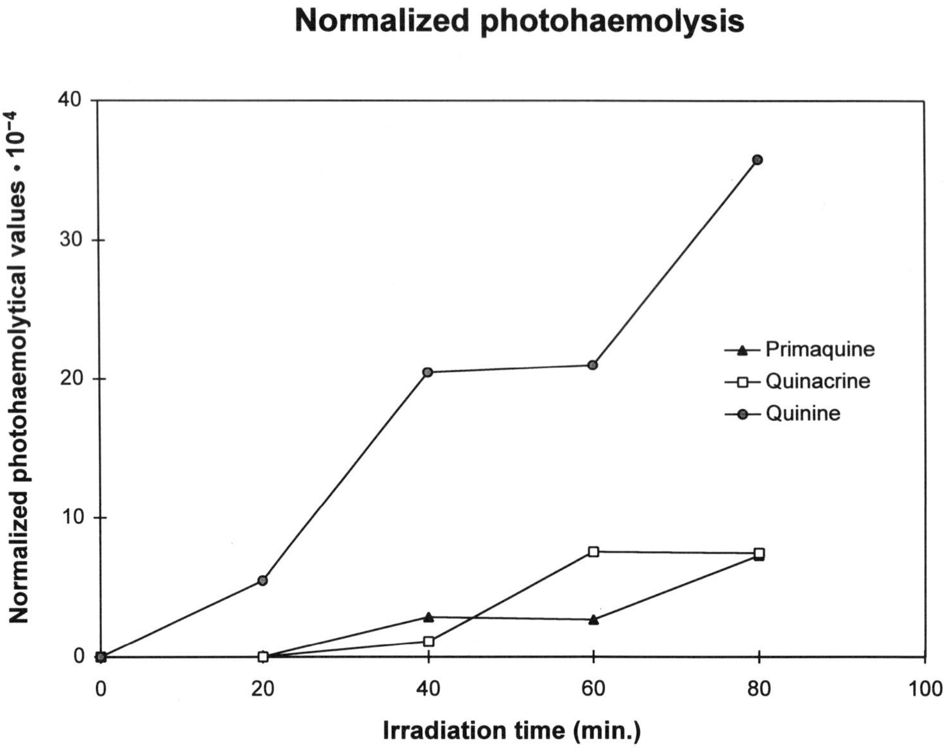

Figure 12.7 Photohaemolytic effect of primaquine, quinacrine and quinine normalized to the amount of light absorbed

cent of the initial concentration to minimize the effect of degradation products formed. This requires methods for quantitation of the drugs in biological media and is not a simple task. In most cases the phototoxic effects observed must be considered as a result of both the parent drug and the photochemical (and hydrolytic) degradation products formed during the experiment.

12.4 Conclusion

This chapter describes the application of an *in vitro* photoassay as a method to estimate the photoreactivity of drugs. The fundamental photochemical behaviour of certain antimalarial compounds has been examined in order to find a basis for the adverse photobiological effects associated with the clinical use of these drugs.

Combining the information obtained about the antimalarial compounds leads so far to the following observations. Intact drugs can after administration reach the skin surface and to a various extent also the eye. The drugs interact with melanin and are likely to accumulate in melanin-rich tissues impairing their clearance from skin and eye. Exposed to direct sunlight the drugs decompose to form some stable photoproducts. Some of the main degradation products are identical to the *in vivo* metabolites and appear to be photoreactive. Free radical intermediates including active oxygen species are generated, leading to cell membrane damage and protein polymerization.

No photosensitizer yields only one type of reactive species. The reactive pathways that the excited molecule follows depend on the nature of both the sensitizer and its environment. This emphasizes the need for using more than one experimental model in the evaluation of drug photoreactivity and to take into account pharmacokinetic parameters and method specificity in the discussion of the normalized results. Observation of drug photoreactivity *in vitro* does not necessarily mean that the same combination of reactions will take place *in vivo*. A certain *in vitro–in vivo* correlation should however be obtained if the *in vitro* photoassay is carefully designed. The assay will then result in a reduction in the number of animal experiments required in the development of new drug substances and drug formulations.

References

BACHEM, A., 1956, Ophthalmic ultraviolet action spectra, *Am. J. Ophthalmol.*, **41**, 969–75.

BOSANQUET, A. G., 1986, Stability of solutions of antineoplastic agents during preparation and storage for in vitro assays. II. Assay Methods, adriamycin and the other antitumor antibiotics. *Cancer Chemother. Pharmacol.*, **17**, 1–10.

EHRENFELD, M., NESHER, R. and MERIN, S., 1986, Delayed-onset chloroquine retinopathy, *Br. J. Ophthalmol.*, **70**, 281–3.

FRAUNFELDER, F. T. and MEYER, S. M., 1989, *Drug-induced Ocular Side Effects and Drug Interactions*, 3rd edn, pp. 58–62. Philadelphia: Lea and Febiger.

KENSLER, T. W. and TAFFE, B. G., 1986, Free radicals in tumor promotion, *Adv. Free Radical Biol. Med.*, **2**, 347–87.

KIRK, B., 1987, The evaluation of a light protective giving set. The photosensitivity of intravenous dacarbazine solutions, *Intensive Ther. Clin. Monit.*, **8**, 78–86.

KOCHEVAR, I. E. and LAMOLA, A. A., 1979, Chlorpromazine and protriptyline phototoxicity: photosensitized oxygen-independent red cell hemolysis, *Photochem. Photobiol.*, **29**, 1177–97.

KRISTENSEN, S., GRINBERG, L. and TØNNESEN, H. H. (in press), Photoreactivity of biologically active compounds. XI. Primaquine and its metabolites as radical inducers, *Int. J. Pharm.*

KRISTENSEN, S., GRISLINGAAS, A.-L., GREENHILL, J. V., SKJETNE, T., KARLSEN, J. and TØNNESEN, H. H., 1993, Photochemical stability of biologically active compounds. V. Photochemical degradation of primaquine in an aqueous medium, *Int. J. Pharm.*, **100**, 15–23.

KRISTENSEN, S., KARLSEN, J. and TØNNESEN, H. H., 1994a, Photoreactivity of biologically active compounds. VI. Photohemolytical properties of antimalarials *in vitro*, *Pharm. Sci. Comm.*, **4**, 183–91.

KRISTENSEN, S., ORSTEEN, A.-L., SANDE, S. A. and TØNNESEN, H. H., 1994b, Photoreactivity of biologically active compounds. VII. Interaction of antimalarial drugs with melanin in vitro as part of phototoxicity screening, *J. Photochem. Photobiol., B:Biol.*, **26**, 87–95.

KRISTENSEN, S., WANG, R.-H., TØNNESEN, H. H., DILLON, J. and ROBERTS, J. E., 1995, Photoreactivity of biologically active compounds. VIII. Photosensitized polymerization of lens proteins by antimalarial drugs *in vitro*, *Photochem. Photobiol.*, **61**, 124–30.

LAURIE, W. A., MCHALE, D. and SAAG, K., 1986, Photoreactions of quinine in aqueous citric acid solution, *Tetrahedron*, **42**, 3711–14.

LAURIE, W. A., MCHALE, D., SAAG, K. and SHERIDAN, J. B., 1988, Photoreactions of quinine in aqueous citric acid solution. Part 2. Some end-products, *Tetrahedron*, **44**, 5905–10.

LINDQUIST, N. G., 1986, Melanin affinity of xenobiotics, *Upsala J. Med. Sci.*, **91**, 283–8.

MCHALE, D., LAURIE, W. A., SAAG, K. and SHERIDAN, J. B., 1989, Photoreactions of quinine in aqueous citric acid solution. Part 3. Products formed in aqueous 2-hydroxy-2-methylpropionic acid, *Tetrahedron*, **45**, 2127–30.

MEGAW, J. M. and DRAKE, L. A., 1986, Photobiology: An overview, in Jackson, E. M. (ed.), *Photobiology of the Skin and Eye*, pp. 1–31. New York: Marcel Dekker.

MOORE, D. E. 1987, Principles and practice of drug decomposition studies, *J. Pharm. Biomed. Anal.*, **5**, 441–53.

MOORE, D. E. and HEMMENS, V. J., 1982, Photosensitization by antimalarial drugs, *Photochem. Photobiol.*, **36**, 71–7.

NEWTON, D. W., FUNG, E. Y. Y. and WILLIAMS, D. A., 1981, Stability of five catecholamines and terbutaline sulfate in 5% dextrose injection in the absence and presence of aminopylline, *Am. J. Hosp. Pharm.*, **38**, 1314–19.

NORD, K., KARLSEN, J. and TØNNESEN, H. H., 1991, Photochemical stability of biological active compounds. IV. Photochemical degradation of chloroquine, *Int. J. Pharm.*, **72**, 11–18.

1994, Photochemical stability of biologically active compounds. IX. Characterization of the spectroscopic properties of the 4-aminoquinolines, chloroquine and hydroxyquinoline, and of selected metabolites by absorption, fluorescence and phosphorescence measurements, *Photochem. Photobiol.*, **60**, 427–31.

PACAKOVA, V., KONAS, M. and KOTVALOVA, V., 1985, Reaction gas chromatography: Study of the photodecomposition of selected substances, *Chromatographia*, **20**, 164–72.

POOLER, J. P. and VALENZENO, D. P., 1982, A method to quantify the potency of photosensitizers that modify cell membranes, *J. Natl. Cancer Inst.*, **69**, 211–15.

ROBERTS, J. E., 1988, Ocular phototoxicity, in Moreno, G., Pottier, R. H. and Truscott, T. G. (eds), *Photosensitization. Molecular, Cellular and Medical Aspects*, pp. 325–30. Berlin: Springer Verlag.

RODIGHIERO, G. and DAL'ACQUA, F., 1976, Biochemical and medical aspects of psoralens, *Photochem. Photobiol.*, **24**, 647–53.

SANDE, S. A. and KARLSEN, J., Evaluation of reaction order. Software in pharmaceutics: III, *Int. J. Pharm.*, **98**, 209–18.

SPIKES, J. D., 1977, Photosensitization, in Smith, K. C. (ed.), *The Science of Photobiology*, pp. 87–112. New York: Plenum Press.

STEVENS, T. J., MARSH, K. L. and BARLTROP, J. A., 1981, Photoperoxidation of unsaturated organic molecules. 21. Sensitizer yields of $O_2\ ^1\Delta_g$, *J. Phys. Chem.*, **85**, 3079–82.

TANENBAUM, L. and TUFFANELLI, D. L., 1980, Antimalarial agents; chloroquine, hydroxychloroquine and quinacrine, *Arch. Dermatol.*, **116**, 587–91.

TAYLOR, R. B., MOODY, R. R., OCHEKPE, N. A., LOW, A. S. and HARPER, M. I. A., 1990, A chemical stability study of proguanil hydrochloride, *Int. J. Pharm.*, **60**, 185–90.

TØNNESEN, H. H. and GRISLINGAAS, A.-L., 1990, Photochemical stability of biologically active compounds. II. Photochemical decomposition of mefloquine in water, *Int. J. Pharm.*, **60**, 157–62.

TØNNESEN, H. H., GRISLINGAAS, A.-L., WOO, S. O. and KARLSEN, J., 1988, Photochemical stability of antimalarials. I. Hydroxychloroquine, *Int. J. Pharm.*, **43**, 215–19.

TØNNESEN, H. H. and KARLSEN, J., 1995, Photochemical degradation of compounds in drug formulations. III. A discussion of experimental conditions, *PharmEuropa*, **7**, 137–41.

TØNNESEN, H. and H. MOORE, D. E., 1991, Photochemical stability of biologically active compounds. III. Mefloquine as a photosensitizer, *Int. J. Pharm.*, **70**, 95–101.

1993, Photochemical degradation of components in drug formulations, *Pharm. Technol.*, **5**, 27, 28, 30, 32–3.

WEBSTER, L. T., 1990, Drugs used in chemotherapy of protozoal infections, in Goodman, A., Gilman, XX., Rall, T. W., Nies, A. S. and Taylor, P. (eds), *The Pharmacological Basis of Therapeutics*, 8th edn, pp. 978–1007. New York: Pergamon Press.

13

Inconsistencies and Deficiencies which exist in the Current Official Regulations concerning the Photolytic Degradation of Drugs

J. C. HUNG

13.1 Introduction

Photolytic degradation is an important limiting factor in the stability of drugs that are sensitive to light. Some light-sensitive drugs are rapidly affected, either by nature's light (especially ultraviolet light) or artificial light (e.g., fluorescent light) and become discoloured or cloudy in appearance, or develop precipitates, while others may slowly undergo photodegradation which may not be visually apparent.

In a photochemical reaction, the light-sensitive drug molecules may be affected directly or indirectly by light, depending upon how the absorbing photon energy is transferred to the drug molecules. With either a direct or indirect light-induced reaction, a drug can only undergo the photodegradation process if the absorbed energy exceeds a threshold. Since ultraviolet radiation has a higher energy level, it is the main cause of many degradation reactions of light-sensitive drugs or drug products. Coloured-glass containers are the most commonly used method to protect light-sensitive drug formulations. Yellow-green glass gives the best protection in the ultraviolet region, whereas amber glass also offers considerable protection from ultraviolet light, but little protection from infrared light.

The photochemical reaction is a very complex process; many variables may be involved in the photolytic degradation kinetics. The velocity of the photochemical reaction may be affected not only by the light source, intensity, and wavelength of the light, but also by the size, shape, composition, and colour of the container. To properly determine the effects (either deleterious or beneficial effects) of light on the quality of a drug substance or drug product, there must be a standard light-stability testing which considers all of the aforementioned variables. Once uniform standard light-stability testing procedures are instituted, one can then establish the proper packaging, storage environment, and expiration date for the light-sensitive drug substance or drug products.

13.2 Standard light-stability testing

Unfortunately, there are no standard procedures, originating either from the United States Pharmacopeia 23 and National Formulary 18 (USP 23/NF 18) or the US Food and Drug Administration (US FDA), for evaluating the photosensitivity of pharmaceuticals. The USP 23/NF 18 lists *Light Transmission* limits for testing light-resistant containers to ensure the proper protection of light-sensitive drugs (Light Transmission, 1995); however, light-stability testing for drugs is not mentioned. The official point of view from the United States Pharmacopeial Convention, Inc. (USPC), is that 'it is not the intent of the USP to give specific guidelines for determining sensitivity to light for all compendial drugs and drug products' (Paul, 1992), and the USPC feels that 'the responsibility for determining the light stability of a particular product belongs to the manufacturer' (Paul, 1992). The USPC further states that there are no specific guidelines in the USP/NF for testing air, pH, moisture, trace metals, and commonly used excipients or solvents on the active ingredient(s), and yet the manufacturer must determine the effects of these conditions (Paul, 1992). Therefore, the USPC thinks that the primary goal of the USP/NF is to provide standards of the listed drugs and not to give specific assay and test procedures for determining compliance with every USP/NF standard; the USPC also believes the obligation for determining the effect of the above-mentioned conditions on a drug's stability, including light sensitivity, is that of the pharmaceutical manufacturer (Paul, 1992). (The USP 23/NF 18 does have test and assay procedures for pH and trace metals such as heavy metals, iron and lead, see 'General Tests and Assays' section of the USP 23/NF 18.)

There is no question that each manufacturer is responsible for determining the light stability, pH, moisture, and so on, for a given drug or drug product. However, because there are no standard evaluation procedures or specifications, different drug manufacturers may evaluate the effect of conditions (e.g. light) differently and may use varying guidelines for judging the results. As an example, if a drug or drug product is packaged in an opaque carton, and it takes three months at 200-foot candles before it discolours, should it carry a 'Protect from light' caution? If another drug or drug product is packaged in an amber glass container and loses 15 per cent potency in three days at 200-foot candles, is 'Protect from light' adequate?

If the drug substance or drug product is susceptible to photolytic degradation, the official USP 23/NF 18 monograph for that drug or drug product will carry a cautionary statement, such as 'Protect from light' or 'Preservation in a light-resistant container'. It is puzzling that if there is no standard or generally accepted light-stability testing method available, by what means did the USPC determine the light sensitivity for the drugs and drug products listed in the USP 23/NF 18?

The USPC feels that the US FDA 'should be contacted for guidance and direction with regard to stability testing' since 'the definition of stability testing is an important part of good manufacturing practice regulations used by FDA' (Paul, 1992). The US FDA is currently working with other countries in an effort to harmonize the standards and regulations among the major pharmaceutical markets (US, Europe and Japan) (Federal Register, 1994). Although this notice on the *International Harmonization; Draft Policy on Standards; Availability* (Federal Register, 1994) does not specifically address the requirements of light-stability testing, it does show that the US FDA continues to participate in international standards activity and is working to harmonize its regulatory requirements with other governments to minimize or eliminate any inconsistent international standard in order to facilitate the availability of safer, more effective, and higher-quality products.

The US FDA does have official guidelines for the evaluation of drug stability (Design and Interpretation of Stability Studies, 1987). One of the required stability assessments includes testing under stressed conditions, e.g., exposure of the drug to various wavelengths of electromagnetic radiation (e.g., 190–780 nm, ultraviolet and visible ranges) on a bulk drug substance. However, there are no details regarding the device and procedures that are officially recognized (e.g., light-source instruments, length of exposure, measurement of discoloration, the intensity of the light source, and the surrounding temperature conditions).

To ensure better and more uniform drug quality, it is critical for both the USP 23/NF 18 and the US FDA to establish standards for testing the effect of light on drug stability. The light type and intensity standards to be used for such testing must be established. Once set, different levels of increasing or decreasing sensitivity to light must also be assigned using duration of exposure, in a manner similar to that which is used for temperature. Since heat from the light source may accelerate the decomposition rate of a drug, an aluminium-foil-wrapped blanket should be used and heat drug sample decomposition should then be applied if necessary. Other recommended specifications and procedures for the determination of the drug's photochemical decomposition stated in this proceedings should also be considered and adopted in the official pharmacopoeia and regulations.

13.3 Official regulations for light-sensitive drugs

Currently, in the US, the monographs in the USP 23/NF 18 and the package inserts for drugs are the only two official resources requiring cautionary statements regarding light sensitivity for drugs or drug products. However, there are many inconsistencies and ambiguities in the legal requirements for the definitions, testing, packaging, storing, and warning labels in the USP 23/NF 18 and between the USP 23/NF 18 and the package inserts.

13.3.1 The deficiencies and inconsistencies in the USP 23/NF 18

Opaque covering. According to the USP 23/NF 18, a light-resistant container protects light-sensitive drug contents from the effects of light by virtue of the specific properties of the material of which it is composed, including any light-resistant coating applied to it (Preservation, Packaging, Storage, and Labeling, 1995). A container intended to provide protection from light or resistant to light must meet the requirements for Light Transmission (1995). A clear and colourless or a translucent container that is made light-resistant by means of an opaque covering is exempt from the requirements for Light Transmission (1995).

Since a container as defined in USP 23/NF 18 is not necessarily in direct contact with the drug (Preservation, Packaging, Storage, and Labeling, 1995), the opaque enclosure would be considered a light-resistant container on the basis of the current USP 23/NF 18 definition. This implies that any drug substance or drug product prone to photolytic degradation can be packaged and stored in a clear and colourless or a translucent container (e.g., ampoule, serum vial or bottle) and maintained in a light-resistant carton or box, thereby fulfilling USP 23/NF 18 requirements (Preservation, Packaging, Storage, and Labeling, 1995). However, the USP 23/NF 18 is not clear as to whether the so-called light-resistant paper carton or box is required to meet the Light Transmission limits (1995). If it is required for such testing, the current testing procedure for Light Transmission (1995) does not have the standard preparation, procedure and limits for the light-resistant paper container.

The USP 23/NF 18 does state that 'Alternatively, a clear and colorless or a translucent container that is made light-resistant by means of an opaque enclosure is exempt from the requirements for light transmission'. Does this mean that the opaque enclosure which is supposed to provide protection and/or resistance from light effect is also exempt from Light Transmission testing in Light Transmission (1995)? The logical thinking would be that a clear and colourless or a translucent container can be exempt from the requirements for Light Transmission (1995) if it is made light resistant by using an opaque enclosure such as a paper carton or a box. However, in order for the paper carton or box to claim light protection or resistance, it must be subject to the same testing procedures and requirements as stated in Light Transmission (1995). Therefore, the USPC must update the Light Transmission testing section (1995) to include the preparation of specimen, procedure, and limits for the 'light-resistant' paper material.

Cautionary statement. When an opaque covering is used to make a clear and colourless or a translucent container light resistant, the USP 23/NF 18 requires '. . . the label of the container bears a statement that the opaque covering is needed until the contents are to be used or administered'. (Preservation, Packaging, Storage, and Labeling, 1995.) The USP 23/NF 18 is not clear which container (i.e., immediate container, outer container or both) should have a cautionary statement for light sensitivity. It would be appropriate to have such a label on both the immediate and outer containers in order to be consistent and thorough to alert the re-packagers or end users regarding the proper storage of the light-sensitive drug or drug products.

For any drugs that may be susceptible to photodegradation, the requirement in the USP 23/NF 18 for the proper packaging and storage of light-sensitive drugs reads, 'Preserve in light-resistant containers' or 'Protect from light'. The USP 23/NF 18 further indicates that preservation in a light-resistant container is intended when the instruction to 'Protect from light' is given in an official monograph for any light-sensitive drug (Preservation, Packaging, Storage, and Labeling, 1995). It appears that these two directions for the packaging and storage of light-sensitive drugs are interchangeable in the USP 23/NF 18. However, it seems that one uniform statement would be more appropriate because it would be help to avoid possible confusion and misinterpretation. Whenever one sees 'Protect from light' in an individual official monograph, either in the USP 23/NF 18 or in the package insert, the manufacturer must assure that the containers meet Light Transmission standards (1995), and the ultimate dispenser or the re-packager needs to be aware that the original package must be retained, or a suitable alternative must be used for proper protection of light-sensitive drugs or drug products.

The other alternative for the cautionary labelling of the light-sensitive drug or drug product is the use of symbols to show light sensitivity, for example, with a numeral, such as 0, 1, 2, and so on, to show the level of light sensitivity. Symbols are much more likely to be observed and followed than words. Such use of symbols has been adopted worldwide for common instruction and for the rising non-English-speaking population of the US; storage and cautionary symbols do obviously meet a practical need.

Light-resistant container for single-use drug. Under the Light-resistant Container section of the USP 23/NF 18 (Preservation, Packaging, Storage, and Labeling, 1995), 'Where an article is required to be packaged in a light-resistant container and if the container is made light resistant by means of an opaque covering, a single-use, unit-dose container or mnemonic pack for dispensing may not be removed from the outer opaque covering prior to dispensing'. (This statement seems to suggest that single-use unit-dose drugs or drugs packaged in a mnemonic pack which are susceptible to photolytic degradation must be stored in a clear and colourless or a translucent container, using an opaque covering for light protection or resistance. However, the intention for such a requirement is not clear.) Does this statement suggest that under the USP 23/NF 18 standards, a drug stored in a single-use unit-dose container or a mnemonic pack which is susceptible to light degradation cannot be removed from the outer opaque covering at any given time? Would it be more appropriate to store such a light-sensitive drug in an immediate container which is light resistant, rather than depending upon the secondary light protection or resistance achieved through the use of an outer opaque covering?

In addition, it is not clear why the USP 23/NF 18 does not have the same restriction for the single-dose container, which is designed for drugs intended for parenteral administration only (Preservation, Packaging, Storage, and Labeling, 1995). For parenteral drugs, it would seem to be more appropriate to store the parenteral drug solution injection in a clear and colourless pre-filled syringe so that proper visual inspection can be performed to observe any colour change, clouding appearance or particulate matter in the drug solution. This type of drug would be more appropriately stored in a clear and colourless or translucent container which would require an opaque enclosure to offer light protection or resistance.

Light exposure. Certain circumstances, such as compounding and/or dispensing of light-sensitive drugs or drug products in a light environment would certainly not be in accordance with full compliance of the requirements which state that the outer covering is not to be removed and discarded until the contents have been used (Preservation, Packaging, Storage, and Labeling, 1995). Since the level of light sensitivity has never been defined, it is uncertain whether a light-sensitive drug substance or drug product can be exposed to light even for a short period of time during visual inspection, re-packaging, compounding or dispensing. In these situations, one has to apply appropriate light-protection or resistance measures (e.g., dim the light resource, or use foil pouches to hold the dispensed drug) in order to minimize the photolytic effects. Short-term spikes due to opening of outer light-resistant container (e.g., paper carton or box) should be viewed as unavoidable and should be acceptable unless the drug substance or drug products are extremely sensitive to light exposure.

13.3.2 USP 23/NF 18 vs package inserts

Light-sensitive cold kits and radiopharmaceuticals. The USP 23/NF 18 is sometimes not in agreement with the light sensitivity cautionary statement listed in the package insert. As my primary work is specialization in nuclear pharmacy, the examples that I will mention in this paper will be strictly related to the radiopharmaceuticals which are used in nuclear pharmacy and nuclear medicine. There are several cold kits and radiopharmaceuticals that are sensitive to light exposure (Table 13.1).

Although the package insert does include a cautionary statement regarding light sensitivity for Microlite® (kit for the preparation of technetium Tc 99 m albumin colloid injection) (Microlite®, 1995), the monograph for technetium Tc 99 m albumin colloid injection in the USP 23/NF 18 fails to mention the light sensitivity (Technetium Tc 99 m albumin colloid injection, 1995). In addition, the

Table 13.1 Light-sensitive cold kits

Cold kit name	Name of radiopharmaceuticals	Manufacturer
DMSA	Technetium Tc 99m succimer (DMSA) injection	Medi-Physics, Inc., Arlington Heights, Illinois, USA
Hepatolite®	Technetium Tc 99m disofenin injection	DuPont Merck Pharmaceutical Co., Billerica, Massachusetts, USA
Microlite®	Technetium Tc 99m albumin colloid injection	DuPont Merck Pharmaceutical Co., Billerica, Massachusetts, USA
TechneScan MAG3®	Technetium Tc 99m mertiatide (MAG3) injection	Mallinckrodt Medical, Inc., St Louis, Missouri, USA
Syringe I[a] of UltraTag® RBC	Technetium Tc 99m labelled red blood cells	Mallinckrodt Medical, Inc., St Louis, Missouri, USA
Reaction vial[b] of OctreoScan®	Indium In-111 pentetreotide	Mallinckrodt Medical, Inc., St Louis, Missouri, USA
Vial A[c] of Neurolite®	Technetium Tc 99m bicisate injection	DuPont Merck Pharmaceutical Co., Billerica, Massachusetts, USA

Notes:
[a]Syringe I contains 0.6 ml of 0.6 mg NaOCl.
[b]Reaction vial contains 10.0 μg pentetreotide, 2.0 mg gentisic acid, 4.9 mg sodium citrate, anhydrous, 0.37 mg citric acid, anhydrous, and 10.0 mg inositol.
[c]Vial A contains 0.9 mg bicisate dihydrochloride (ECD·2HCl), 24 mg mannitol, 0.36 mg edetate sodium, dihydrate, and 72 μg $SnCl_2 \cdot 2H_2O$.

monographs for technetium Tc 99 m disofenin injection in the newly revised USP 23/NF 18 (Technetium Tc 99 m disofenin injection, 1995) does not claim light sensitivity for technetium Tc 99 m disofenin injection as is required by the package insert (Hepatolite®, 1995).

Another commonly observed deficiency of the USP 23/NF 18 is that it still does not contain a complete list of some of the most commonly used technetium Tc 99 m labelled radiopharmaceuticals that are light sensitive (i.e., TechneScan MAG3® for the preparation of technetium Tc 99 m mertiatide injection; syringe I of UltraTag® RBC for the preparation of technetium Tc 99 m labelled red blood cells; and vial A of Neurolite® for the preparation of technetium Tc 99 m bicisate injection). The Neurolite® cold kit was approved by the US FDA on November 28, 1994, for the preparation of technetium Tc 99 m bicisate injection to be used for the localization of stroke (Neurolite®, 1994). The USP 23/NF 18, which was recently revised (published in July 1994 and became official as of January 1, 1995), was not able to include this drug. However, both the TechneScan MAG3® and UltraTag® RBC kits were approved by the US FDA a few years ago (i.e., June 5, 1990, for TechneScan MAG3® and June 10, 1991, for UltraTag® RBC). Thus, it is not clear why the newly revised USP 23/NF 18 does not include these two legend drugs. So far, USP 23/NF 18 only lists one technetium Tc 99 m labelled radiopharmaceutical that is light sensitive, which is technetium Tc 99 m succimer injection (Technetium Tc 99 m succimer injection, 1995).

Interestingly, the product package insert for Choletec® (kit for the preparation of technetium Tc 99 m mebrofenin injection) (Choletec®, 1994), which is an analogue of Hepatolite® (kit for the preparation of technetium Tc 99 m disofenin injection), records no cautionary statement regarding light sensitivity either in the USP 23/NF 18 or in the package insert for technetium Tc 99 m mebrofenin injection (Choletec®, 1994).

Light-sensitive daughter drug. It is common practice in the nuclear pharmacy field to use so-called cold kits (e.g., TechneScan MAG3®, Hepatolite®) to prepare radiopharmaceuticals. In this situation, the cold kit can be referred to as the 'parent drug', whereas the radiopharmaceutical that is prepared with the use of the cold kit can be called the 'daughter drug'. Since the USP 23/NF 18 lists the radioactive drugs in names of the finished radiopharmaceutical products, it would therefore have the requirements and specifications for the 'daughter drugs'. On the other hand, the package inserts for the cold kits usually contain the information not only for the 'parent drugs' (i.e., cold kits), but also for the 'daughter drugs' (i.e., labelled radiopharmaceuticals). However, there is also no additional information which could be determined from the package inserts of the light-sensitive 'parent drugs' (DMSA, 1993; Hepatolite®, 1991; Microlite®, 1991; TechneScan MAG3®, 1992; UltraTag® RBC, 1992; Neurolite®, 1994), regarding whether a 'daughter drug' is also susceptible to light-induced decomposition.

If the drug product in the cold kit is light sensitive, is the radiopharmaceutical prepared from the light-sensitive cold kit also prone to photolytic degradation? What is the time period limit for light exposure which should apply to this type of radiopharmaceutical which is prepared from a light-sensitive 'parent drug'?

The only 'parent and daughter drugs' that the USP 23/NF 18 and the package insert claim light sensitivity is DMSA (kit for the preparation of technetium Tc 99 m succimer injection) (DMSA, 1993; Technetium Tc 99 m succimer injection, 1995).

13.3.3 The deficiencies of package inserts

The package insert for a drug is important official information regarding the drug's description, clinical pharmacology and indications, for example. The light sensitivity statement is usually listed under the 'How Supplied' section or under the 'Description' section of the package insert.

It is interesting to note that although the final language stated in the package insert of the drug must be approved by the US FDA; yet, the guidance and directions with regard to the light-stability and light-sensitivity standards have not been addressed by the US FDA.

Betiatide, which is the main ingredient in the TechneScan® MAG3 reaction vial, is light sensitive, and therefore, must be protected from light (TechneScan MAG3®, 1992). The precautionary statement regarding avoidance of exposure to light can be found in the package insert, on the vial label, and on the label of the cold kit package box containing the US TechneScan MAG3® kit.

However, the warning statement was absent in the original European product packaging (TechneScan® MAG3, 1990). The revised package insert for the European TechneScan® MAG3 (TechneScan® MAG3, 1992) does include a relatively general cautionary statement: TechneScan® MAG3 is to be stored at 2–8 °C in the dark. The language used in this cautionary statement for the European TechneScan® MAG3 is different from the ones used in the USA 'Preserve in light-resistant containers' or 'Protect from light'.

In addition, although the package insert for the US product does include the cautionary statement for light sensitivity, it does not specify the light-sensitivity level of the drug.

The OctreoScan® cold kit was also recently approved by the US FDA for the preparation of indium In-111 pentetreotide for imaging primary and metastatic neuroendocrine tumours bearing somatostatin receptors (OctreoScan®, 1994). Both the OctreoScan® reaction vial pack label and the outer storage carton box bear the cautionary statement 'Protect from light'. Since the label of the 10 ml vial for $^{111}InCl_3$ solution does not contain any cautionary statement regarding light sensitivity, it would indicate that the contents in the OctreoScan® reaction vial (Table 13.1) are light sensitive. Although the reaction vial is packaged appropriately, it is interesting to note that the package insert for the OctreoScan® kit does not include any warning statement regarding light sensitivity of the contents within the reaction vial of the OctreoScan® cold kit (OctreoScan®, 1994).

13.4 The labelling and packaging of light-sensitive cold kits

13.4.1 Labelling light-sensitive cold kits

Most of the light-sensitive cold kits listed in Table 13.1 do comply with the USP 23/NF 18 requirement that the labels on both the immediate and outer containers must state 'Protect from light' (Preservation, Packaging, Storage, and Labeling, 1995). However, there are two light-sensitive cold kits that fail to fully comply with the USP 23/NF 18 labelling requirement (Preservation, Packaging, Storage, and Labeling, 1995). Those two kit formulations are Hepatolite® and Neurolite®.

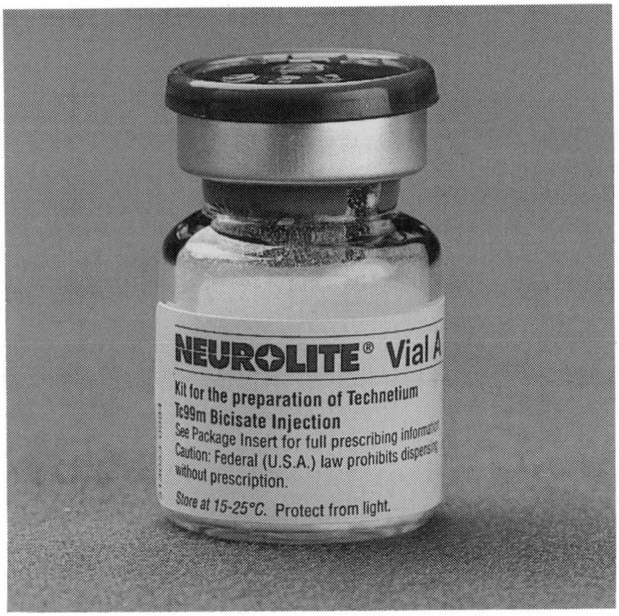

Figure 13.1 Vial A of the Neurolite® cold kits. 'Protect from light' is stated on the vial label (lower right line)

Although the Hepatolite® reaction vial states 'Protect from light' on the label, the outer cold kit box contains no warning statement regarding light sensitivity.

The other cold kit formulation that violates the USP 23/NF 18 labelling requirement (Preservation, Packaging, Storage, and Labeling, 1995) is the Neurolite® kit containing two sets of dual vials (i.e., vials A and B) cold kits (Neurolite®, 1994). Although the label on vial A in the Neurolite® cold kit does state 'Protect from light' (Fig. 13.1), the storage carton containing the light-sensitive vial A in the Neurolite® kit contains no cautionary statement regarding light sensitivity (Fig. 13.2). Because of the light sensitivity of vial A's contents, each of the two sets of Neurolite® cold kits must be kept inside the carton during its storage. The storage carton cannot be removed and discarded until both sets of Neurolite® cold kits have been used.

13.4.2 *Packaging light-sensitive cold kits*

It is common practice for the cold kit manufacturers to package the light-sensitive drugs or drug products in clear and colourless glass vials or syringes. Although this is contrary to the standard practice of protecting light-sensitive drugs in light-resistant (e.g., amber, yellow-green or blue) containers, a clear and colourless or a translucent container does offer some advantages over a coloured container.

Coloured-glass or coloured-plastic containers cost an average of approximately 25 per cent more than the clear and colourless or translucent containers. This may

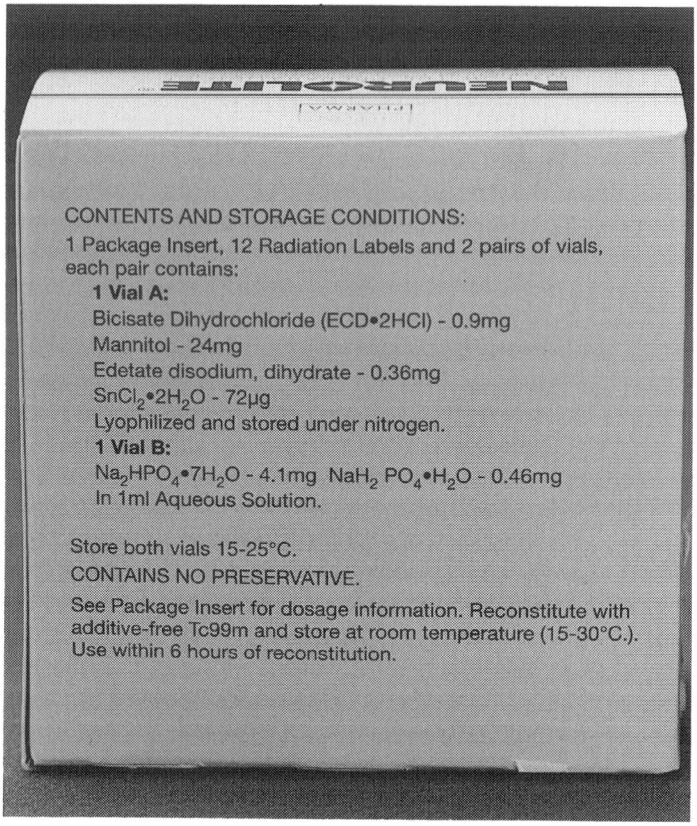

Figure 13.2 The back side of the outer box for the Neurolite® cold kits. No cautionary statement for light sensitivity can be found on the labels of the outer box, except that both vials of Neurolite® and the reconstituted radiopharmaceutical (i.e., technetium Tc 99m bicisate injection) must be stored at 15–25 °C and 15–30 °C, respectively

be due to the more costly material which they are composed of, including any applied coating material in order to make the containers light resistant. The higher cost of the light-resistant containers may also be due to the Light Transmission testing which is required by the USP 23/NF 18 (Light Transmission, 1995). The coloured containers must pass this test in order to qualify for use as light-resistant containers, whereas the clear and colourless or translucent container can be exempt from the requirements of Light Transmission testing if it is made light-resistant by an opaque enclosure (Preservation, Packaging, Storage, and Labeling, 1995).

In addition, with coloured glass or plastic, it is virtually impossible to observe a colour change in a drug formulation and it is difficult to examine injectable drug solution for particulate matter. Therefore, the injectable products raise the most difficult container selection problem for manufacturers because of the need to provide light protection (for light-sensitive drugs or drug products) as well as to allow for visual examination. Often the final inspection is made at the point of use by the dispensers. Many companies, in their attempt to balance these two

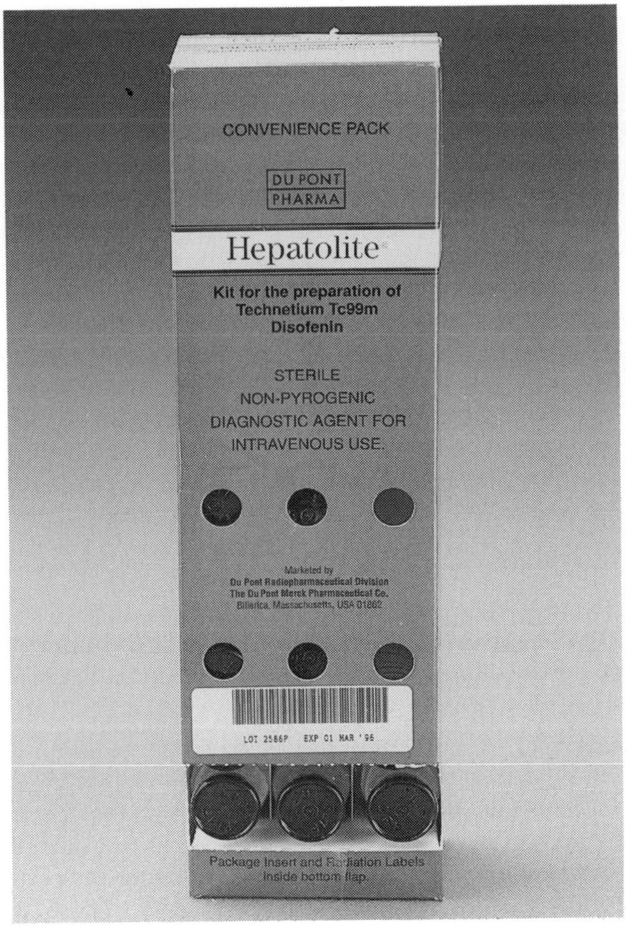

Figure 13.3 The convenience pack for Hepatolite® cold kits

factors, package their injectable products in clear glass containers placed in foil pouches or paper cartons as a secondary means of light protection.

Although the contents of light-sensitive injectable drugs can be transferred to a clear and colourless container (e.g., vial or syringe) to examine the clarity and particulate matter, this approach is not suitable for cold kit formulation in radiopharmaceuticals. The additional step of transferring the entire radioactive solution contained in the light-resistant vial into a colourless and clear syringe or vial increases the risk for degradation of any oxygen-sensitive radiopharmaceutical. The possibility for increased radiation exposure to personnel during the transfer and visual inspection is another concern. Therefore, the light-sensitive cold kits for the preparation of injectable radiopharmaceuticals should be packaged in a clear and colourless or a translucent container. This clear and colourless or translucent container must be protected by a light-resistant opaque covering such as a paper carton or plastic box in order to comply with the USP 23/NF 18 requirements (Preservation, Packaging, Storage, and Labeling, 1995).

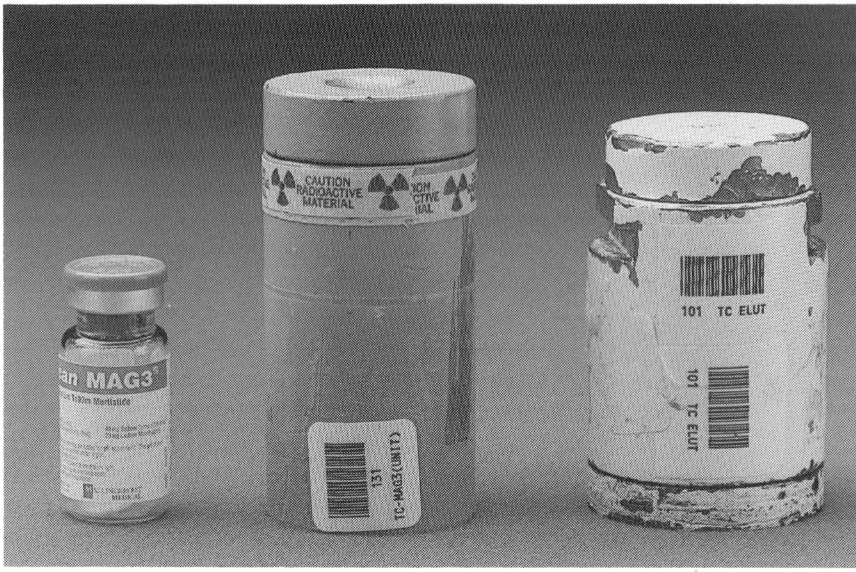

Figure 13.4 The lead container for the storage of technetium Tc 99m labelled radiopharmaceutical (middle) and the lead container for the storage of technetium Tc 99m pertechnetate injection (right) for the radiolabelling of the cold kit (left)

It is also interesting to note that the Hepatolite® reaction vials are stored in a so-called 'convenience pack' carton. This 30-vial convenience pack for Hepatolite® has six 12-mm diameter circular openings for checking the reorder point of the cold kit supply without the need to open the outer box (Fig. 13.3). In addition, the convenience pack also has an 11 × 3 cm round rectangular opening for retrieval of the Hepatolite® cold kits (Fig. 13.3). These openings on the Hepatolite® carton do not provide an adequate light-resistant environment for the Hepatolite® cold kits, which are subject to photolytic degradation. The manufacturer should modify the storage box for the Hepatolite® cold kits in order to provide proper protection for the light-sensitive kit formulation. In the meantime, since there is no indication of the light-sensitive level for the Hepatolite® cold kit, the openings on the Hepatolite® convenience pack should be properly covered to prevent light exposure of the cold kits.

Once the radiopharmaceutical is prepared from a light-sensitive cold kit, it is stored in a lead container to shield from radiation (Fig. 13.4). As shown in Fig. 13.4, the lead container completely encloses the radioactive vial inside the container, offering excellent protection from light exposure for the potentially light-sensitive radioactive drugs. Once the radiopharmaceutical is withdrawn into a syringe, the syringe, except the needle and a portion of the plunger, is completely surrounded with a lead syringe shield (Fig. 13.5). The use of a lead syringe shield for radioactive drugs not only reduces radiation exposure to personnel, but also provides good protection from light exposure for the potentially light-sensitive radiopharmaceuticals. However, it remains unknown whether the lead glass or acrylic that is used in syringes and/or vial shields (most notably the 360° clear-view syringe/vial shields that are usually light yellow in colour) (Fig. 13.5) would meet USP 23/NF 18 Light Transmission testing requirements (1995) for providing

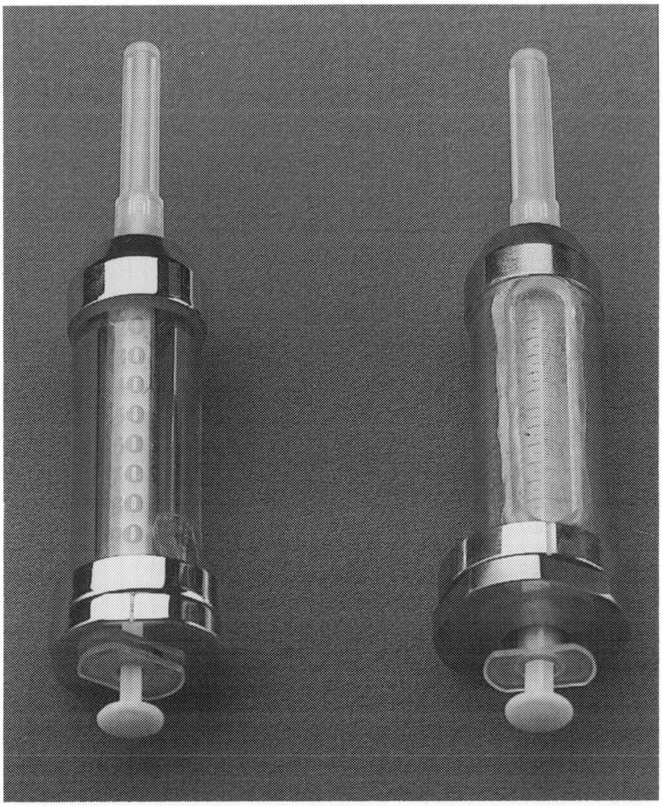

Figure 13.5 Right: lead syringe shield with the lead glass viewing panel; left: lead glass syringe shield with 360° syringe visibility

adequate protection of light-sensitive radiopharmaceuticals from light-exposure degradation.

13.5 Recommendations for the official regulations for light-sensitive drugs

As stated in the Constitution and By-Laws (1995) of the USPC, the primary purpose of the USP 23/NF 18 is '. . . to provide authoritative standards and specifications for materials and substances and their preparations that are used in the practice of the healing arts; they establish titles, definitions, descriptions, and standards for identity, quality, strength, purity, packaging, and labelling and, where practicable, bioavailability, stability, procedures for proper handling and storage, and methods for their examination and formulas for their manufacture or preparation' (Constitution and By-Laws, 1995). It is clear that the main objectives of the USP 23/NF 18 are not only to establish the standards with which the listed drugs or drug products must comply, but also to give specific procedures and guidelines for determining the standards. The USP 23/NF 18 contains a section entitled 'General Tests and Assays' (1995), which provides specific guidelines and information for the performance of certain testing, such as sterility tests, oxygen determination, measurement of pH and so on (General Tests and Assays, 1995). Since light sensitivity is an important factor in the storage of drugs or drug products that are prone to photolytic degradation, it is absolutely necessary to include light-stability testing in the USP/NF.

The Stability Testing of New Drug Substances and Products – the Tripartit Guideline in the USP 23/NF 18 states that light testing should be an integral part of stress testing for drug substances (Stability Testing of New Drug Substances and Products – the Tripartit Guideline, 1995). It also mentions that this guideline has been developed within the Expert Working Group (Quality) of the International Conference on Harmonization (ICH), and it further states that the Expert Working Group is trying to define and standardize the conditions for light-stability testing of active substance and dosage forms (Stability Testing of New Drug Substances and Products – the Tripartit Guideline, 1995). It is my sincere hope that once the Expert Working Group of the ICH has agreed upon a standard light-stability testing, the USPC should include that information in the USP/NF.

The USPC should try to work with the manufacturers and the FDA in order to update their official monographs for drugs and drug products and to standardize the requirements, such as packaging and storage, labelling and standard testings. Up-to-date and consistent official requirements among the USP/NF, US FDA and the package inserts would ensure better compliance from the manufacturers, repackagers, and end users. The other suggestions to enhance compliance are to use uniform cautionary statements, such as 'Protect from light', and/or through the use of symbols (e.g., (○)).

Once the standard light-stability testing is set, using duration of light exposure, the USPC can assign several levels of increasing or decreasing sensitivity to light in a manner similar to that which is used for temperature and humidity (Stability Testing of New Drug Substances and Products – the Tripartit Guideline, 1995). The loss of specific drug potency, any specified degradant exceeding the drug's specification limit, or a drug or drug product exceeding its pH limit can be used as a gauge to determine the shelf-life of the light-sensitive drug or drug products. The USP can then use a combination of symbols (e.g., (☼)) and numerals, such as 0, 1 and 2, to show the light sensitivity and the level of the light sensitivity for the drugs and drug products. Depending on the levels of light sensitivity, the drug manufacturer can use proper precautions to protect the light-sensitive drug from light exposure which occurs during compounding until filling into the final container and can select the proper light-resistant container to store the light-sensitive drugs or drug products. The end user and repackager need only to be concerned with drugs or drug products which may become unacceptable for use within the time period limit permitted after the container is first opened or entered.

The USPC should define the parameters for storage conditions of light-sensitive drugs or drug products. Depending upon the light-sensitivity levels of the drugs or drug products, is it admissible to allow short-term spikes due to opening the light-resistant outer carton for light-sensitive drugs or drug products stored in a clear and colourless or a translucent container? If the drug or drug products is/are very sensitive to light exposure, is it advisable to draw up the solution into a clear and colourless or translucent syringe, thereby causing possible degradation of the drug material? Should the short-term spikes be considered as unavoidable and acceptable for practical reasons? With regard to the time limits for exposure to different light sources, such as ultraviolet light or fluorescent light, it is important for the end user to exercise special caution for the storage, transfer or compounding of light-sensitive drugs or drug products.

The USPC should modify the definitions for light-resistant container for single-use drug so that the definition and requirements have more practical meaning. A clear and colourless or a translucent container has several advantages over the light-resistant coloured container. It is less expensive to manufacture (i.e., approximately 25 per cent cheaper), it is exempt from meeting the light transmission testing requirements as stated in USP 23/NF 18 (Light Transmission, 1995) and it enables visual inspection of the drug or drug products to examine for colour change, clouding appearance, or precipitate formation. If the light-sensitive drug or drug products must be stored in a light-resistant coloured container (e.g., amber vial) to offer assurance of better protection from light sensitivity, should the cold kits for the preparation of the radioactive drugs and the single-dose syringes containing the light-sensitive drugs or drug products be exempted and allowed to be stored in a clear and colourless or a translucent container? Although the contents of light-sensitive cold kits and radiopharmaceuticals can be transferred to a clear and colourless or a translucent container (e.g., vial or syringe) to examine the clarity and particulate matter, this approach is not suitable for cold kit formulations and radiopharmaceuticals due to the possible oxidation effects on these drugs and the additional radiation exposure to personnel during the transfer. The reason for storing the injectable drugs in a pre-drawn single-dose syringe is the ease and convenience of the drug administration. If the drug content in the pre-drawn syringe must be transferred to another clear and colourless container for the final inspection, this would defeat the purpose of the original idea for using the single-use syringe.

The USPC should modify its Light Transmission testing to include procedures and standard limits concerning the use of the light-resistant paper carton or box. The labelling requirements for the secondary or altered light-resistant container, such as a carton or box, must be identical to the labels on the immediate container for the light-sensitive drugs or drug products. There should be no openings on the outer light-resistant box or carton in order to provide an adequate light-resistant environment for those drugs or drug products which are susceptible to photolytic degradation. For a light-resistant outer container which is used to store multiple components of a drug or drug products, if even only one single component of the drug is light sensitive, the label on the outer container is still required to bear a cautionary statement for light sensitivity. This is to ensure that the end user or repackager will be alerted to ensure the proper protection of the light-sensitive drug component.

Acknowledgements

The author would like to thank Ms. Vicki S. Krage for her patience and assistance in the preparation of this paper. This paper was presented in part in the following three articles: Hung, J. C. (1992), Photolytic degradation of drugs, *Am. J. Hosp. Pharm.*, **49**, 2704–5; Hung, J. C. (1993), Photochemical considerations of light-sensitive cold kits and radiopharmaceuticals, *J. Nucl. Med. Technol.*, **21**, 90–1 and Hung, J. C. (1995), The packaging of light-sensitive cold kits, *J. Nucl. Med. Technol.*, in press.

References

CHOLETEC® package insert, 1994, Squibb Diagnostics, Inc., Princeton, New Jersey, USA.

CONSTITUTION and BY-LAWS, 1995, in *United States Pharmacopeia*, 23rd rev., and *National Formulary*, 18th edn, The United States Pharmacopeial Convention, Inc., Rockville, Maryland, USA, p. xxiii.

DESIGN and INTERPRETATION of STABILITY STUDIES (1987), in *Guidelines for Submitting Documentation for the Stability of Human Drugs and Biologics*. Rockville, Maryland, USA: Food and Drug Administration. pp. 8–9.

DMSA package insert, 1993, Medi-Physics, Inc., Arlington Heights, Illinois, USA.

GENERAL TESTS and ASSAYS, 1995, in *United States Pharmacopeia*, 23rd rev., and *National Formulary*, 18th edn, The United States Pharmacopeial Convention, Inc., Rockville, Maryland, USA, pp. 1648–1986.

HEPATOLITE® package insert, 1991, DuPont Merck Pharmaceutical Co., Billerica, Massachusetts, USA.

HUNG, J. C., 1992, Comparison of technetium-99 m MAG3 kit formulations in Europe and the USA, *Eur. J. Nucl. Med.*, **19**, 990–2. Letter.

INTERNATIONAL HARMONIZATION; Draft Policy on Standards; Availability, 1994, *Federal Register*, **59**, 60870–4.

LIGHT TRANSMISSION, 1995, in *United States Pharmacopeia*, 23rd rev., and *National Formulary*, 18th edn. The United States Pharmacopeial Convention, Inc., Rockville, Maryland, USA, pp. 1481–2.

MICROLITE® package insert, 1991, DuPont Merck Pharmaceutical Co., Billerica, Massachusetts, USA.

NEUROLITE® package insert, 1994, DuPont Merck Pharmaceutical Co., Billerica, Massachusetts, USA.

OCTREOSCAN® package insert, 1994, Mallinckrodt Medical, Inc., St Louis, Missouri, USA.

PAUL, W. L., 1992, Photolytic degradation of drugs (reply), *Am. J. Hosp. Pharm.*, **49**, 2704–5. Letter.

PRESERVATION, PACKAGING, STORAGE, and LABELING, 1995, in *United States Pharmacopeia*, 23rd rev., and *National Formulary*, 18th edn, The United States Pharmacopeial Convention, Inc., Rockville, Maryland, USA, pp. 10–11.

TECHNESCAN® MAG3 Data sheet/Directions for use, 1990, Mallinckrodt Medical B.V., Petten, Netherlands.

1992 Mallinckrodt Medical B.V., Petten, Netherlands.

TECHNESCAN MAG3® package insert, 1992, Mallinckrodt Medical, Inc., St. Louis, Missouri, USA.

TECHNETIUM Tc 99 m albumin colloid injection, 1995, in *United States Pharmacopeia*, 23rd rev., and *National Formulary*, 18th edn, The United States Pharmacopeial Convention, Inc., Rockville, Maryland, USA, pp. 1481–2.

TECHNETIUM Tc 99 m disofenin injection, 1995, in *United States Pharmacopeia*, 23rd rev., and National Formulary, 18th edn, The United States Pharmacopeial Convention, Inc., Rockville, Maryland, USA, p. 1483.

TECHNETIUM Tc 99 m succimer injection, 1995, in *United States Pharmacopeia*, 23rd rev., and *National Formulary*, 18th edn, The United States Pharmacopeial Convention, Inc., Rockville, Maryland, USA, pp. 1488–9.

STABILITY TESTING OF NEW DRUG SUBSTANCES and PRODUCTS – the Tripartit Guideline, 1995, in *United States Pharmacopeia*, 23rd rev., and *National Formulary*, 18th edn, The United States Pharmacopeial Convention, Inc., Rockville, Maryland, USA, pp. 1959–63.

ULTRATAG® RBC package insert, 1992, Mallinckrodt Medical, Inc., St Louis, Missouri, USA.

14

Photostability Testing: Design and Interpretation of Tests on Drug Substances and Dosage Forms

N. H. ANDERSON

14.1 Introduction

Photostability testing is an essential part of product development and is needed to ensure satisfactory product quality is maintained during practical usage. A wide diversity of testing procedures has been employed by the pharmaceutical industry (Anderson *et al.*, 1991) (Table 14.1), which is not surprising given (1) the absence of regulatory guidelines; (2) differing assumptions about practical product exposure (Table 14.2) and (3) the lack of published data on daylight (UV and visible) levels inside buildings. The subject has recently been reviewed (Nema *et al.*, 1995) and the principles have been discussed (Thoma and Kerker, 1992; Tønnesen, 1991).

For unstable products, product quality is achieved through suitable labelling and use of a protective pack. This is in contrast to the situation with thermally unstable products where it is necessary for the manufacturer to determine the shelf-life (i.e. period for which the product meets its quality specifications) under defined storage conditions. Thus, whereas for thermal stability testing it is necessary to determine the *rate* of degradation, for photostability testing it is only necessary to determine if the product photostability is sufficient to make a protective pack and warning label unnecessary: this is equivalent to a limit test. In order to determine the relative photostability of different product formulations or the degree of protection afforded by different pack types, it may be necessary to conduct measurements of rate of photodegradation (Moore, 1987; Moore, 1990; Tønnesen, 1991; Sciano, 1990). Such measurements are outside the scope of this chapter.

The proposals set forth in this chapter are largely based on the work of the International Conference on Harmonization (ICH) Photostability Working Group which was formed to provide an international guideline for photostability testing of drug substances and drug products, as part of the ICH process. The resulting ICH Guideline was published as a Step 2 document on 29 November 1995.

UV irradiance is expressed as $W\,m^{-2}$, UV exposure as irradiance x time ($Wh\,m^{-2}$). Visible light is expressed as illuminance (lux) both for practical

Table 14.1 Drug product photostability testing: UK practice

Light source	
Accel. fluorescent	200–1200 lux
exposure time	30–90 days
total exposure	300–1080 klux days
Room fluorescent	1400–6000 lux
exposure time	4–180 days
total exposure	8–900 klux days
Xenon	150–180 klux
exposure time	0.25–30 days
total visible exposure	45–4500 klux days

Table 14.2 Assumptions about practical product exposure

Products very rarely exposed to significant UV light
Blister pack left close to sunny window for up to 30 days
Accumulative exposure to visible light less than 1000 klux/h
Accumulative exposure to visible light up to 10 000 klux/h

convenience and because indoor lighting is normally measured in lux; visible exposure is then lux hours.

14.2 Objectives of testing – photostability under practical usage conditions

It is extremely difficult to establish the actual exposure of pharmaceutical products during practical usage. During discussions of the international harmonization of photostability testing of pharmaceutical products, it has become apparent that there are differences amongst Europe, the US and Japan. In the US and Japan, product exposure to glass filtered daylight is believed to occur rarely, if at all, either in hospitals or pharmacies. In contrast, in Europe, it is recognized that products in hospital pharmacies or during distribution in hospitals, as well as possibly in the home, may be removed from their secondary (outer) carton and be exposed to glass-filtered daylight, possibly over several days (Tønnesen and Karlsen, 1995). Therefore it is important that the susceptibility of products to degradation by glass-filtered daylight or a source simulating glass-filtered daylight is determined. The degree of visible and UV exposure to which products are exposed is not known with any certainty and therefore any test of photostability is based on informed judgement, not on fact. It is, however, possible to give approximate exposure values corresponding to time of exposure to indoor lighting or very close to a sunny window (Table 14.3). Thus, the results of photostability testing of pharmaceutical products should be regarded as essentially qualitative rather than quantitative.

Table 14.3 Simulation of light levels

Exposure (klux h)	Days on sunny window sill	Days 1 m from window	Days under artificial light at 500 lux (24 hr day)
1200	2–3	ca. 12	100
2400	4–6	ca. 24	200
3600	8–12	ca. 36	300
UV exposure $Wh\,m^{-2}$			
200	1–2	ca. 50	
500	3–5	ca. 125	

14.3 Forced photodegradation of drug substance/method validation

In order to develop valid methods for determining the photostability of drug substances and products a forced photodegradation study on the drug substance should be performed. It is easier to develop and largely validate methods for the product photodegradation using the drug substance. The 'forced degradation' experiment on the drug substance should be conducted using a visible light and UV exposure in excess of that used for formal product testing (e.g. by a factor of three- or fivefold) or for exposure to be continued until significant degradation (up to 20 per cent) has occurred. These experiments are usually conducted on both solid drug substance and a solution. The latter is more susceptible to degradation and is particularly relevant to the understanding of the behaviour of solution products. If no degradation is observed, no formal testing of drug substance photostability is required.

According to the Step 4 ICH Guideline on drug substance impurities (ICH, 1995) and the Step 2 ICH Guideline on drug product impurities it is necessary to be able to quantify impurities present in a commercial drug substance or product as low as the 0.1 per cent level (substance and low potency products) and at least to the 1.0 per cent level (high potency products). Thus, the methods developed should be capable of quantifying photodegradants which could be formed at, or above, the 0.1 per cent level in the formal tests. Because the structure of degradants is usually unknown they are estimated by peak area relative to the drug substance using HPLC with UV detection, assuming a response factor of 1 relative to drug substance. It is unnecessary to identify degradants below the Qualification levels defined in Impurities guidelines. Practical details of testing are given below (Section 14.5).

14.4 Formal test for drug substance photostability

The ICH Photostability Guideline recommends that drug substance photostability testing be carried out under the same conditions as those used for product, in order for direct comparisons to be made: the recommended exposure is 1.2 million lux hours and 200 $Wh\,m^{-2}$ UV. During purification and manufacture of the drug substance, total exposure of the substance is extremely unlikely to exceed 100 klux hours of visible light, with no UV exposure and the European Federation of Pharmaceutical Industries (EFPIA) Expert Working Group agreed informally that 100 klux hours' exposure was appropriate for the simulation of exposure during manufacture. Although such a test is not part of the ICH guideline, it may be useful for internal control purposes.

14.5 Drug substance sample presentation, analysis and judgement of results

The following text is taken from the ICH Step 2 guideline.

Sample Presentation

Care should be taken to ensure that the physical characteristics of the samples under test are taken into account and efforts should be made, such as cooling and/or placing the samples in sealed containers, to ensure that the effects of the changes in physical states such as sublimation, evaporation or melting are minimised. All such precautions should be chosen to provide the minimal interference with the irradiation of samples under test. Possible interactions between the samples and any material used for containers or for general protection of the sample should also be considered and eliminated wherever not relevant to the test being carried out.

As a direct challenge for samples of solid drug substance, an appropriate amount of sample should be taken and placed in a suitable glass or plastic dish and protected with a suitable transparent cover, if considered necessary. Solid drug substances, with a particle size distribution representative of material as released for use, should be spread across the container to give a thickness of typically not more than 3 mm. Drug substances which are liquids should be exposed in chemically inert and transparent containers.

Samples may be exposed side-by-side with a validated chemical actinometric system (e.g., quinine for near UV region) to ensure that the specified exposure is obtained, or for the appropriate duration of time when conditions have been monitored using calibrated radiometers/lux meters.

Any protected samples (e.g., wrapped in aluminium foil) used as dark controls should be placed alongside the authentic samples.

Additional comments from the author: Dark control samples may be used to compensate for any temperature effects. Because photodegradation only occurs on the surface of solid samples, the surface area/weight ratio of the material exposed will affect the extent of observed degradation.

Analysis of Samples

At the end of the exposure period, the samples should be examined for any changes in physical properties (e.g., appearance, clarity or colour of solution) and for assay and degradants by a method suitably validated for products likely to arise from photochemical degradation processes.

Where solid drug substance samples are involved, sampling should ensure that a representative portion is used in individual tests. Similar sampling considerations, such as homogenisation of the entire sample, apply to other materials that may not be homogeneous after exposure. The analysis of the exposed sample should be performed concomitantly with that of any protected samples used as dark controls, if these are used in the test.

Judgement of Results

The confirmatory studies should identify precautionary measures needed in manufacturing or in formulation of the drug product, and if light resistant packaging is needed. When evaluating the results of confirmatory studies to determine whether change due to exposure to light is acceptable, it is important to consider the results from the other formal stability studies in order to assure that the drug will be within justified limits at time of use (see the relevant ICH Stability and Impurity Guidelines).

An explanatory comment on the above paragraph is given in Section 14.7.

14.6 Method validation for drug product

Normally the chromatographic method developed for the drug substance will also be valid for the product. However, if during photostability testing on product formulations, it becomes apparent that additional photodegradants are formed, it may be necessary to modify the chromatographic conditions to provide a suitable method.

14.7 Formal test for product photostability

14.7.1 Visible and UV exposure

During the course of ICH discussions on product testing it was agreed that a suitable visible light exposure was 1.2 million lux hours representing some 2–3 days' exposure close to a south-facing sunny window in the summer. A near UV exposure, of 200 Wh m^{-2} between 320 and 400 nm is recommended, corresponding to about 1–2 days close to a sunny window. This exposure recommendation is based on the assumption that products will be exposed to a mixture of glass-filtered natural light and indoor light and not stored where they are exposed to sunlight for any length of time. Where this is not the case, the UV exposure should be increased, for example, up to a maximum of 540 Wh m^{-2}, which is the highest figure considered by the ICH Working Group. UV levels decline rapidly outside the region of direct sunlight inside a room and therefore the ratio of UV/visible exposure recommended is less than that in 'standard' glass-filtered daylight ID-65 (Clarke, 1979; ISO 10 977:1993(E)) for which 1.2 million lux hours corresponds to approximately 540 Wh m^{-2}. UV exposure below 320 nm is unnecessary because the radiation below 320 nm is negligible in glass-filtered daylight. The spectral distribution of standard glass-filtered daylight, ID-65, is shown in Fig. 14.1.

14.7.2 Sequence of testing

For regulatory purposes, the product should first be tested unpacked or in a transparent container, if necessary, for a liquid/semi-solid product (Fig. 14.2). Unstable products are then further tested in primary and secondary (market) packs as necessary, as shown in the figure. When products are stable in the primary pack

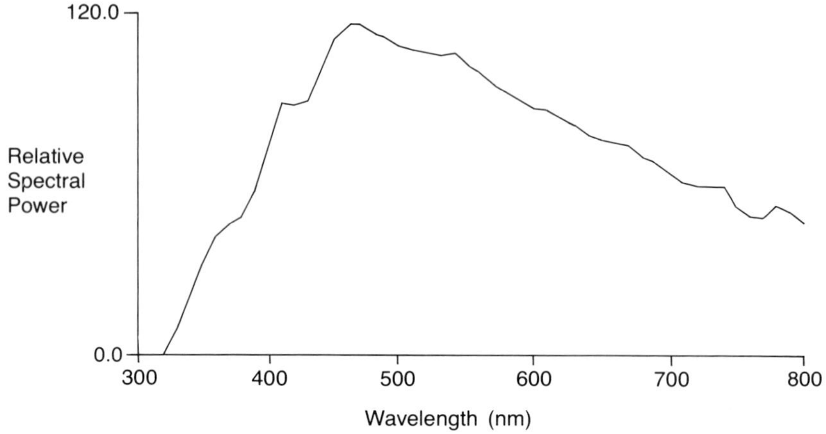

Figure 14.1 Standard indoor indirect daylight ID-65

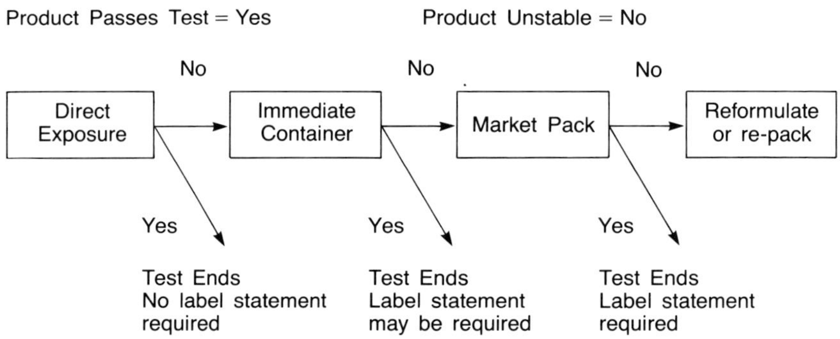

Figure 14.2 Sequential testing of product

but unstable without it, it is necessary to label products to prevent transfer into a less protective primary pack – e.g. by a pharmaceutical wholesaler. It is unnecessary to conduct tests in containers which are completely impenetrable to light, e.g., metal tubes or cans, where these are used for direct dispensing to the patient.

14.7.3 ***In-use testing***

The design of in-use tests such as the stability of reconstituted lyophilized products or products administered through an intravenous drip is excluded from the ICH Guideline but should be based on the same principles and considerations as the formal product test.

14.7.4 Criterion of stability

After each stage of testing, the product should be analysed to determine if it remains within 'limits justified by the Applicant', i.e. within specification. In this case any change is deemed to be acceptable and the product passes this stage of the test.

The ICH Guideline indicates that the overall Judgement of Results should be as follows.

> Depending on the extent of change special labelling or packaging may be required to mitigate exposure to light. When evaluating the results of photostability studies to determine whether change due to exposure to light is acceptable, it is important to consider the results obtained from other formal stability studies in order to assure that the product will be within proposed specifications during the shelf life (see the relevant ICH Stability and Impurity Guidelines).

The specifications set for the product should be in accord with the ICH Step 2 Impurities in New Drug Products Guideline. The limits for any photodegradants will depend on whether or not they have been Qualified. The combined effect of thermal degradation and photodegradation must be taken into account in setting end of shelf-life specifications and/or determining shelf-life. For most products the choice of a suitable pack will prevent photodegradation.

14.7.5 Sources

The source(s) used should be comparable in spectral distribution to those to which products are exposed in practical use, namely glass-filtered daylight and indoor lighting. A source closely simulating 'standard' glass-filtered daylight (Clarke, 1979) will provide a relatively greater UV exposure than most practical situations. It is most important that both the upper (360–400 nm) and lower (320–360 nm) UV ranges contain a significant percentage of the total UV irradiation (e.g. $\geqslant 25$ per cent) and that the UV irradiation extends over the whole band, or nearly so. This is necessary to ensure that products which absorb only in a small part of the UV region e.g. 320–340 nm or 380–400 nm do receive a meaningful UV exposure (Tønnesen and Karlsen, 1995). Some theoretical possibilities are illustrated in Fig. 14.3.

The testing standard DIN 53 487 (DIN 53 387, 1989) requires the UV spectral distribution in the UV to be 3 ± 0.5 per cent (320–360 nm) and 6 ± 1 per cent (360–400 nm) where 100 per cent is the total irradiance over 300–800 nm; this corresponds accurately to standard glass-filtered daylight. This distribution is not essential for pharmaceutical product testing; sources close to or meeting this standard can be used with confidence but will provide a greater UV exposure than necessary (approximately 540 Wh m^{-2} per 1.2 million lux h). Where reproducibility of results between laboratories is important, the spectral distribution of the sources used should be as nearly identical as possible, regardless of whether or not they conform with any particular specification.

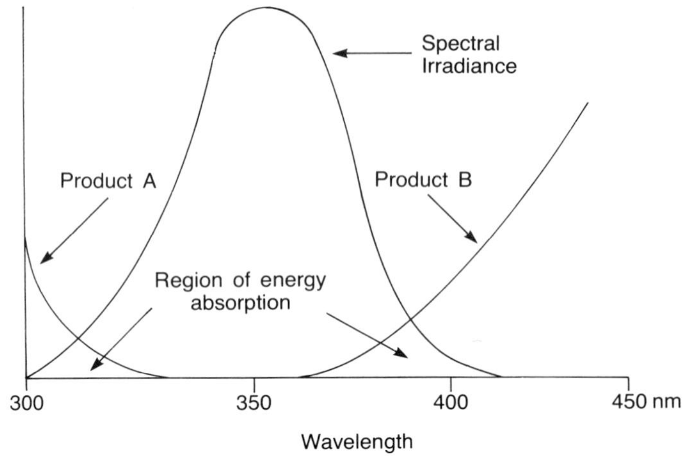

Figure 14.3 Energy absorption of hypothetical products from a near-UV tube

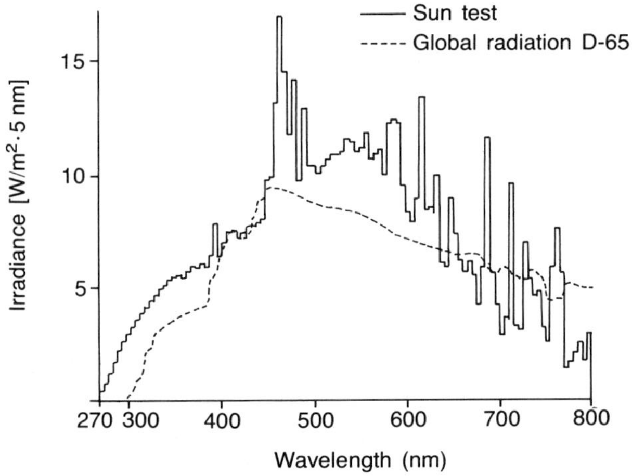

Figure 14.4 Glass-filtered Xenon source (Heraeus Suntest) and glass-filtered daylight

Sources currently used or being evaluated by the pharmaceutical industry are:

1 xenon lamps;
2 metal halide lamps;
3 white fluorescent tubes;
4 artificial daylight tubes;
5 'full spectrum' daylight fluorescent tubes;
6 near-UV fluorescent tubes.

The spectral distribution of these sources is shown in Figs. (14.4) to (14.9). For sources producing significant radiation below 320 nm (principally 1, 2 and 6) a

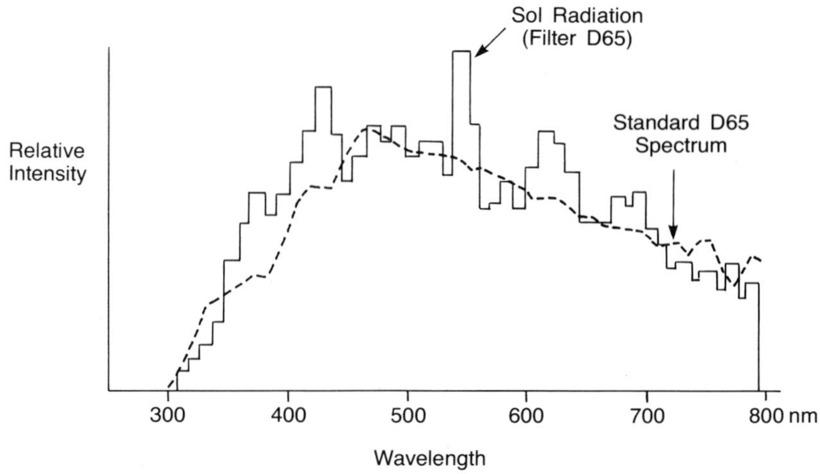

Figure 14.5 Metal halide lamp (sol system)

window glass filter should be used. Some lamp suppliers provide such filters. Fluorescent tubes should be 'burnt in' for 100–200 hr or more since their spectral distribution changes during this period and their output declines noticeably. Output also varies from tube to tube and along the length of tubes so the visible (and UV) levels across the sample area should be 'mapped' to ensure samples are placed at points of equal irradiance. Tubes should be changed at defined intervals (e.g. 5000 or 10 000 hr) or when the output has declined to the lowest level acceptable to the user.

The 'full spectrum' Duro-test Vitalite/Truelite tube emits 8 per cent of its irradiance over 300–800 nm in the UV region, of which only 0.5 per cent is < 320 nm, so that a window glass filter is not essential. The irradiance in the 380–400 nm region is relatively low (2.5 per cent) compared with a glass-filtered xenon source (ca 4 per cent) and the total UV is also slightly lower (8 per cent compared with 9.5 per cent for glass-filtered xenon). However, this and similar 'full spectrum' or artificial daylight tubes do provide an acceptable UV spectral distribution for photostability testing.

For tests simulating indoor light, for which there is no international standard, a typical white or cool white fluorescent tube with a colour temperature of 3500–4300 K is recommended.

For tests simulating glass-filtered daylight, 'artificial daylight' tubes or 'full spectrum' tubes from a specialist manufacturer of tubes designed for this purpose are preferable (Figs 14.7 and 14.8). Tests using sequentially or as a combination, white fluorescent and near-UV tubes are more complicated to conduct and the spectral distribution is less close to glass-filtered daylight (Figs 14.6 and 14.9).

Xenon lamps made by specialist suppliers are the most widely used daylight-simulation sources (Hibbert, 1991), since their spectral distribution is close to solar distribution and changes to only a very small extent during the lifetime of the lamp and is not affected by voltage fluctuations. However, the small illuminated area (typically 25 × 25 cm for a small unit) and high heat output limit its usefulness, although a cooling unit can be fitted. A window glass filter is used to remove radiation below <320 nm. This will also absorb UV at lower wavelengths,

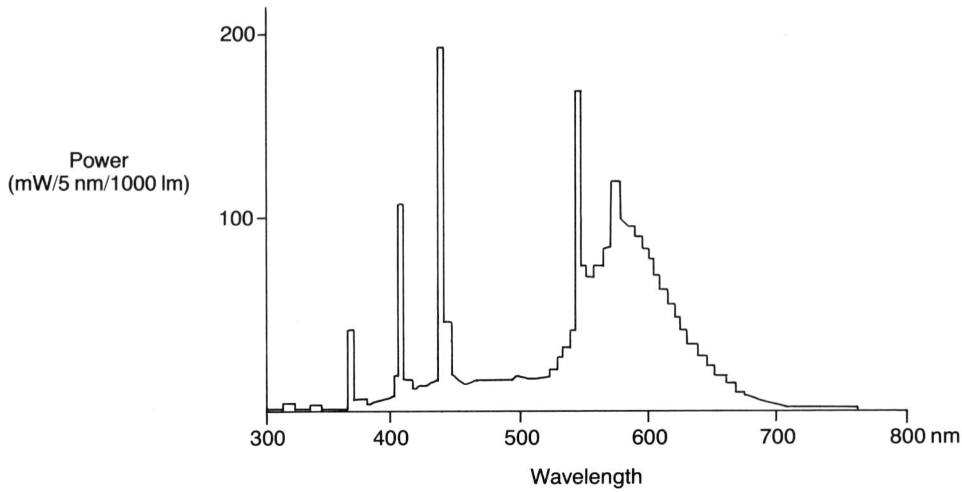

Figure 14.6 White fluorescent tube

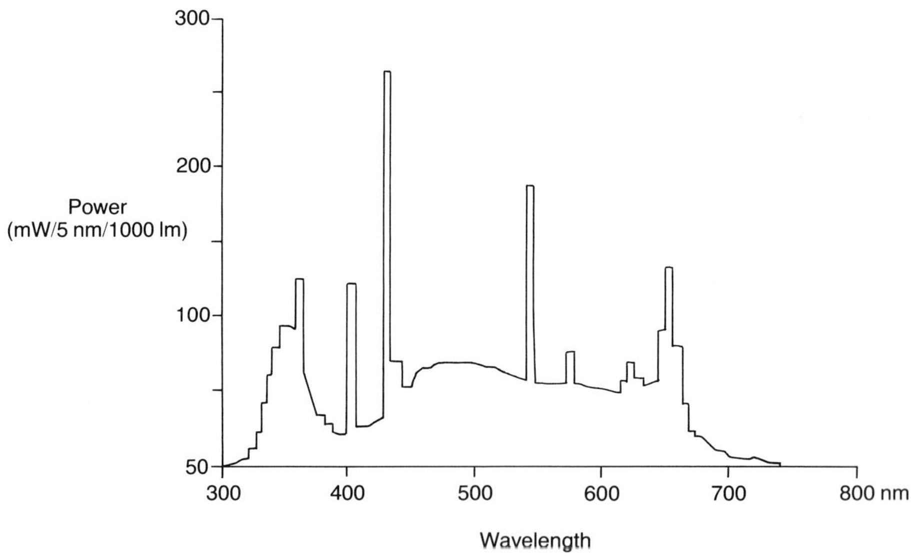

Figure 14.7 Artificial daylight fluorescent tube

particularly 320–360 nm. Specialist metal-halide lamps (with a glass filter) can give a spectral distribution which is as representative of the solar distribution as that of the Xenon source (Fig. 14.4). The UV output declines to a small extent as the source ages. Heat output and the small sample area are again limitations.

If a combination of white fluorescent and near-UV tubes is used, it is important to check that the product does not absorb primarily in the 320–340 nm or 390–420 nm region, where it has been noted that this combination of sources may produce little or no output (Tønnesen and Moore, 1993; Tønnesen and Karlsen, 1995).

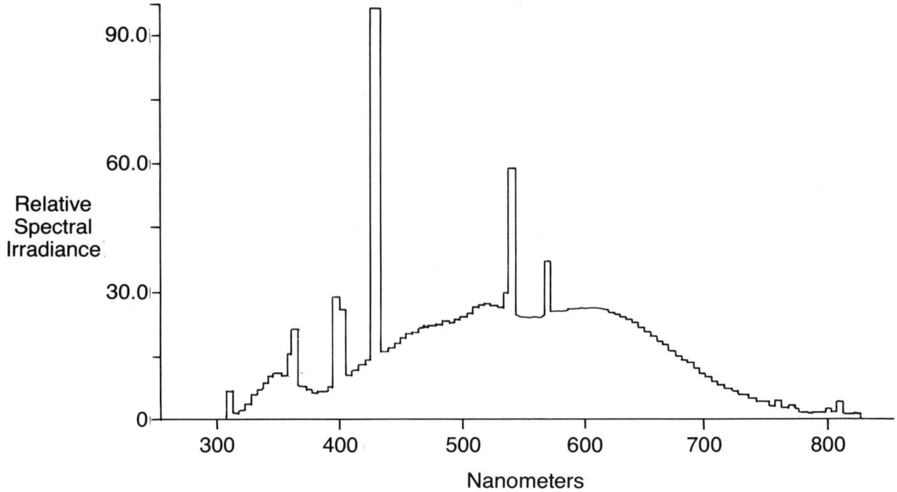

Figure 14.8 Vita-lite 'full spectrum' fluorescent tube

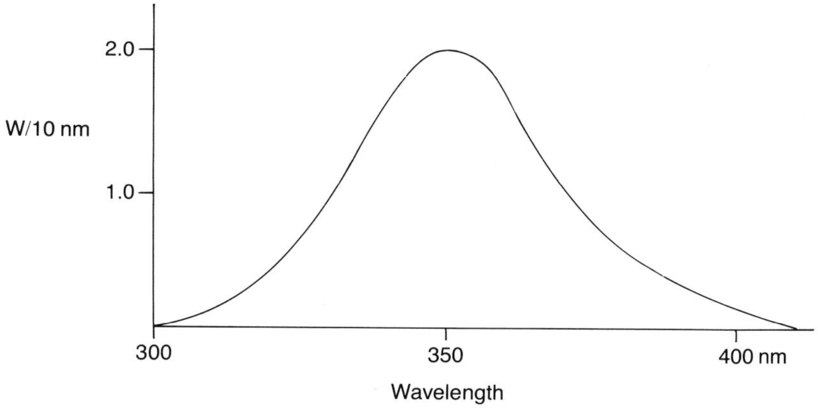

Figure 14.9 Near-UV fluorescent tube

14.7.6 Sample presentation and analysis

The same general considerations apply as for the drug substance (Section 5). The following is taken from the ICH Step 2 draft guideline:

Sample Presentation

Where practicable, when testing samples of the drug product outside of the primary pack, these should be presented in a similar way to the conditions mentioned for the drug substance. The samples should be positioned to provide maximum area of exposure to the light source. For example, tablets, capsules etc., should be spread in a single layer.

If direct exposure is not practical (e.g., due to oxidation of a product), the samples should be placed in a suitable protective inert transparent container (e.g., quartz).

If testing of drug product in the immediate container or as marketed, is needed, the samples should be placed horizontally or transversely with respect to the light source, whichever provides for the most uniform irradiation of the samples. Some adjustment of testing conditions may have to be made when testing large volume containers (e.g., dispensing packs).

Samples may be exposed side-by-side with a validated chemical actinometric system (e.g., quinine for near UV region) to ensure the specified light exposure is obtained or for the appropriate duration of time when conditions have been monitored using calibrated radiometers/lux meters. Any protected samples (e.g., wrapped in aluminium foil) used as dark controls should be placed alongside the authentic sample.

Additional comment from author. Dark control samples should be used to compensate for any temperature (or humidity) effects.

Analysis of Samples

At the end of the exposure period, the samples should be examined for any changes in physical properties (e.g., appearance, clarity or colour of solution, dissolution/disintegration etc.) and for assay and degradants by a method suitably validated for products likely to arise from photochemical degradation processes.

When powder samples are involved, sampling should ensure that a representative portion is used in individual tests. For solid oral dosage form products, testing should be conducted on an appropriately sized composite of, for example, 20 tablets or capsules. Similar sampling considerations, such as homogenisation or solubilisation of the entire sample, apply to other materials that may not be homogeneous after exposure (e.g., creams, ointments, suspensions etc.). The analysis of the exposed sample should be performed concomitantly with that of any protected samples used as dark controls if these are used in the test.

14.8 Number of batches

It is normally only necessary to test one batch of drug substance and drug product during the development phase, and then to confirm the photostability characteristics on a single definitive pilot scale (or larger) batch as described in the parent ICH Stability guideline if the drug substance and product are clearly photostable or photolabile. If the results of the definitive studies are equivocal or borderline, confirmatory testing of up to two additional batches of drug substance and/or product should be conducted. An 'equivocal' result is one where there is doubt as to whether the results indicate a 'pass' or 'fail'.

14.9 Measurement of visible light and UV irradiation

14.9.1 Visible light

A lux meter, calibrated at intervals recommended by the manufacturers, is recommended. Since lux is a measurement of light as perceived by the human eye

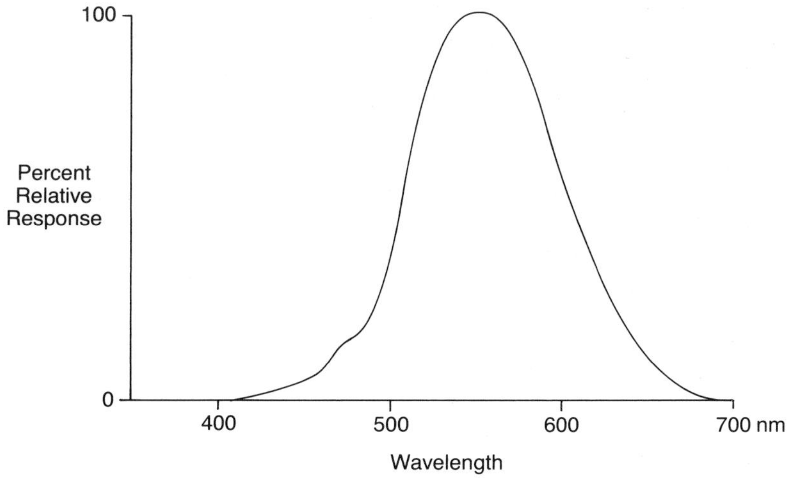

Figure 14.10 CIE photopic curve (lux)

(Fig. 14.10), rather than an absolute measurement, a lux meter is not suitable for comparing the outputs of different types of source. However, it is well suited for checking the evenness of illumination of the sample area and measuring changes in source output with time.

14.9.2 *UV irradiation*

The ICH Working Party has sought to identify a simple and inexpensive means for the average pharmaceutical laboratory to ensure that samples receive the required level of UV exposure. Two methods are recommended.

Radiometry. The UV filter radiometer comprises a detector with a UV wide-band filter which allows UV irradiation of defined wavelengths to reach the sensor (Fig. 14.11). Unfortunately, there are no international standards for filters and so the relative weighting given to the different UV wavelengths depends on the filters, i.e. meters from different manufacturers do not measure the same fraction of irradiance from a given source. Meters should be fitted with a cosine corrector to allow incident radiation from a wide angle of incidence to be measured. The geometrical relationship between the cosine corrector and the filter should be such that the radiation allowed to pass through the filter is not dependent on the angle of incidence. UV filter radiometers may be calibrated to give an absolute measurement of irradiance for a particular type of source – e.g. xenon; fluorescent tube of particular type. These meters provide a simple means of measuring evenness of irradiance across the sample area and changes with time, for a particular source. They cannot be used to compare irradiance between sources, unless calibrated specifically for each source. Recalibration should be conducted in accord with the manufacturer's instructions.

A spectroradiometer is a relatively expensive item of equipment which few pharmaceutical companies would wish to purchase, but is used to measure the absolute spectral irradiance of a source.

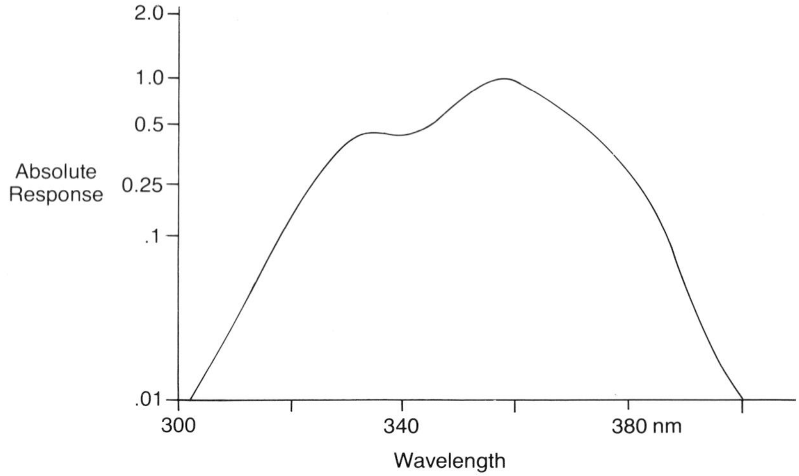

Figure 14.11 UV spectral response for 365 nm radiometer filter

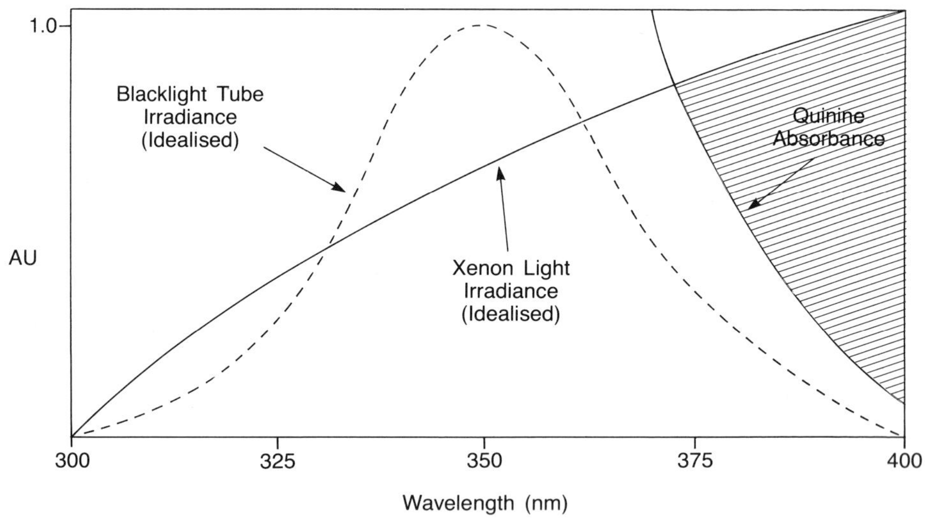

Figure 14.12 Possible limitation of quinine actinometer

Actinometry. The principles of actinometry are well established, but there is no well-established actinometric system which is sensitive only to UV radiation (Kuhn *et al.*, 1989; Piechocki and Walters, 1993). Quinine has been proposed by the Japanese National Institute of Health Sciences and the Japanese Pharmaceutical Manufacturers' Association (Yoshioka *et al.*, 1994). The results of the work of FDA and US laboratories have confirmed its suitability (Brower *et al.*, manuscript in preparation). At the recommended concentration (2 per cent), the quinine solution has an absorbance of ≥ 2 from 320 to 367 nm, so the UV irradiance over this range will be determined using this system. Quinine usually contains a number of alkaloid impurities, principally dihydroquinine; however, the level of the latter does not affect its performance as a UV actinometer.

The monohydrochloride dihydrate salt has been selected as having adequate solubility, and commercially sourced high-purity quinine monohydrochloride is suitable.

The recent work by the FDA and US industry has shown that during UV exposure, the quinine solutions undergo a small fall in pH causing an increase in absorbance at 400 nm, the wavelength recommended for actinometry. In order to make reliable measurements it is therefore important that the vessels used are free from traces of acid, base and buffer salts. The very steep slope of the quinine absorbance curve at 400 nm (>0.01 AU nm^{-1}) means that the monitoring wavelength of 400 nm must be set very accurately, preferably using a diode assay spectrophotometer. Practical details, as described in the ICH guideline, are given below. Quartz cuvettes are a preferable alternative to ampoules.

> Prepare a sufficient quantity of a 2 per cent w/v aqueous solution of quinine monohydrochloride dihydrate (if necessary dissolve by heating). Put *10 ml* of the solution into a 20 ml colourless ampoule,* seal it hermetically, and use this as the sample. Separately, put *10 ml* of the solution into a 20 ml colourless ampoule, seal it hermetically, wrap in aluminium foil to protect completely from light, and use this as the control. Expose the sample and control to the light source for an appropriate number of hours. After exposure determine the absorbances of the sample (A_T) and the control (A_o) at 400 nm using a 1 cm pathlength. Calculate the change in absorbance, $\Delta A = A_T - A_o$.

The quinine actinometric system described above will need to be calibrated for each type of light source used for photostability testing, using the same type of vessel (cuvette or ampoule) as to be used in actinometry. This is because the proportion of the UV irradiance absorbed by the quinine will depend on the spectral irradiance of the source (Fig. 14.12). It can be seen that the shaded UV region of the xenon source is not absorbed by the quinine solution. The calibration will give the absorbance change at 400 nm equivalent to a defined UV exposure from a particular source. For near-UV fluorescent tubes the ICH Guideline indicates that an absorbance change of 0.8 can be taken as corresponding to 200 Wh m^{-2} UV irradiance. Exposure of ampoules or cuvettes of quinine solution alongside test samples or in a separate calibration experiment, can be used to ensure products receive the correct UV exposure.

Quinine actinometry using solutions in cuvettes is only slightly less convenient than UV radiometry for checking the evenness of irradiance across the sample area and for monitoring changes in source output with time. However, it has the advantage of never needing recalibration, unless the type or manufacturer of the source is changed.

10.14 Conclusions

In order to provide a sound basis for taking decisions with regard to labelling and the use of protective packs, pharmaceutical laboratories should determine the susceptibility of their products to simulated glass-filtered daylight, optionally

*Shape and dimensions of Japanese standard ampoule: Stem length: 80.0 ± 1.2 mm; stem diameter: 21.8 ± 0.4 mm; diameter of neck at cutting position: 7.0 ± 0.7 mm.

supplemented with visible light. After much discussion, the ICH Working Party on Photostability has proposed a realistic test with an acceptable level of reproducibility to the Regulatory Agencies, but, which is reasonably straightforward to conduct in Pharmaceutical R&D Laboratories. The guideline represents a major step forward in photostability testing of pharmaceutical products. Once practical experience in the use of the guideline has been gained, opportunities for improving it may be identified.

The Step 2 Guideline will be reviewed by the three regulatory agencies (Europe, Japan, US) and, following modification if necessary, it will be referred to the ICH Steering Committee for adoption. Finally, the guideline will be adopted formally and implemented by the regulatory agencies. The earliest dates for implementation are likely to be for studies initiated after 1st April 1997 (Japan) and for product licence applications submitted after 1st January 1998 (Europe, US).

Acknowledgements

Thanks are due to all the members of the ICH Photostability Working Party, particularly Dr T. Beaumont, for useful discussions. Informal discussions with Professor H. H. Tønnesen and Dr J. Piechocki were appreciated. The contributions of Mrs S. L. Jackson and Mr A. Smail to experimental studies and the preparation of this chapter are gratefully acknowledged.

References

ANDERSON, N. H., JOHNSTON, D., MCLELLAND, M. A. and MUNDEN, P., 1991, Photostability testing of drug substances and drug products in UK pharmaceutical laboratories, *Journal of Pharmaceutical and Biomedical Analysis*, **9**, 443–9.

BROWER, J. F., DREW, H. D., JUHL, W. E. and THORNTON, K. A Proposed Chemical Actinometer System for Pharmaceutical Drug Substance and Drug Product Photostability Studies, manuscript in preparation.

CLARKE, F. J. J., 1979, *Proceedings of 19th Session of CIE*, Kyoto, Japan, p. 75.

DIN 53 387, 1989, Artificial weathering and ageing of plastics and elastomers by exposure to filtered xenon arc radiation.

HIBBERT, M., 1991, Shedding light on stability testing, *Manufacturing Chemist*, **62**, 32–3.

ICH, 1995, International Harmonisation Conference Step 4 Guideline on Impurities in New Drug Substances.

KUHN, H. J., BRASLAVSKY, S. E. and SCHMIDT, R., 1989, Chemical actinometry, *Pure and Applied Chemistry*, **61**, 187–210.

MOORE, D. E., 1987, Principles and practice of drug photodegradation studies, *Journal of Pharmaceutical and Biomedical Analysis*, **5**, 441–53.

1990, Kinetic treatment of photochemical reactions, *International Journal of Pharmaceutics*, **63**, R5–R7.

NEMA, S., WASHKUHN, R. J. and BEUSSINK, D. R., 1995, Photostability testing: an overview, *Pharmaceutical Technology*, **19**, 170–85.

PIECHOCKI, J. T. and WOLTERS, R. J., 1993, Use of actinometry in light-stability studies, *Pharmaceutical Technology*, **18**, 46–52.

SCIANO, J., 1990, *CRC Handbook of Organic Photochemistry, Actinometry* (author Bunce, N. J.), 241–60.

THOMA, K. and KERKER, R., 1992, *Pharmazeutische Industrie*, Photoinstabilität von Arzneimitteln. 1. Mitteilung über das Verhalten von nur im UV-Bereich absorbierenden Substanzen bei der Tageslichtsimulation, **54**, 169–77.

TØNNESEN, H. H., 1991, Photochemical degradation of components in drug formulations. Part I: An approach to the standardization of degradation studies, *Pharmazie*, **46**, 263–5.

TØNNESEN, H. H. and KARLSEN, J., 1995, Photochemical degradation of components in drug formulations, III A discussion of experimental conditions, *Pharmeuropa.*, **7**, 137–41.

TØNNESEN, H. H. and MOORE, D. E., 1993, Photochemical degradation of components in drug formulations, *Pharmaceutical Technology International*, **5**, 27–33.

YOSHIOKA, S., ISHIHARA, Y., TERAZONO, T., TSUNAKAWA, N., MURAI, M., YASUDA, T., KITAMURA, T., KUNIHIRO, Y., SAKAI, K., HIROSE, Y., TONOOKA, K., TAKAYAMA, K., IMAI, F., GODO, M., MATSUO, M., NAKAMURA, K., ASO, Y., KOJIMA, S., TAKEDA, Y. and TERAO, T., 1994, Quinine actinometry as a method for calibrating ultraviolet radiation intensity in light stability testing of pharmaceuticals, *Drug Development and Industrial Pharmacy*, **20**, 2049–62.

15

Mathematical Models for Studies of Photochemical Reactions

S. A. SANDE

15.1 Theoretical basis

Kinetic studies of chemical reactions are based on the general rate equation stating that the reaction rate ($-\delta C/\delta t$) is proportional to the concentrations of the reacting substances raised to the power of the number of reacting molecules of each substance:

$$x \times A + y \times B \ldots \rightarrow \text{Products}$$

and reaction rate:

$$-\frac{\delta C}{\delta t} = k \times [A]^x \times [B]^y \ldots \qquad (15.1)$$

Where $[A]$ is the concentration of substance A.

From equation (15.1) the order of a reaction is defined as the sum of the exponents ($x + y$ ... i.e. the sum of reacting molecules) and plays a major role in kinetic studies of chemical reactions. The reason for this popularity is the combination of information gained from the model, both about the number of molecules involved in the reaction (mechanistic studies) and calculation of how fast the reaction is taking place (kinetic studies). However, as a reaction may take place over a series of steps, where the overall rate is determined by the slowest step, the reaction studied should involve a minimum of steps and usually has to be a simple reaction in order to provide exact information on the mechanism.

Another problem is that parallel or reversible reactions will influence the observed order. Consider for instance the first-order reaction: $R \rightarrow P$ with a rate constant k_1. If R in addition undergoes an unknown second-order reaction with a substance X (rate constant k_2), the total rate of degradation for R is:

$$-\frac{\delta R}{\delta t} = k_1 \times R + k_2 \times R \times X = (k_1 + k_2 \times X) \times R \qquad (15.2)$$

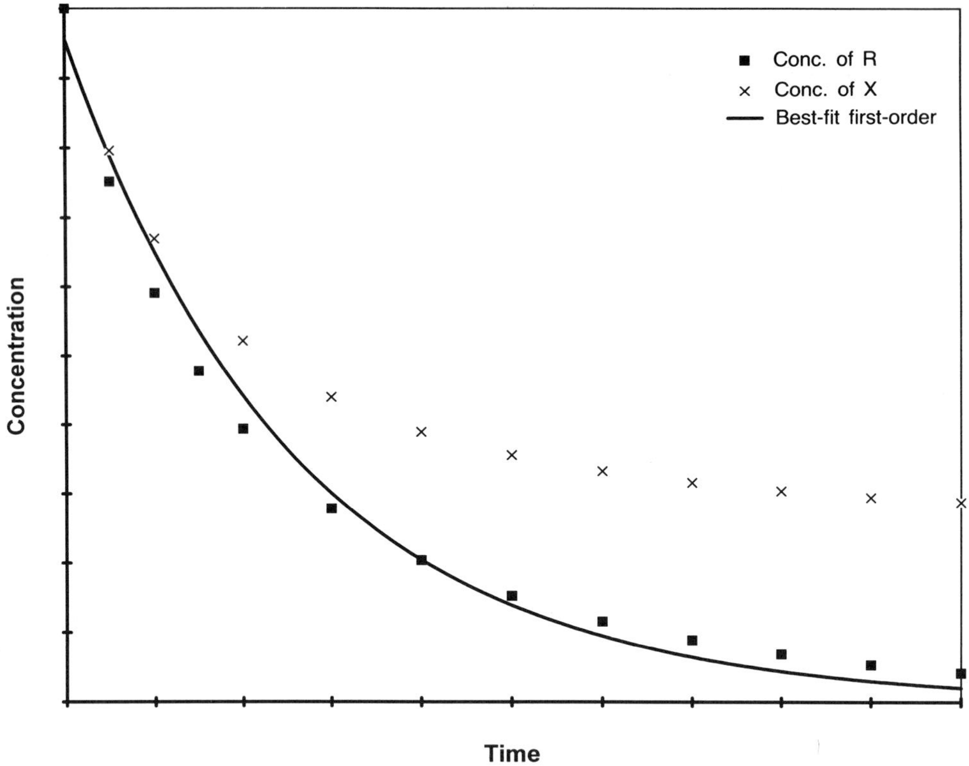

Figure 15.1 Effect of erroneously fitting a first-order model to a degradation with parallel pathways

If $X \gg R$ and remains at an approximately constant concentration during the reaction, the reaction is a pseudo- first-order, and the estimated rate constant is $(k_1 + k_2 \times X)$. A mathematical model describing a first-order reaction is then valid. However, if the initial concentrations of the reactants R and X are of the same magnitude and the first-order model is erroneously fitted to the experimental data, a poor fit is obtained since the calculated k is proportional to X and therefore will vary with time (Fig. 15.1).

For such complex reactions, a thorough analysis of the reacting substances and *all* the reaction products formed are required if a valid estimate for the order of the reaction is to be obtained. In this case the reaction cannot be described by a first-order model.

15.2 A general model valid for all reaction orders

In stability studies, the amount of active substance is usually the major concern. One is consequently less interested in the exact mechanism of the reaction and usually more interested in predicting the shelf-life of the preparation. The exact order of a reaction will therefore be of minor importance as compared with the rate constant or the time required, for example, 10 per cent degradation (Connors *et al.*, 1986).

In these situations a simplified model only taking into account the concentration of one substance may be justified:

$$-\frac{\delta C}{\delta t} = k \times C^x \tag{15.3}$$

where k is a rate constant, *C* is the concentration and *x* is the order of the reaction. We have previously shown and validated a general solution of equation (15.3) (Sande and Karlsen, 1991) showing that under typical conditions for $x \neq 1$:

$$C = (C_0^{1-x} - k \times t \times (1-x))^{1/(1-x)} \tag{15.4}$$

$$\text{with half-life } T_{1/2} = \frac{C_0^{1-x}}{k} \times \frac{2^{x-1} - 1}{(x-1)} \tag{15.5}$$

Where C_0 is the initial concentration (for x = 1 the usual first-order reaction apply: $C = C_0 \times e^{-kt}$)

Equation (15.4) is convenient for estimating shelf-lives of substances showing complex degradation patterns since it is versatile enough to handle several types of reactions.

15.3 Comparison of rate constants

A problem related to varying orders is that for the same raw data the estimated rate constant (*k*) will be a function of x and C_0. For equal reaction order but different initial concentrations, k will be proportional to $C_0^{(1-x)}$. If $C_0 = 1$, k will increase with increasing order, but the actual function will be dependent on several factors, especially the level of the rate constant and the extent to which the degradation is followed.

If the model is fitted to various sets of data producing different values of x, a direct comparison of the estimated rate constants is thus not possible. If, for example, models with x = 0.9 and 1.1 are fitted to data generated from time = 0 to 1 employing a first-order model with $C_0 = 100$ and k = 1, the estimated rate constants will be 1.5 and 0.66 respectively. This difference in magnitude of the estimated rate constants illustrates why a direct comparison of the rate of the reactions is precluded.

A typical example of a reaction likely to produce varying orders where comparison of different rate constants is essential would be a study regarding the effect of an additive on the degradation of an active substance. In this case the correlation between the amount of additive and the rate of degradation cannot be established by a simple comparison of the mean rate constants since both repetitive measurements on the same composition and different levels of additives are likely to result in different orders.

A solution to this problem would be to set the order of the reaction to a predetermined value. Which value to choose would have to be based on previous experiments or knowledge of the reaction mechanisms involved. Another possibility would be to employ the initial half-life or another essential time-value (e.g. shelf-life) for comparison of the reaction rate, since this value will be less dependent on the reaction order. A prerequisite for this method is of course the use of the same initial concentration for the active substance in all the experiments.

15.4 Photochemical reactions

So far all comments have been made on a general basis valid for all chemical reactions, including photochemical reactions. For photochemical reactions in particular the previous concept may be employed provided the incident light intensity is kept constant. The absorbed light is then regarded as a catalyst or a supply of energy, and thus affects the rate constant in the same manner as other environmental conditions (e.g. temperature and humidity). An overview of several types of photochemical degradation is given by Tønnesen (1991).

A reaction scheme for photochemical reactions is presented by Carstensen (1990). The reaction is divided into a two-step process initiated by the absorption of a photon (*p*): $A + p \rightleftharpoons [A^*] \rightarrow$ products. The reversible absorption of the photon creating an activated complex ($[A^*]$) is followed by a non-reversible decomposition into the products.

The overall reaction ($A \rightarrow$ products) of these consecutive steps may be described by a first-order reaction model provided the 'concentration' of *p* is kept constant (constant light intensity). For this model, the rate constant k represents the slowest process, either absorption of photons (usually very fast) or decomposition into products (usually the rate-determining step).

For calculations this is, however, an inconvenient scheme since the gradient of light existing in the volume studied must be taken into account. An approach more suited for kinetic calculations employs both the quantum yield and the Beer-Lambert law (Balzani and Carassiti, 1970). Thus the number of reacting molecules (δN) per time (δt) is given by:

$$-\frac{\delta N}{\delta t} = \Phi \times I_A \qquad (15.6)$$

Where Φ is the quantum yield and I_A is the number of absorbed photons. The concentration of the reacting substance is introduced into the equation via I_A. The Beer-Lambert law describes absorption of light as:

$$A = \varepsilon \times b \times C = -\log\left(\frac{I_T}{I_0}\right) \qquad (15.7)$$

Where A is the absorption, ε is the molar absorptivity, *b* is the length of the light pathway and I_T and I_0 are the transient and incident light intensities respectively.

Substituting μ for $\ln(10) \times \varepsilon$, the number of absorbed photons for a unit volume and time is:

$$I_A = I_0 - I_T = I_0 \times (1 - e^{-\mu N}) \qquad (15.8)$$

Substituting for I_A from equation (15.8) into equation (15.6) produces:

$$-\frac{\delta N}{\delta t} = \Phi \times I_0 \times (1 - e^{-\mu N}) \tag{15.9}$$

A general integration of equation (15.9) and solution for the number of molecules reacted (N) gives:

$$N = \frac{1}{\mu} \times \ln\left[1 + e^{-\mu \Phi I_0 t} \times (e^{-\mu N_0} - 1)\right] \tag{15.10}$$

This equation bears only scarce resemblance to the models described above, and does not necessarily produce an integer reaction order. Only under particular conditions the model may be simplified:

- When the reactant absorbs nearly all the incident light ($I_A \approx I_0$), leading to $-\delta N/\delta t = \Phi \times I_0$, which is a 0-order reaction. (For absorbance > 3 the error is less than 5 per cent)
- When the reactant absorbs only a small amount of the incident light ($I_A \ll I_0$), $I_A \approx I_0 \times \mu \times N$ leading to $-\delta N/\delta t = \Phi \times I_0 \times \mu \times N = k \times N$, which is a first-order reaction. (For absorbance < 0.1 the error is less than 5 per cent)

It may thus be concluded that a photochemical reaction will follow a model based on an integer reaction order only under specific conditions.

It must be noted that the comments so far have been related to simple reaction schemes. In photochemical reactions, complex degradation pathways are frequently found where e.g. the reaction products may alter the original reaction or interact directly with the substance studied (Moore, 1987). In such cases the use of simple integer order models is precluded except for the initial part of the reaction.

In spite of this, several reports on first-order photochemical processes have been made (Lerner *et al.*, 1988; Matsuda and Masahara, 1983). These describe photolysis reactions, whereas oxidation reactions are reported to be zero-order (Asker *et al.*, 1985; Felmeister *et al.*, 1965). The explanations for these findings must be that the combination of conditions (e.g. concentration of other reacting substances) and reaction mechanisms make these simple mathematical models appropriate.

15.5 Determination of reaction order

Differentiating between two or more possible reaction orders is interesting in the study of reaction mechanisms. With the previously described limitations in mind it is therefore of interest to determine which criteria the experimental design has to fulfil in order to ensure a significant differentiation between the competing models (Sande and Karlsen, 1993).

To follow a reaction for too short a period (e.g. until less than 25 per cent decomposed) renders the differentiation between two reaction orders impossible (Taylor and Shivji, 1987; Yang, 1981). As the main parameter affecting the determination of reaction order we therefore studied the influence of the time-span a reaction is followed.

The reason for the importance of this factor is that the differences between the time course of reaction predicted by different orders are small at the initial stages. Other factors that may influence the differentiation between reaction orders are:

- the precision of the analytical method employed;
- the number of samples drawn during the reaction; and
- the magnitude of the difference in reaction order.

  For elementary reactions this would be 1. For complex reactions other figures could be relevant.
- The order of the reaction itself (x).

  The difference between the degradation profiles predicted by second- and third-order models is less than the difference between zero- and first-order models. Consequently, it is more difficult to differentiate between second- and third-order reactions than between zero- and first-order reactions.
- The time schedule of sampling.

  The sampling schedule most frequently used in kinetic studies is a schedule providing equal spacing between the measured concentrations during the experiment. This means frequent sampling at the start of the experiment and less frequent sampling as degradation proceeds. An alternative is sampling at equally spaced times. Both types of sampling schedules are investigated.

Instead of recording the timespan required for significant differentiation between competing models, the more general approach of calculating the percentage remaining at a certain time can be used. Thereby the results will have a general application independent of the actual values for starting concentration and rate constant.

Simulating a first-order degradation with different levels of imposed analytical errors ('analytical precision'), the main results may be summarized as follows:

- Analytical precision is the most important factor in determining a correct reaction order.
- The number of samples affects the estimation of the reaction order to a certain degree.
- The time schedule of sampling has little influence on the estimation of correct reaction order.

The most important result is, however, that even with a very precise method of analysis (SD for the analytical method = 1.0), the reaction has to be followed for almost a half-life in order to get a reliable estimate of the reaction order. This is shown in Fig. 15.2 where the per cent remaining of the reactant necessary for differentiation between zero-, first- and second-order reaction is plotted against number of samples.

It is thus impossible to estimate the exact order of a reaction in cases where the degradation products influence the degradation of the parent compound.

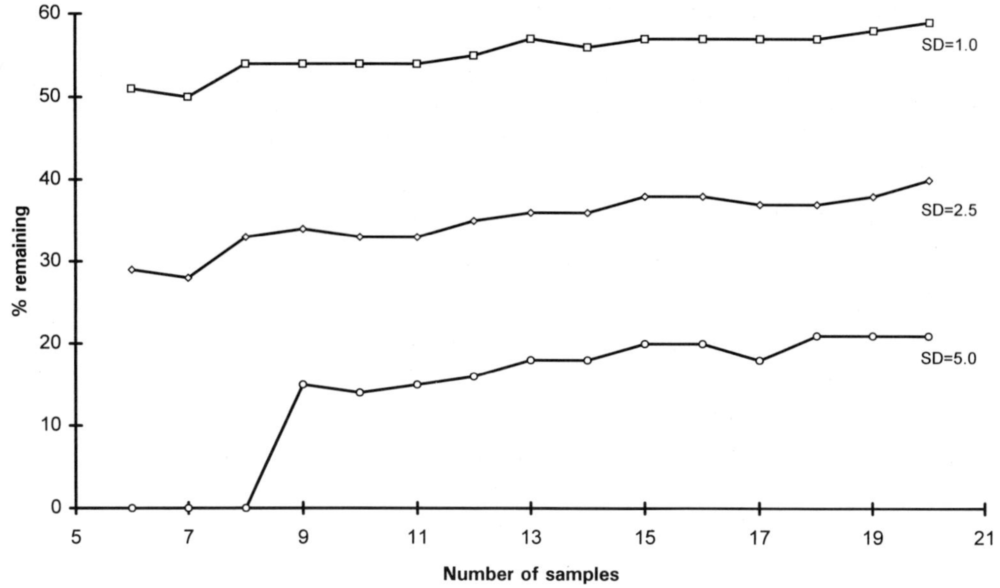

Figure 15.2 The extent to which a reaction must be followed in order to obtain a valid estimate of the reaction order vs. number of samples for different levels of analytical accuracy

15.6 Solid state photochemical reactions

In photochemical solid state reactions the application of a reaction order for characterizing a process is even more difficult than in a solution. Several solid state chemical reactions have been described by Byrn (1982), but also more physico-chemical reactions have been studied (Ho, 1992). Since the molecules do not diffuse, the chances for interaction with reactive groups on other molecules are very small and consequently the supply of energy for the reaction (e.g. from light) becomes the limiting parameter. This usually also leads to alteration in the ratio of the amount of the various products formed, or the formation of altogether new products. The powder bed in which the reaction takes place is usually thicker than a monomolecular layer, leading to varying conditions for the reaction (i.e. the intensity of the light) dependent on the depth in the powder. As stated elsewhere the concept of reaction order is therefore not very useful for solid state reactions (Bamford and Tipper, 1980; De Villiers *et al.*, 1992).

In addition the use of the general model for predicting the rate of a reaction (equation 15.10) is usually of less value since the model predicts that the concentration of the reactant will approach zero as time approaches infinity. In many cases, however, the light will never penetrate the entire powder bed, and therefore a plot of remaining concentration vs time will level off at a value greater than zero.

In Fig. 15.3 a cross-section of a powder bed is shown. The surface area is A and the depth is B. When irradiating this cube with light of a specific wavelength, the intensity will obviously be greater on the top than at the bottom of the sample. The cube may be theoretically segmented into a number of layers for which the

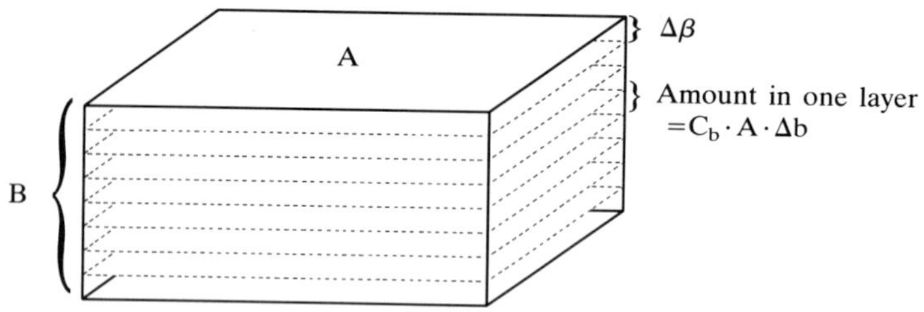

Figure 15.3 Irradiation of an imaginary sectioned powder bed

intensity is (almost) uniform. In practical laboratory work, however, we will not be able to separate our powder into these layers and the measured concentration as a function of time ($C_{tot}(t)$) will be the average of the concentrations for each layer (C(b,t)):

$$C_{tot}(t) = \frac{\sum_{b=0}^{B} (C(b,t) \times A \times \Delta b)}{A \times B} = \frac{1}{B} \int_0^B C(b,\ t)\ \delta b \qquad (15.11)$$

In order to obtain C_{tot} we need an expression for C(b,t):

Since the intensity is uniform throughout each layer (b) the rate of reaction will be:

$$-\frac{\delta N_b}{\delta t} = \Phi \times I_a \times A \qquad (15.12)$$

where Φ is the quantum yield i.e. the ratio between number of reacting molecules and number of absorbed photons (both numbers referring to the same volume and time); I_a is the number of photons absorbed by the reacting substance pr. unit area and A is the area of the irradiated surface.

The number of photons absorbed in the layer will be the difference between the intensity of the light entering and leaving the layer: $I_a = I(b,t) - I(b+1,t)$, where I(b,t) denotes the intensity in layer *b* at time *t* and I(b + 1,t) denotes the intensity in the layer beneath *b* at time *t*. Using the Beer-Lambert law with $\mu_R = \ln(10) \times \varepsilon_R$ (the molar absorptivity of the reactant) as in equation (15.8) will result in the following expression:

$$I_a = I\ (b,t) \times (1 - e^{-\mu_R \Delta b C(b,t)}) \qquad (15.13)$$

Inserting for I_a from Eq. (15.13) into (15.12), and dividing by Avogadro's number and the volume ($N_A \times V$) in order to obtain concentrations instead of number of reacting molecules, results in:

$$-\frac{\delta C(b,t)}{\delta t} = \frac{\Phi \times I\ (b,t) \times (1 - e^{-\mu_R \times \Delta b \times C(b,t)}) \times A}{N_A \times V} \qquad (15.14)$$

Since the volume of the layer (V) equals A × Δb and one is interested in an infinitesimally thin layer (Δb → 0) simplifying the equation is possible:

$$-\frac{\delta C(b,t)}{\delta t} = \frac{\Phi \times I(b,t)}{N_A} \times \lim_{\Delta b \to 0^+} \frac{(1 - e^{-\mu_R \times \Delta b \times C(b,t)}}{\Delta b} \tag{15.15}$$

$$-\frac{\delta C(b,t)}{\delta t} = \frac{\Phi \times I(b,t)}{N_A} \times \mu_R \cdot C(b,t) \tag{15.16}$$

I(b,t) is constant at the surface layer and equal to I_0, but in order to obtain a general equation, an expression for the intensity throughout the powder is needed.

So far only the absorption of light by the reacting substance has needed consideration, but in order to calculate I(b,t), all absorbing species must be considered:

$$-\ln\left(\frac{I(b,t)}{I_0}\right) = b \times (\mu_R \times \bar{C}(b,t) + \mu_P \times \bar{C}_P(b,t) + K') \tag{15.17}$$

where $\bar{c}$ and $\bar{c}_P$ are the average concentration of the reactant and degradation product respectively, μ_R and μ_P are the corresponding ln-based molar absorptivities and K′ is the light absorption by all other substances. It is important to notice that $\bar{c}$ and $\bar{c}_P$ are the average concentrations *above* the layer in question, and since the average concentration equals

$$\frac{1}{b}\int_0^b C(b,t)\,\delta b \tag{15.18}$$

we may express I(b,t) as:

$$I(b,t) = I_0 \times e^{-(\mu_R \times \int_0^b C(b,t)\,\delta b + \mu_P \times \int_0^b C_P(b,t)\,\delta b + K)} \tag{15.19}$$

Inserting equation (15.19) in equation (15.16) gives:

$$-\frac{\delta C(b,t)}{\delta t} = \frac{\Phi}{N_A} \times I_0 \times e^{-(\mu_R \times \int_0^b C(b,t)\,\delta b + \mu_P \times \int_0^b C_P(b,t)\,\delta b + K)} \times \mu_R \times C(b,t) \tag{15.20}$$

We have now managed to obtain an, admittedly rather complicated, differential equation describing the change in concentration for one single layer of a solid sample. A part solution of a corresponding equation is given by Treushnikov and Yanin (1990), but for the purpose of simulations, equation (15.20) will be adequate.

Restrictions and prerequisites of the model. Some comments are justified on the estimation of the parameters. Molar absorptivity of a compound (μ_R and μ_P) is usually determined in solution. The absorptivities should preferably be estimated in a non-polar solvent to determine μ for a non-dissociated molecule.

The value for μ determined in solution is usually employed also for the compound in solid state, but it should be noted that diffraction (included in K) is much more significant in the solid state (Paulsen, 1990). This will lead to a greater absorption in the upper layers and a greater reduction in the intensity over Δb than predicted solely on the basis of the higher concentrations present in the solid state.

The actual influence on the result will be interrelated with the method used for determination of the quantum yield. This determination is difficult in the solid state since few techniques are available. Since diffraction will be important also for this determination, a higher quantum yield may be expected, which, to some degree, compensates for the diffraction in the sample.

The quantum yield is not expected to be dependent on the wavelength as long as the light possesses sufficient energy to excite the molecule into the S1 state.

The model is developed on the basis of one uniform crystal of solid. The presence of pores in the material or use of powder is, however, not expected to influence the average degradation. For the individual layers, pores or voids will produce diverging concentrations, but unless the reaction changes the volume of the pores, the differences over the entire powder bed will level out. Since the diffraction from surfaces differs from diffraction inside the sample this factor is affected by pores and internal surfaces. For porous material, control of the diffraction therefore becomes increasingly important.

In these calculations, the use of monochromatic light has been assumed. When other sources of light are employed the overlap between emission and absorption spectra must be taken into account. For the computations this means that the effects of the individual wavelengths have to be summed. The possible presence of sensitizers will also affect the effective molar absorptivity (Arney *et al.*, 1989).

Simplification of the model, pure substance. When a pure substance is irradiated, the only absorbing species are the reactant and the possible degradation products. This allows a simplification of equation (15.20) since at $t = 0$ the concentration of absorbing substances in all layers are equal (C_0), and at later times C_p may be calculated from:

$$C(b,t) + C_P(b,t) = C_0 \quad \rightleftarrows \quad C_P(b,t) = C_0 - C(b,t) \qquad (15.21)$$

Substituting for C_p in equation (15.20) and rearranging give:

$$-\frac{\delta C(b,t)}{\delta t} = \frac{\Phi}{N_A} \times I_0 \times e^{-[(\mu_R - \mu_P)\times \int_0^b C(b,t)\,\delta b + \mu_P \times C_0 \times b]} \times \mu_R \times C(b,t) \qquad (15.22)$$

If the reacting substance is a large molecule and the photolysis does not affect the chromophore system of the substance, it is not unlikely that both the reactant and the degradation product have the same molar absorptivity (μ) at the specific wavelength. In that case equation (15.22) may be further simplified:

$$-\frac{\delta C(b,t)}{\delta t} = \frac{\Phi}{N_A} \times I_0 \times e^{-\mu \times C_0 \times b} \times \mu \times C(b,t) = k \times C(b,t) \qquad (15.23)$$

This expresses a first-order reaction, but it only applies for a single layer, so again, the change in total concentration measured as a function of time does not necessarily follow any particular reaction order model.

15.7 Simulation

Integration of equation (15.22) and a subsequent general solution for C(b,t) does not appear to be possible. A general expression for the total concentration (equation (15.11)) is therefore also not available. Instead a numeric evaluation of the expressions is possible using either a spreadsheet or a specially designed computer program.

At the top layer of the solid sample the light intensity is constant and equal to I_0. A general solution of equation (15.22) is thus possible:

$$C(0,t) = C_0 \times e^{-\Phi \times I_0 / N_A \times \mu_R \times t} \tag{15.24}$$

For the following layers, the concentration at each time is calculated from:

$$C(b,t+1) = C(b,t) - \frac{\Phi}{N_A} \times I_0 \times e^{-[\Delta b \times (\mu_R - \mu_P) \times \Sigma_{b=0}^{b-1} C(b,t) + \mu_P \times C_0 \times b]} \times \mu_R \times C(b,t) \times \Delta t \tag{15.25}$$

Where Δb is the thickness of each layer and Δt is the timespan over which the rate of reaction is considered constant. The initial concentration for each layer is $C(b,0) = C_0$. Having calculated the concentration for each layer, the total concentration of the sample may be calculated by averaging the concentrations over all layers for each specific time according to:

$$C_{tot}(t) = \frac{\Delta b}{B} \times \sum_{b=0}^{B} C(b,t) \tag{15.26}$$

The two domains over which the degradation rate and light intensity are considered constant (Δt and Δb) have to be chosen carefully in order to obtain valid results. The actual values are dependent on the values of the other parameters and should be chosen on the basis of the change in concentration and light intensity over the interval. These changes should be less than a predetermined value (1–10 per cent depending on the accuracy needed). Employing a too high value for Δt and Δb produces unstable calculations (oscillation) or negative concentrations.

Equation (15.25) consists of one physical constant (N_A), five parameters (Φ, I_0, μ_R, μ_P, C_0) and two variables (b,t). The aim of the following simulations is to clarify how the five parameters influence the rate of degradation. This is done by calculating the concentrations in each layer for varying values of the parameters. The average concentration (C_{tot}) as a function of time and the concentration in each layer at the time where $C_{tot} = C_0/2$ (concentration cross-section) are subsequently examined in order to evaluate the influence of the individual parameters.

It should be noted that the actual values employed for the parameters during the simulations are of no practical value taken directly and are merely reported to indicate the range over which the parameters have been varied.

Since this simulation is dealing with pure substance, C_0 is determined by crystal structure and powder density, and is thus more or less outside our control. In these simulations C_0 is therefore kept constant at 100 (per cent). On a more general basis it is worth noting that the model is actually describing a change in number of molecules, and C_0 and total depth of the powder will therefore be interrelated. The parameters μ_R and μ_P are also interrelated. Variation in the sum and ratio of these parameters will therefore provide most information. I_0 and Φ are interchangeable and are therefore varied together during the simulation.

15.8 Results of simulation

Depth of the powder bed. This parameter is not a direct part of the model, and increasing the depth has of course no influence on the upper layers of the sample. The measured total concentration is, however, affected since a thicker powder bed will apparently result in a slower degradation. The light will penetrate only fractions of a millimetre in a solid sample, so as long as the total depth of the sample is larger than the depth penetrated by the light, only apparent changes are made to the time-profile of C_{tot}.

Quantum yield (Φ) and incident light intensity (I_0). Increase in the factor $\Phi \times I_0/N_A$ produces the expected increase in degradation rate, as can be seen in Fig. 15.4. The concentration cross-section at the time of 50 per cent total degradation shows no differences indicating lack of mechanistic differences. These parameters may consequently be viewed solely as a scaling parameter for the time axis.

Sum of molar absorptivity of the reactant (μ_R) and molar absorptivity of the product (μ_P). In these simulations the ratio between μ_R and μ_P is kept constant (= 1). In Fig. 15.5 it may be seen that for low values of absorptivity, increased absorption by the parent compound produces an increase in the rate of degradation. This may be explained by the increased amount of photons providing energy for the degradation. As the molar absorptivity is further increased, light penetration into the lower layers is blocked and the total degradation rate will decrease. It should be noted that for the top layer a constantly increased degradation rate is observed. Also noteworthy is the change in profile from a near-linear degradation for the lower to a nearly exponential decay for the higher absorptivities.

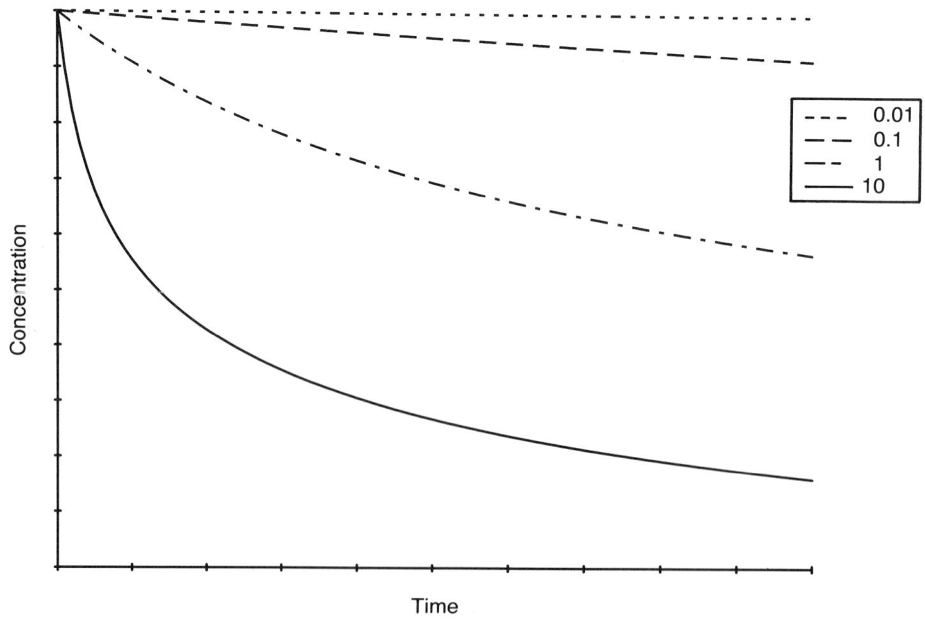

Figure 15.4 Total concentration as a function of irradiation time for several values of $\Phi \times I_0/N_A$

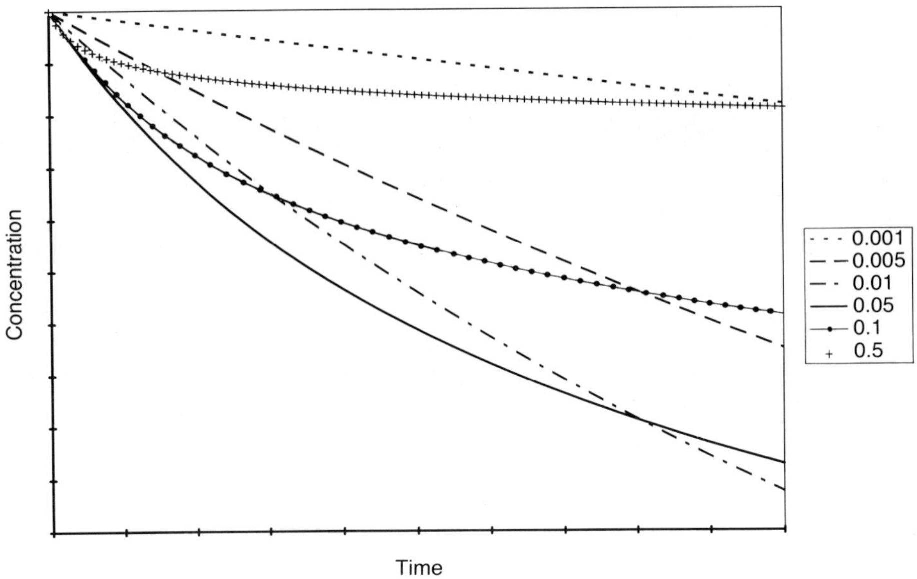

Figure 15.5 Total concentration as a function of irradiation time for several equal values of molar absorptivity of reactant and product (μ_R and μ_P)

The effect of increased absorption on the cross-section illustrates the same concept (Fig. 15.6). For low absorptivities the degradation is uniform throughout the powder (nearly horizontal profile) with a gradually increasing difference as the molar absorptivity increases. Eventually a front of degraded substance is created virtually moving through the powder bed as time proceeds. Above this front nearly all substance is degraded, and below nearly nothing.

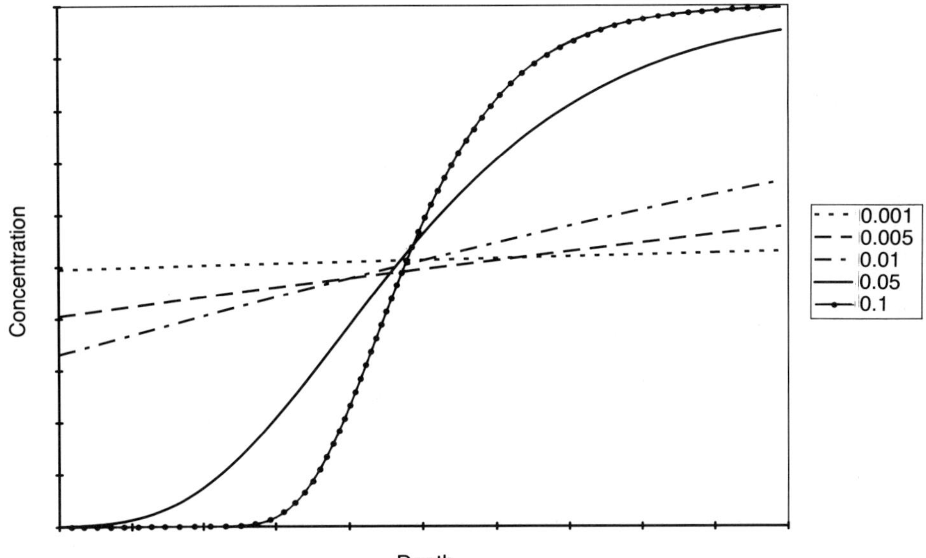

Figure 15.6 Concentration cross-section at time = $T_{1/2}$ for several equal values of molar absorptivity of reactant and product (μ_R and μ_P)

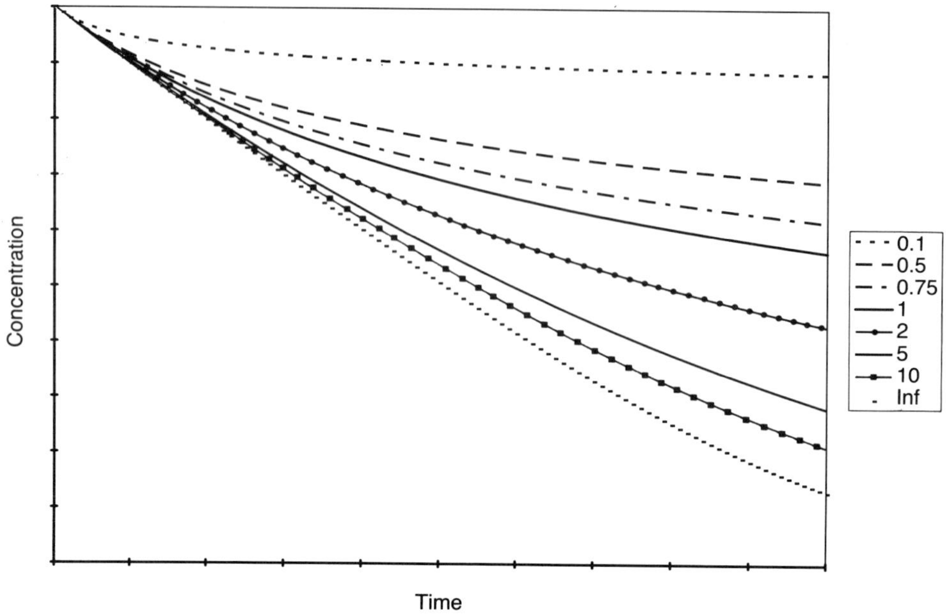

Figure 15.7 Total concentration as a function of irradiation time for several ratios of molar absorptivity of reactant and product (μ_R/μ_P)

Ratio of molar absorptivity of the reactant (μ_R) and molar absorptivity of the product (μ_P). In this simulation μ_R is kept constant and μ_P decreased to obtain increasing ratios. This increase produces an increasingly rapid reduction in C_{tot} as demonstrated in Fig. 15.7. This is natural since a relative increase in the amount

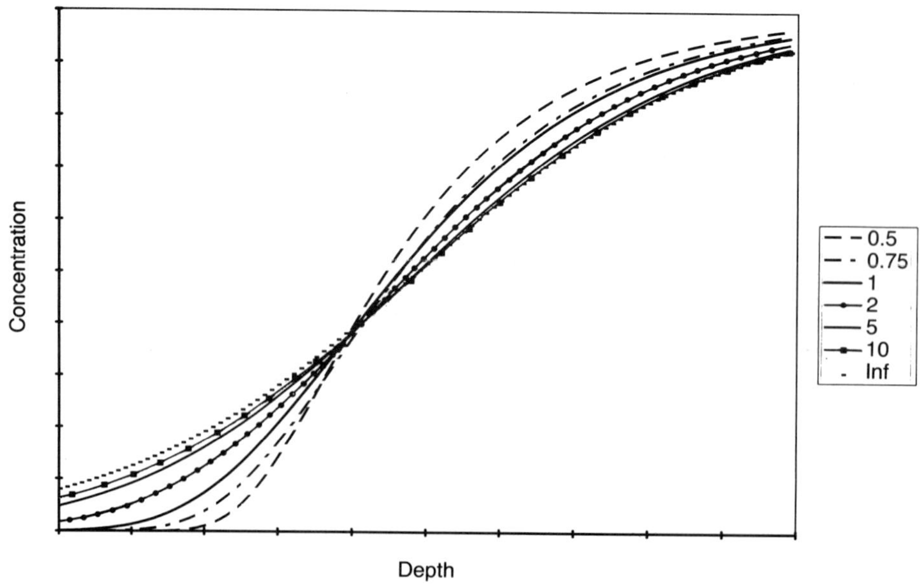

Figure 15.8 Concentration cross-section for time = $T_{1/2}$ for several ratios of molar absorptivity of reactant and product (μ_R/μ_P)

of absorbed light by the reacting substance will increase the number of reacting molecules.

According to Fig. 15.8, increase in the ratio leads to a deeper penetration of the degradation into the powder. This is the same trend as observed for the sum of μ_R and μ_P, and since μ_R is kept constant and μ_P decreased this may explain the observed effect. The ratio thus seems to show only scarce influence on the cross-section.

15.9 Effect of temperature

The effect of temperature has not been considered in these simulations. The reason is that temperature is not expected to affect the absorption of photons as such and no energy in addition to the photon is required for the reaction to take place (activation energy = 0).

The temperature will of course affect subsequent chemical degradations in the usual manner as described by the Arrhenius equation. If secondary thermal reactions are taking place, a temperature effect on the overall reaction quantum yield would be expected. If the reaction is taking place in solution, an indirect temperature dependence caused by, for example, change in solvent viscosity may also be expected (Balzani and Carassiti, 1970). Since light-irradiation without an increase in temperature is difficult to achieve, this change in the experimental conditions must be taken into account. It is, however, not to be expected that the

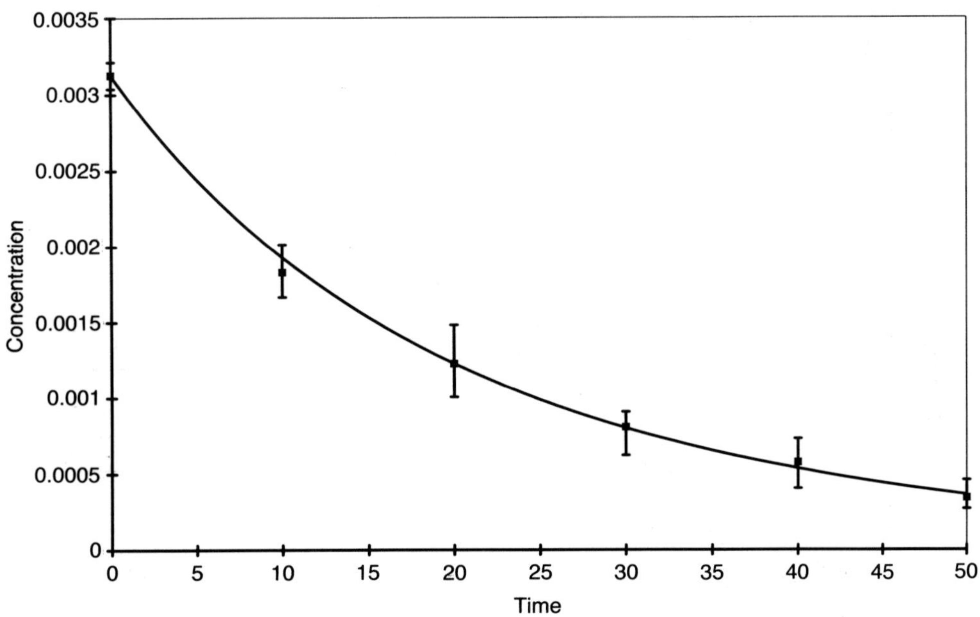

Figure 15.9 Measured values for concentration of crystal form I of chloroquine as a function of irradiation time (mean, maximum and minimum, n = 4) with estimated function according to equation (15.23)

effects on the reaction rate would differ from increase in temperature by other means. Thus the two types of influences may be treated as separate entities.

15.10 Experimental example

To demonstrate a practical application of the described model, the transition from one crystal form of chloroquine (CQ) to another was measured by differential scanning calorimetry. Samples of CQ were irradiated for 0 to 50 h in a Suntest CPS (Heraeus Instruments GmbH, D-Hanau) employing a windowglass filter. Since the samples consisted of pure substance and there were no chemical changes (constant μ), equation (15.23) was employed for the calculations. C_0 was calculated from the powder density and molecular weight of CQ (320). I_0 for the individual wavelengths was estimated from a plot of spectral radiation provided by the manufacturer, and corrected for the transmission of the windowglass. μ was estimated from the absorbance spectrum of CQ in the non-polar solvent EPA (diethyl ether, isopentane and ethanol 5:5:2) with 0.5 per cent 1 mol/l HCl. The quantum yield was estimated from the data.

As can be seen from Fig. 15.9 the estimated profile is in good accordance with the measured data.

Acknowledgements

I would like to thank Prof. H. H. Tønnesen and senior lecturer K. Nord for invaluable discussions and information provided during my work with this chapter. I will also express my gratitude towards Prof. T. Waaler for valuable comments during preparation of this paper. Finally, I would like to thank H. Andersen for providing the experimental results employed in demonstrating the practical use of the mathematical model.

References

Arney, J. S., Dowler, J. A. and Ruder, D. L., 1989, Radiation attenuation effects in photosensitive coatings of microcapsules, *J. Imag. Sci.*, **33**, 184–8.

Asker, A. F., Canady, D. and Cobb, C., 1985, Influence of DL-methionine on the photostability of ascorbic acid solutions, *Drug. Dev. Ind. Pharm.*, **11**, 2109–25.

Balzani, V. and Carassiti, V., 1970, *Photochemistry of Coordination Compounds.* London: Academic Press, pp. 11–13.

Bamford, C. H. and Tipper, C. F. H. (eds), 1980, Reactions in the solid state Comprehensive Chemical Kinetics, Vol. 22, pp. 41–109. Amsterdam: Elsevier.

Byrn, S. R., 1982, *Solid-state Chemistry of Drugs*. London: Academic Press, pp. 259–82.

Carstensen, J. T. (ed.), 1990, Drug stability: Principles and practices, *Drugs and the pharmaceutical sciences*, Vol. 43, pp. 99–101. New York: Marcel Dekker Inc.

Connors, K. A., Amidon, G. L. and Valentino, J. S., 1986, *Chemical Stability of Pharmaceuticals*, New York: Wiley, p. 59.

De Villiers, M. M., van der Watt, J. G. and Lötter, A. P., 1992, Kinetic study of the solid-state photolytic degradation of two polymorphic forms of furosemide, *Int. J. Pharm.*, **88**, 275–83.

FELMEISTER, A., SCHAUBMAN, R. and HOWE, H., 1965, Dismutation of a semiquinone free radical of chlorpromazine, *J. Pharm. Sci.*, **54**, 1589–93.

HO, W., 1992, Fundamental mechanisms of surface photochemistry, *Res. Chem. Interm.*, **17**, 27–38.

LERNER, D. A., BONNEFORD, G., FABRE, H., MANDROU, B. and DEBOUCHBERG, M. S., 1988, Photodegradation paths of cefotaxime, *J. Pharm. Sci.*, **77**, 699–703.

MATSUDA, Y. and MASAHARA, T., 1983, Photostability of solid-state ubidecarenon at ordinary and elevated temperatures under exaggerated UV irradiation, *J. Pharm. Sci.*, **72**, 1198–203.

MOORE, D., 1987, Principles and practice of drug photodegradation studies, *J. Pharm. Biomed. Anal.*, **5**, 441–53.

PAULSEN, M.-H., 1990, Lysfordeling i hud i bølgeområdet rundt 500 nm. (English summary), SIS report, National Institute of Radiation Hygiene, Oslo, Norway.

SANDE, S. A. and KARLSEN, J., 1991, Curve fitting of stability data by personal computer, *Int. J. Pharm.*, **73**, 147–56.

1993, Evaluation of reaction order, *Int. J. Pharm.*, **98**, 209–18.

TAYLOR, R. B. and SHIVJI, A. S. H., 1987, A critical appraisal of drug stability testing methods, *Pharm. Res.*, **4**, 177–80.

TØNNESEN, H. H., 1991, Photochemical degradation of components in drug formulations, *Pharmazie*, **46**, 263–5.

TREUSHNIKOV, V. M. and YANIN, A. M., 1990, A mathematical description of the kinetics of photochemical processes in solids, *J. Sci. Appl. Photo. Cinema.*, **32**, 351–9.

YANG, W., 1981, Errors in the estimation of the activation energy and the projected shelf life in employing an incorrect kinetic order in an accelerated stability test, *Drug. Dev. Ind. Pharm.*, **7**, 717–38.

16

The Application of Photoacoustic Spectroscopy to the Photodegradation of Drugs

R. S. DAVIDSON

16.1 Introduction

The problem of monitoring the photodegradation of drugs is very similar to that of monitoring the degradation of other molecular species, e.g. dyes and polymers. In every case we need to find a 'signal' that will enable the degradation process to be monitored. Usually, such a signal is related to some molecular property, for example, a chromophore being destroyed or created. In the case of photooxidation processes the rate of absorption of oxygen can be used to follow the course of the reaction. Sometimes we have to rely upon physical changes such as embrittlement or softening which, although related to molecular changes, contain information due to other processes, e.g., a change in surface properties and therefore renders interpretation of the data more difficult.

What type of processes can be monitored spectroscopically? Since we are dealing with *photo*degradation, changes in photophysical properties such as the quantum yields of fluorescence and phosphorescence, fluorescence and phosphorescence lifetimes may be helpful. Photodegradation, in common with other degradation processes, may involve molecular rearrangement, oxidation or reduction, elimination reactions, polymerization and depolymerization processes. Where these processes involve extended chromophores, fluorescence and ultraviolet/visible (UV/VIS) electronic absorption spectroscopy can prove to be very useful. In many cases, the absorption bands are broad and featureless and this makes interpretation of spectra and hence unravelling the changes that have occurred at the molecular level, difficult. If the degradation process involves loss of a substituent such as halogen or molecular rearrangement, infrared spectroscopy (IR) is likely to be more useful.

Where does photoacoustic spectroscopy (PAS) fit into this scenario? PAS is not an alternative to UV/VIS and IR spectroscopy. PAS covers the UV, visible, near-IR and IR regions and offers an alternative way of detecting the signal. All absorption spectroscopies rely on measuring the intensity of the incident, and the transmitted radiation and reflectance techniques rely upon measuring the intensity

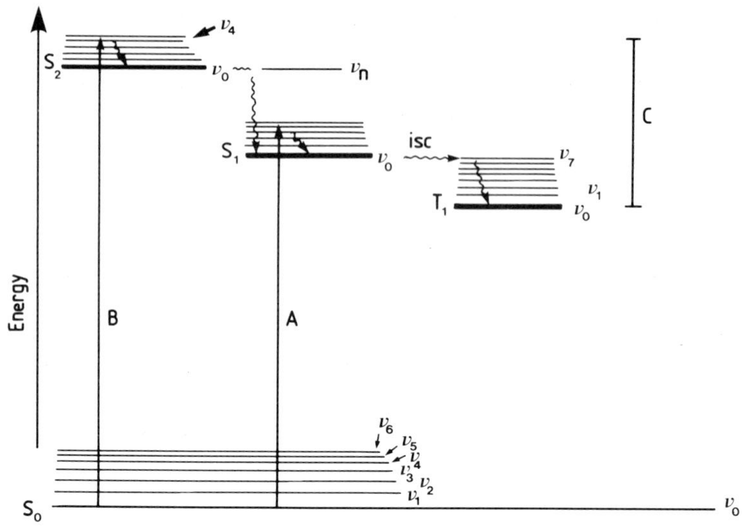

Figure 16.1 Jablonski diagram. C = heat lost following internal conversion and intersystem crossing after populating S_2 at the fourth vibrational level

of the incident and reflected radiation. Photoacoustic spectroscopy is very different.

In 1881 Alexander Graham Bell (Bell, 1881) found that if solid matter was placed inside a glass tube which had a hearing-tube (a form of microphone) attached an audible signal could be detected if the material in the tube was illuminated with a rapidly interrupted beam of sunlight. Of particular importance to the subject being discussed here, Bell found that the loudest signals are produced from substances in a loose, porous, spongy condition and from those that have the darkest or most absorbent colours.

What is the origin of the photoacoustic effect? Absorption of UV/visible radiation results in the population of excited states. For most compounds in the solid state or in fluid solution, excitation produces a vibrationally excited singlet state, which loses energy as *heat* to give the excited singlet state in its lowest vibrational level (Fig. 16.1). This may decay to the ground state by emission of radiation (fluorescence) or radiationless decay (i.e. the energy is lost as heat). Alternatively or concomitantly, the excited singlet state may undergo an isothermal process (intersystem crossing) to give a vibrationally excited singlet state which loses energy as *heat* to give the lowest triplet state in its lowest vibrational level. This state may relax to the ground state by emission of light (phosphorescence) or by radiationless decay (i.e. the energy is lost as *heat*). We have identified from the Jablonski diagram (Fig. 16.1) where heat is evolved. If chemical reaction occurs from an excited state, the amount of heat generated will be directly related to the energy requirements of the reaction.

By examining the Jablonski diagram in detail we can see that not only are there several different processes that degrade light into heat but also that the evolution of the signal due to loss of heat occurs on a timescale that reflects the timescale of the photophysical effects. Thus in principle we should be able to measure the lifetime of excited singlet and triplet states, the rate of intersystem crossing and the rate of chemical reactions. In the last few years enormous progress has been made in following the fate of excited states by photoacoustic spectroscopy (Jabben *et al.*, 1984). If, rather than following the heat loss as a function of time, we measure the total heat output, then this value together with other determined at a series of wavelengths will give us an absorption spectrum. Since the absorption spectrum is not obtained by measuring the intensity of the non-absorbed radiation, the PA signal should not be so dependent upon light-scattering effects. For this reason PAS was solid on the basis that unlike related spectroscopies, no sample preparation was necessary (Rosencwaig, 1973). As we shall see later, this claim is not entirely true.

PAS has been used for recording near-infrared spectra. Such spectra usually offer a good fingerprint spectrum but it can be difficult to make unequivocal assignments to absorption bands. Another problem is that the presence of atmospheric moisture can severely affect the quality of the spectra obtained.

PAS can be very usefully applied to acquiring infrared spectra. The use of a PA cell in conjunction with an FTIR spectrometer can yield excellent spectra for samples which are difficult to handle by other sampling techniques e.g. surface coatings on paper (Davidson *et al.*, 1987). Absorption of IR radiation leads to motion of groups, for example carbonyl stretching which leads to degradation of the absorbed energy to heat.

16.2 The Photoacoustic signal – samples as films

For recording the spectra of samples in the ground state, as opposed to the excited state, the heat evolution process is monitored by a microphone as Bell did in 1881. The cell design and sensitivity of the microphones have undergone substantial improvement over the intervening years. A typical cell design is shown in Fig. 16.2.

When the incident radiation is absorbed by the sample and radiationless processes ensue, heat is generated at different depths in the sample. If we consider the idealized situation shown in Fig. 16.3, heat is generated at sites 1, 2 and 3.

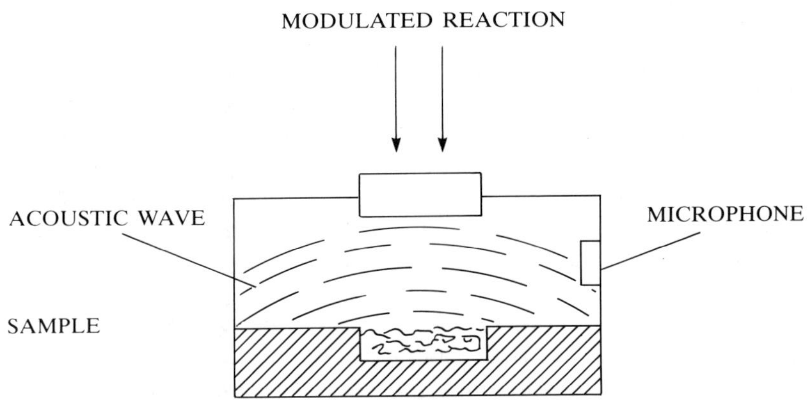

Figure 16.2 Schematic drawing of a photoacoustic cell

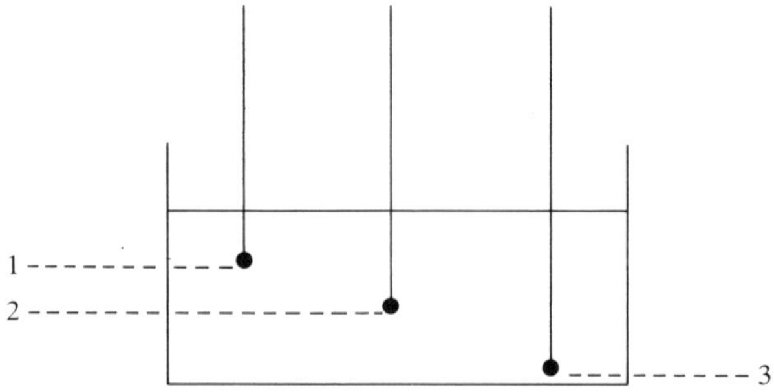

Figure 16.3 An idealized situation where heat is evolved at three different depths

The heat signal generated from site 1 will be modulated since the incident radiation is modulated. Thus, from site 1 a modulated thermal wave is generated and when this reaches the surface of the sample the temperature of the surface changes. The periodic variation in the temperature at the surface of the sample results in the generation of an acoustic wave in the adjacent gas and this wave propagates through the volume of the gas to the microphone where a signal is produced. A similar train of events occurs for the heat evolved at sites 2 and 3.

If for the sake of simplicity we consider our sample to be a homogeneous solid and that the incident light reaches the bottom of the sample, the thermal waves from sites 1, 2 and 3 will arrive at progressively longer intervals with the magnitude of the separation interval being dependent upon the thermal diffusivity of the sample. Thus, for a signal to arrive from site 3, it is necessary for the thickness of the sample to be less than $2\pi\mu_S$ where μ_S is the thermal diffusion length. When this is the case, high modulation frequencies will favour seeing signals from site 1 and low modulation frequencies will favour seeing signals from site 3. An example of this situation is given in Fig. 16.4 which shows the spectra obtained from a laminate formed from covering a piece of blue paper with sellotape. When a high modulation frequency is used the materials near the surface of the sellotape are seen. It is only at low modulation frequencies that the spectrum of the underlying blue layer can be seen.

Clearly, the thermal properties of the sample determine the maximum depth from which a signal can be recovered. The simple relationship shown in equation 16.1 illustrates how the depth examined (μ_S) may be varied by adjusting the modulation frequency w.

$$\mu_S = \frac{2x}{\omega} \tag{16.1}$$

where μ_S is depth examined; x is the thermal diffusivity and ω = modulation frequency

Keeping with our homogenous sample, the situation could arise that for certain wavelengths the light does not penetrate to the bottom of the sample. Needless to say we will not gain information from where the light has not penetrated! However, will we gain information for that part of the sample which was interrogated by the light? If the thermal diffusion length (μ_S) is greater than the

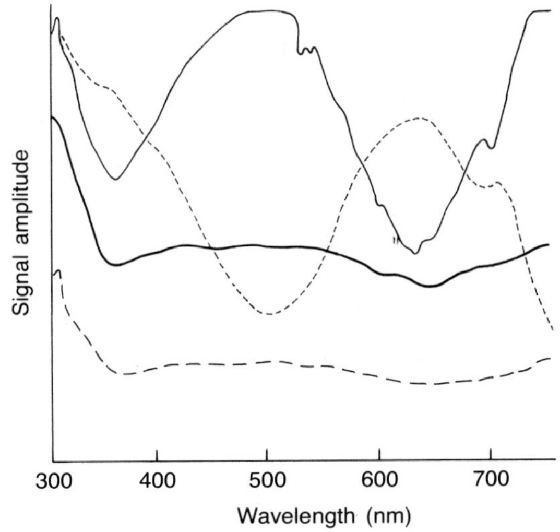

Figure 16.4 PA spectrum of sellotape on the surface of coloured foil recorded at various chopping frequencies — = 240 Hz; — = 80 Hz; — — — = 40 Hz; - - - = 10 Hz

depth penetrated by the light (l_β) the answer is no. Under these circumstances we are considering an opaque sample and when all the light is absorbed over a range of frequencies the thermal signal will be very similar and under these circumstances we are in a situation known as 'saturation', and consequently, although there is a thermal signal, no spectral information is obtained. If μ_s is less than l_β a spectrum can be recorded.

We have for a solid homogeneous sample the situations depicted in Fig. 16.5.

An extremely important practical point that is worth bearing in mind is that as the modulation frequency is increased, so the signal intensity decreases (the

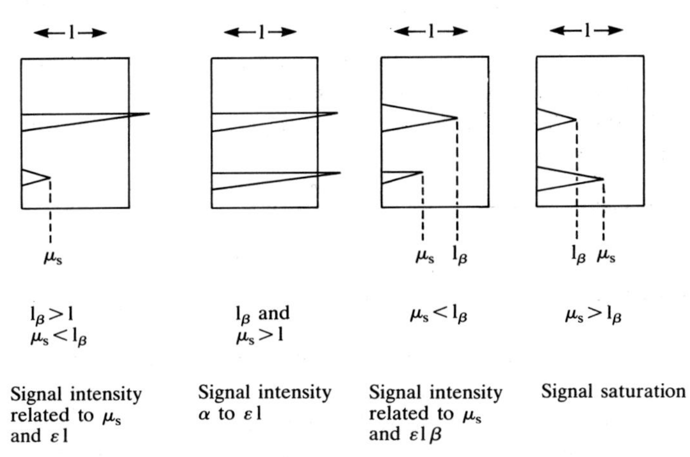

Figure 16.5 Effect of varying thermal and optical depth upon the PAS signal

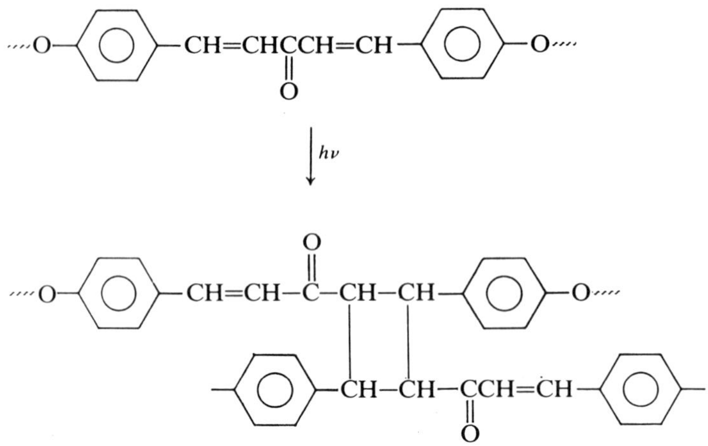

Figure 16.6 Photocrosslinking of a chalcone

light is resident on the sample for less time and hence less heat is evolved). The quality of a PA signal is improved by increasing the intensity of the light beam since this increases the heat output.

To illustrate these points let us have a look at a photopolymerizable system (Davidson and Lowe, 1989). When the chalcone (Fig. 16.6) is irradiated it undergoes a [2 + 2] cycloaddition reaction thereby leading to crosslinking and hence polymerization.

A PA spectrum of a 2.5-μm-thick film is shown in Fig. 16.7. The spectrum exhibits no spectral features such as a λ_{max} around 350 nm which is what you would

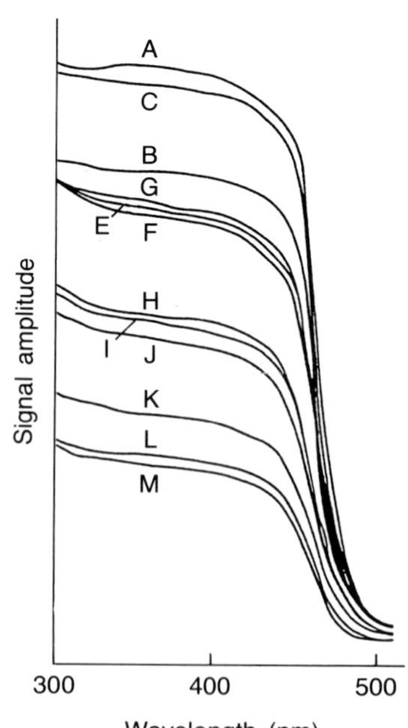

Figure 16.7 PA spectrum of polymer film (2.5 μm). Chopping frequencies: 10 Hz (A,B,C); 80 Hz (E,F,G); 160 Hz (H,I,J); 240 Hz (K,L,M)

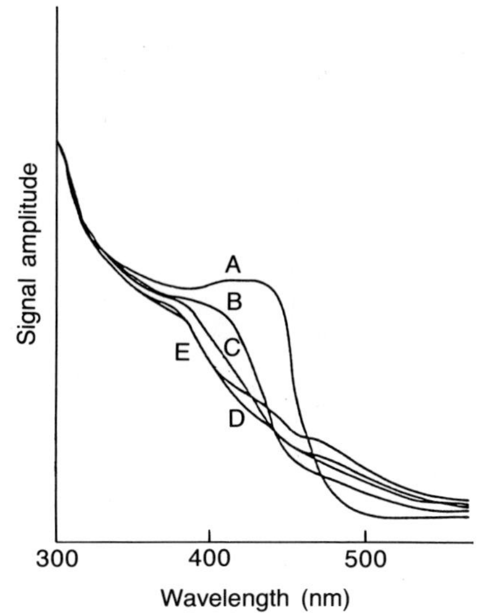

Figure 16.8 Effect of irradiation upon polymer film (2.5 μm) in the PA cell and monitored by PAS. Irradiation time (A) 0 min, (B) 10 min, (C) 20 min, (D) 30 min, (E) 60 min

expect. Instead, the spectra show an absorption edge and then it 'levels' off, i.e. we are observing signal saturation. When the chopping frequency is increased from 10 to 240 Hz no improvement in the spectral information is achieved and at the higher chopping frequency the signal is weaker. On irradiation of these films

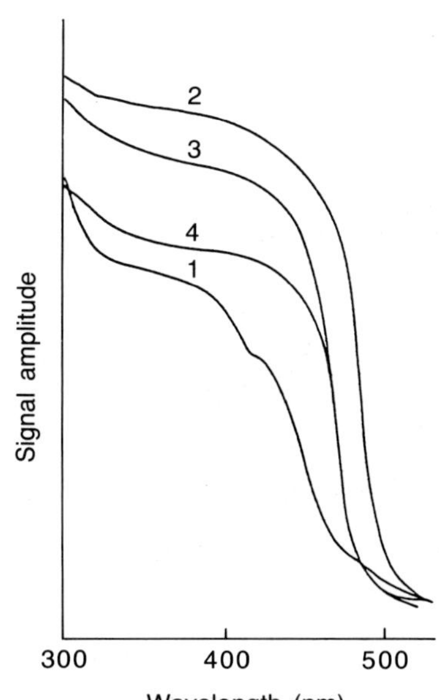

Figure 16.9 PA spectrum of thick polymer film (35 μm) irradiated for 15 min. (1) Irradiated face, chopping frequency 240 Hz; (2) irradiated face, chopping frequency 10 Hz; (3) unirradiated face, chopping frequency 240 Hz; (4) unirradiated face, chopping frequency 10 Hz

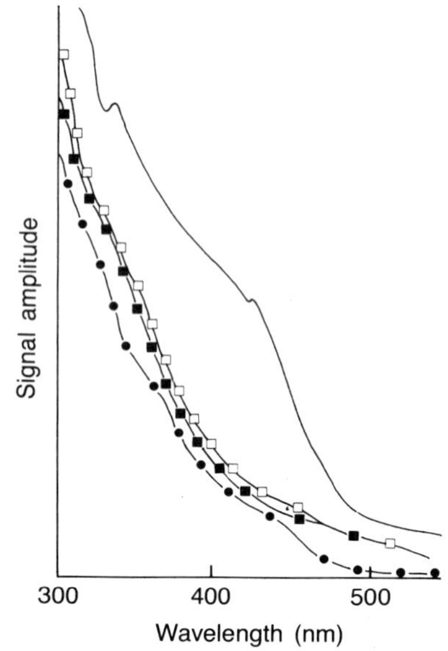

Figure 16.10 PA spectrum of a thin film of polymer irradiated in the PA cell for different periods. —— = unirradiated; -□-□- = irradiated for 0.5 min; -■-■- = irradiated for 0.75 min; -●-●- = irradiated for 1.5 min

double bonds are consumed with consequent reduction in absorption intensity and possibly a moving to the blue of the absorption maximum. The PA spectra of irradiated films (Fig. 16.8) show the band edge progressively moving to shorter wavelengths and the over-irradiation causing some unwanted reactions which manifest themselves as yellowing (curve E, Fig. 16.8). If a similar experiment is carried out but with a much thicker film (35 μm) the effect of radiation upon the illuminated face can only be seen when a high modulation frequency is employed (curve 1, Fig. 16.9). With a lower modulation frequency (10 Hz) the illuminated face shows the same spectral characteristics as the non-irradiated face. When an ultra-thin film of the polymer is laid down, its PA spectrum shows a little more detail than the thicker film (Fig. 16.10). The effects of radiation upon this film are much easier to see and one would conclude from the spectra that in all probability both double bonds in the chalcone chromophore are being destroyed. Attempts to follow this reaction by FTIR-PAS were frustrated by the signals due to the carbonyl group exhibiting saturation. It will have been noted that the situation of signal saturation arises when $\mu_s < l_\beta < l$. When recording an infrared spectrum this can occur for some bands and not others since the extinction coefficients for the bands (ε values) can vary enormously and hence the depth of light penetration also changes.

16.3 The photoacoustic signal – samples in powder form

So far we have been concerned with a sample in slab form, for example a piece of coal or a section of polymer. This seems to be a far cry from the more usual pharmaceutical preparation.

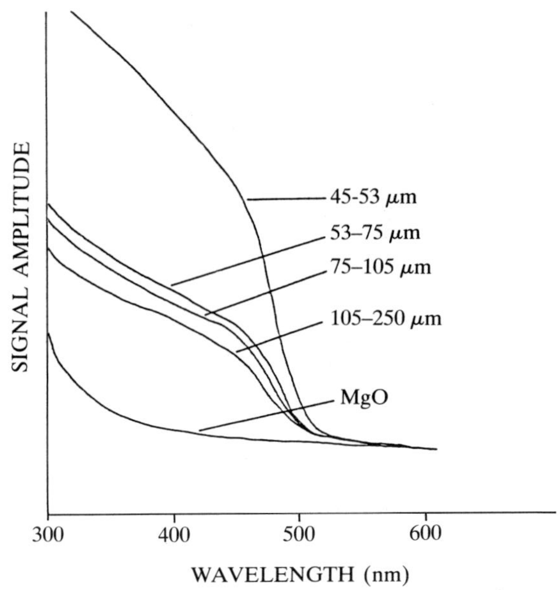

Figure 16.11 PA spectra of potassium chromate particles

Let us consider particulate materials, e.g. samples in powder form. Going back to the observation by Bell that the best signals are observed for loose, porous and spongy samples, we can expect materials in powder form to give good signals.

What factors should influence the intensity of the PA signal of a powder? The extinction coefficients (ε) will be very important and will in part determine whether or not a signal is in or out of saturation. Particle size which relates to the surface

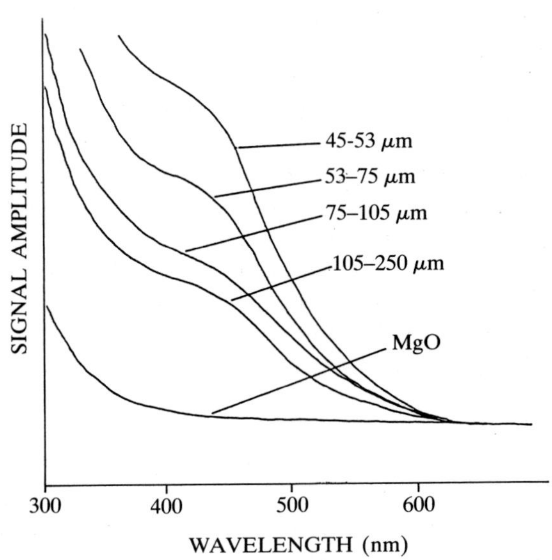

Figure 16.12 PA spectra of potassium ferricyanide particles

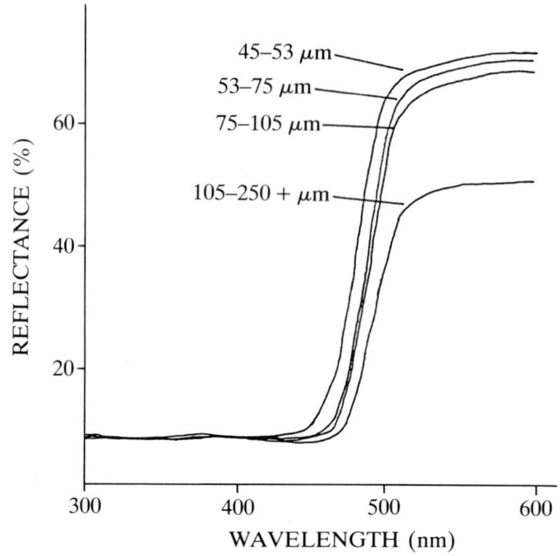

Figure 16.13 Reflectance spectra of potassium chromate particles

area of the particles and the porosity of the particles will affect the heat transfer process (Burgraff and Leyden, 1981).

Does PA spectroscopy offer any advantage over diffuse reflectance (DR) spectroscopy for the recording of spectra of powder sample? A common feature is that in both cases the signal amplitude is dependent upon the light absorption. For weakly to moderately absorbing materials the extent of reflectance increases with decreasing particle size and as a consequence the effective depth of penetration decreases; hence there is also a decrease in the fraction of light

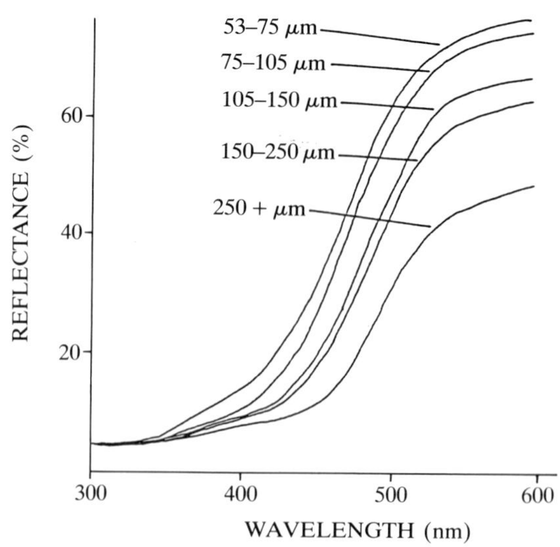

Figure 16.14 Reflectance spectra of potassium ferricyanide particles

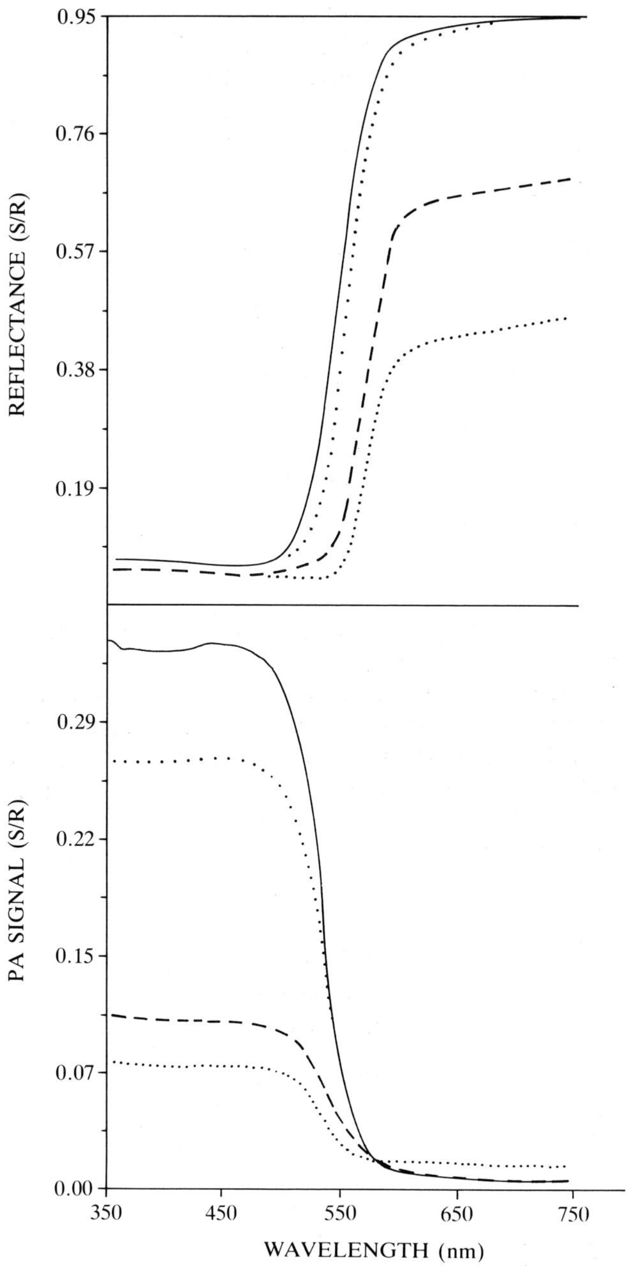

Figure 16.15 PA and DR spectra of potassium dichromate crystals of various particle sizes (x) (——) $x < 50$ μm; (···) $100 > x > 50$ μm; (--) $250 > x > 100$ μm; (–––) $500 > x > 250$ μm

available for absorption. Thus, both the PA and DR signals should *decrease* in intensity if this is the only factor to be taken into account (Childers *et al.*, 1986). As the particle size becomes larger the scattering coefficient becomes smaller and hence the light can penetrate deeper into the sample and thereby increases the amount of light absorbed. Internal reflection within the larger crystals can also lead to an increase in the amount of light absorbed. These effects can lead to a distorted or saturated DR spectrum but may enhance a PA spectrum. When strongly absorbing materials are used, DR spectra are less affected by light scattering and on the whole the spectra are independent of particle size.

The PA signal is dependent upon the amount of light absorbed, the efficiency with which the heat is released and the efficiency of transformation of the heat into an acoustic wave. With very small particles the thermal diffusion length is probably greater than the particle size and hence the heat produced can be transformed into a measurable signal. With strongly absorbing particles this may not necessarily be the case and hence the PA signal may not show much if any increase in signal with decrease in particle size.

These effects can be demonstrated by comparing the DR and PA signal from powdered samples of potassium chromate ($\varepsilon \times 10^3 = 2.5\ \mathrm{L mol^{-1} cm^{-1}}$ at 370 nm) and potassium ferricyanide ($\varepsilon \times 10^3 = 0.8\ \mathrm{L mol^{-1} cm^{-1}}$ at 420 nm). PA spectra for the materials are shown in Figs 16.11 and 16.12 (Davidson *et al.*, 1984). These spectra (recorded at 10 Hz) show quite clearly that the signal amplitude increases with decreasing particle size. The corresponding diffuse reflectance spectra (Figs 16.13 and 16.14) show that as the particle size increases so the signal amplitude increases. For the more weakly absorbing potassium ferricyanide there is less loss of spectral information than the more strongly absorbing potassium chromate. In

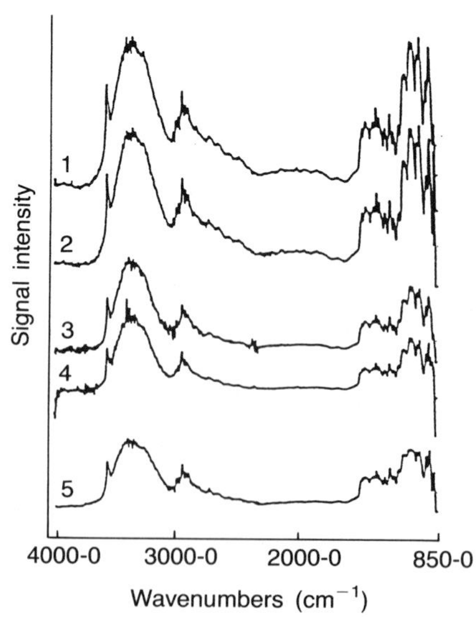

Figure 16.16 Mid-infrared Fourier transform photoacoustic spectra of sucrose powders with different particle sizes: 1, 33 μm; 2, 41.5 μm; 3, 69 μm; 4, 82.5 μm; 5, 186 μm. Photoacoustic intensities were ratioed to that of a finely divided carbon black material to correct for the output spectrum of the source

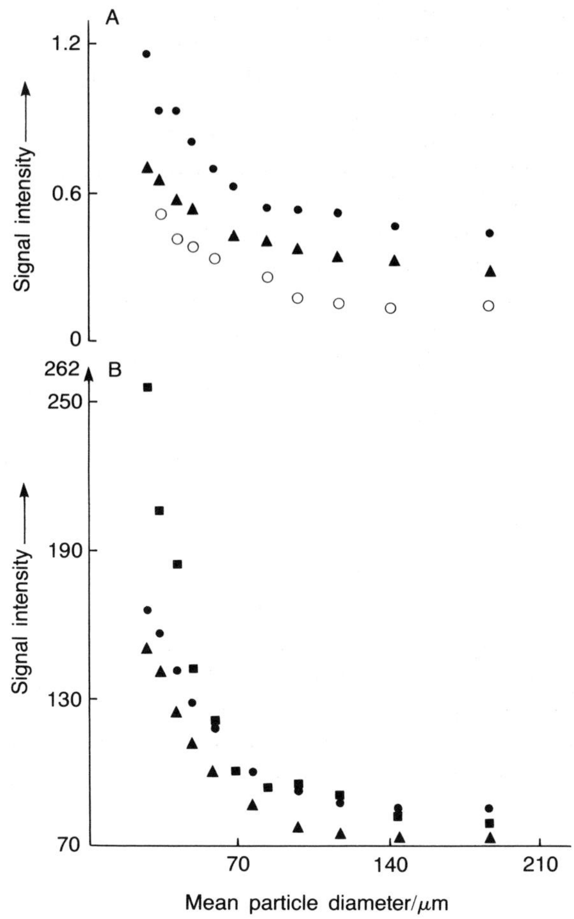

Figure 16.17 (A) Near-infrared dispersive PAS intensities at $4762\,cm^{-1}$ of (●) carbon black, (▲) pure sucrose, and (○) a mixture of sucrose and potassium bromide containing 40 per cent sucrose. (B) Mid-infrared FTIR-PAS intensities at (■) $2940\,cm^{-1}$ for carbon black, (●) $2940\,cm^{-1}$ for sucrose, and (▲) $1430\,cm^{-1}$ for sucrose

these cases the PA spectra are suffering less from saturation than the DR spectra. This is also seen with the more strongly absorbing potassium dichromate ($\varepsilon \times 10^3 = 3.5\,L mol^{-1} cm^{-1}$ at 362 nm) (Fig. 16.15) (Childers *et al.*, 1986).

The efficiency of heat transfer from solid to gas is of the utmost importance if the best results are to be obtained from PAS, demonstrated by a study on sucrose – carbon black and sucrose potassium bromide mixtures. Figure 16.16 shows that the FTIR PAS signal decreases with increasing particle size. A similar situation was obtained for sucrose, sucrose/potassium bromide mixture and carbon black samples when examined by near-IR PAS (Fig. 16.17) (Belton *et al.*, 1987). This particular study was extended by showing that an increase in porosity of the sample also led to increased signal intensity (Fig. 16.18).

A possible way of increasing porosity is to adsorb material onto a highly porous support (King *et al.*, 1982). Potassium dichromate is a strong absorber and when

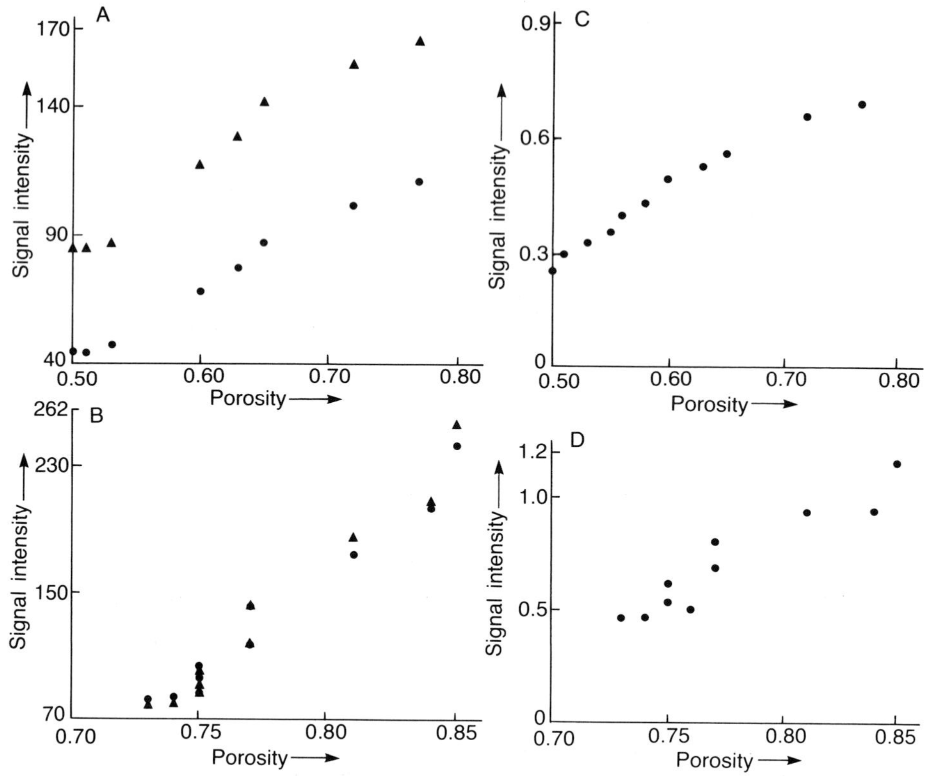

Figure 16.18 Variation of the intensity of photoacoustic signals with porosity: (A) mid-infrared FTIR-PAS at (▲) 2940 cm^{-1} in sucrose and (•) 1430 cm^{-1} in sucrose; (B) mid-infrared FTIR-PAS at (▲) 2940 cm^{-1} in carbon black and (•) 1430 cm^{-1} in carbon black; (C) near-infrared dispersive PAS at 4762 cm^{-1} in sucrose; (D) near-infrared dispersive PAS at 4762 cm^{-1} in carbon black

diluted by grinding with barium sulphate, alumina and lithium fluoride relatively poor spectra were obtained. However, when silica and magnesium oxide were used (Figs. 16.19 and 16.20) good spectral resolution was achieved. Since good spectra were obtained, the relationship between signal intensity and the concentration was studied. As can be seen from Fig. 16.21 a non-linear relationship was found for silica but this was not the case for magnesium oxide. If PAS is to be used quantitatively it is absolutely necessary that a calibration graph be obtained.

Not surprisingly PAS has proved useful for studying the modification of silicas by covalently linking organic species to the surface (Davidson *et al.*, 1981). Thus the two HPLC support materials shown in Fig. 16.22 have been studied and from the spectra it can be shown that there is site-to-site interaction between the amino groups and the picramido groups as evidenced by the appearance of a charge transfer band at 410 nm.

PA spectroscopy has been used for quantitative TLC (Rosencwaig and Hall, 1975; Castleden *et al.*, 1979). When fluorescein is adsorbed onto silica plates it is claimed that a linear relationship exists between the PA signal amplitude and

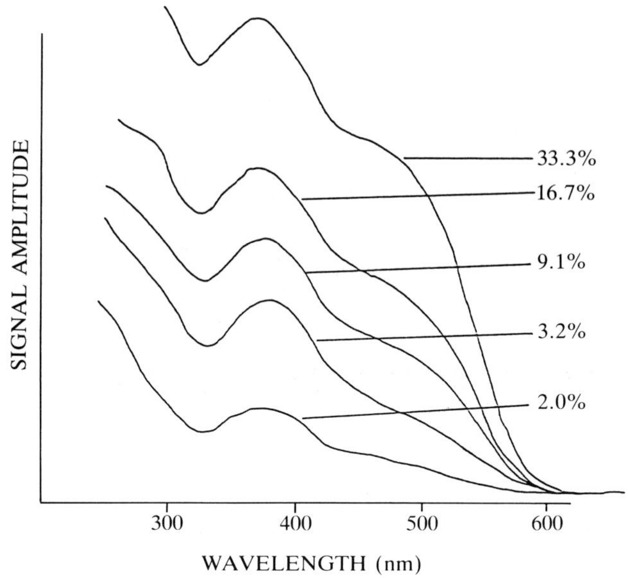

Figure 16.19 PA spectra of potassium dichromate diluted by grinding with silica

the amount of fluorescein. With the advent of open-ended PA cells (Takamoto *et al.*, 1992) PA spectroscopy in conjunction with TLC analysis may become very useful in monitoring the photodegradation of a variety of materials including drugs.

Compounds are often adsorbed onto supports for a variety of reasons. When fabrics are dyed it is hoped that the dye will not only be adsorbed but also that

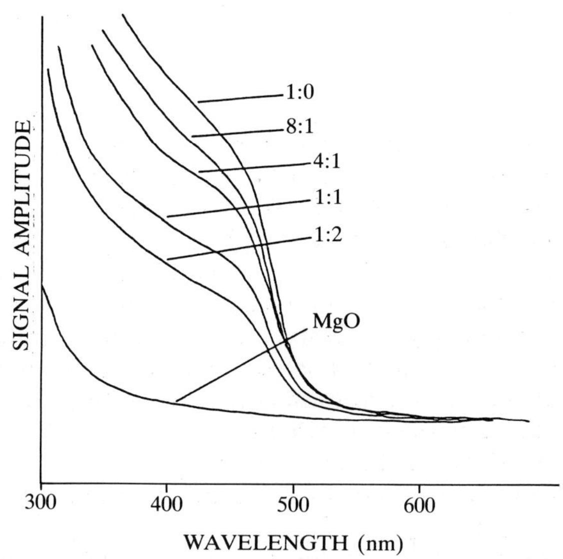

Figure 16.20 PA spectra of potassium chromate powder diluted with magnesium oxide (expressed as weight ratio potassium chromate : magnesium oxide)

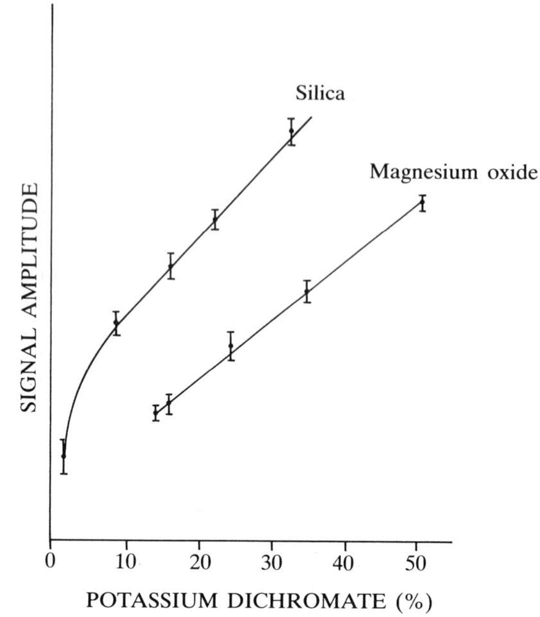

Figure 16.21 Potassium dichromate diluted by grinding: the effect of dilution on the PA signal

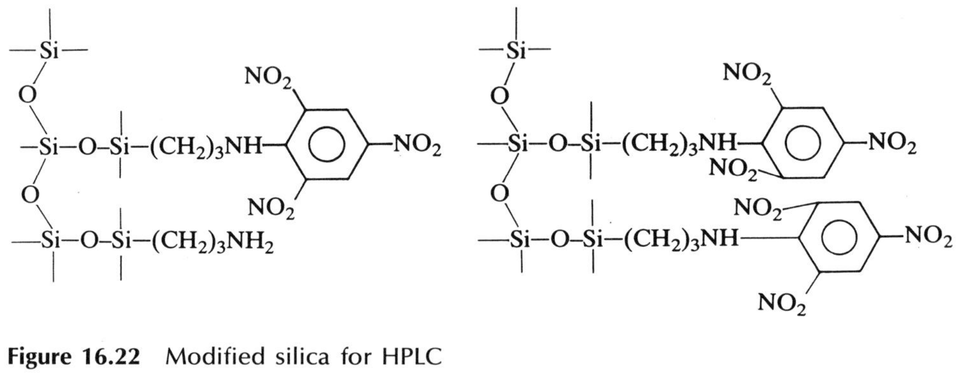

Figure 16.22 Modified silica for HPLC

it will penetrate the fibre. How does PAS cope with this difficult situation? Figures 16.23 and 16.24 compare the DR and PA spectra of wool dyed with Blue 50 and Blue 177 (Davidson *et al.*, 1983). These spectra show that PA spectra are less prone to distortion due to problems associated with internal reflection. Figure 16.25 shows the PA spectra of Acid Blue 224 applied to wool. It is very obvious that there is not a simple relationship between the amount of dye applied and the intensity of the signal. This could be due to the way in which the dye is taken up by the wool. To probe this point further, the dye was adsorbed onto silica and the results are shown in Fig. 16.26. In the case of this dye and also Acridine Orange a linear dependence between signal intensity and dye loading was not found

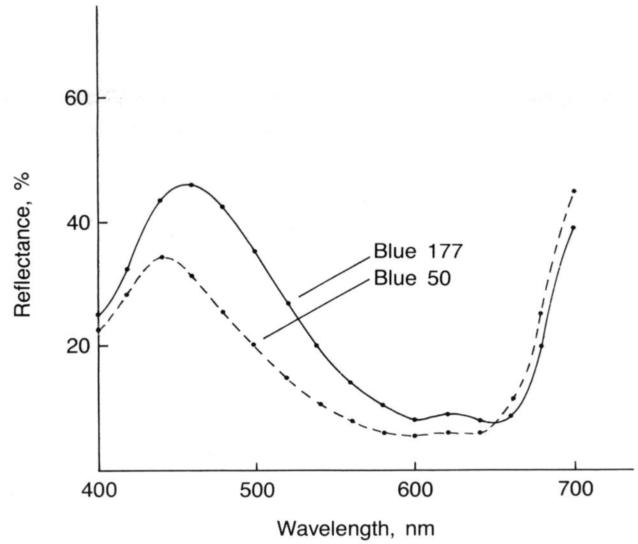

Figure 16.23 The reflectance spectra of C.I. Reactive Blue 177 and Blue 50 on wool at 1 per cent on the weight of fibre (o.w.f.).

whereas it was in the case of Rose Bengal (Fig. 16.27) (Davidson, 1983). Could the non-linearity be caused by aggregation of the dye? The dyes were applied to ion exchange resins either by a covalent bond (Lanasol Blue) or an ionic bond (Acridine Orange and Rose Bengal). As can be seen from Fig. 16.28 a linear relationship was not found. We conclude that light-scattering effects play an important and as yet an unquantifiable part.

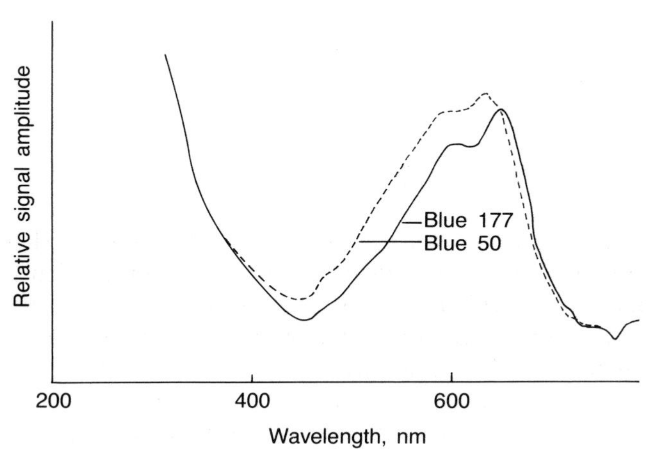

Figure 16.24 The photoacoustic spectra of C.I. Reactive Blue 177 and Blue 50 on wool at 1 per cent o.w.f.

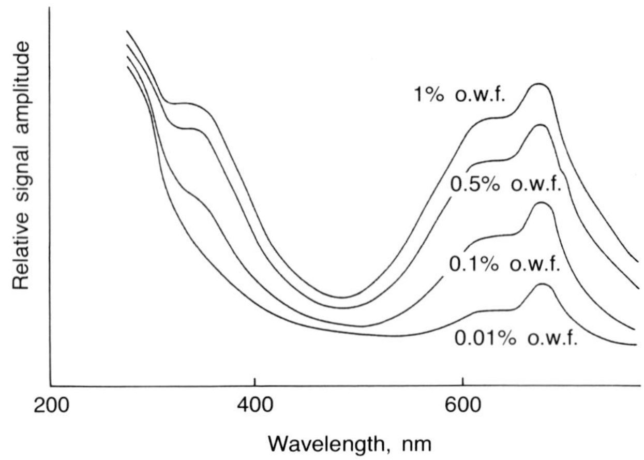

Figure 16.25 The photoacoustic spectra of C.I. Acid Blue 224 at different dye concentrations on wool

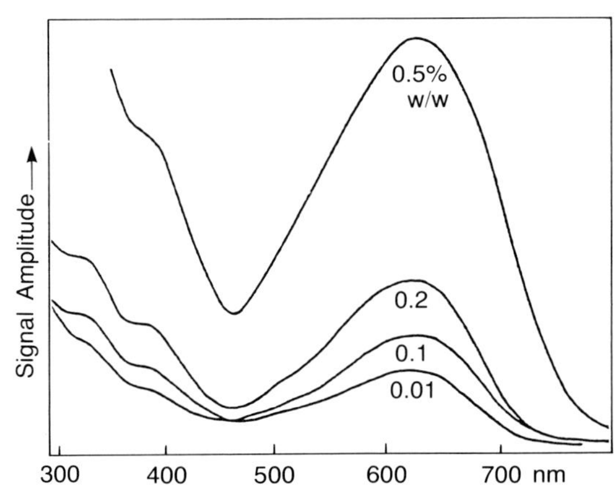

Figure 16.26 PA spectra of Lanasol Blue on silica

16.4 Application of PAS to studying biological samples

PAS has been used to study a wide variety of biological samples. The PA spectrum of an intact green leaf shows all the optical characteristics of the chloroplasts in the leaf, e.g. the chlorophyll absorption between 600–700 nm (Cahen *et al.*, 1978). A variety of algae and marine phytoplankton have been studied and in every case the presence of the coloured pigments can be detected (O'Hara *et al.*, 1983). A quite remarkable feature of PA spectroscopy is that it can also be used to detect whether algae are dead or alive. If the PA spectrum of the algae is recorded whilst the algae are being irradiated with c.w. broad-band light, the signal intensity at low modulation frequency is affected by the photosynthetic process since this produces oxygen, thereby altering the gas composition around the sample and

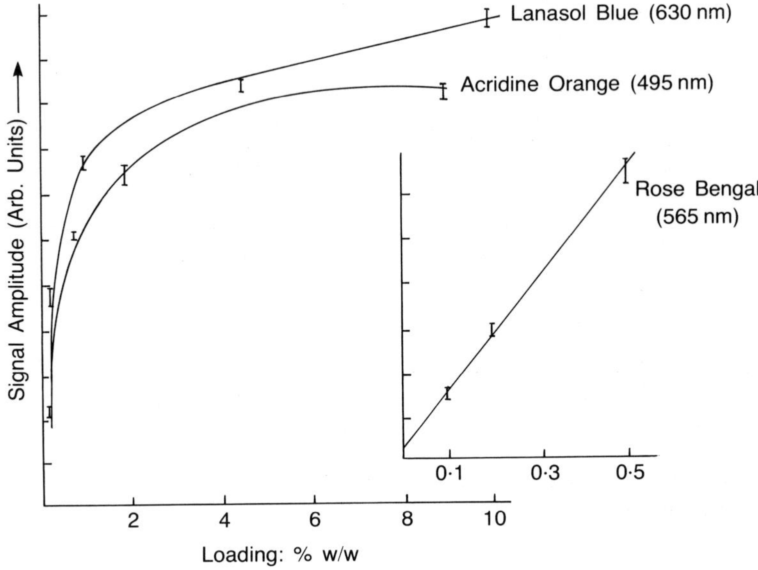

Figure 16.27 Plots of signal amplitude versus concentration of dye for samples of dyes adsorbed on silica

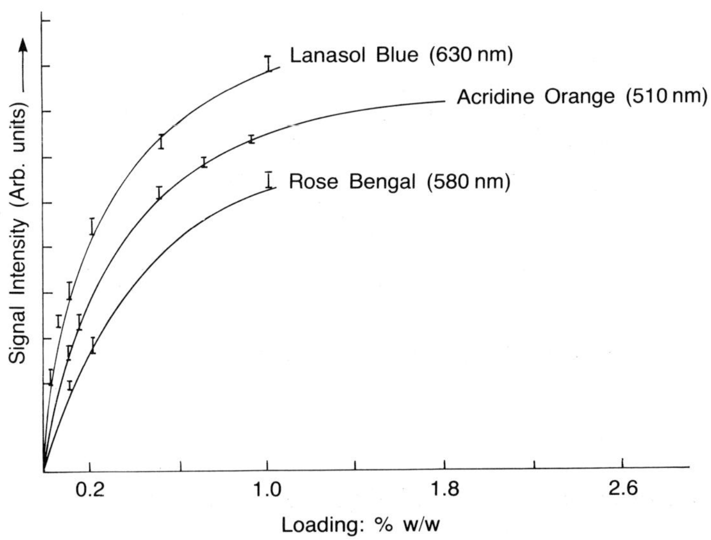

Figure 16.28 Plots of signal intensity versus dye concentration for samples of dyes adsorbed on ion-exchange resins

hence the production of the acoustic wave. Photosynthetic activity in tobacco leaves has also been detected by photoacoustic spectroscopy (Bults *et al.*, 1982). The sensitivity of the intensity of the PA signal to the composition of the gas in contact with a solid has been used to monitor the photo-oxidative destruction of rubrene (Gray and Baird, 1978).

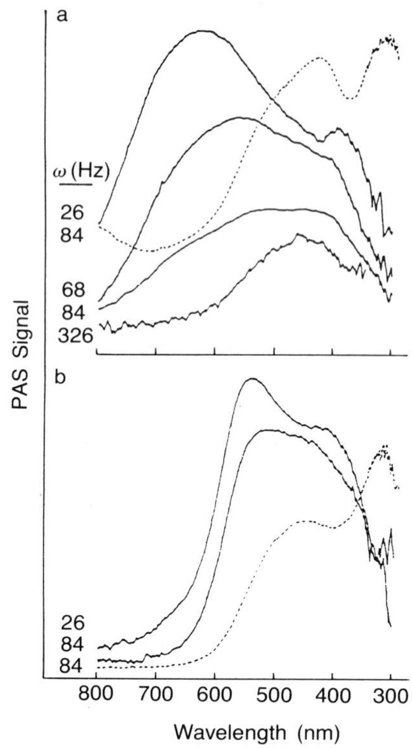

Figure 16.29 Photoacoustic absorption spectra of native (*a*) and denatured (*b*) lobster shell as a function of modulation frequency ω (indicated with each spectrum) and phase ϕ

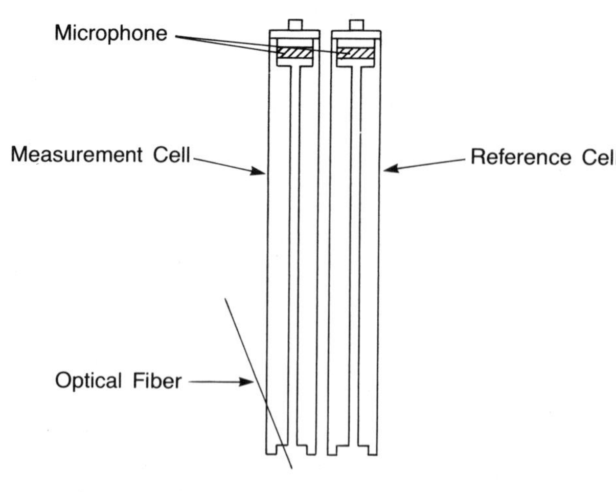

Figure 16.30 Diagram of a portable and double open-ended PA cell for *in vivo* measurement

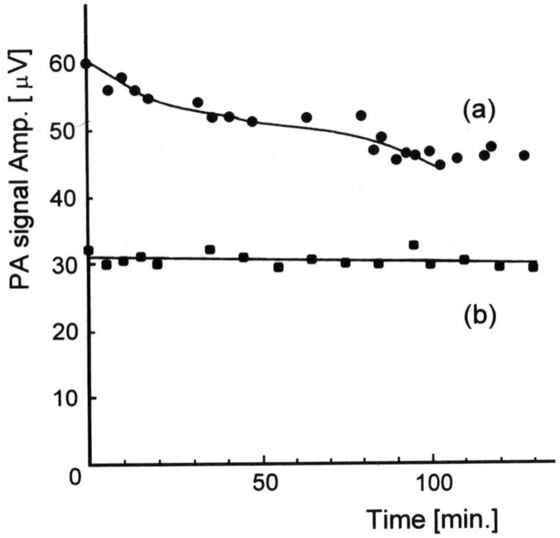

Figure 16.31 PA signal versus time when measuring human *in vivo* percutaneous absorption: (a) sample 3 per cent shilkonin ointment; (b) sample vehicle

Of particular interest to seafood gourmets is of course the lobster (Mackenthun *et al.*, 1979). The colour of the lobster is due to a single carotenoid pigment astaxanthin which is bound to specific proteins. The colour cannot be assigned to a single complex. A mature live lobster yielded a small sample (~4 mm^2) of its shell which was bright blue in colour. Figure 16.29 shows the PA spectrum of the sample. First, one sees the epicuticle with a λ_{max} of 425 nm. On reducing the modulation frequency from 326 to 26 Hz it can be seen that the spectrum of the carotenoid complex alters. If the shell fragment is denatured by treatment with boiling water one finds that the spectrum of the epicuticle is similar to that recorded for the native sample. The shape and position of the carotenoid absorption bands have been changed quite dramatically by the denaturing process. The change in shape of the protein has given spectra more akin to that of the carotenoid.

An open-ended photoacoustic cell has been designed which allows the *in vivo* percutaneous absorption of shilkonin (Takamoto *et al.*, 1992). The cell is shown in Fig. 16.30 and was carefully constructed so as to maximize the signal-to-noise ratio. As can be seen from Fig. 16.31, the take-up of the drug can be measured very effectively.

16.5 Application of PAS to the measurement of quantum yields of fluorescence

The quantum yields of fluorescence of solid samples are notoriously difficult to determine. In the case of sodium salicylate values range from 0.25 to 0.99. Using fluorimetry, problems arise due to the many corrections which have to be applied to the measured value, for example, corrections for experimental geometry, variation of detector sensitivity as a function of wavelength and non-ideality of absolute reflectance standards. Photoacoustic spectroscopy has been used to determine quantum yields in a reliable fashion (Adams *et al.*, 1980, 1981).

The absolute luminescence quantum efficiency Q of a compound is given by the following expression:

$$Q = \frac{\lambda_F}{\lambda_o}\left(1 - \frac{A_F}{A_{NF}} \times \frac{K_{NF}}{K_F} \times \frac{P_{Abs(NF)}}{P_{Abs(F)}}\right)$$

where λ_F and λ_o are the mean emission and excitation wavelengths; A_F and A_{NF} is the photoacoustic signal amplitudes for luminescent and non-luminescent references irradiated with identical intensity of radiation at λ_o; K_F and K_{NF} is the constants which account for the surface area and thermal properties of the samples and instrumental arrangement; P_{abs} and $P_{abs(NF)}$ is the photoacoustic response factors for the absorption of incident radiant power.

Under conditions of photoacoustic signal saturation (assuming identical reflectance for the luminescent and non-luminescent samples), $P_{abs} = P_{abs(NF)}$ are identical. Therefore

$$Q = \frac{\lambda_F}{\lambda}\left(1 - \frac{A_F}{A_{NF}} \times \frac{K_{NF}}{K_F}\right)$$

How can we operate under saturation conditions? A very thin layer of the sample is applied to a thick substrate. Thus all the light is absorbed in the sample layer but the sample and reference will have similar thermal diffusivities which are similar to that of the substrate.

For strongly absorbing chromophores (e.g. $\varepsilon = 10\,000\ \mathrm{Lmol^{-1}\,cm^{-1}}$) the optical depth corresponds to 3×10^{-5}cm (0.3 μm), therefore a layer of 1 μm evenly deposited on a substrate of 20 μm thicker will absorb most of the incident radiation whilst providing a composite thermal diffusivity that clearly resembles that of the pure substrate. Since the thermal properties are the same for sample and reference $K_F = K_{NF}$ therefore

$$Q = \frac{\lambda_F}{\lambda}\left(1 - \frac{A_F}{A_{NF}}\right)$$

In this way Q_F for sodium salicylate was determined as 0.56 (using Congo Red as the non-fluorescent reference).

To determine the quantum yield of fluorescence of a sample in solution is much simpler (Adams *et al.*, 1977):

$$P = P_{Abs}\,\beta$$

where P is the PAS signal magnitude; P_{Abs} is the radiant power absorbed and β is the efficiency factor for converted absorbed power to heat.

The PAS signal (P_{NF}) for a non-fluorescent sample is measured using the same wavelength and optical density as for the fluorescent sample. If both values are determined using the same solvent, the solution will possess identical thermal characteristics and therefore P_{Abs} can be calculated.

$$\beta = 1 - Q + Q\left(\frac{\nu_o - \nu_F}{\nu_o}\right)$$

Q is the quantum efficiency; ν_o is the frequency of excitation; ν_F = mean frequency of fluorescent radiation;

$$\text{therefore } P_F = P_{Abs}\left(1 - Q + Q\,\frac{(\nu_o - \nu_F)}{\nu_o}\right)$$

for a non-fluorescent compound $\beta = 1$ therefore $P_{NF} = P_{Abs}$ for the sample and non-fluorescent standard having *identical thermal characteristics*.

How can we achieve these conditions? We record P_F for the sample. The fluorescence of the sample is then quenched by adding a suitable quencher. The signal for this sample corresponds to P_{NF}.

In the case of quinine bisulphate the quenching can be achieved by adding chloride ions. Alternatively, if Q_F is known, this method can be used to measure the rate constant for fluorescence quenching.

16.6 Some potential uses of PAS to the study of the photostability of drugs

From the foregoing discussion it is clear that PAS has many potential applications for the study of the photostability of drugs. These include:

1 Recording changes in UV/visible, near-IR and IR spectra of solid samples.
2 As in 1 but recording the spectra whilst the samples are subjected to continuous irradiation.
3 Following spectral changes in turbid and translucent samples, e.g. creams.
4 Measurement of spectral changes in tissue, e.g. skin, both *in vitro* and *in vivo*.
5 Determining whether changes are purely surface phenomena.
6 Following rates of oxygen consumption.
7 Determination of Q_F and changes in fluorescence intensity during irradiation of solid samples.

Hopefully it will not be long before we see the application of photoacoustic spectroscopy making a positive contribution to an understanding of the photoprocess involved in the photodegradation of drugs.

Acknowledgements

The author greatly acknowledges the following organizations for permission to reproduce figures: American Chemical Society (Figs 16.11–16.21 and 16.31), Society of Dyers and Colourists (Figs 16.23–16.25), Elsevier (Figs 16.4 and 16.7–16.10) and Macmillan Journals Ltd (Fig. 16.29).

References

Adams, M. J., Highfield, J. G. and Kirkbright, G. F., 1977, Determination of absolute flourescence quantum efficiency of quinine bisulphate in aqueous medium using optoacoustic spectrometry, *Anal. Chem.*, **49**, 1850–2.

1980, Determination of absolute quantum efficiency of luminescence of solid materials employing photoacoustic spectroscopy, *Anal. Chem.*, **52**, 1260–4.

1981, Determination of the absolute quantum efficiency of sodium salicylate using photoacoustic spectroscopy, *Analyst*, **106**, 850–4.

BELL, A. G., 1881, Upon the production of sound by radiant energy, *Philosophical Mag.*, **11** (Ser. 5), 510–26.

BELTON, P. S., WILSON, R. H. and SAFFA, A. M., 1987, Effects of particle size on quantitative photoacoustic spectroscopy using a gas microphone cell, *Anal. Chem.*, **59**, 2378–82.

BULTS, G., HORWITZ, B. A., MALKIN, S. and CAHEN, D., 1982, Photochemistry and gas exchange, *Biochim. Biophys. Acta*, **679**, 452–65.

BURGGRAF, L. W. and LEYDEN, D. E., 1981, Quantitative photoacoustic spectroscopy of intensity light-scattering thermally thick samples, *Anal. Chem.*, **53**, 759–64.

CAHEN, D., MALKIN, S. and LERNER, E., 1978, Photoacoustic spectroscopy of chloroplast membranes. Listening to photosynthesis, *FEBS Letters*, **91**, 339–42.

CASTLEDEN, C. L., ELLIOTT, C. M., KIRKBRIGHT, G. F. and SPILLANE, D. E. M., 1979, Quantitative examination of thin layer chromatography plates by photoacoustic spectroscopy, *Anal. Chem.*, **51**, 2152–3.

CHILDERS, J. W., RÖHL, R. and PALMER, R. A., 1986, Direct comparison of the capabilities of photoacoustic and diffuse reflectance spectroscopies in the ultraviolet, visible and near-infra red regions, *Anal. Chem.*, **58**, 2629–36.

DAVIDSON, R. S., 1983, The contribution of photoacoustic and photothermal spectroscopy of surface chemistry: adsorbed species, catalysis and corrosion, *Journal de Physique*, **44**, C6-267–74.

DAVIDSON, R. S., ELLIS, R., WILKINSON, S. and SUMMERSGILL, C., 1987, A study of the polymerisation of acrylates using electron beam radiation, *Eur. Polym. J.*, **23**, 105–8.

DAVIDSON, R. S. and KING, D., 1984, Effect of particle size on photoacoustic signal amplitude, *Anal. Chem.*, **56**, 1409–11.

DAVIDSON, R. S., KING, D., DUFFIELD, P. A. and LEWIS, D. M., 1983, Photoacoustic spectroscopy for the study of the adsorption of dyes on wool fabrics, *Journal Soc. Dyers and Col.*, **99**, 123–6.

DAVIDSON, R. S., LOUGH, J. W., MATLIN, S. A. and MORRISON, C. L., 1981, Photoacoustic spectroscopic evidence for site-site interactions in a bifunctional surface bonded phase, *J. Chem. Soc. Chem. Commun.*, 517–18.

DAVIDSON, R. S. and LOWE, C., 1989, Use of UV/visible photoacoustic spectroscopy to study the photoinduced crosslinking of oligomers containing chalcone units, *Eur. Polym. J.*, **25**, 159–65.

GRAY, R. C. and BARD, A. J., 1978, Photoacoustic spectroscopy applied to systems involving gas evolution or consumption, *Anal. Chem.*, **50**, 1262–5.

JABBEN, M., HEIHOFF, K., BRASLAVSKY, S. E. and SCHAFFNER, K., 1984, Studies on phytochrome photoconversions *in vitro* with laser-induced optoacoustic spectroscopy, *Photochem. and Photobiol.*, **40**, 361–7.

KING, D., DAVIDSON, R. S. and PHILLIPS, M., 1982, Effects of concentration and sample preparation in photoacoustic spectroscopy of powdered samples, *Anal. Chem.*, **54**, 2191–4.

MACKENTHUN, M. L., TOM, R. D. and MOORE, T. A., 1979, Lobster shell carotenoprotein organisation *in situ* studied by photoacoustic spectroscopy, *Nature*, **279**, 265–6.

O'HARA, E. P., TOM, R. D. and MOORE, T. A., 1983, Determination of the *in vivo* absorption and photosynthetic properties of lichen Acarospora Schleicheri using photoacoustic spectroscopy, *Photochem. and Photobiol.*, **38**, 709–15.

ROSENCWAIG, A., 1973, Photoacoustic spectroscopy of biological materials, *Science*, **181**, 657–8.

ROSENCWAIG, A. and HALL, S., 1975, Thin layer chromatography and photoacoustic spectrometry, *Anal. Chem.*, **47**, 548–9.

TAKAMOTO, R., NAMBA, R., MATSUOKA, M. and SANADA, T., 1992, Human *in vivo* percutaneous absorptimetry using the laser photoacoustic method, *Anal. Chem.*, **64**, 2661–3.

Useful terms and expressions in the photoreactivity testing of drugs

(The list is not comprehensive)

Actinometer A chemical system or physical device which determines the number of photons in a beam integrally or per unit time. This name is commonly applied to devices used in the ultraviolet and visible wavelength ranges. Solutions of iron (III) oxalate can be used as a chemical actinometer while thermopiles are examples of physical devices giving a reading that can be correlated to the number of photons detected.

Action spectrum A plot of the reciprocal of the number of incident photons required to produce a given effect compared with the wavelength of the radiation employed.

Black-body A body which completely absorbs radiation of any wavelength falling upon it at any angle.

Black-body radiator A black-body which emits, in every direction and at any wavelength, the maximum possible radiant energy, as compared with other temperature radiators of the same temperature, geometrical shape and dimension.

Black-light lamp Fluorescent lamp that emits ultraviolet radiation in a broad band from 320 to 380 nm.

Candle Equivalent to light produced by a spermaceti candle 7/8 inch in diameter burning at rate of 120 grams per hour.

Candlepower See luminous intensity.

Conversion spectrum A plot of a quantity related to the absorption (absorbance, cross-section, etc.) multiplied by the quantum yield for the considered process against a suitable measure of photon energy, such as frequency, υ, wavenumber, σ, or wavelength, λ.

Cut-off filter An optical device which only permits the transmission of radiation of wavelengths that are longer than or shorter than a specified wavelength.

Deactivation Any loss of energy by an excited molecular entity.

Dose The energy or amount of photons absorbed per unit area or unit volume by an irradiated object during a particular exposure time. Dose can also be used in the sense of the energy or amount of photons per unit area or unit volume received by an irradiated object during a particular exposure time.

Efficiency spectrum A plot of the biological or chemical change or response per absorbed photon vs wavelength.
Electronically excited state A state of an atom or molecular entity which has greater electronic energy than the ground state of the same entity.
Emission Radiative deactivation of an excited state.
Footcandle Unit of intensity of illumination, obtained when a source of 1 candlepower illuminates a screen 1 ft away. 1 footcandle = 10.76 lux.
Illuminance (E_v) The quotient of the luminous flux elements ($d\Phi_v$) divided by the irradiated surface element (dA). Unit: ($lm\,m^{-2}$) = lux.
Inner filter effect During a light irradiation experiment the term refers to a sample with a high optical density, resulting in a significant reduction in light intensity at the centre of the cuvette compared with an infinitely dilute solution.
Irradiance (E) The radiant flux or radiant power (P) *incident* on an infinitesimal element of surface containing the point under consideration divided by the area of the element (S) (dP/dS), simplified: E = P/S when the radiant power is constant over the surface area considered. The SI unit is $W\,m^{-2}$.
Isosbestic point A wavelength, wavenumber, or frequency at which absorption coefficients are equal, i.e. the total absorbance of a sample at this wavelength does not change during a chemical reaction or physical change of the sample.
Lumen Unit of luminous flux falling on a square centimetre at a distance of one centimetre from one international candle, cell cavity, passageway or opening.
Luminance Luminous flux per unit solid angle leaving element of surface in a given direction, divided by the area of orthogonal projection on the plane perpendicular to this direction.
Luminosity factor Ratio of total luminous flux to total energy emitted by a light source at a given wavelength.
Luminous flux Total visible energy emitted by a source per unit time.
Luminous intensity Amount of luminous flux emitted by a point source of light per solid angle, compared with a standard candle.
Lux Measure of illumination of a surface, equal to 0.092 902 footcandle or 1.000 lumen per square metre.
Photoacoustic effect Generation of heat after absorption of radiation, due to radiationless deactivation or chemical reaction.
Photoallergy An acquired immunologic reactivity dependent on antibody or cell-mediated hypersensitivity.
Photochemical reaction A chemical reaction caused by absorption of ultraviolet, visible, or infrared radiation.
Photochemotherapy The combination of a photoactive chemical and light. The interaction of the light and the chemical produces a synergistic effect.
Photodynamic effect Photoinduced damage requiring the simultaneous presence of light, photosensitizer and molecular oxygen.
Photometry The measurement of quantities associated with light, i.e. based on the average apparent intensity of a light source as viewed by a normal light-adapted human eye. Photometric units report light intensity in terms of the illuminance, e.g. candlepower (lumen/ft^2) or lux (lumen/m^2). Photometric units are only appropriate for visible radiation.
Photooxidation Oxidation reactions induced by light, e.g. the loss of one or more electrons from a chemical species as a result of photoexcitation of that species or the reaction of a substance with oxygen under the influence of light.

Photooxygenation Incorporation of molecular oxygen into a molecular entity. There are three common mechanisms:

Type I — The reaction of triplet molecular oxygen with radicals formed photochemically.

Type II — The reaction of photochemically produced singlet molecular oxygen with molecular entities to give rise to oxygen containing molecular entities.

Type III — Mechanism proceeds by electron transfer producing superoxide anion as the reactive species.

Photophysical process Photoexcitation and subsequent events which lead from one to another state of a molecular entity through radiation and radiationless transitions. No chemical change results.
Photopolymerization Polymerization processes requiring a photon for the propagation step.
Photoreduction Reduction reactions induced by light, e.g. addition of one or more electrons to a photoexcited species or the photochemical hydrogenation of a substance.
Photosensitivity A broad term used to describe an adverse reaction to light, which may be phototoxic or photoallergic in nature.
Photosensitization The process by which a photochemical or photophysical alteration occurs in one molecular entity as a result of initial absorption of radiation by another molecular entity called a photosensitizer.
Photosensitized oxidation Two mechanisms named Type I and Type II (see *Photooxygenation*).

Type I — Substrate or solvent reacts with the sensitizer excited state (either singlet or triplet sens*) to give radicals or radical ions, respectively, by hydrogen atom or electron transfer, leading to oxygenated products.

Type II — The excited sensitizer reacts with oxygen to form singlet molecular oxygen which then reacts with substrate to form the products.

Photothermal effect An effect produced by photoexcitation resulting partially or totally in the production of heat.
Phototoxicity The conversion of an otherwise nontoxic chemical to one directly toxic to tissues after the absorption of electromagnetic radiation.
Quantum counter A medium emitting with a quantum yield independent of the excitation energy over a defined spectral range, e.g. concentrated rhodamine 6G solution between 300 and 600 nm.
Quantum yield (Φ) The number of defined events which occur per photon absorbed by the system.
Quencher A molecular entity that deactivates (quenches) an excited state of another molecular entity, either by energy transfer, electron transfer, or by a chemical mechanism.
Quenching The deactivation of an excited molecular entity intermolecularly by an external environmental influence (such as a quencher) or intramolecularly by a substituent through a nonradiative process.

Radiance (L) The radiant power (P) leaving or passing through a surface element (S) in a given direction from the source, divided by the projection dS cos Q of the surface element, where Q is the angle between the direction of radiation and the normal to the surface. SI units $W\,m^{-2}$ (parallel beam) or $W\,m^{-2}\,sr^{-1}$ (divergent beam).
Radiant energy (Q) The total energy emitted, transferred or received as radiation in a defined period of time, i.e. the product of radiant power (P) and time (t) when the radiant power is constant over the time considered. The SI unit is J.
Radiant (energy) flux (Φ) Same as radiant power (P). Power emitted, transferred, or received as radiation. The SI unit is $J\,s^{-1} = W$.
Radiant exposure (H) The irradiance, E, integrated over the time of irradiation. SI unit is Jm^{-2}.
Radiant intensity (I) Radiant (energy) flux or radiant power, P, per unit solid angle, ω. The SI unit is $W\,sr^{-1}$.
Radiant power (P) See radiant (energy) flux.
Radiometry The measurement of quantities associated with radiant energy. The radiometric unit of intensity is irradiance.
Self-quenching Quenching of an excited atom or molecular entity by interaction with another atom or molecular entity of the same species in the ground state.
Singlet molecular oxygen The oxygen molecule (dioxygen), O_2, in an excited singlet state. The ground state of O_2 is a triplet.
Singlet state A state having a total electron spin quantum number equal to 0.
Temperature radiation Every body which has a temperature higher than 0 K emits radiation due to its own temperature.
Thermopile Radiation-measuring instrument consisting of a number of thermocouples connected together in series. Measures the incident total radiant flux (calibrated in microwatts). The irradiance is obtained by dividing the measured value by the effective sensitive receiving surface area.
Triplet state A state having a total electron spin quantum number of 1.

Relevant literature on photostability testing of actual drug substances and drug formulations

(The list is not comprehensive)

Adrenaline

DE MOL, N. J., BEIJERSBERGEN VAN HENEGOUWEN, G. M. J. and GERRITSMA, K. W., 1979, Photochemical decomposition of catecholamines – II. The extent of aminochrome formation from adrenaline, isoprenaline and noradrenaline induced by ultraviolet light, *Photochem. Photobiol.*, **29**, 479–82.

HOEVENAARS, P. C. M., 1965, Stabiliteit van adrenaline in injectievloeistoffen, *Pharm. Weekbl.*, **100**, 1151–62.

NEWTON, D. W., YIN YEE FUNG, E. and WILLIAMS, D. A., 1981, Stability of five catecholamines and terbutaline sulfate in 5% dextrose injection in the absence and presence of aminophylline, *Am. J. Hosp. Pharm.*, **38**, 1314–19.

WOLLMANN, H. and GRÜNERT, R., 1984, Einfluss des sichtbaren Lichtes auf die Haltbarkeit von Isoprenalin-, Epinephrin- und levarteren Öllösungen in unterschiedlicher Behältnissen, *Pharmazie*, **39**, 161–3.

Amidopyrin

REISCH, J. and FITZEK, A., 1967, Über die zersetzung von wässrigen amidopyrinlösungen unter dem einfluss von licht und γ-strahlen, *Dtsch. Apoth.-Ztg.*, **107**, 1358–9.

Amidinohydrazones

SCHLEUDER, M., RICHTER, P. H., KECKEIS, A. and JIRA, TH., 1993, Antiarrhythmisch wirksame Amidinohydrazone substituierter Benzophenone, *Pharmazie*, **48**, 33–7.

Adriamycin

TAVOLONI, N., GUARINO, A. M. and BERK, P. D., 1980, Photolytic degradation of adriamycin, *J. Pharm. Pharmacol.*, **32**, 860–2.

Amiloride

HAMOUDI, H. I., HEELIS, P. F., JONES, R. A., NAVARATNAM, J. S., PARSONS, B. J., PHILIPPS, G. O., VANDENBURG, M. J. and CURRIE, W. J. C., 1984, A laser flash photolysis and pulse radiolysis study of amiloride in aqueous and alcoholic solution, *Photochem. Photobiol.*, **40**, 35–9.

Aminophenazone

REISCH, J. and ABDEL-KHALEK, M., 1979, Zur fotooxidation von kristallinem aminophenazon, *Pharmazie*, **34**, 408–10.

Aminophylline

BOAK, L. R., 1987, Aminophylline stability, *Can. J. Hosp. Pharm.*, **40**, 155.

Aminosalicylic acid

JENSEN, J., CORNETT, C., OLSEN, C. E., TJØRNELUND, J. and HANSEN, S. H., 1992, Identification of major degradation products of 5-aminosalicylic acid formed in aqueous solutions and in pharmaceuticals, *Int. J. Pharm.*, **88**, 177–87.

Amiodarone

LI, A. S. W. and CHIGNELL, C. F., 1987, A spin trapping study of the photolysis of amiodarone and desethylamiodarone, *Photochem. Photobiol.*, **45**, 191–7.

PAILLOUS, N. and VERRIER, M., 1988, Photolysis of amiodarone, an antiarrhythmic drug, *Photochem. Photobiol.*, **47**, 337–43.

Amodiaquine

OWOYALE, J. A., 1989, Amodiaquine less sensitive than chloroquine to photochemical reactions, *Int. J. Pharm.*, **56**, 213–15.

Amonafide

SÁNCHEZ, M. A. C., SUÁRES, A. I. T. and SANZ, M. P., 1989, Estabilidad de disoluciones de amonafide frente a la luz y la temperatura, *Cienc. Ind. Farm.*, **8**, 104–9.

Amphotericin B

BLOCK, E. R. and BENNETT, J. E., 1973 Stability of amphotericin B in infusion bottles, *Antimicrob. Agents Chemother.*, **4**, 648–9.

GALLELLI, J. F., 1967, Assay and stability of amphotericin B in aqueous solution, *Drug Intel.*, **1**, 103–5.

LEE, M. D., HESS, M. M., BOUCHER, B. A. and APPLE, A. M., 1994, Stability of amphotericin B in 5% dextrose injection stored at 4 or 25° C for 120 hours, *Am. J. Hosp. Pharm.*, **51**, 394–6.

SHADOMY, S., BRUMMER, D. L. and INGROFF, A. V., 1973, Light sensitivity of prepared solutions of amphotericin B, *Am. Rev. Respir. Dis.*, **107**, 303–4.

Amsacrine

CARTWRIGHT-SHAMOON, J. M., MCELNAY, J. C. and D'ARCY, P. F., 1988, Examination of sorption and photodegradation of amsacrine during storage in intravenous kuvette administration sets, *Int. J. Pharm.*, **42**, 41–6.

Azapropazon

REISCH, J., EKIZ-GÜCER, N., TAKÀCS, M., GUNAHERATH, G. M. and KAMAL B., 1989, Photochemische studien, 53. Mitt. Über die photoisomerisierung des azapropazons, *Arch. Pharm.*, **322**, 295–6.

Azathioprine

HEMMENS, V. J. and MOORE, D. E., 1986, Photochemical sensitization by azathioprine and its metabolites – II. Azathioprine and nitroimidazoale metabolites, *Photochem. Photobiol.*, **43**, 257–62.

1986, Photochemical sensitization by azathioprine and its metabolites – 1.6-Mercaptopurine, *Photochem. Photobiol.*, **43**, 247–55.

MOORE, D. E., CHIGNELL, C. F., SIK, R. H. and MOTTEN, A. G., 1986, Generation of radical anions from metronidazole, misonidazole and azathioprine by photoreduction in the presence of EDTA, *Int. J. Radiat. Biol.*, **50**, 885–91.

Aztreonam

FABRE, H., IBROK, H. and LERNER, D. A., 1992, Photodegradation kinetics under UV light of aztreonam solutions, *J. Pharm. Biomed. Anal.*, **10**, 645–50.

Barbituric acid

BARTÓN, H. J., BOJARSKI, J. and ZUROWSKA, A., 1986, Stereospecificity of the Photoinduced Conversion of Methylphenobarbital to Meptienytoin, *Arch. Pharm.*, **319**, 457–61.

BARTON, H., MOKROSZ, J., BOJARSKI, J. and KLIMCZAK, M., 1980, Photochemical Degradation of Barbituric Acid Derivatives. Part 1: Products of Photolysis and Hydrolysis of Pentobarbital, *Pharmazie*, **35**, 155–8.

JOCHYM, K., BARTON, H. and BOJARSKI, J., 1988, Photochemical degradation of barbituric acid derivatives. Part 8: Photolysis of sodium salts of barbiturates in solid state, *Pharmazie*, **43**, 623–4.

MOKROSZ, J. and BOJARSKI, J., 1980, Photochemical Degradation of Barbituric Acid Derivatives. Part 3: Rate Constants of Photolysis of Barbituric and Thiobarbituric Acid Derivatives, *Pharmazie*, **35**, 768–73.

MOKROSZ, J., KLIMCZAK, M., BARTON, H. and BJOARSKI, J., 1980, Photochemical degradation of barbituric acid derivatives. Part 2: Kinetics of Pentobarbital Photolysis, *Pharmazie*, **35**, 205–8.

MOKROSZ, J., ZUROWSKA, A. and BJOARSKI, J., 1982, Photochemical degradation of barbituric acid derivatives. Part 4: Kinetics and TLC investigations of photolysis of proxibarbal, *Pharmazie*, **37**, 832–5.

PALUCHOWSKA, M. H. and BOJARSKI, J., 1988, Photochemical Formation of Primidone from 2-Thiophenobarbital, *Arch. Pharm.*, **321**, 343–4.

REISCH, J., MÜLLER, M. and MÜNSTER, 1984, Über die photostabilität offizineller arznei- und hilfsstoffe I: Barbiturate, *Pharm. Acta Helv.*, **59**, 56–61.

Benorylate

CASTELL, J. V., GOMEZ-L., M. J., MIRABET, V., MIRANDA, M. A. and MORERA, I. M., 1987, Photolytic Degradation of Benorylate: Effects of the Photoproducts on Cultured Hepatocytes, *J. Pharm. Sci.*, **76, 374–8.**

Benoxaprofen

MOORE, D. W. and CHAPPUIS, P. P., 1988, A comparative study of the photochemistry of the non-steroidal anti-inflammatory drugs, naproxen, benoxaprofen and indomethacin, *Photochem. Photobiol.*, **47**, 173–80.

NAVARATNAM, S., HUGHES, J. L., PARSONS, B. J. and PHILLIPS, G. O., 1985, Laser flash and steady-state photolysis of benoxaprofen in aqueous solution, *Photochem. Photobiol.*, **41**, 375–80.

RESZKA, K. and CHIGNELL, C. F., 1983, Spectroscopic Studies of Cutaneous Photosensitizing Agents – IV. The Photolysis of Benoxaprofen, an Anti-Inflammatory Drug with Phototoxic Properties, *Photochem. Photobiol.*, **38**, 281–91.

Benzamide

NYQVIST, H. and WADSTEN, T., 1986, Preformulation of solid dosage forms: Light stability testing of polymorphs as a part of a preformulation program, *Acta Pharm. Technol.*, **32**, 130–2.

Benzocaine

CHINGPAISAL, P., FLETCHER, G. and DAVIS, D. J. G., 1977, The effect of CTAB on the radiation sensitivity of benzocaine to hydrated electrons and hydroxyl radicals in aqueous solution, *J. Pharm. Pharmacol.*, **29**, 47P.

FLETCHER, G. and DAVIES, D. J. G., 1974, The effect of surfactants on the radiation sensitivity of benzocaine in aqueous solution, *J. Pharm. Pharmacol.*, **26**, 82P.

Benzodiazepine

REISCH, J., EKIZ-GÜCER, N. and TEWES, G., 1992, Photochemische studien, 63. Photostabilität einiger 1,4-benzodiazepine in kristallinem zustand, *Liebigs Ann. Chem.*, **69–70.**

Benzoquinones

KALLMAYER, H. J. and FRITZEN, W., 1987, 2-amino-3,5,6-tribromo-1,4-benzochinone und ihre labilität am tageslicht, *Arch. Pharm.*, **320**, 769–75.

1992, Photoreaktivität einiger 2-amino-3,5,6,-trichloro-1,4-benzochinone, *Pharm. Acta Helv.*, **67**, 210–3.

Benzydamine

VARGAS, F., RIVAS, C., MACHADO, R. and SARABIA, Z., 1993, Photodegradation of benzydamine: Phototoxicity of an isolated photoproduct on erythrocytes, *J. Pharm. Sci.*, **82**, 371–2.

Betamethasone

THOMA, K., KERKER, R. and WEISSBACH, C., 1987, Untersuchung des Einflusses von Bestrahlungsmethoden auf die Photostabilität von Betamethason, *Pharm. Ind.*, **49**, 961–3.

Bleomycin

DOUGLAS, K. T., 1983, Photoactivity of bleomycin, *Biomed. Pharmacother.*, **37**, 191–3.

DOUGLAS, K. T., RATWATTE, H. A. M. and THAKRAR, N., 1983, Photoreactivity of Bleomycin and its Implications, *Bull. Cancer*, **70**, 372–80.

THAKRAR, N. and DOUGLAS, K. T., 1981, Photolability of Bleomycin and its Complexes, *Cancer Lett.*, **13**, 265–8.

Bupivacaine

TU, Y.-H., STILES, M. L. and ALLEN, L. V., 1990, Stability of fentanyl citrate and bupivacaine hydrochloride in portable pump reservoirs, *Am. J. Hosp. Pharm.*, **47**, 2037–40.

Butibufen

CASTELL, J. V. and GÓMEZ-LECHÓN, M. J., 1992, Phototoxicity of non-steroidal anti-inflammatory drugs: in vitro testing of the photoproducts of butibufen and flurbiprofen, *J. Photochem. Photobiol. B: Biol.*, **13**, 71–81.

MATSUDA, Y., AKAZAWA, R., TERAOKA, R. and OTSUKA, M., 1994, Pharmaceutical evaluation of carbamazepine modifications: Comparative study for photostability of carbamazepine polymorphs by using fourier-transformed reflection-absorption infrared spectroscopy and colorimetric measurement, *J. Pharm. Pharmacol.*, **46**, 162–7.

Carbisocaine

BEZAKOVA, Z., BACHRATÀ, M., BLESOVÀ, M. and BOROVANSKY, A., 1986, Studium lokalnych anestetik. LXXXIII Stabilita karbizokainiumchloridu a pentakainiumchloridu, *Farm. Obz.*, **55**, 195–203.

Carmustine

FREDRIKSSON, K., LUNDGREN, P. and LANDERSJØ, L., 1986, Stability of carmustine – kinetics and compatibility during administration, *Acta Pharm. Suec.*, **23**, 115–24.

Cefotaxime

LERNER, D. A., BONNEFOND, G., FABRE, H., MANDROU, B. and SIMEON DE BOUCHBERG, M., 1988, Photodegradation paths of cefotaxime, *J. Pharm. Sci.*, **77**, 699–703.

Cephaeline

TESHIMA, D., OTSUBO, K., HIGUCHI, S., HIRAYAMA, F., UEKAMA, K. and AOYAMA, T., 1989, Effects of cyclodextrins on degradations of emetine and cephaeline in aqueous solution, *Chem. Pharm. Bull.*, **37**, 1591–4.

Cephalexine

OLIVEIRA, A. G. and PETROVICK, P. R., 1984, Photochemical stability of cephalexine monohydrated suspension in mineral oil. Degradation kinetics and half-life period, *Rev. Cienc. Farm.*, **6**, 63–6.

Cephradine

SIGNORETTI, E. C., ONORI, S., VALVO, L., FATTIBENE, P., SAVELLA, A. L., DE SENA, C. and ALIMONTI, S., 1993, Ionizing radiation induced effects on cephradine. Influence on sample moisture content, irradiation dose and storage conditions, *Drug Dev. Ind. Pharm.*, **19**, 1693–708.

Chalcone

UTSUKI, T., IMAMURA, K., HIRAYAMA, F. and UEKAMA K., 1993, Stoichiometry-dependent changes of solubility and photoreactivity of an antiulcer agent, 2-carboxymethoxy-4,4-bis (3-methyl-2-butenyloxy)chalcone, in cyclodextrin inclusion complexes, *Eur. J. Pharm. Sci.*, **1**, 81–7.

Chloramphenicol

BEIJERSBERGEN VAN HENEGOUWEN, G. M. J., 1991, New trends in photobiology, *J. Photochem. Photobiol. B: Biol.*, **10**, 183–210.

DE VRIES, H., BEIJERSBERGEN VAN HENEGOUWEN, G. M. J. and HUF, F. A., 1984, Photochemical decomposition of chloramphenicol in a 0.25% eyedrop and in a therapeutic intraocular concentration, *Int. J. Pharm.*, **20**, 265–71.

MUBARAK, S. I. M., STANFORD, J. B. and SUGDEN, J. K., 1982, Some aspects of the photochemical degradation of chloramphenicol, *Pharm. Acta Helv.*, **57**, 226–30.

1983, Photochemical reactions of chloramphenicol with diols, *Pharm. Acta Helv.*, **58**, 343–7.

REICH, J. and WEIDMAN, K. G., 1971, Photo- und Radiolyse des Chloramphenicols, *Arch. Pharm.*, **304**, 911–19.

SHIH, I. K., 1971, Photodegradation products of chloramphenicol in aqueous solution, *J. Pharm. Sci.*, **60**, 1889–90.

THOMA, VON K. and KERKER, R., 1992, Photoinstabilität von arzneimotteln. 1. Mitteilung über das verhalten von nur im UV-bereich absorbierenden substanzen bei der tageslichtsimulation, *Pharm. Ind.*, **54**, 169–77.

Chlordiazepoxide

BEIJERSBERGEN VAN HENEGOUWEN, G. M. J., 1991, New trends in photobiology, *J. Photochem. Photobiol. B: Biol.*, **10**, 183–210.

CORNELISSEN, P. J. G., BEIJERSBERGEN VAN HENEGOUWEN, G. M. J. and

Gerritsma, K. W., 1979, Photochemical decomposition of 1,4-benzodiazepines. Chlordiazepoxide, *Int. J. Pharm.*, **3**, 205–20.

Moore, D. E. and Tamat, S. R., 1980, Photosensitization by drugs: photolysis of some chloride-containing drugs, *J. Pharm. Pharmacol.*, **32**, 172–7.

Chloroquine

Aaron, J. J. and Fidanza, J., 1982, Photochemical analysis studies – III. A fluorimetric and ultraviolet spectrophotometric study of the photolysis of chloroquine on silica-gel thin layers, and its analytical application, *Talenta.*, **29**, 383–9.

Nord, K., Karlsen, J. and Tønnesen, H. H., 1991, Photochemical stability of biologically active compounds. IV. Photochemical degradation of chloroquine, *Int. J. Pharm.*, **72**, 11–18.

1994, Photochemical stability of biologically active compounds. IX. Characterization of the spectroscopic properties of the 4-aminoquinolines chloroquine and hydroxychloroquine, and of selected metabolites by absorption, fluorescence and phosphorescence measurements, *Photochem. Photobiol.*, **60**, 427–31.

Odusote, M. O. and Nasipuri, R. N., 1988, Effect of pH and storage conditions on the stability of a novel chloroquine phosphate syrup formulation, *Pharm. Ind.*, **50**, 367–9.

Owoyale, J. A., 1989, Decomposition profile of ultraviolet-irradiated chloroquine, *Int. J. Pharm.*, **52**, 179–81.

Chlorpromazine

Beijersbergen van Henegouwen, G. M. J., 1991, New trends in photobiology, *J. Photochem. Photobiol. B: Biol.*, **10**, 183–210.

Davies, A. K., Navaratnam, S. and Phillips, G. O., 1976, Photochemistry of chlorpromazine (2-chloro-N-(3-dimethyl-aminopropyl)phenothiazine) in propan-2-ol solution, *J. Chem. Soc. Perkin Trans.*, **2**, 25–9.

Hall, R. D., Buettner, G. R., Motten, A. G. and Chignell, C. F., 1987, Near infrared detection of singlet molecular oxygen produced by photosensitization with chlorpromazine, *Photochem. Photobiol.*, **46**, 295–300.

Huang, C. L. and Sands, F. L., 1964, The effect of UV irradiation on chlorpromazine, *J. Chromatogr.*, **13**, 246–9.

1967, Effect of ultra-violet irradiation on chlorpromazine. II Anaerobic condition, *J. Pharm. Sci.*, **56**, 259–64.

Matsuda, Y. and Masahara, R., 1980, Comparative evaluation of coloration of photosensitive solid drugs under various light sources, *Yakugaku Zasshi*, **100**, 953–7.

Moore, D. E. and Tamat, S. R., 1980, Photosensitization by drugs: photolysis of some chloride-containing drugs, *J. Pharm. Pharmacol.*, **32**, 172–7.

Motten, A. G., Buettner, G. R. and Chignell, C. F., 1985, Spectroscopic Studies of Cutaneous Photosensitizing Agents – VIII. A spin-trapping study of light induced free radicals from chlorpromazine and promazine, *Photochem. Photobiol.*, **42**, 9–15.

Rosenthal, I., Ben-Hur, E., Prager, A. and Riklis, E., 1978, Photochemical reactions of chlorpromazine; chemical and biochemical implications, *Photochem. Photobiol.*, **28**, 591–4.

Chlortetracycline

Moore, D. E. and Tamat, S. R., 1980, Photosensitization by drugs: photolysis of some chloride-containing drugs, *J. Pharm. Pharmacol.*, **32**, 172–7.

Cianidanol

AKIMOTO, K., INOUE, K. and SUGIMOTO, I., 1985, Photostability of several crystal forms of cianidanol, *Chem. Pharm. Bull.*, **33**, 4050–3.

AKIMOTO, K., NAKAGAWA, H. and SUGIMOTO, I., 1985, Photo-stability of cianidanol in aqueous solution and in the solid state, *Drug Devel. Ind. Pharm.*, **II**, 865–89.

Cinoxacin

VARGAS, F., RIVAS, C. and CANUDAS, N., 1994, Photosensitized lipid peroxidation by cinoxacin and its photoproducts. Involvement of a derived peroxide in its phototoxicity, *Pharmazie*, **49**, 742–5.

Cisplatin

CUBELLS, M. P., JANE, M. T., ANTUNEZ, X. D. and DE BOLOS CAPDEVILA, J., 1992, Degradación fotoquimica del cisplatino, *Circ. Farm.*, **314**, 149–54.

GREENE, R. F., CHALLEJI, D. C., HIRANAKA, P. K. and GALLELLI, J. F., 1979, Stability of cisplatin in aqueous solution, *Am. J. Hosp. Pharm.*, **36**, 38–43.

HONDA, D. H., JANSEN, J. R., MINOR, D. R. and GOOD, J. W., 1979, Preprinted Physician's order form for intravenous cisplatin therapy, *Am. J. Hosp. Pharm.*, **36**, 742–3.

MACKA, M., BORÀK, J., SEMÈNKOVÀ, L. and KISS, F., 1994, Decomposition of cisplatin in aqueous solutions containing chlorides by ultrasonic energy and light, *J. Pharm. Sci.*, **83**, 815–18.

STEWART, C. F. and FLEMING, R. A., 1990, Compatibility of cisplatin and fluorouracil in 0.9% sodium chloride injection, *Am. J. Hosp. Pharm.*, **47**, 1373–7.

VILLAREJO, A. D., REVILLA, P., ALONSO, P. and DOADRIO, A., 1986, Estudio de la estabilidad fotoquimica del cis-dicloro diamino Pt (II) por cromatografia liquida, *An. R. Acad. Farm.*, **52**, 265–78.

ZIESKE, P. A., KOBERDA, M., HINES, J. L., KNIGHT, C. C., SRIRAM, R., RAGHAVAN, N. V. and RABINOW, B. E., 1991, Characterization of cisplatin degradation as affected by pH and light, *AJHP*, **48**, 1500–6.

Clofazimine

SINGH, S., KHANNA, M. and SARIN, J. P. S., 1987, *In vivo* evaluation of tablet containing clofazimine and dapsone, *Indian J. Farm. Sci.*, **49**, 236–8.

Clonazepam

WAD, N., 1986, Degradation of clonazepam in serum by light confirmed by means of a high performance liquid chromatographic method, *Ther. Drug Monit.*, **8**, 358–60.

Colchicine

HABIB, M. J. and ASKER, A. F., 1989, Influence of certain additives on the photostability of colchicine solutions, *Drug Dev. Ind. Pharm.*, **15**, 1904–9.

1989, Influence of certain additives on the photostability of colchicine solutions, *Drug. Dev. Ind. Pharm.*, **15**, 845–9.

Contraceptives

BEIJERSBERGEN VAN HENEGOUWEN, G. M. J., 1991, New trends in Photobiology, *J. Photochem. Photobiol. B: Biol.*, **10**, 183–210.

Dacarbazine

BAIRD, G. M. and WILLOUGHBY, M. L. N., 1978, Photodegradation of dacarbazine, *Lancet*, **2**, 681.

Horton, J. K. and Stevens, M. F. G., 1981, A new light on the photodecomposition of the antitumour drug DTIC, *J. Pharm. Pharmacol.*, **33**, 808–11.

Islam, M. S. and Asker, A. F., 1994, Photostabilization of dacarbazine with reduced glutathione, *J. Pharm. Sci. Technol.*, **48**, 38–40.

Kirk, B., 1987, The evaluation of a light-protective giving set. The photosensitivity of intravenous dacarbazine solutions, *Intensive Ther. Clin. Monit.*, **8**, 78–86.

Dapsone

Singh, S., Khanna, M. and Sarin, J. P. S., 1987, *In vivo* evaluation of tablet containing clofazimine and dapsone, *Indian J. Farm. Sci.*, **49**, 236–8.

Daunorubicin

Maniez-Devos, D. M., Baurain, R., Lesne, M. and Trouet, A., 1986, Degradation of doxorubicin and daunorubicin in human and rabbit biological fluids, *J. Pharm. Biomed. Anal.*, **4**, 353–65.

Thoma, K. and Klimek, R., 1991, Photostabilization of drugs in dosage forms without protection from packaging materials, *Int. J. Pharm.*, **67**, 169–75.

Wood, M. J., Irwin, W. J. and Scott, D. K., 1990, Photodegradation of doxorubicin, daunorubicin and epirubicin measured by high-performance liquid chromatography, *J. Clin. Pharm. Therapeut.*, **15**, 291–300.

Demeclocycline

Moore, D. E. and Tamat, S. R., 1980, Photosensitization by drugs: photolysis of some chloride-containing drugs, *J. Pharm. Pharmacol.*, **32**, 172–7.

Dichloracetamids

Reich, J. and Weidman, K. G., 1971, Notiz zur Photochemie des Dichloracetamids, *Arch. Pharm.*, **304**, 920–2.

Diclofenac

Moore, D. E., Roberts-Thomson, S., Zhen, D. and Duke, C. C., 1990, Photochemical studies on the anti-inflammatory drug diclofenac, *Photochem. Photobiol.*, **52**, 685–90.

Digitoxin

Ekiz-Gücer, N. and Reisch, J., 1991, Photochemische studien, 61. Photoisomerisierung von digitoxin in kristallinem zustand, *Liebigs Ann. Chem.*, 1105–6.

Diltiazem

Suleiman, M. S., Adbulhameed, M. E., Najib, N. M. and Muti, H. Y., 1989, Effect of ultraviolet radiation on the stability of diltiazem, *Int. J. Pharm.*, **50**, 71–3.

Diphenhydramine

Beijersbergen van Henegouwen, G. M. J., 1991, New trends in photobiology, *J. Photochem. Photobiol. B: Biol.*, **10**, 183–210.

Beijersbergen van Henegouwen, G. M. J., van de Zijde, H. J., van de Griend, J. and de Vries, H., 1987, Photochemical decomposition of diphenhydramine in water, *Int. J. Pharm.*, **35**, 259–62.

Dipyridamole

Ameer, B., Callahan, R. J. and Dragotakes, S. C., 1989, Preparation and stability of an oral suspension of dipyridamole, *J. Pharm. Technol.*, **5**, 202–5.

Dithranol

MÜLLER, K., WIEGREBE, W. and YOUNES, M., 1987, Formation of active oxygen species by dithranol. III. Dithranol, active oxygen species and lipid peroxidation *in vivo*, *Arch. Pharm.*, **320**, 59–66.

Dobutamin

PRAMAR, Y., DAS GUPTA, V., GARDNER, S. N. and YAU, B., 1991, Stabilities of dobutamine, dopamine, nitroglycerin and sodium nitroprusside in disposable plastic syringes, *J. Clin. Pharm. Ther.*, **16**, 203–7.

Dopamine

DANDURAND, K. R. and STENNETT, D. J., 1985, Stability of dopamine hydrochloride exposed to blue-light phototherapy, *Am. J. Hosp. Pharm.*, **42**, 595–7.

NEWTON, D. W., YIN YEE FUNG, E. and WILLIAMS, D. A., 1981, Stability of five catecholamines and terbutaline sulfate in 5% dextrose injection in the absence and presence of aminophylline, *Am. J. Hosp. Pharm.*, **38**, 1314–19.

PRAMAR, Y., DAS GUPTA, V., GARDNER, S. N. and YAU, B., 1991, Stabilities of dobutamine, dopamine, nitroglycerin and sodium nitroprusside in disposable plastic syringes, *J. Clin. Pharm. Ther.*, **16**, 203–7.

Dothiepin

TAMMILEHTO, S. and TORNIAINEN, K., 1989, Photochemical stability of dothiephin in aqueous solutions, *Int. J. Pharm.*, **52**, 123–8.

Doxorubicin

ASKER, A. F. and HABIB, M. J., 1988, Effect of glutathione on photolytic degradation of doxorubicin hydrochloride, *J. Parent. Sci. Technol.*, **42**, 153–6.

HABIB, J. M. and ASKER, A. F., 1989, Photostabilization of doxorubicin hydrochloride with radioprotective and photoprotective agents: potential mechanism for enhancing chemotherapy during radiotherapy, *J. Parent. Sci. Technol.*, **43**, 259–61.

MANIEZ-DEVOS, D. M., BAURAIN, R., LESNE, M. and TROUET, A., 1986, Degradation of doxorubicin and daunorubicin in human and rabbit biological fluids, *J. Pharm. Biomed. Anal.*, **4**, 353–65.

WOOD, M. J., IRWIN, W. J. and SCOTT, D. K., 1990, Photodegradation of doxorubicin, daunorubicin and epirubicin measured by high-performance liquid chromatography, *J. Clin. Pharm. Ther.*, **15**, 291–300.

Emetine

TESHIMA, D., OTSUBO, K., HIGUCHI, S., HIRAYAMA, F., UEKAMA, K. and AOYAMA, T., 1989, Effects of cyclodextrins on degradations of emetine and cephaeline in aqueous solution, *Chem. Pharm. Bull.*, **37**, 1591–4.

Epirubicin

WOOD, M. J., IRWIN, W. J. and SCOTT, D. K., 1990, Photodegradation of doxorubicin, daunorubicin and epirubicin measured by high-performance liquid chromatography, *J. Clin. Pharm. Therapeut.*, **15**, 291–300.

Ergotamine

PUTTEMANS, M., DE COCK Jr, R., HOOGEWIJS, G., DRYON, L. and MASSART, D. L., 1985, Separation and quantitative determination by HPLC of ergotamin and its stereo-isomer ergotaminine in pharmaceutical preparations, *J. Pharm. Belg.*, **40**, 387–94.

THOMA, K. and KLIMEK, R., 1991, Photostabilization of drugs in dosage forms without protection from packaging materials, *Int. J. Pharm.*, **67**, 169–75.

Etopside

MCLEOD, H. L. and RELLING, M. V., 1992, Stability of etoposide solution for oral use, *Am. J. Hosp. Pharm.*, **49**, 2784–5.

Famotidine

UNDERBERG, W. J. M., KOOMEN, J. M. and BEIJNEN, J. H., 1988, Stability of famotidine in commonly used nutritional infusion fluids, *J. Parent. Sci. Technol.*, **42**, 94–7.

Fentanyl

ALLEN, L. V., STILES, M. L. and TU, Y.-H., 1990, Stability of fentanyl citrate in 0.9% sodium chloride solution in portable infusion pumps, *Am. J. Hosp. Pharm.*, **47**, 1572.

TU, Y.-H., STILES, M. L. and ALLEN, L. V., 1990, Stability of fentanyl citrate and bupivacaine hydrochloride in portable pump reservoirs, *Am. J. Hosp. Pharm.*, **47**, 2037–40.

Fibrates

VARGAS, F. and CANUDAS, N., 1993, Photolability and photohemolytic studies of fibrates, *Pharmazie*, **48**, 900–4.

VARGAS, F., RIVAS, C. and CANUDAS, N., 1993, Formation of a perbenzoic acid derivative in the photodegradation of fenofibrate: Phototoxicity studies on erythrocytes, *J. Pharm. Sci.*, **82**, 590–1.

Flordipine

NARURKAR, A. N., SHEEN, P.-C., BERNSTEIN, D. F. and AUGUSTINE, M. A., 1986, Studies on the light stability of flordipine tablets in amber blister packaging material, *Drug. Devel. Ind. Pharm.*, **12**, 1241–7.

Flucytosine

BIONDI, L. and NAIRN, J. G., 1986, Stability of 5-fluorouracil and flucytosine in parenteral solutions, *Can. J. Hosp. Pharm.*, **39**, 60–4.

Flunitrazepam

BUSKER, R. W., BEIJERSBERGEN VAN HENEGOUWEN, G. M. J., KWEE, B. M. C. and WINKENS, J. H. M., 1987, Photobinding of flunitrazepam and its major photo-decomposition product N-desmethylflunitrazepam, *Int. J. Pharm.*, **36**, 113–20.

GIVENS, R. S., GINGRICH, J. and MECKLENBURG, S., 1986, Photochemistry of flunitrazepam: a product and model study, *Int. J. Pharm.*, **29**, 67–72.

Fluorouracil

BIONDI, L. and NAIRN, J. G., 1986, Stability of 5-fluorouracil and flucytosine in parenteral solutions, *Can. J. Hosp. Pharm.*, **39**, 60–4.

LORILLON, P., CORBEL, J. C., MORDELET, M.-F., BASLE, B. and GUESNIER, L. R., 1992, Photosensibilité du 5-fluoro-uracile et du méthotrexate dans des perfuseurs translucides ou opaques, *J. Pharm. Clin.*, **11**, 285–95.

MILOVANOVIC, D. and NAIRN, J. G., 1980, Stability of fluorouracil in amber glass bottles, *Am. J. Hosp. Pharm.*, **37**, 164–5.

STEWART, C. F. and FLEMING, R. A., 1990, Compatibility of cisplatin and fluorouracil in 0.9% sodium chloride injection, *Am. J. Hosp. Pharm.*, **47**, 1373–7.

Flurbiprofen

CASTELL, J. V. and GÓMEZ-LECHÓN, M. J., 1992, Phototoxicity of non-steroidal anti-inflammatory drugs: *in vitro* testing of the photoproducts of butibufen and flurbiprofen, *J. Photochem. Photobiol., B: Biol.*, **13**, 71–81.

Foscarnet

WOODS, K., STEINMANN, W., BRUNS, L. and NEELS, J. T., 1994, Stability of foscarnet sodium in 0.9% sodium chloride injection, *Am. J. Hosp. Pharm.*, **51**, 88–90.

Furnidipine

NÙNEZ-VERGARA, L. J., SUNKEL, C. and SQUELLA, J. A., 1994, Photodecomposition of a new 1,4-dihydropyridine: furnidipine, *J. Pharm. Sci.*, **83**, 502–7.

Furosemide

BUNDGAARD, H., NØRGAARD, T. and NIELSEN, T. M., 1988, Photodegradation and hydrolysis of furosemide and furosemide esters in aqueous solutions, *Int. J. Pharm.*, **42**, 217–24.

KERREMANS, A. L. M., TAN, Y., VAN GNNEKEN, C. A. M., and GRIBNAU, F. W. J., 1982, Specimen handling and high-performance liquid chromatographic determination of furosemide, *J. Chromatogr.*, **229**, 129–39.

MATSUDA, Y. and MASAHARA, R., 1980, Comparative evaluation of coloration of photosensitive solid drugs under various light sources, *Yakugaku Zasshi*, **100**, 953–7.

MOORE, D. E. and SITHIPITAKS, V., 1983, Photolytic degradation of frusemide, *J. Pharm. Pharmacol.*, **35**, 489–93.

MOORE, D. E. and TAMAT, S. R., 1980, Photosensitization by drugs: photolysis of some chlorine-containing drugs, *J. Pharm. Pharmacol.*, **32**, 172–7.

ROWBOTHAM, P. C., STANFORD, J. B. and SUGDEN, J. K., 1976, Some aspects of the photochemical degradation of frusemide, *Pharm. Acta Helv.*, **51**, 304–7.

THOMA, K. and KLIMEK, R., 1991, Photostabilization of drugs in dosage forms without protection from packaging materials, *Int. J. Pharm.*, **67**, 169–75.

VILLIERS, DE, M. M., WATT, VAN DER, J. G. and LÖTTER, A. P., 1992, Kinetic study of the solid-state photolytic degradation of two polymorphic forms of furosemide, *Int. J. Pharm.*, **88**, 275–83.

1993, Influence of the cohesive behaviour of small particles on the solid-state photolytic degradation of furosemide, *Drug Dev. Ind. Pharm.*, **19**, 383–94.

YAHYA, A. M., MCELNAY, J. C. and D'ARCY, P. F., 1986, Photodegradation of frusemide during storage in burette administration sets, *Int. J. Pharm.*, **31**, 65–8.

Haloperidol

THOMA, K. and KLIMEK, R., 1991, Photostabilization of drugs in dosage forms without protection from packaging materials, *Int. J. Pharm.*, **67**, 169–75.

Hexachlorophane

MOORE, D. E. and TAMAT, S. R., 1980, Photosensitization by drugs: photolysis of some chlorine-containing drugs, *J. Pharm. Pharmacol.*, **32**, 172–7.

Hydralazine

HALASI, S. and NAIRN, J. G., 1990, Stability of hydralazine hydrochloride in parenteral solutions, *Can. J. Hosp. Pharm.*, **43**, 237–41.

Hydroxychloroquine

TØNNESEN, H. H., GRISLINGAAS, A.-L., WOO, S. O. and KARLSEN, J., 1988, Photochemical stability of antimalarials. I. Hydroxychloroquine, *Int. J. Pharm.*, **43**, 215–19.

Hypochlorite

VINCENT-BALLEREAU, F., MERVILLE, C. and LAFLEURIEL, M. T., 1989, Sodium hypochloride as a disinfectant for injection materials in third world rural dispensaries, *Int. J. Pharm.*, **50**, 87–8.

Ibuprofen

CASTELL, J. V., GÒMEZ-L., M. J., MIRANDA, M. A. and MOREA, I. M., 1987, Photolytic degradation of ibuprofen, *Photochem. Photobiol.*, **46**, 991–6.

PANDIT, J. K., PAL, R. N. and MISHRA, B., 1989, Effect of formulation variables and storage conditions on the release rate of ibuprofen solid dosage forms, *East. Pharm.*, **17**, 133–7.

Imipramin

MATSUDA, Y. and MASAHARA, R., 1980, Comparative evaluation of coloration of photosensitive solid drugs under various light sources, *Yakugaku Zasshi*, **100**, 953–7.

Indapamide

DAVIS, R., WELLS, C. H. J. and TAYLOR, A. R., 1979, Photolytic decomposition of indapamide, *J. Pharm. Sci.*, **68**, 1063–4.

Indomethacin

DABESTANI, R., SIK, R. H., DAVIS, D. G., DUBAY, G. and CHIGNELL, C. F., 1993, Spectroscopic studies of cutaneous photosensitizing agents. XVIII. Indomethacin, *Photochem. Photobiol.*, **58**, 367–73.

EKIZ-GÜCER, N. and REISCH, J., 1991, Photostabilität von indometacin in kristallinem zustand (2). Photochemische studien, 58. Mitt, *Pharm. Acta Helv.*, **66**, 66–7.

MATSUDA, Y., ITOOKA, T. and MITSUHASHI, Y., 1980, Photostability of indomethacin in model gelatine capsules: effects of film thickness and concentration of titanium dioxide on the coloration and photolytic degradation, *Chem. Pharm. Bull.*, **28**, 2665–71.

MATSUDA, Y. and MASAHARA, R., 1980, Comparative evaluation of coloration of photosensitive solid drugs under various light sources, *Yakugaku Zasshi*, **100**, 953–7.

MOORE, D. W. and CHAPPUIS, P. P., 1988, A comparative study of the photochemistry of the non-steroidal anti-inflammatory drugs, naproxen, benoxaprofen and indomethacin, *Photochem. Photobiol.*, **47**, 173–80.

THOMA, VON K. and KERKER, R., 1992, Photoinstabilität von arzneimotteln. 1. Mitteilung über das verhalten von nur im UV-bereich absorbierenden substanzen bei der tageslichtsimulation, *Pharm. Ind.*, **54**, 169–77.

Isoprenaline

DE MOL, N. J., BEIJERSBERGEN VAN HENEGOUWEN, G. M. J. and GERRITSMA, K. W., 1979, Photochemical decomposition of catecholamines – II. The extent of aminochrome formation from adrenaline, isoprenaline and noradrenaline induced by ultraviolet light, *Photochem. Photobiol.*, **29**, 479–82.

WOLLMANN, H. and GRÜNERT, R., 1984, Einfluss des sichtbaren Lichtes auf die Haltbarkeit von Isoprenalin-, Epinephrin- und levarterenollösungen in unterschiedlicher Behältnissen, *Pharmazie*, **39**, 161–3.

Isopropylaminophenazone

REISCH, J., EKIZ, N. and GÜNERI, T., 1986, Photo- und strahlenchemische studien, 46. Mitt. Zur photochemie des isopropylaminophenazons im kristallinen zustand und wässriger lösung, *Arch. Pharm.*, **319**, 973–7.

Isoproterenol

NEWTON, D. W., YIN YEE FUNG, E. and WILLIAMS, D. A., 1981, Stability of five catecholamines and terbutaline sulfate in 5% dextrose injection in the absence and presence of aminophylline, *Am. J. Hosp. Pharm.*, **38**, 1314–19.

Ketoconazole

KUMER, K. P., OKONOMAH, A. D. and BRADSHAW, W. G., 1991, Stability of ketoconazole in ethanolic solutions, *Drug Dev. Ind. Pharm.*, **17**, 577–80.

Ketoprofen

BOSCÀ, F., MIRANDA, M. A., CARGANICO, G. and MAULEÒN, D., 1994, Photochemical and photobiological properties of ketoprofen associated with the benzophenone chromophore, *Photochem. Photobiol.*, **60**, 96–101.

Ketorolac tromethamine

GU, L., CHIANG, H.-S. and JOHNSON, D., 1988, Light degradation of ketorolac tromethamine, *Int. J. Pharm.*, **41**, 105–13.

Lactucin

SCHENK, G., HUKE, M. and NAUMANN, T., 1964, Untersuchungen über die Photochemische Umwandlung des lactucins, *Archiv.*, **297**, 703.

Levothyroxine

KATO, Y., SAITO, M., KOIZUMI, H. and TOYODA, Y., 1986, Determination and stability of levothyroxine sodium injections prepared in hospital, *Jap. J. Hosp. Pharm.*, **12**, 253–6.

Meclofenamic acid

PHILIP, J. and SZULCZEWSKI, D. H., 1973, Photolytic decomposition of N-(2,6-dichloro-m-tolyl) anthranilic acid (meclofenamic acid), *J. Pharm. Sci.*, **62**, 1479–82.

Mefloquine

EPLING, G. A. and YOON, U. C., 1982, Photochemical reactivity of mefloquine, a new synthetic antimalarial, *Chemistry Letters*, **2**, 211–14.

TØNNESEN, H. H. and GRISLINGAAS, A.-L., 1990, Photochemical stability of biologically active compounds. II. Photochemical decomposition of mefloquine in water, *Int. J. Pharm.*, **60**, 157–62.

TØNNESEN, H. H. and MOORE, D. E., 1991, Photochemical stability of biologically active compounds. III. Mefloquine as a photosensitizer, *Int. J. Pharm.*, **70**, 95–101.

Menadione

ASKER, A. F. and HABIB, M. J., 1989, Photostabilization of menadione sodium bisulfite by glutathione, *J. Parent. Sci. Technol.*, **43**, 204–7.

6-Mercaptopurine

HEMMENS, V. J. and MOORE, D. E., 1984, Photo-oxidation of 6-Mercaptopurine in aqueous solution, *J. Chem. Soc. Perkin Trans.*, **2**, 209–11.

Methadone

DENSON, D. D., CREWS, J. C., GRUMMICH, K. W., STIRM, E. J. and SUE, C. A., 1991, Stability of methadone hydrochloride in 0.9% sodium chloride injection in single-dose plastic containers, *Am. J. Hosp. Pharm.*, **48**, 515–17.

Methaqualone

PÖHLMANN, H., THEIL, F.-P. and OTTO, H., 1986, Nachweis eines Oxaziridins als Zwischenprodukt der fotochemischer Umsetzung von Methaqualon-1-oxid., *Pharmazie*, **41**, 390–4.

STEVENS, M. F. G. and GUNN, B. C., 1972, Photolysis of a methaqualone metabolite, *J. Pharm. Pharmacol.*, **24**, 141P.

THEIL, F. P., PÖHLMANN, H., PFEIFER, S. and FRANKE, P., 1985, Fotochemische Reaktivität von Methaqualon-1-oxid., *Pharmazie*, **40**, 328–31.

Methotrexate

CHAHIDI, C., GIRAUD, M., AUBAILLY, M., VALLA, A. and SANTUS, R., 1986, 2,4-diamino-6-pteridinocarboxaldehyde and an azobenzene derivative are produced by UV photodegradation of methtrexate, *Photochem. Photobiol.*, **44**, 231–3.

CHAHIDI, C., MORLIERE, P., AUBAILLY, M., DUBERTRET, L. and SANTUS, R., 1983, Photosensitization by methotrexate photoproducts, *Photochem. Photobiol.*, **38**, 317–22.

CHATTERJI, D. C. and GALLELLI, J. F., 1978, Thermal and photolytic decomposition of methotrexate in aqueous solutions, *J. Pharm. Sci.*, **67**, 526–31.

DYVIK, O., GRISLINGAAS, A.-L., TØNNESEN, H. H. and KARLSEN, J., 1986, Methotrexate in infusion solutions – a stability test for the hospital pharmacy, *J. Clin. Hosp. Pharm.*, **II**, 343–8.

LORILLON, P., CORBEL, J. C., MORDELET, M.-F., BASLE, B. and GUESNIER, L. R., 1992, Photosensibilité du 5-fluoro-uracile et du méthotrexate dans des perfuseurs translucides ou opaques, *J. Pharm. Clin.*, **11**, 285–295.

Methyldopa

NEWTON, D. W., YIN YEE FUNG, E. and WILLIAMS, D. A., 1981, Stability of five catecholamines and terbutaline sulfate in 5% dextrose injection in the absence and presence of aminophylline, *Am. J. Hosp. Pharm.*, **38**, 1314–19.

Metoprolol

NASA, S. L. and YADAV, S., 1989, Microencapsulation of metoprolol tartrate using phase separation conservation technique, *East. Pharm.*, **32**, 133–4.

Metronidazole

CANO, S. B. and GLOGIEWICZ, F. L., 1986, Storage requirements for metronidazole injection, *Am. J. Hosp. Pharm.*, **43**, 2983.

ELFATIH, I. A., KARIM IBRAHIM, K. E. and ADAM, M. E., 1991, Studies on the photochemical decomposition of metronidazole, *Int. J. Pharm.*, **76**, 261–4.

GODFREY, R. and EDWARDS, R., 1991, A chromatographic and spectroscopic study of photodegraded metronidazole in aqueous solution, *J. Pharm. Sci.*, **80**, 212–18.

HABIB, M. J. and ASKER, A. F., 1989, Complex formation between metronidazole and sodium urate: effect on photodegradation of metronidazole, *Pharm. Res.*, **6**, 58–61.

MOORE, D. E. and WILKINS, B. J., 1990, Common products from gamma-radiolysis and ultraviolet photolysis of metronidazole, *Radiat. Phys. Chem.*, **36**, 547–50.

MOORE, D. E., CHIGNELL, C. F., SIK, R. H. and MOTTEN, A. G., 1986, Generation of radical anions from metronidazole, misonidazole and azathioprine by photoreduction in the presence of EDTA, *Int. J. Radiat. Biol.*, **50**, 885–91.

THOMA, VON K. and KERKER, R., 1992, Photoinstabilität von arzneimotteln. 1. Mitteilung über das verhalten von nur im UV-bereich absorbierenden substanzen bei der tageslichtsimulation, *Pharm. Ind.*, **54**, 169–77.

Midazolam

ANDERSIN, R., OVASKAINEN, J. and KALTIA, S., 1994, Photochemical decomposition of midazolam. III. Isolation and identification of products in aqueous solutions, *J. Pharm. Biomed. Anal.*, **12**, 165–72.

ANDERSIN, R. and TAMMILEHTO, S., 1989, Photochemical decomposition of midazolam. II. Kinetics in ethanol, *Int. J. Pharm.*, **56**, 175–9.

BLEASEL, M. D., PETERSON, G. M. and JESTRIMSKI, K. W., 1993, Stability of midazolam in sodium chloride infusion packs, *Aust. J. Hosp. Pharm.*, **23**, 260–2.

SELKÄMAA, R. and TAMMILEHTO, S., 1989, Photochemical decomposition of midazolam. I. Isolation and identification of products, *Int. J. Pharm.*, **49**, 83–9.

Misonidazole

MOORE, D. E., CHIGNELL, C. F., SIK, R. H. and MOTTEN, A. G., 1986, Generation of radical anions from metronidazole, misonidazole and azathioprine by photoreduction in the presence of EDTA, *Int. J. Radiat. Biol.*, **50**, 885–91.

Mitomycin C

BEIJNEN, J. H., GIJN, R. VAN and UNDERBERG, W. J. M., 1990, Chemical stability of the antitumor drug mitomycin C in solutions for intravesical instillation, *J. Parent. Sci. Technol.*, **44**, 332–5.

Mitonafide

TORRES SUÀREZ, A. I. and CAMACHO, M. A., 1994, Photolability evaluation of the new cytostatic drug mitonafide, *Arzneim.-Forsch./Drug Res.*, **44** (I), Nr. 1, 81–3.

Molsidomine

THOMA, VON K. and KERKER, R., 1992, Photoinstabilität von arzneimitteln. 1. Mitteilung über das verhalten von nur im UV-bereich absorbierenden substanzen bei der tageslichtsimulation, *Pharm. Ind.*, **54**, 169–77.

1992, Photostabilität von arzneimitteln. 6. Mitteilung: Untersuchungen zur photostabilität von molsidomin., *Pharm. Ind.*, **54**, 630–8.

VANDENBOSSCHE, G. M. R., DE MUYNCK, C., COLARDYN, F. and REMON, J. P., 1993, Light stability of molsidomine in infusion fluids, *J. Pharm. Pharmacol.*, **45**, 486–8.

Morphine

GRASSBY, P. F. and HUTCHINGS, L., 1993, Factors affecting the physical and chemical stability of morphine sulphate solutions stored in syringes, *Int. J. Pharm.*, Pr., 2, 39–43.

LANDERSJØ, L. and HÖK, L., 1986, Stabilitet av morfin i Infusor, *Sjukhusfarmaci*, **nr. 5**, s. 85–7.

POGGI, G. L., 1991, Compatibility of morphine tartrate admixtures in propylene syringes, *Austral. J. Hosp. Pharm.*, **21**, 316.

Nalidixic acid

FERNANDEZ, E., PEÑA, W., VINET, R. and HIDALGO, M. E., 1986, Kinetics of photodegradation of nalidixic acid, *An. Quim., Ser. A*, **82**, 96–8.

MOORE, D. E., HEMMENS, V. S. and YIP, H., 1984, Photosensitization by drugs: nalidixic and oxolinic acids, *Photochem. Photobiol.*, **39**, 57–61.

TATE, T. J., DIFFEY, B. L. and DAVIS, A., 1980, An ultraviolet radiation dosimeter based on the photosensitising drug, nalidixic acid, *Photochem. Photobiol.*, **31**, 27–30.

Naproxen

ALBINI, A., 1991, Fotochimica di sostanze farmaceutiche, *Boll. Chim. Farmaceutico*, **130**, 393.

BOSCÀ, F., MIRANDA, M. A., VANÒ, L. and VARGAS, F., 1990, New photodegradation for naproxen, a phototoxic non-steroidal anti-inflammatory drug, *J. Photochem. Photobiol. A: Chem.*, **54**, 131.

MOORE, D. E., 1987, Principles and practice of drug photodegradation studies, *J. Pharm. Biomed. Anal.*, **5**, 441–53.

MOORE, D. W. and CHAPPUIS, P. P., 1988, A comparative study of the photochemistry of the non-steroidal anti-inflammatory drugs, naproxen, benoxaprofen and indomethacin, *Photochem. Photobiol.*, **47**, 173–80.

Neocarzinostatin

BURGER, R. M., PEISACH, J. and HORWITZ, S. B., 1978, Effect of light and oxygen on neocarzinostatin stability and DNA-cleaving activity, *J. Biol. Chem.*, **253**, 4830–2.

Nicardipine

BONFERONI, M. C., MELLERIO, G., GIUNCHEDI, P., CARAMELLA, C. and CONTE, U., 1992, Photostability evaluations of nicardipine HCL solutions, *Int. J. Pharm.*, **80**, 109–17.

Nifedipine

AL-TURK, W. A., MAJEED, I. A., MURRAY, W. J., and OTHMAN, S., 1988, Some factors affecting the photodecomposition of nifedipine, *Int. J. Pharm.*, **41**, 227–30.

BÈCHARD, S. R., QURAISHI, O. and KWONG, E., 1992, Film coating: effect of titanium dioxide concentration and film thickness on the photostability of nifedipine, *Int. J. Pharm.*, **87**, 133–9.

BEIJERSBERGEN VAN HENEGOUWEN, G. M. J., 1991, New trends in photobiology, *J. Photochem. Photobiol. B: Biol.*, **10**, 183–210.

DESAI, D. S., ABDELNASSER, M. A., RUBITSKI, B. A. and VARIA, S. A., 1994, Photostabilization of uncoated tablets of sorivudine amd nifedipine by incorporation of synthetic iron oxides, *Int. J. Pharm.*, **103**, 69–76.

FELTKAMP, H., MEYLE, E. and TOPPEL, G., 1989, Nifedipin – Qualität und Chargenkonformität retardierter Nifedipin-Generika., *Deutsche Apotheker Zeitung*, **129**, 2697–704.

GIBBS, N. K., TRAYNOR, N. J., JOHNSON, B. E. and FERGUSON, J., 1992, *In vitro* phototoxicity of nifedipine: Sequential induction of toxic and non-toxic photoproducts with UVA radiation, *J. Photochem. Photobiol., B: Biol.*, **13**, 275–88.

LOCATI, D. and DONDI, G., 1986, Nuova metodica gas-cromatografica per la determinazione della nifedipina nei liquidi perfusionali, *Boll. Chim. Farm.*, **125**, 254–8.

LOGAN, B. K. and PATRICK, K. S., 1990, Photodegradation of nifedipine relative to nitrendipine evaluated by liquid and gas chromatography, *J. Chromatogr.*, **529**, 175–81.

MAJEED, I. A., MURRAY, W. J., OTHMAN, S. and AL-TURK, W. A., 1987, Spectrophotometric study of the photodecomposition kinetics of nifedipine, *J. Pharm. Pharmacol.*, **39**, 1044–6.

MARCINIEC, B. and RYCHCIK, W., 1994, Kinetic analysis of nifedipine photodegradation in the solid state, *Pharmazie*, **49**, 894–7.

Matsuda, Y., Teraoka, R. and Sugimoto, I., 1989, Comparative evaluation of photostability of solid-state nifedipine under ordinary and intensive light irradiation conditions, *Int. J. Pharm.*, **54**, 211–21.

Müller, B. W. and Albers, E., 1992, Complexation of dihydropyridine derivatives with cyclodextrins and 2-hydroxypropyl-β-cyclodextrin in solution, *Int. J. Pharm.*, **79**, 273–88.

Ogawa, S., Itagaki, Y., Hayase, N., Takemoto, I., Kasahara, N., Akutsu, S. and Inagaki, S., 1990, Photostability of nifedipine in powder, obtained by crushing tablet, granule of fine-granule, *Jpn. J. Hosp. Pharm.*, **16**, 189–97.

Pietta, P., Rava, A. and Biondi, P., 1981, High-performance liquid chromatography of nifedipine, its metabolites and photochemical degradation products, *J. Chromatogr.*, **210**, 516–21.

Sadana, G. S. and Ghogare, A. B., 1991, Quantitative proton magnetic resonance spectroscopic determination of nifedipine and its photodecomposition products from pharmaceutical preparations, *J. Pharm. Sci.*, **80**, 895–8.

Sadana, G. J. and Ghogare, A. B., 1991, Mechanistic studies on photolytic degradation of nifedipine by use of ^{1}H-NMR and ^{13}C-NMR spectroscopy, *Int. J. Pharm.*, **70**, 195–9.

Sugimoto, I., Tohgo, K., Sasaki, K., Nakagawa, H., Matsuda, Y. and Masahara, R., 1981, Wavelength dependency of the photodegradation of nifedipine tablets, *Yakugaku Zasshi*, **101**, 1149–63.

Teraoka, R., Matsuda, Y. and Sugimoto, I., 1988, Quantitative design for photostabilization of nifedipine by using titanium dioxide and/or tartrazine as colourants in model film coating systems, *J. Pharm. Pharmacol.*, **41**, 293–7.

Testa, R., Dolfini, E., Reschiotto, C., Secchi, C. and Biondi, P. A., 1979, GLC determination of nifedipine, a light sensitive drug, in plasma, *Il Farmaco Ed. Pr.*, **34**, 463–73.

Thoma, K. and Klimek, R., 1985, Untersuchungen zur Photoinstabilität von Nifedipin, *Pharm. Ind.*, **47**, 207–15.

1985, Untersuchungen zur Photoinstabilität von Nifedipin. 2. MIH.: Einfluss von Milieubedingungen, *Pharm. Ind.*, **47**, 319–27.

1991, Untersuchungen zur Photoinstabilität von Nifedipin, *Pharm. Ind.*, **53**, 388–96.

1991, Photostabilization of drugs in dosage forms without protection from packaging materials, *Int. J. Pharm.*, **67**, 169–75.

Thoma, von K. and Kerker, R., 1992, Photostabilität von arzneimitteln. 2. Mitteilung über das verhalten von im sichtbaren bereich absorbierenden substanzen bei der tageslichtsimulation, *Pharm. Ind.*, **54**, 287–93.

1992, Photoinstabilität von arzneimitteln. 3. Mitteilung: Photostabilität und stabilisierung von nifedipin in arzneiformen., *Pharm. Ind.*, **54**, 359–65.

1992, Photostabilität von arzneimitteln. 4. Mitteilung: Untersuchung zu den zersezungsprodukten von nifedipin, *Pharm. Ind.*, **54**, 465–8.

Tucker, F. A., Minty, P. S. B. and MacGregor, G. A., 1985, Study of nifedipine photodecomposition in plasma and whole blood using capillary gas-liquid chromatography, *J. Chromatogr.*, **342**, 193–8.

Vargas, F., Rivas, C. and Machado, R., 1992, Photodegradation of Nifedipine under Aerobic Conditions: Evidence of Formation of Singlet oxygen and Radical Intermediate, *J. Pharm. Sci.*, **81**, 399–400.

Nimodipine

ZANOCCO, A. L., DIAZ, L., LÒPEZ, M., NUNEZ-VERGARA, L. J. and SQUELLA, J. A., 1992, Polarographic study of the photodecomposition of nimodipine, *J. Pharm. Sci.*, **81**, 920–4.

Nitrazepam

ROTH, H. J. and ADOMEIT, M., 1973, Photochemie des Nitrazepams., *Arch Pharm.*, **306**, 889–97.

Nisoldipine

MÜLLER, B. W. and ALBERS, E., 1992, Complexation of dihydropyridine derivatives with cyclodextrins and 2-hydroxypropyl-β-cyclodextrin in solution, *Int. J. Pharm.*, **79**, 273–88.

Nitrendipine

LOGAN, B. K. and PATRICK, K. S., 1990, Photodegradation of nifedipine relative to nitrendipine evaluated by liquid and gas chromatography, *J. Chromatogr.*, **529**, 175–81.

MÜLLER, B. W. and ALBERS, E., 1992, Complexation of dihydropyridine derivatives with cyclodextrins and 2-hydroxypropyl-β-cyclodextrin in solution, *Int. J. Pharm.*, **79**, 273–88.

SQUELLA, J. A., ZANOCCO, A., PERNA, S. and NUÑEZ-VERGARA, L. J., 1990, A polarographic study of the photodegradation of nitrendipine, *J. Pharm. Biomed. Anal.*, **8**, 43–7.

Nitrobenzaldehydes

REICH, J. and WEIDMANN, K. G., 1971, Photochemie des p-nitrobenzaldehyds, *Arch. Pharm.*, **304**, 906–10.

Nitrofurantoin

BEIJERSBERGEN VAN HENEGOUWEN, G. M. J., 1991, New trends in photobiology, *J. Photochem. Photobiol. B: Biol.*, **10**, 183–210.

Nitrofurazone

QUILLIAM, M. A., MCCARRY, B. E., HOO, K. H., MCCALLA, D. R. and VIATEKUNAS, S., 1987, Identification of the photolysis products of nitrofurazone irradiated with laboratory illumination, *Cand. J. Chem.*, **65**, 1128–32.

THOMA, K. and KLIMEK, R., 1991, Photostabilization of drugs in dosage forms without protection from packaging materials, *Int. J. Pharm.*, **67**, 169–75.

THOMA, VON K. and KERKER, R., 1992, Photostabilität von arzneimitteln. 2. Mitteilung über das verhalten von im sichtbaren bereich absorbierenden substanzen bei der tageslichtsimulation, *Pharm. Ind.*, **54**, 287–93.

Nitroglycerin

LUDWIG, D. J. and UEDA, C. T., 1978, Apparent stability of nitroglycerin in dextrose 5% in water, *Am. J. Hosp. Pharm.*, **35**, 541–4.

PRAMAR, Y., DAS GUPTA, V., GARDNER, S. N. and YAU, B., 1991, Stabilities of dobutamine, dopamine, nitroglycerin and sodium nitroprusside in disposable plastic syringes, *J. Clin. Pharm. Therapeu.*, **16**, 203–7.

STUREK, J. K., SOKOLOSKI, T. D., WINSLEY, W. T. and STACH, P. E., 1978, Stability of nitroglycerin injection determined by gas chromatography, *Am. J. Hosp. Pharm.*, **35**, 537–41.

Nitroprusside

DAVIDSON, S. W. and LYALL, D., 1987, Sodium nitroprusside stability in light-protective administration sets, *Pharm. J.*, **239**, 599–601.

EDWARDS, R., 1986, Stability and light sensitivity of sodium nitroprusside infusions, *Aust. J. Hosp. Pharm.*, **16**, 145.

FRANK, M. J., JOHNSON, J. B. and RUBIN, S. H., 1976, Spectrophotometric determination of sodium nitroprusside and its photodegradation products, *J. Pharm. Sci.*, **65**, 44–8.

LEEUWENKAMP, O. R., VAN BENNEKOM, W. P., VAN DER MARK, E. J. and BULT, A., 1984, Nitroprusside, antihypertensive drug and analytical reagent. Review of (photo) stability, pharmacology and analytical properties, *Pharm. Weekbl. Sci. Ed.*, **6**, 129–40.

LEEUWENKAMP, O. R., VAN DER MARK, E. J., VAN BENNEKOM, W. P. and BULT, A., 1985, Investigation of the photochemical and thermal degradation of aqueous nitroprusside solutions using liquid chromatography, *Int. J. Pharm.*, **24**, 27–41.

MAHONY, C., BROWN, J. E., STARGET, W. W., VERGHESE, C. P. and BJORNSSON, T. D., 1984, *In vitro* stability of sodium nitroprusside solutions for intravenous administration, *J. Pharm. Sci.*, **73**, 838–9.

PRAMAR, Y., DAS GUPTA, V., GARDNER, S. N. and YAU, B., 1991, Stabilities of dobutamine, dopamine, nitroglycerin and sodium nitroprusside in disposable plastic syringes, *J. Clin. Pharm. Therapeu.*, **16**, 203–7.

SAUNDERS, A., 1986, Stability and light sensitivity of sodium nitroprusside infusions, *Aust. J. Hosp. Pharm.*, **16**, 55–6.

VESEY, C. J. and BATISTONI, G. A., 1977, The determination and stability of sodium nitroprusside in aqueous solutions (determination and stability of SNP), *J. Clin. Pharm.*, **2**, 105–17.

Noradrenaline

DE MOL, N. J., BEIJERSBERGEN VAN HENEGOUWEN, G. M. J. and GERRITSMA, K. W., 1979, Photochemical decomposition of catecholamines – II. The extent of aminochrome formation from adrenaline, isoprenaline and noradrenaline induced by ultraviolet light, *Photochem. Photobiol.*, **29**, 479–82.

Norepinephrine

NEWTON, D. W., YIN YEE FUNG, E. and WILLIAMS, D. A., 1981, Stability of five catecholamines and terbutaline sulfate in 5% dextrose injection in the absence and presence of aminophylline, *Am. J. Hosp. Pharm.*, **38**, 1314–19.

Norfloxacin

NANGIA, A., LAM, F. and HUNG, C. T., 1991, A stability study of aqueous solution of norfloxacin, *Drug Dev. Ind. Pharm.*, **17**, 681–94.

Olaquindox

BEIJERSBERGEN VAN HENEGOUWEN, G. M. J., 1991, New trends in photobiology, *J. Photochem. Photobiol. B: Biol.*, **10**, 183–210.

Oxolinic acid

MOORE, D. E., HEMMENS, V. S. and YIP, H., 1984, Photosensitization by drugs: nalidixic and oxolinic acids, *Photochem. Photobiol.*, **39**, 57–61.

Pentacaine

BEZAKOVA, Z., BACHRATÀ, M., BLESOVÀ, M. and BOROVANSKY, A., 1986, Studium lokalnych anestetik, LXXXIII Stabilita karbizokainiumchloridu a pentakainiumchloridu, *Farm. Obzor.*, **55**, 195–203.

LAU, T. VAN, ČIŽMARIK, J. and SOVIAR, K., 1991, Decomposition of the pentacaine base, *Pharmazie*, **46**, 356.

Perazine

PAWELCZYK, E. and MARCINIEK, B., 1977, Kinetics of drug decomposition. Part 47. Effect of substituents on photochemical stability of perazine derivatives, *Pol. J. Pharmacol. Pharm.*, **29**, 143–9.

Perphenazine

MATSUDA, Y. and MASAHARA, R., 1980, Comparative evaluation of coloration of photosensitive solid drugs under various light sources, *Yakugaku Zasshi*, **100**, 953–7.

Phenazone

MARCINIEC, B., 1983, Photochemical decomposition of phenazone derivatives. Part 2: Kinetics of decomposition in the solid state, *Pharmazie*, **38**, 848–50.

1985, Photochemical decomposition of phenazone derivatives. IV. Kinetics of photolysis and photooxidation in solutions, *Acta Polon. Pharm.*, **XLII,** 448–54.

REISCH, J., EKIZ, N. and GUNERI, T., 1986, Fotoabbau von 4-isopropyl-2, 3-dimethyl-1-phenyl-3-pyrazolin-5-on (propyphenazon) 49. Mitteilung: Foto- und strahlenchemische studien, *Pharmazie*, **41**, 287–8.

Phenothiazines

BEIJERSBERGEN VAN HENEGOUWEN, G. M. J., 1991, New trends in Photobiology, *J. Photochem. Photobiol. B: Biol.*, **10**, 183–210.

Phenylbutazone

FABRE, F., HUSSAM-EDDINE, N., LERNER, D. and MANDROU, B., 1984, Autooxidation and hydrolysis kinetics of the sodium salt of phenylbutazone in aqueous solution, *J. Pharm. Sci.*, **73**, 1709–13.

Phenylephrine

CHAFETZ, L. and CHOW, L. H., 1988, Photochemical hydroxylation of phenylephrine to epinephrine (adrenaline), *J. Pharm., Biomed. Anal.*, **6**, 511–14.

Phenytoin

MATSUDA, Y. and MASAHARA, R., 1980, Comparative evaluation of coloration of photosensitive solid drugs under various light sources, *Yakugaku Zasshi*, **100**, 953–7.

Pilocarpine

PLAIZIER-VERCAMMEN, J. A., 1993, Parametereffecten op de stabiliteit en activiteit van pilocarpine collyria, een overzicht, *Farm. Tijdschr. Belg.*, **70**, 2–10.

Pralidoxime

HOLCOMBE, D. G., 1986, Short papers in pharmaceutical analysis. Stability of pralidoxime mesylate injections, *Anal. Proc.*, **23**, 320–1.

Primaquine

BROSSI, A., GESSNER, W., HUFFORD, C. D., BAKER, J. K., HOMO, F., MILLET, P. and LANDAU, I., 1987, Photooxidation products of primaquine. Structure, antimalarial activity and hemolytic effects, *FEBS Letters*, **223**, 77–81.

KRISTENSEN, S., GRISLINGAAS, A.-L., GREENHILL, J. V., SKJETNE, T., KARLSEN, J. and TØNNESEN, H. H., 1993, Photochemical stability of biologically active compounds. V. Photochemical degradation of primaquine in an aqueous medium, *Int. J. Pharm.*, **100**, 15–23.

Proguanil

OWOYALE, J. A. and ELMARAKBY, Z. S., 1989, Effect of sunlight, ultraviolet irradiation and heat on proguanil, *Int. J. Pharm.*, **50**, 219–21.

TAYLOR, R. B., MOODY, R. R., OCHEKPE, N. A., LOW, A. S. and HARPER, M. I. A., 1990, A chemical stability study of proguanil hydrochloride, *Int. J. Pharm.*, **60**, 185–90.

Promazine

HALL, R. D., BUETTNER, G. R., MOTTEN, A. G. and CHIGNELL, C. F., 1987, Near infrared detection of singlet molecular oxygen produced by photosensitization with chlorpromazine, *Photochem. Photobiol.*, **46**, 295–300.

MOTTEN, A. G., BUETTNER, G. R. and CHIGNELL, C. F., 1985, Spectroscopic studies of cutaneous photosensitizing agents – VIII. A spin-trapping study of light induced free radicals from chlorpromazine and promazine, *Photochem. Photobiol.*, **42**, 9–15.

Promethazine

MATSUDA, Y. and MASAHARA, R., 1980, Comparative evaluation of coloration of photosensitive solid drugs under various light sources, *Yakugaku Zasshi*, **100**, 953–7.

Propionic acid

VARGAS, F., RIVAS, C., MIRANDA, M. A. and BOSCA, F., 1991, Photochemistry of the non-steroidal anti-inflammatory drugs, propionic acid-derived, *Pharmazie*, **46**, 767–71.

Psoralen

HEARST, J. E., 1982, Psoralen photochemistry and nucleic acid structure, *Biorg. Khim.*, **8**, 945–53.

Pyrazinamide

CAIN, M. J., SPENCER, N. R., DUSCI, L. and ONG, R., 1986, Stability of Pyrazinamide Infusion, *Aust. J. Hosp. Pharm.*, **16**, 217–18.

Quinidine

PÖHLMANN, H., THEIL, F.-P. and PFEIFER, S., 1986, *In-vitro* und *in-vivo*-Untersuchungen zur fotochemischer Reaktivität der Chinin- und Chinidin – N – Oxide, *Pharmazie*, **41**, 859–62.

Quinine

LAURIE, W. A., MCHALE, D. and SAAG, K., 1986, Photoreactions of quinine in aqueous citric acid solutions, *Tetrahedron*, **42**, 3711–14.

LAURIE, W. A., MCHALE, D., SAAG, K. and SHERIDAN, J. B., 1988, Photoreactions of quinine in aqueous citric acid solution. Part 2. Some end-products, *Tetrahedron*, **44**, 5905–10.

MCHALE, D., LAURIE, W. A., SAAG, K. and SHERIDAN, J. B., 1989, Photoreactions of quinine in aqueous citric acid solution. Part 3. Products formed in aqueous 2-hydroxy-2-methylpropionic acid, *Tetrahedron*, **45**, 2127–30.

PÖHLMANN, H., THEIL, F.-P. and PFEIFER, S., 1986, *In-vitro* und *in-vivo*-Untersuchungen zur fotochemischer Reaktivität der Chinin- und Chinidin – N – Oxide, *Pharmazie*, **41**, 859–62.

Quinolones

TIEFENBACHER, E.-M., HAEN, E., PRZYBILLA, B. and KURZ, H., 1994, Photodegradation of some quinolones used as antimicrobial therapeutics, *J. Pharm. Sci.*, **83**, 463–7.

Ranitidin

Williams, M. F., Hak, L. J. and Dukes, G., 1990, *In vitro* evaluation of the stability of ranitidine hydrochloride in total parenteral nutrient mixtures, *Am. J. Hosp. Pharm.*, **47**, 1574–9.

Salicylanilides

Davies, A. K., Hilal, N. S., McKellar, J. F. and Phillips, G. O., 1975, Photodegradation of salicylanilides, *Brit. J. Derm.*, **92**, 143–7.

Salicylic acid

Mills, A., Holland, C. E., Davies, R. H. and Worsley, D., 1994, Photomineralization of salicylic acid: a kinetic study, *J. Photochem. Photobiol. A: Chem.*, **83**, 257–63.

Sorivudine

Desai, D. S., Abdelnasser, M. A., Rubitski, B. A. and Varia, S. A., 1994, Photostabilization of uncoated tablets of sorivudine and nifedipine by incorporation of synthetic iron oxides, *Int. J. Pharm.*, **103**, 69–76.

Spironolactone

Mathur, L. K. and Wickman, A., 1989, Stability of extemporaneously compounded spironolactone suspensions, *Am. J. Hosp. Pharm.*, **46**, 2040–2.

Steroids

Ekiz-Gücer, N., Reisch, J. and Nolte, G., 1991, Photostabilität von kreuzkonjugierten glucocorticoiden in kristallinem zustand. Photochemische studien, 60. Mitteilung, *Eur. J. Pharm. Biopharm.*, **37**, 234–7.

Ekiz-Gücer, N., Zappel, J. and Reisch, J., 1991, Photostabilität von kontrazeptiven steroiden in kristallinem zustand, *Pharm. Acta Helv.*, **66**, 2–4.

Hamlin, W. E., Chulski, T., Johnson, R. H. and Wagner, J. G., 1960, A note on the photolytic degradation of anti-inflammatory steroids, *J. Am. Pharm. Asso.*, **49**, 253–7.

Lin, C.-T., Perrier, P., Clay, G. G., Sutton, P. A. and Byrn, S. R., 1982, Solid-state photooxidation of 21-cortisol tert-butylacetate to 21-cortisone tert-butylacetate, *J. Org. Chem.*, **47**, 2978–81.

Lou Stiles, M., Allen, Jr, L. V., Resztak, K. E. and Prince, S. J., 1993, Stability of octreotide acetate in polypropylene syringes, *Am. J. Hosp. Pharm.*, **50**, 2356–8.

Reisch, J., Ekiz, N. and Takàcs, M., 1989, Photochemische studien, 52. Mitt.: Untersuchungen zur photostabilität von testosteron und methyltestosteron im kristallinen zustand, *Arch. Pharm.*, **322**, 173–5.

Reisch, J., Ekiz-Gücer, N., Takàcs, M. and Henkel, G., 1989, Photochemische studien, 54. Photodimerisierung von testosteronpropionat in kristallinem zustand und kristallstruktur von testosteronpropionat, *Liebigs Ann. Chem.*, 595–7.

Reisch, J., Henkel, G., Ekiz-Gücer, N. and Nolte, G., 1992, Photochemische studien, 62. Untersuchungen zur photostabilität von halometason und prednicarbat in kristallinem zustand sowie kristallstruktur von halometason, *Liebigs Ann. Chem.*, 63–7.

Reisch, J., Iranshahi, L. and Ekiz-Gücer, N., 1992, Photochemische studien, 64. Photostabilität von glucocorticoiden in kristallinem zustand, *Liebigs Ann. Chem.*, 1199–200.

REISCH, J., TOPALOGLU, Y. and HENKEL, G., 1986, Über die photostabilität offizineller arznei- und hilfsstoffe, II: Die gekreuzt- konjugierten corticoide des europäischen arzneibuches, *Acta. Pharm. Technol.*, **32**, 115–21.

REISCH, J., ZAPPEL, J., HENKEL, G. and EKIZ-GÜCER, N., 1993, Photochemische studien, 65. Mitt. Untersuchungen zur photostabilität von ethisteron und norethisteron sowie deren kristallstrukturen, *Monatshefte für chemie*, **124**, 1169–75.

REISCH, J., ZAPPEL, J., RAGHU, A. and RAO, R., 1995, Photochemical studies part 68. Solid state photochemical investigations of methylprednisolone, prednisolone and triamcinolone acetonide, *Acta Pharmaceutica Turcica*, **XXXVII**, 13–17.

REISCH, J., ZAPPEL, J., RAO, A. R. R. and HENKEL, G., 1994, Dimerisation of levonorgestrel in solid state ultraviolet light irradiation, *Pharm. Acta Helv.*, **69**, 97–100.

SEDEE, A. G. J., BEIJERSBERGEN VAN HENEGOUWEN, G. M. J. and BLAAUWGEERS, H. J. A., 1983, Isolation, identification and densitometric determination of norethisterone-4β,5β-epoxide after photochemical decomposition of norethisterone, *Int. J. Pharm.*, **15**, 149–58.

SEDEE, A. G. J., BEIJERSBERGEN VAN HENEGOUWEN, G. M. J., DE VRIES, H., GUIJT, W. and HAASNOOT, C. A. G., 1985, The photochemical decomposition of the progestronic 19-nor-steroid, norethisterone, in aqueous medium, *Pharm. Weekbl. Sci. Ed.*, **7**, 194–201.

TAKÀCS, M., EKIZ-GÜCER, N., REISCH, J. and GERGELY-ZOBIN, A., 1991, Lichtempfindlichkeit von corticosteroiden in kristallinem zustand. *Photochemiche studien*, 59. Mitt. *Pharm. Acta Helv.*, **66**, 137–40.

THOMA, VON K. and KERKER, R., 1992, Photoinstabilität von arzneimitteln. 1. Mitteilung über das verhalten von nur im UV-bereich absorbierenden substanzen bei der tageslichtsimulation, *Pharm. Ind.*, **54**, 169–77.

1992, Photostabilität von arzneimitteln. 5. Mitteilung: Untersuchungen zur photostabilisierung von glukokortikoiden, *Pharm. Ind.*, **54**, 551–4.

Succinylcholine

SCHMUTZ, C. W. and MÜHLEBACH, S. F., 1991, Stability of succinylcholine chloride injection, *Am. J. Hosp. Pharm.*, **48**, 501–6.

Sulfacetamide

LAND, E. J., NAVARATNAM, S., PARSONS, B. J. and PHILLIPS, G. O., 1982, Primary processes in the photochemistry of aqueous sulfacetamid: a laser flash photolysis study, *Photochem. Photobiol.*, **35**, 637–42.

REICH, J. and NIEMEYER, D. H., 1972, Photochemische studien am sulfanilamid, sulfacetamid und 4-aminobenzoesäure-äthylester, *Arch. Pharm. (Weinheim)*, **305**, 135–40.

TAMMILEHTO, S., 1989, Photodecomposition of sulphacetamide in water-ethanol solutions, *Acta Pharm. Fenn.*, **98**, 225–30.

Sulfamethoxazole

ZHOU, W. and MOORE, D. E., 1994, Photochemical decomposition of sulfamethoxazole, *Int. J. Pharm.*, **110**, 55–63.

Sulfathiazole

ASKER, A. F. and LAROSE, M., 1987, Influence of uric acid on photostability of sulfathiazole sodium solutions, *Drug Dev. Ind. Pharm.*, **13**, 2239–48.

Sulfisomidine

MATSUDA, Y., INOUYE, H. and NAKANISHI, R., 1978, Stabilization of Sulfisomidine Tablets by Use of Film Coating Containing UV Absorber: Protection of Coloration and Photolytic Degradation from Exaggerated light, *J. Pharm. Sci.*, **67**, 196–201.

Sulphanilamide

PAWLACZYK, J. and TUROWSKA, W., 1976, Photochemistry of photodynamic compounds. VII. Identification of some products of photolysis of sulphanilamide aqeuous solutions, *Acta Polon. Pharm.*, **XXXIII**, 505–9.

REISCH, J. and NIEMEYER, D. H., 1972, Photochemische studien am sulfanilamid, sulfacetamid und 4-aminobenzoesäure-äthylester, *Arch. Pharm. (Weinheim)*, **305**, 135–40.

Sulpyrine

MATSUDA, Y. and MASAHARA, R., 1980, Comparative evaluation of coloration of photosensitive solid drugs under various light sources, *Yakugaku Zasshi*, **100**, 953–7.

Suprofen

CASTELL, J. V., GOMEZ-LECHON, M. J., GRASSA, C., MARTINEZ, L. A., MIRANDA, M. A. and TARREGA, P., 1994, Photodynamic lipid peroxidation by the photosensitizing nonsteroidal, antiinflammatory drugs suprofen and tiaprofenic acid, *Photochem. Photobiol.*, **59**, 35–9.

Suramin

BEIJNEN, J. H., GIJN, R. VAN, HORENBLAS, S. and UNDERBERG, W. J. M., 1990, Chemical stability of suramin commonly used infusion solutions, DICP, *The Annals of Pharmacotherapy*, **24**, 1056–8.

Tauromustine

BETTERIDGE, R. F., CULVERWELL, A. L. and BOSANQUET, A. G., 1989, Stability of tauromustine (TCNU) in aqueous solutions during preparation and storage, *Int. J. Pharm.*, **56**, 37–41.

Terbutaline

GLASCOCK, J. C., DIPIRO, J. T., CADWALLADER, D. E. and PERRI, M., 1987, Stability of terbutaline sulfate repackaged in disposable plastic syringes, *Am. J. Hosp. Pharm.*, **44**, 2291–3.

NEWTON, D. W., YIN YEE FUNG, E. and WILLIAMS, D. A., 1981, Stability of five catecholamines and terbutaline sulfate in 5% dextrose injection in the absence and presence of aminophylline, *Am. J. Hosp. Pharm.*, **38**, 1314–19.

Terfenadine

CHEN, T.-M., SILL, A. D. and HOUSMYER, A. L., 1986, Solution stability study of terfenadine by high-performance liquid chromatography, *J. Pharm. Biomed. Anal.*, **4**, 533–9.

Tetracyclines

ALBINI, A., 1991, Fotochimica di sostanze farmaceutiche, *Boll. Chim. Farmaceutico.*, **130**, 393.

ASKER, A. F. and HABIB, M. J., 1991, Effect of certain additives on photodegradation of tetracycline hydrochloride solutions, *J. Parent. Sci. Technol.*, **45**, 113–15.

WIEBE, J. A. and MOORE, D. E., 1977, Oxidation photosensitized by tetracyclines, *J. Pharm. Sci.*, **66**, 186–9.

Theophylline

BAAK, L. R., 1987, Aminophylline stability, *The Canadian J. Hosp. Pharm.*, **40**, 155.

Thiazide

MOORE, D. E. and MALLESCH, J. L., 1991, Photochemical interaction between triamterene and hydrochlorthiazide, *Int. J. Pharm.*, **76**, 187–90.

MOORE, D. E. and TAMAT, S. R., 1980, Photosensitization by drugs: photolysis of some chlorine-containing drugs, *J. Pharm. Pharmacol.*, **32**, 172–7.

TAMAT, S. R. and MOORE, D. E., 1983, Photolytic decomposition of hydrochlorthiazide, *J. Pharm. Sci.*, **72**, 180–3.

ULVI, V., 1987, Tiatsididiureettien kemiasta, hajoamisesta ja fotobiologiasta. *Acta Pharm. Fennica*, **96**, 147–56.

ULVI, V. and TAMMILEHTO, S., 1989, Photodecomposition studies on chlorothiazide and hydrochlorothiazide, *Acta Pharm. Nord.*, **1**, 195–200.

1990, High-performance liquid chromatographic method for studies on the photodecomposition kinetics of chlorothiazide, *J. Chromatogr.*, **507**, 151–6.

Thioridazine

EAP, C. B., KOEB, L. and BAUMANN, P., 1993, Artifacts in the analysis of thioridazine and other neuroleptics, *J. Pharm. Biomed. Anal.*, **11**, 451–7.

PAWELCZYK, E. and MARCINIEC, B., 1974, Chemical characterization of decomposition products of drugs, *Pharmazie*, **29**, 585–9.

Thiorphan

GIMENEZ, F., POSTAIRE, E., PROGNON, P., LE HOANG, M. D., LECOMTE, J. M., PRADEAU, D. and HAZEBROUCQ, G., 1988, Study of thiorphan degradation, *Int. J. Pharm.*, **43**, 23–30.

Thiothixene

THOMA, K. and KLIMEK, R., 1991, Photostabilization of drugs in dosage forms without protection from packaging materials, *Int. J. Pharm.*, **67**, 169–75.

Tiaprofenic acid

BOSCÁ , F., MIRANDA, M. A. and VARGAS, F., 1992, Chemistry of tiaprofenic acid, a nonsteroidal antiinflammatory drug with phototoxic side effects, *J. Pharm. Sci.*, **81**, 181–2.

CASTELL, J. V., GOMEZ-LECHON, M. J., GRASSA, C., MARTINEZ, L. A., MIRANDA, M. A. and TARREGA, P., 1994, Photodynamic lipid peroxidation by the photosensitizing nonsteroidal, antiinflammatory drugs suprofen and tiaprofenic acid, *Photochem. Photobiol.*, **59**, 35–9.

CASTELL, J. V., GOMEZ-LECHON, M. J., HERNANDEZ, D., MARTINEZ, L. A. and MIRANDA, M. A., 1994, Molecular basis of drug phototoxicity: Photosensitized cell damage by the major photoproduct of tiaprofenic acid, *Photochem. Photobiol.*, **60**, 586–90.

Tinidazole

SALOMIES, H., 1991, Structure elucidation of the photolysis and hydrolysis products of tinidazole, *Acta Pharm. Nord.*, **3**, 211–14.

Tretinoin

BRISAERT, M. G., EVERAERTS, G. R., BULTINCK, G. R. and PLAIZIER-VERCAMMEN, J. A., 1992, Onderzoek naar de chemische stabiliteit van tretinoine in dermatologische preparaten, *Farm. Tijdschr. Belg.*, **69**, 194–9.

Triamterene

MOORE, D. E. and MALLESCH, J. L., 1991, Photochemical interaction between triamterene and hydrochlorthiazide, *Int. J. Pharm.*, **76**, 187–90.

Trimethoprim

Bergh, J. J., Breytenbach, J. C. and Wessels, P. L., 1989, Degradation of trimethoprim, *J. Pharm. Sci.*, **78**, 348–50.

Ubidecarenone

Matsuda, Y. and Masahara, R., 1983, Photostability of solid-state ubidecarenone at ordinary and elevated temperatures under exaggerated UV irradiation, *J. Pharm. Sci.*, **72**, 1198–203.

Matsuda, Y. and Teraoka, R., 1985, Improvement of the photostability of ubidecarenone microcapsules by incorporating fat-soluble vitamins, *Int. J. Pharm.*, **26**, 289–301.

Takeuchi, H., Sasaki, H., Niwa, T., Hino, T., Kawashima, Y., Uesugi, K. and Ozawa, H., 1992, Improvement of the photostability of ubidecarenone in the formulation of a novel powered dosage form termed redispersible dry emulsion, *Int. J. Pharm.*, **86**, 25–33.

Vesnarinone

Zhan, X., Yin, G. and Liu, S., 1995, Kinetic study on the photostability of solid vesnarinone and the equivalent relationship between daylight and lamplight, *J. Pharm. Sci.*, **84**, 624–6.

Vinblastine

Black, J., Buechter, D. D. and Thurston, D. F., 1988, Stability of vinblastine sulphate when exposed to light, *Drug Intell. Clin. Pharm.*, **22**, 634–5.

Black, J., Buechter, D. D., Chinn, J. W., Gard, J. and Thurston, D. E., 1988, Studies on the stability of vinblasting sulfate in aqueous solution, *J. Pharm. Sci.*, **77**, 630–4.

Vitamin A

Allwood, M. C. and Plane, J. H., 1986, The wavelength-dependent degradation of vitamin A exposed to ultraviolet radiation, *Int. J. Pharm.*, **31**, 1–7.

Bluhm, D. P., Summers, R. S., Lowes, M. M. J. and Dürheim, H. H., 1991, Lipid emulsion content and vitamin A stability in TPN admixtures, *Int. J. Pharm.*, **68**, 277–88.

1991, Influence of container on vitamin A stability in TPN admixtures, *Int. J. Pharm.*, **68**, 281–3.

Smith, J. L., Canham, J. E. and Wells, P. A., 1988, Effect of phototherapy light, sodium bisulfite, and pH on vitamine stability in total parenteral nutrition admixtures, *J. Parent. Enter. Nutrit.*, **12**, 394–402.

Vitamin B

Ahmad, I., David, H. and Rapson, C., 1990, Multicomponent spectrophotometric assay of riboflavine and photoproducts, *J. Pharma. Biomed. Anal.*, **8**, 217–23.

Asker, A. F. and Habib, M. J., 1990, Effect of certain stabilizers on photobleaching of riboflavin solutions, *Drug Dev. Ind. Pharm.*, **16**, 149–56.

Chen, M. F., Boyce, H. W. and Triplett, L., 1983, Stability of the B vitamins in mixed parenteral nutrition solution, *JPEN. J. Parenter. Enteral Nutr.*, **7**, 462–4.

Habib, M. J. and Asker, F., 1991, Photostabilization of riboflavin by incorporation into liposomes, *J. Parenter. Sci. Technol.*, **45**, 124–7.

Martens Von Harrie, J. M., 1989, Stabilität wasserlöslicher vitamine in verschiedenen infusionsbeuteln, *Krankenhauspharmazie*, **10**, 359–61.

OSTREA, E. M., GREEN, C. D. and BALUN, J. E., 1982, Decomposition of TPN solutions exposed to phototherapy, *J. Pediatr.*, **100**, 669–70.

SMITH, J. L., CANHAM, J. E. and WELLS, P. A., 1988, Effect of phototherapy light, sodium bisulfite, and pH on vitamine stability in total parenteral nutrition admixtures, *JPEN J. Parenter. Enteral. Nutr.*, **12**, 394–402.

Vitamin C

MARTENS VON HARRIE, J. M., 1989, Stabilität wasserlöslicher vitamine in verschiedenen infusionsbeuteln, *Krankenhauspharmazie*, **10**, 359–61.

SIDHU, D. S. and SUGDEN, J. K., 1992, Effect of food dyes on the photostability of aqueous solutions of L-ascorbic acid, *Int. J. Pharm.*, **83**, 263–6.

SMITH, J. L., CANHAM, J. E. and WELLS, P. A., 1988, Effect of phototherapy light, sodium bisulfite, and pH on vitamine stability in total parenteral nutrition admixtures, *J. Parenter. Enteral. Nutr.*, **12**, 394–402.

Vitamin D

HÜTTENRAUCH, R., FRICKE, S. and MATTHEY, K., 1985, Zur Notwendigkeit des Lichtschutzes für wässerige vitamin-D-Lösungen, *Pharmazie*, **41**, 742.

Vitamin E

SMITH, J. L., CANHAM, J. E. and WELLS, P. A., 1988, Effect of phototherapy light, sodium bisulfite, and pH on vitamine stability in total parenteral nutrition admixtures, *JPEN. J. Parenter. Enteral. Nutr.*, **12**, 394–402.

Vitamin K

MARCINIEC, B. and WITKOWSKA, D., 1988, Fotochemiczny rozklad Witaminy K_4 W facie statej, *Acta Polon. Pharm.*, **XLV**, 528–34.

NISHIKAWA, M., SUZUKI, K., MORITA, H., KAWAKAGE, I. and FUJII, K., 1986, Stability of vitamin K2 syrup after dispensing, *Jap. J. Hosp. Pharm.*, **12**, 257–60.

TERAOKA, R. and MATSUDA, Y., 1993, Stabilization-oriented preformulation study of photolabile menatetrenone (vitamin K_2), *Int. J. Pharm.*, **93**, 85–90.

WILSON, R. M., WALSH, T. F. and GEE, S. K., 1980, Laser photochemistry: the wavelength dependent oxidative photodegradation of vitamin K analogs, *Tetrahedron Lett.*, **21**, 3459–63.

Warfarin

REISCH, J. and ZAPPEL, J., 1993, Photochemische studien, 70. Photostabilitätsuntersuchungen an natrium-warfarin in kristallinem zustand, *Sci. Pharm.*, **61**, 283–6.

Reviews

ALBINI, A., 1991, Fotochimica di sostanze farmaceutiche, *Boll. Chim. Farmaceutico.*, **130**, 393–7.

ANDERSON, N. H., JOHNSTON, D., MCLELLAND, M. A. and MUNDEN, P., 1991, Photostability testing of drug substances and drug products in UK pharmaceutical laboratories, *J. Pharm. Biomed. Anal.*, **9**, 443–9.

BEIJERSBERGEN VAN HENEGOUWEN, 1977, Relevantie van fotochemi. binnen de farmacie, *Pharmaceutisch Weekblad*, **112**, 1101–16.

BENVENUTO, J. A., ANDERSON, R. W., KERKOF, K., SMITH, R. G. and LOO, T. L., 1981, Stability and compatibility of antitumor agents in glass and plastic containers, *Am. J. Hosp. Pharm.*, **38**, 1914–18.

BOSANQUET, A. G., 1985, Stability of solutions of antineoplastic agents during preparation and storage for *in vitro* assays. General considerations, the nitrosoureas and alkylating agents, *Cancer Chemother. Pharmacol.*, **14**, 83–95.

1986, Stability of solutions of antineoplastic agents during preparation and storage for *in vitro* assays. II. Assay methods, adriamycin and the other antitumor antibiotics, *Cancer Chemother. Pharmacol.*, **17**, 1–10.

CARSTENSEN, J. T., 1974, Stability of solids and solid dosage forms, *J. Pharm. Sci.*, **63**, 1–14.

1990, Drug stability, Marcel Dekker, Inc., New York, pp. 288–9.

GREENHILL, J. V., 1995, Photodecomposition of drugs, Encyclopedia of Pharmaceutical Technology (Swarbrick, J. and Boyland, J. C., eds) Marcel Dekker, New York, Vol. 12, 105–35.

GREENHILL, J. V. and MCLELLAND, M. A., 1990, Photodecomposition of drugs. *Progress in Medicinal Chemistry* (Ellis, G. P., West, G. B., eds), Elsevier, Amsterdam, Vol. 27, 51–121.

GRÜNERT, R. and WOLLMANN, H., 1982, Die einwirkung von ultraviolettem und sichtbarem licht auf arzneistoffe aus der reihe der phenylalkylamine im hinblick auf ihre stabilität in plastbehältnissen, *Pharmazie*, **37**, 798–9.

KRÄMER, VON I. and FRANK, P., 1987, Lichtschutzbedürftigkeit photoinstabiler arzneimittel zur infusion, *Krankenhauspharmazie*, **8**, 137–44.

MENDENHALL, D. W., 1984, Stability of parenterals, *Drug Development and Industrial Pharmacy*, **10** (8&9), 1297–342.

MOORE, D. E., 1987, Photochemistry in medicine. Techniques for analysis of photodegradation pathways of medicinal compounds, *Trends in Analytical Chemistry*, **6**, 234–8.

NEMA, S., WASHKUHN, R. J. and BEUSSINK, D. R., 1995, Photostability testing: An overview, *Pharm. Technol.*, 170–85.

NYQVIST, H., 1984, Light stability testing of tablets in the xenotest and the fadeometer, *Acta Pharm. Suec.*, **21**, 245–52.

OPPENLÄNDER, T., 1988, A comprehensive photochemical and photophysical assey exploring the photoreactivity of drugs, *Chimia*, **42**, 331–42.

PARKASH, A. K., SUGDEN, J. K., YUCAN, Z. and MINGXIN, W., 1993, A rapid method for screening non steroidal anti-inflammatory agents for hydroxyl radical scavenging, *Int. J. Pharm.*, **92**, 151–5.

PAUL, W. L., 1992, Photolytic degradation of drugs, *Am. J. Hosp. Pharm.*, **49**, 2704–5. Letter.

SHAHJAHAN, M., 1989, Ultraviolet light absorbers and their uses in pharmacy. *The Eastern Pharmacist*, **32**, 37–41.

STRUHAR, M. and SYNKOVA, A., 1986, Uloha svetla pri rozklade lieciv (The part of light in the breakdown of drugs), *Farm. Obzor.*, **55**, 211–16.

TAKÀCS, M., REISCH, J., GERGELY-ZOBIN, A. and GÜCER-EKIZ, N., 1990, Zur untersuchung der lichtempfindlichkeit von festsubstanzen, *Sci. Pharm.*, **58**, 289–97.

THOMA, K., MARSCHALL, M. and SCHUBERT, O. E., 1991, Augentropfen physikalische und chemische instabilität nach anbruch, *Deutsche Apotheker Zeitung*, **34**, 1739–43.

THOMA, K. and STRUVE, M., 1986, Einfluss hydrophiler lösungsmittel auf die instabilität von arzneistoffen, *Pharm. Ind.*, **48**, 179–83.

TØNNESEN, H. H., 1991, Photochemical degradation of components in drug formulations. Part 1: An approach to the standardization of degradation studies, *Pharmazie*, **46**, 263–6.

TØNNESEN, H. H. and KARLSEN, J., 1995, Photochemical degradation of components in drug formulations, *Pharmeuropa*, **7**, 137–41.

TØNNESEN, H. H. and MOORE, D. E., 1993, Photochemical degradation of components in drug formulations, *Pharma. Technol.*, **5**, 27–8, 30–3.

VON HANSJØRG SAUER, 1990, Aufbewahrung von Zytostatika-Lösungen, *Krankenhauspharmazie*, **11**, 373–5.

Index